Lecture Notes in Artificial Intelligence   11580

Subseries of Lecture Notes in Computer Science

More information about this series at http://www.springer.com/series/1244

Dylan D. Schmorrow · Cali M. Fidopiastis (Eds.)

Augmented Cognition

13th International Conference, AC 2019
Held as Part of the 21st HCI International Conference, HCII 2019
Orlando, FL, USA, July 26–31, 2019
Proceedings

 Springer

Editors
Dylan D. Schmorrow
Soar Technology Inc.
Orlando, FL, USA

Cali M. Fidopiastis
Design Interactive, Inc.
Orlando, FL, USA

ISSN 0302-9743   ISSN 1611-3349 (electronic)
Lecture Notes in Artificial Intelligence
ISBN 978-3-030-22418-9   ISBN 978-3-030-22419-6 (eBook)
https://doi.org/10.1007/978-3-030-22419-6

LNCS Sublibrary: SL7 – Artificial Intelligence

This Springer imprint is published by the registered company Springer Nature Switzerland AG
The registered company address is: Gewerbestrasse 11, 6330 Cham, Switzerland

Foreword

The 21st International Conference on Human-Computer Interaction, HCI International 2019, was held in Orlando, FL, USA, during July 26–31, 2019. The event incorporated the 18 thematic areas and affiliated conferences listed on the following page.

A total of 5,029 individuals from academia, research institutes, industry, and governmental agencies from 73 countries submitted contributions, and 1,274 papers and 209 posters were included in the pre-conference proceedings. These contributions address the latest research and development efforts and highlight the human aspects of design and use of computing systems. The contributions thoroughly cover the entire field of human-computer interaction, addressing major advances in knowledge and effective use of computers in a variety of application areas. The volumes constituting the full set of the pre-conference proceedings are listed in the following pages.

This year the HCI International (HCII) conference introduced the new option of "late-breaking work." This applies both for papers and posters and the corresponding volume(s) of the proceedings will be published just after the conference. Full papers will be included in the *HCII 2019 Late-Breaking Work Papers Proceedings* volume of the proceedings to be published in the Springer LNCS series, while poster extended abstracts will be included as short papers in the HCII 2019 *Late-Breaking Work Poster Extended Abstracts* volume to be published in the Springer CCIS series.

I would like to thank the program board chairs and the members of the program boards of all thematic areas and affiliated conferences for their contribution to the highest scientific quality and the overall success of the HCI International 2019 conference.

This conference would not have been possible without the continuous and unwavering support and advice of the founder, Conference General Chair Emeritus and Conference Scientific Advisor Prof. Gavriel Salvendy. For his outstanding efforts, I would like to express my appreciation to the communications chair and editor of *HCI International News,* Dr. Abbas Moallem.

July 2019   Constantine Stephanidis

HCI International 2019 Thematic Areas and Affiliated Conferences

Thematic areas:

- HCI 2019: Human-Computer Interaction
- HIMI 2019: Human Interface and the Management of Information

Affiliated conferences:

- EPCE 2019: 16th International Conference on Engineering Psychology and Cognitive Ergonomics
- UAHCI 2019: 13th International Conference on Universal Access in Human-Computer Interaction
- VAMR 2019: 11th International Conference on Virtual, Augmented and Mixed Reality
- CCD 2019: 11th International Conference on Cross-Cultural Design
- SCSM 2019: 11th International Conference on Social Computing and Social Media
- AC 2019: 13th International Conference on Augmented Cognition
- DHM 2019: 10th International Conference on Digital Human Modeling and Applications in Health, Safety, Ergonomics and Risk Management
- DUXU 2019: 8th International Conference on Design, User Experience, and Usability
- DAPI 2019: 7th International Conference on Distributed, Ambient and Pervasive Interactions
- HCIBGO 2019: 6th International Conference on HCI in Business, Government and Organizations
- LCT 2019: 6th International Conference on Learning and Collaboration Technologies
- ITAP 2019: 5th International Conference on Human Aspects of IT for the Aged Population
- HCI-CPT 2019: First International Conference on HCI for Cybersecurity, Privacy and Trust
- HCI-Games 2019: First International Conference on HCI in Games
- MobiTAS 2019: First International Conference on HCI in Mobility, Transport, and Automotive Systems
- AIS 2019: First International Conference on Adaptive Instructional Systems

Pre-conference Proceedings Volumes Full List

1. LNCS 11566, Human-Computer Interaction: Perspectives on Design (Part I), edited by Masaaki Kurosu
2. LNCS 11567, Human-Computer Interaction: Recognition and Interaction Technologies (Part II), edited by Masaaki Kurosu
3. LNCS 11568, Human-Computer Interaction: Design Practice in Contemporary Societies (Part III), edited by Masaaki Kurosu
4. LNCS 11569, Human Interface and the Management of Information: Visual Information and Knowledge Management (Part I), edited by Sakae Yamamoto and Hirohiko Mori
5. LNCS 11570, Human Interface and the Management of Information: Information in Intelligent Systems (Part II), edited by Sakae Yamamoto and Hirohiko Mori
6. LNAI 11571, Engineering Psychology and Cognitive Ergonomics, edited by Don Harris
7. LNCS 11572, Universal Access in Human-Computer Interaction: Theory, Methods and Tools (Part I), edited by Margherita Antona and Constantine Stephanidis
8. LNCS 11573, Universal Access in Human-Computer Interaction: Multimodality and Assistive Environments (Part II), edited by Margherita Antona and Constantine Stephanidis
9. LNCS 11574, Virtual, Augmented and Mixed Reality: Multimodal Interaction (Part I), edited by Jessie Y. C. Chen and Gino Fragomeni
10. LNCS 11575, Virtual, Augmented and Mixed Reality: Applications and Case Studies (Part II), edited by Jessie Y. C. Chen and Gino Fragomeni
11. LNCS 11576, Cross-Cultural Design: Methods, Tools and User Experience (Part I), edited by P. L. Patrick Rau
12. LNCS 11577, Cross-Cultural Design: Culture and Society (Part II), edited by P. L. Patrick Rau
13. LNCS 11578, Social Computing and Social Media: Design, Human Behavior and Analytics (Part I), edited by Gabriele Meiselwitz
14. LNCS 11579, Social Computing and Social Media: Communication and Social Communities (Part II), edited by Gabriele Meiselwitz
15. LNAI 11580, Augmented Cognition, edited by Dylan D. Schmorrow and Cali M. Fidopiastis
16. LNCS 11581, Digital Human Modeling and Applications in Health, Safety, Ergonomics and Risk Management: Human Body and Motion (Part I), edited by Vincent G. Duffy

http://2019.hci.international/proceedings

13th International Conference on Augmented Cognition (AC 2019)

Program Board Chair(s): **Dylan D. Schmorrow**
and Cali M. Fidopiastis, *USA*

- Brendan Allison, USA
- Amy Bolton, USA
- Micah Clark, USA
- Martha Crosby, USA
- Fausto De Carvalho, Portugal
- Daniel Dolgin, USA
- Sven Fuchs, Germany
- Rodolphe Gentili, USA
- Scott S. Grigsby, USA
- Katy Hancock, USA
- Monte Hancock, USA
- Frank Hannigan, USA
- Robert Hubal, USA
- Kurtulus Izzetoglu, USA
- Øyvind Jøsok, Norway
- Ion Juvina, USA
- Benjamin Knott, USA
- Benjamin Knox, Norway
- Julie Marble, USA
- Chang S. Nam, USA
- Banu Onaral, USA
- Sarah Ostadabbas, USA
- Lesley Perg, USA
- Robinson Pino, USA
- Mannes Poele, The Netherlands
- Lauren Reinerman-Jones, USA
- Stefan Sütterlin, Norway
- Suraj Sood, USA
- Ayoung Suh, Hong Kong, SAR China
- Georgios Triantafyllidis, Denmark
- Christian Wagner, Hong Kong, SAR China
- Melissa Walwanis, USA
- Quan Wang, USA
- Martin Westhoven, Germany

The full list with the Program Board Chairs and the members of the Program Boards of all thematic areas and affiliated conferences is available online at:

http://www.hci.international/board-members-2019.php

HCI International 2020

The 22nd International Conference on Human-Computer Interaction, HCI International 2020, will be held jointly with the affiliated conferences in Copenhagen, Denmark, at the Bella Center Copenhagen, July 19–24, 2020. It will cover a broad spectrum of themes related to HCI, including theoretical issues, methods, tools, processes, and case studies in HCI design, as well as novel interaction techniques, interfaces, and applications. The proceedings will be published by Springer. More information will be available on the conference website: http://2020.hci.international/

General Chair
Prof. Constantine Stephanidis
University of Crete and ICS-FORTH
Heraklion, Crete, Greece
E-mail: general_chair@hcii2020.org

http://2020.hci.international/

Contents

Brain-Computer Interfaces and Electroencephalography

Augmented Learning

Cognitive Modeling, Perception, Emotion and Interaction

Creating Affording Situations with Animate Objects

Chris Baber[(⊠)] ⓘ, Sara Al-Tunaib, and Ahmed Khattab

University of Birmingham, Birmingham B15 2TT, UK
c.baber@bham.ac.uk

Abstract. In this paper, we report the design, development and initial evaluation of the concept of animate objects (as a form of Tangible User Interface, TUI) to support the cueing of action sequences. Animate objects support user actions through the provision of affordances, and developments in actuator technologies allow those devices to change their physical form. Each object communicates (in a small-scale 'Internet of Things') to cue other objects to act, and these, in turn, provide cues to the user. The set of cues creates affording situations which can be understood in terms of Forms of Engagement. A user trial was conducted in which participants had to either perform a simulated tea-making task or search for letters hidden under objects to spell words. The first task involved a familiar sequence (and we also included an unfamiliar sequence), and the second task involved either real or false words. We show that, when cues provided by the objects correspond to a 'logical' task sequence (i.e., familiar and/or easily interpretable), they aid performance and when these cues relate to 'illogical' task sequences, performance is impaired. We explain these results in terms of forms of engagement and suggest applications for rehabilitation.

Keywords: Tangible User Interface · Animate objects · Affordance

1 Introduction

1.1 Tangible User Interfaces

Tangible User Interfaces (TUIs) employ physical objects to collect data from, or display information to, users [1–3]. With the decreasing cost of sensors, actuators and processors, and access to 3D printing, it is easy to design and build all manner of things that have the appearance of familiar objects combined with the capability to sense and respond to user activity, and communicate with other objects. The networking of objects, the recognition of user activity and prediction of user intent, and the use of intent prediction to cue specific actions create new challenges for Ergonomics. In a previous paper, we described the underlying concepts relating to activity and intent, and the use of the Blynk protocol for managing networked communications between smart objects [4]. We also, in that paper, reported development of an initial set of objects and conducted a user trial which showed that, when the handle on a jug rose

© Springer Nature Switzerland AG 2019
D. D. Schmorrow and C. M. Fidopiastis (Eds.): HCII 2019, LNAI 11580, pp. 3–13, 2019.
https://doi.org/10.1007/978-3-030-22419-6_1

automatically, most (22/23 participants in the trial) people would use the hand adjacent to the handle to pick it up rather than their dominant hand – even if, during interview after the experiment, they did not notice that they had done this. We believe that this finding implied that their behaviour was influenced by that of the object, and that the influence could well be occurring at a pre-conscious level. Encouraged by this, we have developed more objects and an experimental protocol that allows us to explore such interactions.

In this paper, familiar objects are fitted with sensors (to support 'awareness and monitoring') and with actuators and other means of display to provide cues to a user. Of particular interest are questions relating to the interpretation of an appropriate action in response to the cues presented by the object. So, if one is performing a familiar task, will the cues be irritating or distracting? If the cues are erroneous (in that they cue a sequence that is different to the one that you intended), will you continue to respond to the cues? In order to hypothesize why users might either ignore cues or respond erroneously, we propose that the interactions between users and smart objects can be considered in terms of affording situations.

1.2 Affording Situations

Gibson [5] introduced the term affordance into psychology, suggesting that we perceive the world in terms of opportunities for action. What an object affords is determined by the physical properties of the objects (e.g., shape, orientation, size), by the action capabilities of the agent, and by the intention that the use of this object will support. This means that 'affordance' cannot simply be a property of an object (so it does not make sense to simply state that a 'cup affords drinking'). Rather, one needs to be situate the use of the object in the context of an ongoing goal-driven activity being performed by an individual with sufficient ability to use *that* object to achieve *that* goal. Thus, an affordance is the relationship between an individual's ability to act and the opportunities provided to that person in the given situation in pursuit of a given goal. That is, a cup of particular dimensions can be grasped by a person with particular physical abilities (e.g., hand size, motor skills etc.) in the context of performing a task with a particular goal: a person laying the table will pick up the cup differently than a person who intends to drink from it; picking up a cup that is full to the brim will be different from picking up a half empty cup. These simple observations lead to the proposal that affordance emerges from the interactions between person and object in a given environment and in pursuit of a given goal and this relationship is captured by the idea of Forms of Engagement, illustrated by Fig. 1 and developed in [6–8].

In Fig. 1, interaction (between person and object) involves several Forms of Engagement. Responding to the specific features of an object in an environment in order to perform an action defines the 'affordance' (indicated by a dotted line around *Environmental* and *Motor Engagement*). This results in a change of state of the object, which the person attends to through *Perceptual Engagement* (i.e., interpreting visual, tactile, kinaesthetic, auditory etc. information as feedback from the performance of an action). So, this proposes two forms of perception by the person: in *environmental engagement* the person responds to features of the object that can support actions (i.e., perception-action coupling) and in *perceptual engagement*, the person responds to the changing state of the object (and environment) as an action is performed. As the object (and environment) change, so this can produce new opportunities for *environmental engagement*. In order to act on an object, there is a need to respond to the 'information' (in a Gibsonian sense) that it conveys. In other words, people are able to 'see' aspects of the object, or the environment, in terms of an action that they want, and are able, to perform in pursuit of a goal. We further separate *Motor Engagement* (the ability to perform an action) from *Morphological Engagement*, the disposition of the person, e.g., in terms of the size of the hand. We have a two-way link between these because hand shaping will be influenced by subsequent actions.

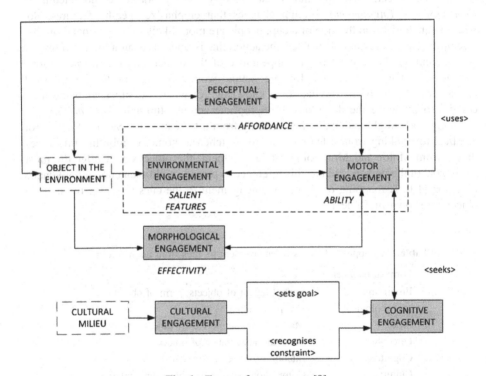

Fig. 1. Forms of engagement [9]

The role of *Cognitive Engagement*, in this description, is two-fold. First, it provides high-level management on ongoing actions (checking for lapses, slips, mistakes etc.) and second, to manage the actions in terms of an overarching goal (e.g., in terms of the Anticipative Actions of the Adaptive Control Model). Of particular interest to this paper, is the extent to which these goals might not be fully-defined prior to performing an action but might evolve in response to the opportunities presented by the objects.

Finally, the notion of an 'acceptable' goal (or 'acceptable' actions to perform on objects) could relate to the culture in which one is acting and is characterized as *Cultural Engagement*.

1.3   Specifying Animate Objects

Having proposed that interaction comprises Forms of Engagement, one can relate these to the possible inferences that animate objects could make (about user intention) as they are being interacted with. At the most basic level, sensors on the object could provide data to characterize the motion, orientation, position, etc. of the object. These data would define the *motor engagement* with which the person was interacting with the object. The object, assuming that it can modify its appearance, could encourage *Environmental Engagement* through changes that emphasize specific features. So, when a handle rises on the side of a cup, people are more likely to use the hand on that side of the cup to pick it up [9]. Thus, the action that is selected in anticipation of motor engagement, could be cued by the appearance of the object, and the *morphological engagement* that is necessary for the action would correspond to the physical appearance of the object. Having some knowledge of where the object is being used could also influence the definition of appropriate actions, through *Cultural Engagement*, e.g., one might anticipate that drinking from a tea cup in the Savoy Hotel is not identical to drinking from a tea cup in one's kitchen at home. Combining inferences drawn from Motor and Morphological Engagement, the object could infer the most likely goal of the person, and use this inference to provide additional cues and guidance of action [10]. A mapping of Forms of Engagement to the cues provided by animate objects in shown in Table 1.

Table 1. Mapping of forms of engagement to cues from animate objects.

Form of engagement	Cues
Environmental	Spatial layout of objects; form of objects
Morphological	Grasp permitted by object
Motor	Manipulation supported by object
Perceptual	State and appearance of object
Cognitive	Intention (of user or 'system')
Cultural	Conventions governing object manipulation

Let us assume that the Animate Object looks like something familiar which has been fitted with sensors [9]. For example, Fig. 2 shows a jug with its sensor unit (developed for the CogWatch project[1]).

---

[1] https://www.birmingham.ac.uk/research/impact/original/cogwatch.aspx.

Fig. 2. Jug and sensor unit

On the one hand, a jug is an object that we 'know' how to use, but on the other hand, this is an alien object that is capable to doing things that we do not, necessarily, fully understand. The jug could, for example, be part of a system that monitors our daily cream intake and the system could have a 'intention' of ensuring that we drink a specified quantity of liquid, or it might be part of a system that has the 'intention' of reducing our caffeine intake (using the pouring of cream as a proxy for drinking coffee). One way in which such 'intentions' could be communicated to the user would via the objects themselves (through lights, sounds, movement etc.), providing feedback and cues to the person. In this way, the form of the objects could display their function and we can create 'affording situations' in which the appropriate action is cued by the objects that the person needs to use.

Relating this to Forms of Engagement and the cues provided by Animate Objects, we could suggest that the motion of an object (part or whole) would be compatible with a movement to be made by the person, e.g., if the handle on the object moved, then one might expect the person to move their hand to that handle. Alternatively, a light on the object turned on (or the object made a noise), one would expect the person's attention to turn to that object. From this, Cue Compatibility draws on environmental and perceptual Forms of Engagement, and can influence Morphological and Motor Engagement. Cognitive Engagement relates to selecting which action to perform, and Cultural Engagement constrains selection of action in terms of 'normal' behavior. From this one can propose a set of tasks that reflect the relationship between the Task that the person ought to perform and the State of the Object (Table 2).

Table 2. Mapping user action to object state

Action	State (object after action)	Pre-condition (object prior to action)
Pick up	Raised	$Jug.Lowered = true \wedge Jug.Handle = up \wedge Jug.LED = on \wedge Jug.Audio = \text{'pick\_me\_up'}$
Put down	Lowered	$Jug.Raised = true$
Move	Location changed	$Jug.Raised = true$
Pour	Object tilted	$Jug.Raised = true \wedge Jug.Location(proximal.cup) = true$

1.4 Constructing Animate Objects

The objects used in this study were designed and 3D printed to look like familiar, everyday objects, e.g., mugs, jugs, kettle, drawer handle, spoon handle. They were designed to incorporate electronics, i.e., sensors, a microprocessor (Arduino), Wi-Fi communications. Communications were managed using the Blynk Bridge Application, which allowed each object to communicate on a network and also allowed an Apple iPhone to be used to send commands to each object and log their status. Figure 3 shows some of the objects used in this study. Each object is powered by a rechargeable power pack and all objects have multi-colour Light Emitting Diode (LED) strips. Commands (from the iPhone or from other objects via the Blynk Bridge network) can turn the LEDs on or off, or change the colour of LED on the objects. Some objects had small motors that could vibrate the object (e.g., the spoon) or lower and raise the handles (e.g., jug, kettle), and small audio chip to replay recorded voice messages.

Fig. 3. Animate objects arranged for the 'tea-making' task

2   User Trial

The aim of the user trial was to explore the interaction between cues offered by Animate Objects and actions provided by users. A basic question was whether cueing from Animate Objects could over-ride participants' expectations of familiar tasks. So, participants were asked to perform two tasks. That is, whether the top-down control of action implied by cognitive engagement could be over-ridden by environmental engagement. In the first task ('simulated tea-making') we assumed that participants would have a well-learned 'script' that defined a sequence of tasks. In this case, the cues from the Objects should have little impact (other than, perhaps, slowing the participants as the waited for the next cue). Conversely, if the Objects cued a sequence that was unfamiliar or illogical, participants would be likely to ignore the cues and rely on their prior knowledge. In a second task ('letter finding'), participants could not rely on prior knowledge. In this case, individual letters were hidden under the Objects and participants had to lift each Object to find the letter in order to spell a 3-letter word. In this case, good performance would involve lifting only correct Objects.

2.1   Participants

Twenty participants (Female = 7, Male = 13, age = 26±, 3.2 years) were recruited from Undergraduate students in the School of Electronic, Electrical and Systems Engineering. None of those recruited had any previous experience interacting with the previous project.

2.2   Procedure

The design of evaluation trials was approved by the School of Engineering, University of Birmingham ethics committee and followed the Declaration of Helsinki. Once they had signed an informed consent form, each participant was shown a printed copy of instructions, to be read prior to commencing their experiments. The instructions were removed from the participants after they had been read and understood. This was done to create a situation to resemble natural interaction with objects.

Three restrictions were imposed on the participants:

1. There was an upper time limit of one minute
2. Only the use of one hand was permitted.
3. Participants were asked to imagine there was water in the jugs in the tea making task.

Two sets of tasks were used. The first set relied on participants being able to define a goal and use this to complete a familiar task sequence. We used simulated tea-making, in which participants acted out the tasks involved in making tea. In the second set, we required participants to learn an unfamiliar sequence. In this case, letters were hidden under each object and participants had to lift the object to find the letters and then spell out a 3-letter word. In this condition, one sequence of cues resulted in the real word 'TIN' and another sequence resulted in the false word 'TEI'. The order in which participants performed these conditions were counter-balanced to minimise learning effects.

Prior to each task, participants placed both hands on the table in front of the Objects. In the no cue condition (for 'simulated tea making'), participants were asked to interact with the Objects as if they were making a cup of tea. In the cue conditions (for both tasks), Object states were controlled by the Experimenter according to a defined script. This allowed motor, audio and LEDs to be turned on or off, so that the State of the Objects changed to provide cues to the participants. The Dependent Variable in the experiment was Total Task Time, which began when one or both hands were lifted and ended when the participant declared that the task was complete.

2.3   Results

The corpus of data is collected from 320 trials, together with 10.5 h of video (which was analyzed for timing of actions; 2 analysts viewed the video and agreed (r = 0.92) on timings), and 40 feedback forms completed by participants. The results are organized by the dependent variables: total time taken and recall of LED colors.

As there were three conditions in which tea-making was performed, a one-way Analysis of Variance (no cues x familiar x unfamiliar) was run. This showed a significant main effect [$F_{(2, 59)} = 8.79$, $p = 0.0005$]. Subsequent pairwise comparison, using paired t-tests, showed a difference between no cues and familiar [$t(19) = 2.23$, $p < 0.05$], between no cues and unfamiliar [$t(19) = 3.6$, $p < 0.005$], and between and familiar and unfamiliar [$t(19) = 5.8$, $p < 0.0001$]. This is illustrated by Fig. 4.

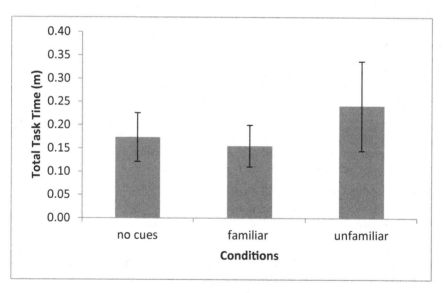

Fig. 4.   Average time to complete tea-making tasks

For the spelling task, there were two conditions (real word and false word). A paired t-test showed a significant difference [$t(19) = 2.65$, $p = 0.05$]. This is illustrated by Fig. 5.

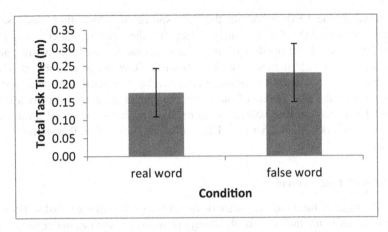

Fig. 5. Average time to complete 'spelling' task

3 Discussion

As might be expected, participants took longer to complete the tea-making task when they were prompted to follow an unfamiliar sequence. More interestingly, participants were faster when they were prompted (in the tea-making task) than when there were no cues. This suggests that, even with familiar task sequences, the provision of cues could be beneficial (at least, in terms of task completion time). For the 'spelling' task, performance on the 'real word' was faster than on the 'false word' – as if participants were trying to align the sequence with a goal but struggling to determine what this might be.

From observation, and discussion with participants during debrief, the use of cues in the unfamiliar tea-making sequence or for the false word task caused participants to interrupt the task sequence or led to them becoming confused. They might, for example, ask for clarification, e.g., 'is it meant to do that?' before continuing with the task. Thus, in the tea-making task, the vibrating spoon was a distraction from the intended sequence but led to a pause while the participant noticed it and then decided whether or not to respond.

3.1 Cues for Action

The cues that were employed in the experiment (audio, LED, moving parts) were attended to and commented upon by participants. Whether or not the cues were responded to depended on when the cues occurred and how the participant would incorporate this into the task sequence that was being followed. If the cue accorded with a logical sequence, it was followed, but if not then it was questioned (or, at least, responded to more slowly).

In the 'spelling' task (in which there might not be so clear a script to define the logical sequence), the cues were more likely to be followed. However, even in this task, participants would seek to make sense of what sequence the cues were guiding them towards. Further, the LEDs were not the only source of information used. Several participants mentioned that the position of objects (for the 'spelling' task) was felt to be important (even though the position of the objects was randomized across trials and so did not provide such information). We infer from this that the association of LED to object to letter was not made as consistently as we had expected, particularly if participants regarded the task as one of managing the spatial arrangement of objects. On reflection, if the only reliable source of information available is spatial arrangement (because they are not certain what the LEDs are telling them), then this strategy makes some sense.

3.2   Forms of Engagement

The role of animate objects in user activity varied across the different tasks. Thus, it is incorrect to assume that the cues would always provoke a response, but equally wrong to assume that these are over-ridden by prior knowledge. More interestingly, it appears that the cues were more beneficial when they support a logical (or familiar) sequence of tasks, and exacerbated confusion on illogical (or unfamiliar) sequences. According to the notion of Forms of Engagement, environmental engagement (in terms of the state of the object) can guide a sequence of actions, with perceptual engagement (as a way of interpreting the state of objects) being mediated by cognitive engagement (as a way of defining a logical sequence) but only when the sequence begins to feel wrong.

3.3   Affording Situations

The notion of 'affording situations' is intended to highlight the importance of context in understanding affordance; it is not simply a matter of saying a 'jug affords lifting or pouring'. While these actions are, of course, possible with a jug, in order to know that this particular jug *could* be lifted or poured at this moment in time, one needs to know the capabilities of the users. Furthermore, in order to know whether either lifting or pouring is appropriate, one also needs to know the intention of the person (or the intention that is plausible for the predicted task sequence). Our aim is to develop technologies that can recognize and interpret contextual factors (based on the analysis of data defining the Forms of Engagement) and use these to discern the plausibility of actions in a given sequence. We can then adapt the animate objects to encourage (or discourage) specific actions.

3.4   Implications for Medical Aids

While the focus of this paper has been on the design and testing of animate objects, it is clear that there could be scope for exploring this concept in the medical domain. Connecting smart objects into a 'rehabilitation internet of things' [11] could offer benefit for monitoring patient activity. Having such objects present cues to guide users can encourage patients to develop or practice actions, particularly if the cues are auditory [12]. Prior work, in the CogWatch project, suggests that adapting familiar objects can reduce anxiety in stroke patients undergoing rehabilitation and also that this can help recall of previously known sequences (lost as a result of stroke). It is a moot point as to whether the type of animate object developed in this project (in which moving parts are intended to cue specific morphological and motor engagement) will prove beneficial for rehabilitation and this is a matter for future research.

3.5   Conclusions

In this paper, we continue our exploration of the ways in which everyday objects can be modified to become animate, and consider how such objects can encourage specific user actions. We demonstrate that people will respond to cues if these correspond to their expectation and understanding of the task. This suggests that the response is mediated by expectation and interpretation. However, relating this mediation to the notion of forms of engagement, we propose that this implies different levels of control. That is, 'affordance' is not simply a matter of responding to the opportunity to act, but is contextualized in terms of the definition of the context. When the context is clear, there appears to be seamless combination of cue response and user activity, such that the cues actually speed-up performance. When the context is unclear or ambiguous (i.e., a cue that is not related to the task being performed, or an outcome that is not clearly understood), response to the cues *could* be helpful but people expend cognitive effort in making sense of these.

References

1. Hornecker, E., Buur, J.: Getting a grip on tangible interaction: A framework on physical space and social interaction. In: CHI 2006, pp. 437–446. ACM, New York (2006)
2. Ishii, H., Ullmer, B.: Tangible bits: towards seamless interfaces between people, bits and atoms. In: CHI 1997, pp. 234–241. ACM, New York (1997)
3. Ishii, H.: The tangible user interface and its evolution. Commun. ACM **51**, 32–36 (2008)
4. Baber, C., Khattab, A., Hermsdörfer, J., Wing, A., Russell, M.: Coaching through smart objects. In: Proceedings of the 11th ACM/EAI International Conference on Pervasive Computing Technologies for Healthcare, pp. 298–306 (2017)
5. Gibson, J.J.: The Ecological Approach to Visual Perception. Houghton Mifflin, Boston (1986)
6. Baber, C.: Cognition and Tool Use. CRC Press, Boca Raton (2003)
7. Baber, C.: Cognitive aspects of tool use. Appl. Ergon. **37**, 3–15 (2006)
8. Baber, C.: Designing smart objects to support affording situations. Front. Psychol. **9**, 292–302 (2018)
9. Baber, C., Khattab, A., Russell, M., Hermsdörfer, J., Wing, A.: Creating affording situations: coaching through animate objects. Sensors **17**, 2308 (2017)
10. Jean-Baptiste, E.M., et al.: Intelligent assistive system using real-time action recognition for stroke survivors. In: 2014 IEEE International Conference on Healthcare Informatics (ICHI), pp. 39–44, September 2014
11. Dobkin, B.H.: A rehabilitation internet of things in the home to augment motor skills and exercise training. Neurorehabilitation Neural Repair **31**, 217–227 (2016)
12. Bienkiewicz, M.N., Gulde, P., Schlegel, A., Hermsdorfer, J.: The use of ecological sounds in facilitation of tool use in apraxia. In: 2nd International Conference on Neurorehabilitation (ICR 2014) (2014)

FUN*ii*: The Physio-Behavioural Adaptive Video Game

Alexis Fortin-Côté[1]([✉]), Nicolas Beaudoin-Gagnon[2], Cindy Chamberland[1],
Frédéric Desbiens[3], Ludovic Lefebvre[3], Jérémy Bergeron[3],
Alexandre Campeau-Lecours[2], Sébastien Tremblay[1], and Philip L. Jackson[1]

[1] School of Psychology, Université Laval, Quebec City, Canada
`alexis.fortin-cote.1@ulaval.ca`
[2] Department of Mechanical Engineering, Université Laval, Quebec City, Canada
[3] Ubisoft Québec, Quebec City, Canada

Abstract. This paper investigates the use of physio-behavioural detection of fun to model players' preferences in real-time in the context of an adaptive game. To do so, a Physiological and Behavioural Model of Fun (PBMF), previously trained on 218 players, was used to model players' preferences based n gameplay events. As a proof-of-concept, we leverged the PBMF to generate a simple player's preference profile tailored to our test-bench game: *Assassin's Creed: Odyssey*, an open-world, action-adventure game. This model associated every player to one of 3 predetermined stereotypical types of player, namely Fight, Stealth and Explore, which are closely tied to mechanics of the *Assassin's Creed* series. Using the inferred preferences, we compared an adaptive vs a non-adaptive version of the same game and tested whether the adaptive version was perceived as more fun than the non-adaptive version by the 39 participants of this study. The results point to the creation of an accurate player's preference profiles during a baseline mission, with profile matching both a "ground truth" Fun Trace – a continuous, subjective rating of a player's fun – and a self-reported profile with an accuracy of 69% and 72% respectively. This, however, did not translate into a measurable difference in reported fun between the adaptive version of the game and the non-adaptive version in neither Fun Trace ratings nor questionnaire answers. Theses findings support that stereotypical preference modelling can be achieved successfully through a physio-behavioural model of fun, but that further investigation on adaptation strategies to those preferences are needed in order to reach the adaptive game's promise of maximizing player's enjoyment.

Keywords: Affective computing · Bio-feedback · Video game · Physiological signals

1 Introduction

Adaptive video game research aims at creating games that adapt to players in order to create more enjoyable, engaging gaming experiences. In the line of affec-

© Springer Nature Switzerland AG 2019
D. D. Schmorrow and C. M. Fidopiastis (Eds.): HCII 2019, LNAI 11580, pp. 14–28, 2019.
https://doi.org/10.1007/978-3-030-22419-6_2

tive computing research, most studies so far have focused on using player's emotional state, physiologically-inferred using fuzzy or supervised learning models, as the target of adaptation. Instead of using basic emotions [1] in the adaption loop, this paper investigates the use of fun levels trough a combined Physiological and Behavioural Model of Fun (PMBF) to model the player's gameplay preferences.

In the literature, Ravaja et al. [2] were the first to show that instantaneous video game events (e.g., gameplay or story elements) could elicit phasic psychophysiological responses indexing emotional valence and arousal, thus highlighting the potential of physiological measures for gaining insight into the player's experience. Therefore, most affective gaming researches have used physiological data as a means of assessing emotional states of players in relation to game content. For example, Gilleade and Dix [3] explored frustration in adaptive games using physiological indicators of emotional arousal. The authors argued that monitoring (and eventually manipulating) the player's frustration level could lead to the development of more complex emotional gaming experiences. Interestingly, other authors later suggested that in-game frustration could also increase player engagement, possibly resulting in an overall more satisfying gaming experience [4]. Also, Martínez et al. [5] studied the generality of physiological features of heart rate and skin conductance as predictors of a player's affective state. They showed that heat rate (HR) and skin conductance (SC) features could be used to predict affective states through different game genres and game mechanics. Finally, Nogueira et al. [6] proposed a model to investigate relationships between emotions (represented as valence and arousal) and game events using the fuzzy physiological model of valence and arousal proposed by Mandryk and Atkins [7].

While some authors have worked on integrating physiological data as direct inputs to a game (i.e. biofeedback games, e.g. [8]), most research using physiological signals have focused on ways of integrating emotions to game design. Those researches have focused mainly on specific emotional states, such as anxiety or fear, that could have more or less straightforward applications in adaptive games. For instance, Liu et al. [9] and Rani et al. [10] used peripheral physiological signals (ECG, EDA, EMG, etc.) to model the player's anxiety level. In accordance with the concept of challenge-skill balance of the flow theory [11–13], they used anxiety as a tool indicating when dynamic difficulty adjustments were required in a game. Similar work has been carried out by Chanel et al. [14] who used both central and peripheral physiological signals for the same purpose. Other authors have employed physiological (ECG and EEG) and behavioural data in order to monitor suspense level in an adaptive survival horror game [15].

Although some authors in the affective gaming community have proposed frameworks to integrate emotions in the design of adaptive games, such as Hudlicka [16,17] and Tijs et al. [18], more research is required to bring this technology to industry-ready levels. Indeed, little is known on how emotion-driven adaptations should be carried out, since different players will most probably react in dissimilar ways to a same adaptation. Thus, emotion-driven adaptations also require player modelling [19], which constitute an ongoing research topic.

Alternatively, recent studies have focused on developing physiological models of the player's fun level. Using *Assassin's Creed* video game series, it has been demonstrated in previous phases of the FUN*ii* project that fun variations were detectable through players' physiological signals and behavioural cues (i.e. physio-behavioural measures). Using the continuous fun rating (Fun Trace) proposed in [20], Clerico et al. [21] and Fortin-Cote et al. [22] have trained a supervised machine learning model that detects player's fun changes throughout a game session, making possible the continuous monitoring of the player's fun. To train this model, they used the FUNii Database, which contains the physiological and behavioural data along with subjective Fun Traces of 218 players, totalling over 400 game sessions.

Monitoring fun using physio-behavioural modalities rather than post-game questionnaires provides continuous assessment of player experience in real-time, without disturbing gameplay. Furthermore, modelling player enjoyment directly from physiology, instead of modelling emotions from physiology, as in most works in the literature, circumvents the problem of having to provide a model that maps emotions to player enjoyment afterwards. Finally, monitoring the player's fun level gives insight into what game events likely yielded increases or decreases of a player's fun during a game session. This kind of information can then be used to build, in real-time, a model of players' preferences.

The FUNii project aims at developing an adaptive gaming system that uses a physio-behavioural model of fun (heart rate, respiratory activity, skin conductance, eye-tracking and head movements) to detect and maximize players' enjoyment in real-time. The first phase of this project was conducted in [20–22] and aimed at designing a PBMF and involved training supervised classification models on over 200 video game players' physio-behavioural data. This paper presents the second and third phase of the FUNii project, which focus on integrating and testing the effectiveness of the system inside an adaptive game.

This paper investigates the use of fun detection to model player's preferences in real-time using physio-behavioural modalities in the context of an adaptive game. The goal of this paper is two-fold. First, we test the reliability of a Physiological and Behavioural Model of Fun (PBMF), trained on 218 players in previous works, to model player's preferences using gameplay events and according to a predetermined stereotypical model of players. Secondly, as a proof-of-concept we use the inferred preferences to tailor the gaming experience and test which of two versions of the same game (an adaptive and a non-adaptive one) is perceived as more fun by the players.

2 Method

Participants were invited to play missions of *Assassin's Creed: Odyssey* that were custom-built by Ubisoft Québec. They first played a baseline mission followed by two variants of a second mission, one predicted by the model to be the player's preferred one and the other, the least preferred one among three possibilities. Their level of fun was measured subjectively afterwards using both questionnaires and the Fun Trace tool in order to determine whether the player's preferred variant of the mission could be correctly identified by the PBMF and gameplay events.

2.1 Experimental Design

The participants played a baseline mission during which they were exposed to a mix of three different styles of gameplay, namely Fight, Exploration and Stealth gameplay styles within the same game (*Assassin's Creed: Odyssey*). Each style respectively implied to: fight one or more enemies (Fight), explore the game world to discover cues or just wander (Exploration), and sneak around enemies while trying to remain unseen (Stealth). Those gameplay styles were detected by the game itself and were based on ingame events such as entering combat, moving crouched and the like. If the game character was not doing an action associated to fight or stealth, it was classified within the exploration category. While playing, fun increases were detected based on physiological signals in real time using our PBMF previously trained on 218 participants during prior phases of the project [21, 22]. At the end of the baseline mission, the player's preference levels for each gameplay styles were compiled by a process further detailed in Sect. 2.6. Those preferences levels allowed us to infer the most and the least preferred gameplay style. Following the baseline mission, players were asked to play two variants of a second mission in a counterbalanced order: one tailored to their most preferred gameplay style and one tailored to their least preferred one.

2.2 Participants

A group of 39 (5 women) participants aged between 20 and 28 years old (M: 23, SD: 2.5) were recruited through Université Laval' student email list as well as Ubisoft Québec player database. Selected participants reported having no diagnosed mental illness, cognitive, neurological or nervous system disorder, nor any uncorrected visual impairment. They also needed to have played the previous instalment of the *Assassin's Creed* series: *Origin*. This was required so that all participants would be already familiar with the game controls and new mechanics introduced in this opus, which are to a great extent the same as in *Assassin's Creed: Odyssey*, and would not have to learn them before the experiment.

2.3 Material

Participants played a custom-built version of the most recent opus of the series *Assassin's Creed: Odyssey*, which was not released to the public at the time of the experiment. A total of 4 custom missions were developed by Ubisoft Québec developers: a baseline mission and three variants of a second mission, namely

Table 1. Missions tested in the experiment along with their descriptions. Each variant was designed to fit one of the 3 stereotypical preference profiles, while Baseline mission presented a balanced mix of fight, stealth and exploration game events.

Mission name	Description
Baseline	In this mission the player is tasked with stealing an object from a camp somewhere in the game world. To this end, the player has to explore the environment to find the camp, sneak around (Stealth) and fight his/her way to the object. It therefore contains all of the three types of gameplay studied
Fight variant	The player is tasked with finding a spy. Finding direction to the spy is not difficult and the challenge resides in defeating groups of enemies, as well a stronger enemy ("Boss") guarding the spy
Stealth variant	The player is tasked with finding a spy, but his location is unknown and the player has to sneak into a heavily guarded fortress to find a map containing the location of the spy. Fighting is required only if the player is detected
Exploration variant	The player is tasked with finding a spy, but through a trail of clues that leads to exploration of the player's surroundings. Little to no fighting is required

a fight tailored mission, a stealth tailored mission and an exploration tailored mission. Summary of the mission's objectives are presented in Table 1.

Physiological and behavioural measures were recorded during every mission by a *Biopac MP150* system at a sampling rate of 100 Hz and the *Smart Eye Pro* eye-tracking system at a sampling rate of 60 Hz. Measurements details are presented in Table 2. Also, a webcam was used to record video of the participant and the *OBS Studio* screen capture software was used to record gameplay.

2.4   Fun Assessment

For this experiment, 3 methods allowed to assess subjective fun and gameplay style preferences during each mission: Fun Trace, fun assessment questionnaire, and gameplay preference questionnaire. First, Fun Trace, which is a continuous rating (analogue scale from −100 to 100) of fun throughout the whole game session, was recorded after a mission playthrough. The Fun Trace homemade software, which is similar to *GTrace* [24], shows participants their gameplay recordings while also presenting a scrolling analogue trace as a visual feedback of their fun annotation. Participants controlled the Fun Trace through a physical control knob: the *PowerMate USB* from *Griffin technology*. Figure 1 presents the application as well as the control knob.

One thing to note is that when participants turned the knob to a value below 0, the scale turns red, while it is green otherwise, making a clear demarcation between positive and negative values. Also, the concept of "fun" itself was

Table 2. Physiological measures recorded during experiment. ECG, RSP and EDA were recorded using a *Biopac MP150* (with sampling rate of 100 Hz), while a *Smart Eye Pro* system (60 Hz) was used for ET and HM.

Physiological signal	Details
Cardiac activity (ECG)	Recorded using an electrocardiogram in lead II configuration
Respiratory activity (RSP)	Recorded using a respiration belt transducer placed around the participant's chest
Electrodermal activity (EDA)	Recorded on the left thenar and hypothenar eminences of the left hand using exosomatic recording with direct current [23]
Eye-tracking (ET)	Measurements include pupil size, blinks and fixations durations and saccades counts along with onscreen gaze intersections
Head movements (HM)	Recorded in six degrees-of-freedom

deliberately left undefined (to gain insight into participants' own conceptions of "fun"). During Fun Trace recording, the playback speed of the video is set to 1.5× in order to minimize task boredom, which could affect validity of the Fun Trace ratings.

Fig. 1. The fun trace application

The second subjective fun assessment was through questionnaires using a six-point Likert scale. The first fun related question was asked after each mission:

Question 1. *How pleasant was this mission?*

The second question was asked at the end of the two mission variants:

Question 2. *What version of the mission did I prefer?*

The responses ranged from 1 to 6, 1 signifying that the first mission was strongly preferred and 6 signifying that the second mission was strongly preferred. The use of a six-point Likert scale forced participants to make a categorical choice between the two variants.

Finally, to get a self-reported measure of the gameplay style preferences (fight, stealth and exploration) of the participants, they were asked, at the end of the study, the following three questions to answer on a six-point Likert scale:

Question 3. *I prefer the fighting gameplay style, meaning that I prefer direct confrontations with the enemy.*

Question 4. *I prefer the stealthy gameplay style, meaning that I prefer moving stealthily while avoiding direct confrontations.*

Question 5. *I prefer the exploration gameplay style, meaning that I prefer to explore the world around to find the best path and hidden treasure.*

2.5 Experiment Protocol

The total duration of the experiment ranged between 2 h and 30 min to 3 h and there was a 20\$ compensation for participating in the study. Participants were first welcomed and invited to sign the required agreements. Electrodes for the *Biopac MP150* system were placed before participants sat at the computer. They were asked to fill out a profile questionnaire which included questions about their gaming habits, self-reported skill level and favourite types of game. The calibration phase of the *Smart Eye Pro* software was then performed. Afterwards, baseline physio-behavioural signals were recorded for 30 s and participants were asked to remain still while looking at a fixation cross. Participants were then presented with a tutorial that served only as a quick refresher since they were already familiar with the last instalment of the *Assassin's Creed* series, where controls were very similar. Participants were then presented with a training mission, where they had to fulfill a set of goals that insure that they possessed the minimal abilities to succeed in the following missions. A schematic representation of the following phases of the experiment is shown in Fig. 2 to help visualize the process. Following this mission, participants played the baseline mission, where they experienced the three types of gameplay. Subsequently, they used the Fun Trace app to generate a Fun Trace for the baseline mission, which was used as a "ground truth" to assess the validity of the inference afterwards. They then played a first variant of the second mission –the variant they prefer (counterbalanced)–, responded to **Question** 1 and used the Fun Trace app. They then played the second variant of the second mission, responded to **Question** 1 again and used the Fun Trace app for the last time. Finally, they were asked to fill the gameplay style preferences questionnaire including **Questions** 2–5. Participants were then debriefed and given monetary compensation.

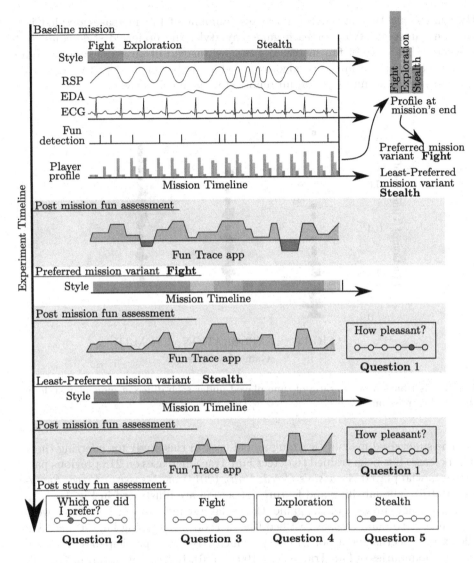

Fig. 2. A schematic representation of the experimental procedure.

2.6   Profiles Generation

Participants' profiles were generated through detection of their fun increases during the baseline mission using the PBMF. This mission gave players the opportunity to experiment with each of the three gameplay styles. Figure 3 displays a summary of the relative amount of time participants experienced each style. One thing to note is that time spent under the fight style was lower than exploration and stealth and that is a consequence of the game architecture: fight sequences are inherently shorter than the two other gameplay style sequences.

By contrasting the rate of fun increases (amount of fun increases divided by time in gameplay style) for each gameplay style, the preferred style could be predicted. The rate of fun increases was used instead of an average of the fun level because of the inference algorithm, which is better at detecting discrete increases over absolute level of fun reported with the Fun Trace.

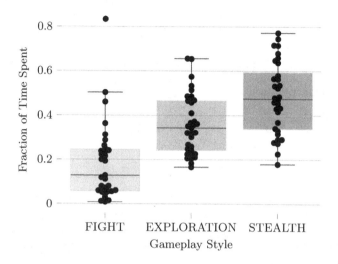

Fig. 3. Distribution of the ratio of time played under each different gameplay style in the baseline mission.

The supervised machine learning model used in this study for inferring these fun increases has been trained to detect Fun Trace increases on 218 previous participants who played *Assassin's Creed: Unity* (2014) or *Assassin's Creed: Syndicate* (2015). The labelling and feature extraction are illustrated in Fig. 4.

For a single mission of the game, the 20 largest increases of Fun Trace were identified. Two main reasons justified this method. First, using largest increases alleviates concerns about border effects, which arise when participants hit one of the two boundaries of Fun Trace (i.e. −100 or 100). Indeed, when this happened, participants were forced to increase or decrease Fun Trace, which introduced noise. Second, using increases of Fun Trace instead of the value itself is explained by the fact that human preferences are arguably more ordinal in nature than cardinal [25], meaning that relative levels of Fun Trace in a game session (e.g., going from low to high Fun Trace value) is more likely to capture relevant information about a player's experience than absolute Fun Trace value. With those increases identified, a 20 s temporal window of the physio-behavioural signals was extracted around the increases to capture its physio-behavioural signature. Examples of constant (no changes) Fun Trace and decrease in the Fun Trace were extracted in a similar fashion. We ended up with 7623 labelled samples, 2496 of which corresponds to Fun Trace increases, while the remaining one were examples of constant (2706) or decrease (2421) in the Fun Trace. Temporal and spectral statistics were extracted from each of the physio-behavioural signal as to compile a vector of 201 features from each sample. Using inter-participant, meaning no samples from a single participant was used in the training of the model used to make prediction on his/her subset of data, K-Fold cross validation was used to train the model, tune the hyper parameters and select the most accurate model with the Scikit-learn library [26]. In this case, the most accurate model, with an F1 score of 65% was an extreme gradient boosting classifiers (XGBoost implementation [27]) compared to an F1 score of 56% for a Stratified Dummy Classifier.

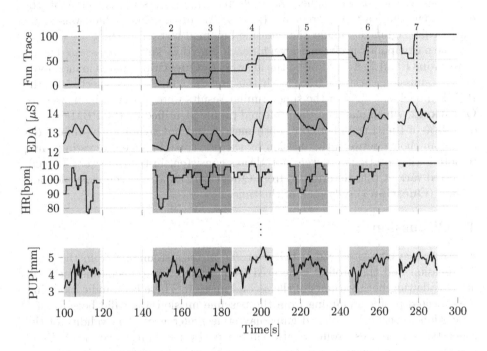

Fig. 4. Examples of the fun trace labelling and corresponding physio-behavioural signals. A total of 7 significant increases in the Fun Trace are represented by a vertical dotted line. The 20 s windows extracted around fun increases are represented by shaded regions in each of the plot. The sample of physio-behavioural signals shown is electrodermal activity (EDA), heart rate (HR) and pupil dilatation (PUP).

3   Results

3.1   Validation of the Generated Profile on the Baseline Mission

In order to validate the generated profile, the profile computed by the real-time algorithm was compared to the one computed using the actual Fun Trace of the baseline mission. Using the amount of fun increases in the "ground-truth" Fun Trace for the preferred and the least preferred variants, as determined by the model, the observed agreement was 69% ($p = 0.03$) between the two. This means that the actual Fun Trace correctly showed more fun increases in the preferred profile 69% of the time. This is statistically significant under the one-side binomial test ($\alpha = 0.05$), where the null hypothesis is that the detected profile is random (probability of success 0.5). To further consolidate that the detected profile was a valid one, it was possible to compare it to the self-reported gameplay style preferences provided by the participant on a six-point Likert scale at the end of the experiment (**Questions** 3–5). Results revealed that 76% of the time, participants rated higher the preferred gameplay style detected by the model than the least preferred one ($p = 0.004$) of the time, which was again statistically significant under the one-side binomial test ($\alpha = 0.05$).

The participants' preference level for each played variants were assessed using **Questions** 1 and 2. According to the answers to **Question** 1, participants' most pleasant mission variant agreed with the preferred mission as selected by the PBMF model only 52% of the time. Similar results were observed in answers to **Question** 2, which matched the preferred profile identified by the PBMF model only 48% of the time. Therefore, both metrics did not significantly differ from a random choice between the mission by the participants. One interesting thing to note is that participants were not always consistent in their answers. Indeed, preferred variant (has determined from **Question** 2) matched the most pleasant variant (**Question** 1) only 69% of the time.

4   Discussion

A discrepancy was observed between profile metrics stemming from the baseline mission and the profiles generated from subjective appreciation questionnaires following both variants of the second mission. There is an indication that the detected profile stemming from the baseline mission was valid because of a concordance with self-reported gameplay style preferences. The validity of the generated preferences profile is also supported by the Fun Trace of the baseline mission. This suggests valid inference of player's preference profile from the baseline mission, and therefore supports that our PBMF can be used to model players' preferences in the context of a predetermined stereotypical preferences model. The discordance with the fun reports of second mission variants could stem from the adaptation strategy or the measure of the response to the adaptation, similar to those raised by Fuchs [28]. The adaptation strategy could be at fault in that the categorisation of the variant of each type might be too coarse. Indeed, while each variant was designed to favour its corresponding style, the game did not enforce a particular way to play. For example, in the stealth variant of the second mission a participant might still tried to fight its way through the level, which is difficult (if not impossible) and prompted the experimenter to redirect the participant to the more streamlined path.

A failure in the measure of the response to the game's adaptation could also be at fault. Fun Traces are subjective, their temporal resolution is higher than questionnaires', and therefore allow for a much more precise inspection of differences in game experience between each gameplay style, something that was not reflected by answers to the questionnaire.

Furthermore, the overall improvement of game experience caused by tailoring a single mission to inferred preferences might not be important enough to be measurable with a six-point Likert scale, especially considering that participants rated the mission between 3 or 5 in most cases (85% of the time). Another possibility is that the 3-class preferences model that we used in this paper, even though being strongly tied to our test-bench game mechanics, is too simplistic to properly orient the adaptation process. Indeed, it is not straightforward that a player's preferences are fixed throughout the game [29], and it is even less straightforward that they unfold only in one gameplay dimension (e.g. a player that enjoys Fight might as well enjoy Stealth, even if it is to a lesser extent). Thus, a adaptive game using the same PBMF would most likely benefit from using a game-agnostic model of player types [30].

Finally, there is also the possibility that a single adaptive mission is not long enough to measure fun increases, but that a sequence of multiple missions tailored to the evolving player profile might generate an adaptive game that is perceived as more fun overall. This would necessitate further study, including multiple, longer, play sessions with the same players as they progress through several tailored missions.

5 Conclusion

This research is a step towards integrating real-time player modelling, using objective measurements of players experience through physio-behavioural data, to the design of an adaptive video game. Using real-time prediction of the fun level of player can help steer the game towards the player preferences and even adapt to changing preferences during gameplay. While the real-time generated profile seems accurate under two different metrics (the "ground truth" Fun Trace as well as self-reported profile), the adaptation strategy did not provide measurable improvements in enjoyment of the subsequent mission. Further work might include investigation into which types of adaptation strategies might show a measurable improvement in enjoyment. A simpler game that allows easier adaptation to different styles might provide opportunities to test more real-time adaptation and leverage further benefits from real-time profile generation. For example, becoming tired of a particular gameplay style could be detected by the use of a rolling average of the profiles allowing for varying preferences inside a mission and more fluid adaptations.

Acknowledgement. This project was funded by NSERC-CRSNG, Ubisoft Québec and Prompt. Additional thanks to Nvidia for providing a video card for deep learning analysis through their GPU Grant Program.

References

1. Ekman, P.: An argument for basic emotions. Cogn. Emot. **6**(3–4), 169–200 (1992). https://doi.org/10.1080/02699939208411068
2. Ravaja, N., Saari, T., Salminen, M., Laarni, J., Kallinen, K.: Phasic emotional reactions to video game events: a psychophysiological investigation. Media Psychol. **8**(4), 343–367 (2006). https://doi.org/10.1207/s1532785xmep0804_2. ISSN 1521-3269, 1532-785X
3. Gilleade, K.M., Dix, A.: Using frustration in the design of adaptive videogames. In: Proceedings of the 2004 ACM SIGCHI International Conference on Advances in Computer Entertainment Technology, ACE 2004, pp. 228–232. ACM, New York. ISBN 978-1-58113-882-5. https://doi.org/10.1145/1067343.1067372
4. Boulton, A., Hourizi, R., Jefferies, D., Guy, A.: A little bit of frustration can go a long way. In: Winands, M.H.M., van den Herik, H.J., Kosters, W.A. (eds.) ACG 2017. LNCS, vol. 10664, pp. 188–200. Springer, Cham (2017). https://doi.org/10.1007/978-3-319-71649-7_16
5. Perez Martínez, H., Garbarino, M., Yannakakis, G.N.: Generic physiological features as predictors of player experience. In: D'Mello, S., Graesser, A., Schuller, B., Martin, J.-C. (eds.) ACII 2011. LNCS, vol. 6974, pp. 267–276. Springer, Heidelberg (2011). https://doi.org/10.1007/978-3-642-24600-5_30
6. Nogueira, P.A., Aguiar, R., Rodrigues, R., Oliveira, E.: Computational models of players' physiological-based emotional reactions: a digital games case study. In: 2014 IEEE/WIC/ACM International Joint Conferences on Web Intelligence (WI) and Intelligent Agent Technologies (IAT), vol. 3, pp. 278–285, August 2014. https://doi.org/10.1109/WI-IAT.2014.178
7. Mandryk, R.L., Atkins, M.S.: A fuzzy physiological approach for continuously modeling emotion during interaction with play technologies. Int. J. Hum.-Comput. Stud. **65**(4), 329–347 (2007). ISSN 10715819, https://doi.org/10.1016/j.ijhcs.2006.11.011, http://linkinghub.elsevier.com/retrieve/pii/S1071581906001881
8. Nacke, L.E., Kalyn, M., Lough, C., Mandryk, R.L.: Biofeedback game design: using direct and indirect physiological control to enhance game interaction. In: Proceedings of the SIGCHI Conference on Human Factors in Computing Systems, CHI 2011, pp. 103–112. ACM, New York (2011). ISBN 978-1-4503-0228-9. https://doi.org/10.1145/1978942.1978958
9. Liu, C., Agrawal, P., Sarkar, N., Chen, S.: Dynamic difficulty adjustment in computer games through real-time anxiety-based affective feedback. Int. J. Hum.-Comput. Interact. **25**(6), 506–529 (2009). https://doi.org/10.1080/10447310902963944. ISSN 1044-7318, 1532-7590
10. Rani, P., Sarkar, N., Liu, C.: Maintaining optimal challenge in computer games through real-time physiological feedback. In: Proceedings of the 11th International Conference on Human Computer Interaction, vol. 58 (2005)
11. Csikszentmihalyi, M.: FLOW: The Psychology of Optimal Experience, p. 6. Harper Perennial, New York (1990)
12. Nakamura, J., Csikszentmihalyi, M.: The Concept of Flow. Flow and the Foundations of Positive Psychology, pp. 239–263. Springer, Dordrecht (2014). https://doi.org/10.1007/978-94-017-9088-8_16
13. Chen, J.: Flow in games (and everything else). Commun. ACM **50**(4), 31–34 (2007). https://doi.org/10.1145/1232743.1232769. ISSN 0001-0782
14. Chanel, G., Rebetez, C., Bétrancourt, M., Pun, T.: Emotion assessment from physiological signals for adaptation of game difficulty. IEEE Trans. Syst. Man Cybern.-Part A: Syst. Hum. **41**(6), 1052–1063 (2011)

15. Vachiratamporn, V., Legaspi, R., Moriyama, K., Numao, M.: Towards the design of affective survival horror games: an investigation on player affect. In: 2013 Humaine Association Conference on Affective Computing and Intelligent Interaction, pp. 576–581, September 2013. https://doi.org/10.1109/ACII.2013.101

16. Hudlicka, E.: Affective game engines: motivation and requirements. In: Proceedings of the 4th International Conference on Foundations of Digital Games, FDG 2009, pp. 299–306. ACM, New York (2009). ISBN 978-1-60558-437-9. https://doi.org/10.1145/1536513.1536565

17. Nogueira, P.A., Rodrigues, R., Oliveira, E., Nacke, L.E.: Guided emotional state regulation: understanding and shaping players' affective experiences in digital games. In: Proceedings of the Ninth AAAI Conference on Artificial Intelligence and Interactive Digital Entertainment, AIIDE 2013, pp. 51–57. AAAI Press, Boston (2013). ISBN 978-1-57735-607-3. http://dl.acm.org/citation.cfm?id=3014712.3014721

18. Tijs, T., Brokken, D., IJsselsteijn, W.: Creating an emotionally adaptive game. In: Stevens, S.M., Saldamarco, S.J. (eds.) ICEC 2008. LNCS, vol. 5309, pp. 122–133. Springer, Heidelberg (2008). https://doi.org/10.1007/978-3-540-89222-9_14

19. Yannakakis, G.N., Spronck, P., Loiacono, D., André, E.: Player modeling. In: Lucas, S.M., Mateas, M., Preuss, M., Spronck, P., Togelius, J. (eds.) Artificial and Computational Intelligence in Games, volume 6 of Dagstuhl Follow-Ups, pp. 45–59. Schloss Dagstuhl-Leibniz-Zentrum fuer Informatik, Dagstuhl (2013). ISBN 978-3-939897-62-0. https://doi.org/10.4230/DFU.Vol6.12191.45, http://drops.dagstuhl.de/opus/volltexte/2013/4335

20. Chamberland, C., Grégoire, M., Michon, P.-E., Gagnon, J.-C., Jackson, P.L., Tremblay, S.: A cognitive and affective neuroergonomics approach to game design. In: Proceedings of the Human Factors and Ergonomics Society Annual Meeting **59**(1), 1075–1079 (2015). ISSN 1541-9312. https://doi.org/10.1177/1541931215591301

21. Clerico, A., et al.: Biometrics and classifier fusion to predict the fun-factor in video gaming. In: 2016 IEEE Conference on Computational Intelligence and Games (CIG), pp. 1–8. IEEE (2016). http://ieeexplore.ieee.org/abstract/document/7860418/

22. Fortin-Cote, A., et al.: Predicting video game players' fun from physiological and behavioural data. In: Future of Information and Communication Conference (2018)

23. Boucsein, W.: Electrodermal Activity, 2nd edn. Springer, New York (2012). ISBN 978-1-4614-1125-3, 978-1-4614-1126-0

24. Cowie, R., Sawey, M., Doherty, C., Jaimovich, J., Fyans, C., Stapleton P.: Gtrace: general trace program compatible with emotionml. In: 2013 Humaine Association Conference on Affective Computing and Intelligent Interaction (ACII), pp. 709–710, September 2014. https://doi.org/10.1109/ACII.2013.126

25. Yannakakis, G.N., Cowie, R., Busso, C.: The ordinal nature of emotions: an emerging approach. IEEE Transactions on Affective Computing, 1 (2018). ISSN 1949-3045. https://doi.org/10.1109/TAFFC.2018.2879512

26. Pedregosa, F., et al.: Scikit-learn: machine learning in python. J. Mach. Learn. Res. **12**, 2825–2830 (2011)

27. Chen, T., Guestrin C.: XGBoost: a scalable tree boosting system. In: Proceedings of the 22nd ACM SIGKDD International Conference on Knowledge Discovery and Data Mining, KDD 2016, pp. 785–794. ACM, New York (2016). ISBN 978-1-4503-4232-2. https://doi.org/10.1145/2939672.2939785

28. Fuchs, S.: Session overview: adaptation strategies and adaptation management. In: Schmorrow, D.D., Fidopiastis, C.M. (eds.) AC 2018. LNCS (LNAI), vol. 10915, pp. 3–8. Springer, Cham (2018). https://doi.org/10.1007/978-3-319-91470-1_1

29. Kotsia, I., Zafeiriou, S., Fotopoulos, S.: Affective gaming: a comprehensive survey. In: 2013 IEEE Conference on Computer Vision and Pattern Recognition Workshops, pp. 663–670, June 2013. https://doi.org/10.1109/CVPRW.2013.100
30. Hamari, J., Tuunanen, J.: Player types: a meta-synthesis. Trans. Digit. Games Res. Assoc. **1**(2) (2014). ISSN 2328–9422, 2328–9414. https://doi.org/10.26503/todigra.v1i2.13, http://todigra.org/index.php/todigra/article/view/13

Deriving Features for Designing Ambient Media

Kota Gushima$^{(\boxtimes)}$, Shuma Toyama, Yukiko Kinoshita,
and Tatsuo Nakajima

Department of Computer Science and Engineering,
Waseda University, Tokyo, Japan
{gushi,toyama,yukiko-kinoshita,
tatsuo}@dcl.cs.waseda.ac.jp

Abstract. In this study, we focus on an information presentation method to reduce cognitive load. For example, a typical method of presenting information in a useful manner is Augmented Reality, which overlays information on the real world; however, this approach causes high cognitive load because it is too loud compared to the real world. Thus, we focus on an ambient media approach proposed by Hiroshi Ishii that works on the background of human consciousness to reduce the cognitive load. We aim to find expressions that are midway between information superposition and ambient media. In this study, we focus on the visual and auditory senses, which are mainly used for information presentation, and learn the features necessary to design ambient media through our case studies. We develop two case studies: HoloList and Ambient Table. We implement and evaluate these systems to learn the features required to design the ambient media.

Keywords: Augmented reality · Mixed reality · Ambient media ·
Notification · Head-mounted display

1 Introduction

A great deal of information is processed by people in their daily lives. In the real world, physical data such as temperature, humidity, position information, and sound are available. People use sensory organs to detect these data and process it as information. In recent years, with the development of digital technology, more information is being transformed into forms that people can feel; consequently, people live with various information. For example, by digitizing various information such as weather forecasts and stock prices, people can analyze a wider range of information. As a result, they can learn trends and new facts. In addition, sensors are being introduced to various places using IoT technology, and the range where data can be acquired has also expanded [3].

The method of information presentation using digital technology has been developed over time. When information presentation using a computer changed from the character-based interface (CUI) to the graphical user interface (GUI), more people were able to handle computers; consequently, discussion on information provision using digital technology has continued. In addition, augmented reality (AR) has been

D. D. Schmorrow and C. M. Fidopiastis (Eds.): HCII 2019, LNAI 11580, pp. 29–40, 2019.
https://doi.org/10.1007/978-3-030-22419-6_3

developed, and it is now possible to seamlessly present information to the real world. A typical method of presenting information in AR overlays the information on the real world. The approach used with AR is primarily visual, but it is also used for audio. For example, a car navigation system using voice is similar to AR, in that it overlays sounds. While this approach supplies much information to people, there is a possibility that the cognitive load will rise beyond ideal levels.

Ambient media has been researched as an information-providing method that is integrated into the environment while having a low cognitive load. Proposed by Hiroshi Ishii, this media works on the background of human consciousness [6]. An example would be to display information by projecting an abstract image on a wall [10]. In this study, we focus on the visual and auditory senses, which are mainly used for information acquisition. For visual ambient media, we focus on mixed reality (MR) technology to provide information in daily life and we developed a prototype named HoloList. For auditory ambient media, we developed Ambient Table, which tells users the positions of objects on a table. Through these case studies, we learn the features required to design modern ambient media.

2 Related Work: Ambient Media

Ambient Media uses intermediary media such as sound, light, air flow, or water movement to interact with people's perception of the background [6]. Ambient Media research began in the 1990s, and currently, there are several related products. As an early example, Pinwheel [5] was developed in 1998. It provides digital information with wind and is a system in which wind turbines spin according to the value of information. Another example is ambientROOM [7], which extends the real space of the user's physical environment with information technology with, e.g., sound, as an interface of digital contents to provide information passively.

A modern example, Class Beacons [2], is a system in which devices are placed in a classroom desk and the color of the device changes according to the position of the teacher. Other Ambient Media on a display include Sideshow [4] and Scope [11]. CityCell [8], which puts emphasis on interactions, has stacked hexagonally shining objects in public spaces, changed the way they are stacked, and collaborated with smartphones to change how light is emitted.

3 Visual Case Study: HoloList

3.1 Design and Purpose

The prototype HoloList used in this research is a HoloLens application developed with the Unity game engine[1] and C#. In this application, as shown in Fig. 1, the user can interact with the system with an index finger. The purpose of the case study was to investigate methods for providing information using visual virtual objects.

---

[1] https://unity3d.com/

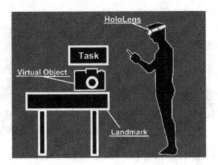

Fig. 1. Concept of HoloList

HoloList is an application that manages a task schedule by linking a virtual object with the task. In HoloList, the task display format can be switched between three modes. Details of each mode are shown in Fig. 2.

In Mode 1, the display format is a planar object showing the task contents in text, as shown in Fig. 2a. In this mode, it is assumed that the user places the virtual object on a flat surface, such as an indoor desk and a wall, and the visual effect on the real space by the virtual object is small.

In Mode 2, a 3D model is arranged that informs the user of a task, and the task contents are displayed by text when the user gazes at the model, as shown in Fig. 2b. In this mode, it is assumed that the user selects 3D models prepared for the task contents in the application and places them at a position where real objects such as indoor desks and shelves are present. As a feature of this mode, if the user does not gaze it, the contents of the task are not displayed, thus reducing the amount of information that is visible simultaneously. There are four 3D models that can be selected in this application: a camera, a book, a medicine bottle, and a laptop computer.

In Mode 3, task contents are displayed in text by gazing at the 3D model that performs a specific animation, as shown in Fig. 2c. In this mode, it is possible to select the type of animation to be performed. As a feature of this mode, the 3D model is animated, so it is easy to distinguish it from the real object.

3.2 Preliminary Evaluation

In this case study, we conducted preliminary experiments on the display type of virtual objects. The purpose was to obtain knowledge for improving applications of virtual object provision methods by evaluating each mode of application and comparing them.

User Enactments [9] was adopted as the evaluation method for this project. User Enactments is a method to evaluate the degree of accomplishment when a subject performs a task, following a scenario created to evaluate an application using a new technology. The scenario used in this experiment is shown below.

Alan performed schedule management with HoloList to spend the holidays effectively. Alan set task contents, 3D model, and corresponding animation as follows.

(a)　　　　　　　　　　　(b)　　　　　　　　　　　(c)

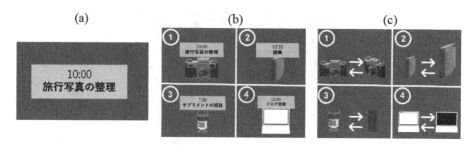

Fig. 2. Design of HoloList. (a): Mode 1. (b): Mode 2. (c): Mode 3.

- 10:00: Organize Travel Photo - Camera - Object rotates
- 12:30: Updating Blog - Object size changes
- 19:30: Supplement intake - object color changes
- 22:30: Reading - book - 3D model of display part blinks

The participants were seven males and one female aged 20 to 24 who have experience using HMD-based AR/VR/MR applications. The participants were instructed to perform tasks using HoloList for each mode according to the scenario and to answer a questionnaire and undergo an interview after the task was finished.

The user browsed the four virtual objects placed in each mode. The user evaluated through a questionnaire and an interview whether the mode was able to achieve the scenario. The 3D models handled in Modes 2 and 3 are shown in Fig. 2. Figure 3 shows the results of the questionnaire for evaluating HoloList in each mode.

3.3 Review of Design

3.3.1 Stepwise Display of Information

Regarding the method of displaying information, it was found that displaying the text via the visualized object is preferred over directly arranging the text, as shown in Fig. 3b. The reason is that the displayed text has less affinity with the real object, and the information not focused upon by the user is seen. However, according to comments from the interviews, the interaction method using gazing should be changed, because the current gaze-based interaction method causes unintentional action.

3.3.2 Information Presentation by Animation

For the animation of the virtual object in Mode 3, many opinions were obtained from comments. From the positive opinions, the animation can be used to distinguish between real objects and virtual objects. From the negative opinions, depending on the type of animation, the user's consciousness is always directed towards the object. From these opinions, it is assumed that verifying whether the degree of attention in the type of animation of the virtual object can be presented is necessary.

3.3.3 Mode 4: Representing Start Time of the Task by Animation

Mode 4 was created based on the above considerations. In this mode, the contents are displayed in text while the user gazes at the animated 3D model and gestures to bend an

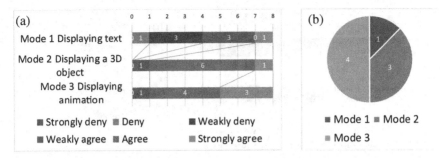

Fig. 3. (a): Do you think that this display format is suitable for the presented scenario? (b): Which mode do you think is most suitable for the scenario?

index finger. The intent of the gesture input is to avoid displaying tasks unintentionally by setting the user's conscious behavior to present the information. Additionally, unlike Mode 3, all 3D objects are animated to rotate, and the rotation speed increases as the time of the task becomes closer. With this function, the user can understand the priority of the task from the animation.

3.4 Experiment

The application was improved based on the review and the experiment was conducted again. The purpose of this experiment was to perform common tasks in the application before and after the improvement and to verify if the results improved.

In this experiment, the evaluation was performed using the same scenario as in the preliminary experiment. The subjects were the seven males and one female who participated in the preliminary experiment. We evaluated the use of Mode 3 and Mode 4 of HoloList to verify the change of interaction for displaying information from the 3D object and the possibility of presenting the information by animation. The task contents were the same as in the preliminary experiment, and a questionnaire and an interview were conducted after the completion of the task. For the speed of animation in Mode 4, we set the time of the experiment to perform the task at 7 AM virtually, and the system reflects the difference between 7 AM and the start time of each task shown in the scenario.

Figures 4 and 5 show the results of the questionnaire for each mode. From Fig. 4b, Mode 4 is evaluated higher than Mode 3. The participants felt that the presentation of information in the animation, which is the object of Mode 4, is effective. Indeed, as shown in Fig. 5a, six participants were impressed by the book object in Mode 3. However, from the comments, that was regarded as a bad impression, and it seems that the use of such an animation with a big influence on the perspective of the user should be avoided.

For the display format, from the interview, we found that many users favorably evaluated the display information provided by the intentional gesture. However, there are comments that users are able to change the display format arbitrarily. It is thought that the display format needs to be designed to be customizable, according to the user's preference.

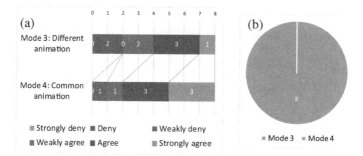

Fig. 4. Result of the experiment. (a): Which mode do you think is most suitable for the scenario? (b): Do you think that this display format is suitable for the presented scenario?

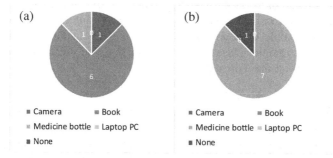

Fig. 5. Result of the question "which object is most impressive?" (a): in Mode 3. (b): in Mode 4.

4 Auditory Case Study: Ambient Media

4.1 Overview

We developed a system called Ambient Table as an information presentation system using Ambient Music. This system presents the situation on a desk with music and aims to present information naturally by hearing. Specifically, it uses the following two scenarios.

Scenario 1: *Alice has a 5-year-old daughter, who is compatible with part-time jobs and housework. Her daughter has recently become interested in drawing. While her daughter is drawing in the living room, Alice started office work on a dining table. After some time, she noticed that the music became noisy. The music indicated that her daughter was spreading crayons and papers in the living room. She stopped working and started to take care of her.*

Scenario 2: *Bob is a college student who lives in a shared house with five people. Recently they became his friends and he chatted with them in a shared space. One day, Bob faced an exam, so he was studying in his room. At night, Bob felt a little hungry, so he decided to make instant noodles in the kitchen. At that time, he heard the music. It seems that things were scattered on the table of the shared space. He noticed that everyone went to sleep without cleaning up. The desk could not be seen because it was dark, but he could recognize the terrible situation by hearing the music. Finally, he started to clean up the shared desk.*

To create a system that reproduces this scenario, we decided to present the positional information of the object on the desk as music.

4.2 Implementation

We used OpenCV, which is a Python library, and Sonic Pi to develop this prototype. In the prototype, users worked on edged paper as a table, as shown in Fig. 6. The system uses a web camera to capture the situation on the table. The image is binarized using OpenCV to detect the outline of an object within the edges. From the contours of the object, the system calculates the size and centroid of the object. Next, the system considers each object as a musical note and creates music as if it were drawing a musical score on the table. The vertical position on the table corresponds to the pitch of a note on the staff (i.e., the sound's pitch), the horizontal position corresponds to the timing of a note, and the size of the object corresponds to the duration of the sound. To express the sound, we use Sonic Pi [1], a tool to perform live coding, which is a new method of creating music in real time by coding.

4.3 Experiment

When using audio for information presentation, one problem is that it takes time to present the information. For voice guidance, as the amount of information increases, it takes more time to present all the information. I the situation on the table changes constantly, it cannot be reflected in real time if the music speed is slow. On the other hand, if the music speed is fast, the situation on the table can be reflected and the change of sounds becomes easier to understand; however it is difficult to produce these sounds with ambience. Therefore, we evaluated how to balance the speed of presentation of information and the sound's ambient quality. In the experiment, we also evaluated the feasibility of the scenario to provide new value, by incorporating music in daily life.

The experiment was conducted with a total of 16 people, comprising 13 males and 3 females, who read the scenarios and answered a questionnaire. Next, we introduced the system and confirmed they knew how to use it. They assembled puzzles for 2 min while listening to the music generated by the system. After performing this task, they answered a questionnaire. This procedure was one set of the evaluation, and three of these sets were conducted with different speeds of the music. Finally, we interviewed them and asked them to comment freely on this system.

4.4 Results

The experimental results are shown in Fig. 7. Additional answers obtained from the interview are summarized as follows.

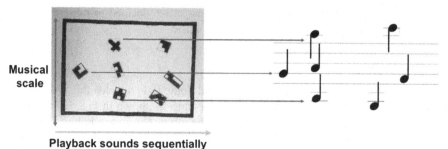

Fig. 6. Concept of Ambient Table system

Regarding the scenario:

- *It can be used instead of vision, I do not need information where I can see it. I want it to compensate for information I do not see.*
- *Auditory information is always coming, so it is nonsense to express an unpleasant state with unpleasant music.*
- *Ambience and alertness cannot coexist. I think I do not listen to music even if I stop concentrating, or I listen to music and concentrate at the same time. Information becomes noisy when the information is given as sound.*
- *If you want various types of information, you do not know what information corresponds to which element. Mapping is important. The amount of information decreases if the number of elements decrease, to understand the media more easily.*
- *I think it will be affected by the surrounding sound.*

Regarding the system:

- *Responses were divided into people who felt that the fast tempo was bad because it felt like being hurried, and people who felt the change was enjoyable.*
- *Many people noticed a change in the number of sounds and the pitch of the sound. There were a few people who noticed the tone.*
- *The slow tempo music was often not liked, because they felt it was noisy.*
- *Because dissonance was anxious, it may be good to change it so that the music becomes beautiful when the situation is improved.*

4.5 Discussion

From the scenario's questionnaire, if the necessity for information is low, the system is not required unless it enhances the aesthetics as music. On the other hand, if the necessity for information is high and it is necessary to convey information properly, we believe that the system will be used even if the aesthetic is lost.

From the system's questionnaire, the same result as the prediction was obtained; the change on the table is easy to understand as information when it occurred with a fast playback speed. However, many people can concentrate on the task while listening to the middle playback speed. We expected that people can concentrate with the slow playback speed, but they actually feel it is not ambient. According to the interview,

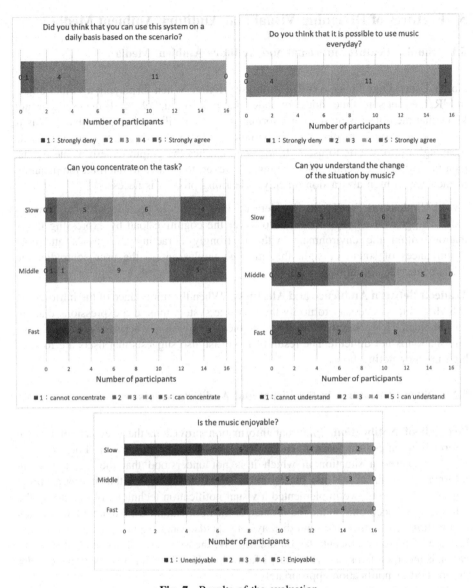

Fig. 7. Results of the evaluation

people may be calm listening to the slow ambient music; however, in reality they feel that the slow music is redundant and noisy. Because people usually work with low ambient background music, people can adjust to loud music with concentration. Based on the results, we think that to the use of ambient music is unnecessary, because it is often preferred to use music that can indicate the change.

5 Features of Designing Visual and Auditory Ambient Media

5.1 Similar Features in Visual and Auditory Ambient Media

Notify Users That There Are Information Media. When displaying a virtual object in MR, the user sometimes does not notice that it is a virtual object. If the user does not know that the object is presenting information, he/she will overlook the object. This is also true for sound media; it is important to clearly indicate that sounds provide meaningful information to users. If someone teaches the output pattern of the media, then the user can recognize the meaning. However, to be able to interpret the meaning of media with high abstraction quickly, a learning process is necessary.

Avoid Unpleasant Feedback. Ambient media's aim is to provide information naturally. The goal of ambient media is to lower the cognitive load by expressing information within the environment without strongly attracting the user's attention. Disturbances of sound or sight that may be unpleasant to the user contradict the fundamental idea of ambient media.

Tradeoff Between Ambience and Alertness. When the importance of the information is high and it is necessary to notify the user, the system must use expressions that are not ambient. However, making a non-ambient expression when the user needs the information is not difficult. The result of HoloList also suggests that users are fond of high urgency notifications.

5.2 Different Features in the Visual and Auditory Senses

Strength of Notification. In visual information provision, the user cannot receive information unless the user intentionally gazes at the information. For example, HoloList caused a situation in which it is not understood that the object was an information media in the first place, unless it was first clarified that it was a virtual object. By updating, we implemented a visual notification technique of increasing the rotation speed according to the importance of the notification. On the other hand, visual notifications on smartphones and displays are standard and the notifications work well because, if the user concentrates on the display, then he/she will notice it. If a notification is incorporated in a part of daily life, such as HoloList, it is necessary to consider the method of notification appropriately.

In auditory information provision, systems can provide information to the user semi-forcibly. Therefore, it is necessary to avoid expressions that are particularly unpleasant when providing information using sound.

Metaphor. Visual metaphors are common in everyday life. There are various forms of metaphors in daily life, such as pictograms and icons. HoloList provided information by using media that becomes a metaphor. However, there are few cases of auditory metaphors. Currently, notifications such as electronic sounds only link notification sounds and content types and are not metaphors. However, there is a possibility that a metaphor could be implemented using music, rather than just a sound. For example, it might be possible to realize an auditory metaphor by sampling the sound of water flowing and tying it to content related to water.

6 Conclusion and Future Work

In this study, we reveal features for designing visual and auditory ambient media. To clarify the features, we developed two case studies: a visual case study and an auditory case study. In the visual case study, we developed an MR application using HoloLens, named HoloList, to provide information on tasks related to positions in real space. In the auditory case study, we developed a system using ambient media, named Ambient Table. This system indicates the situation on a table using ambient music. From these case studies, we found that the user forgives that their attention is drawn strongly when the content is important to him/her. This is a common feature for both visual and auditory media.

In this study, we found features for designing ambient media. In future work, we will focus on how to develop new ambient media based on these features we found. Finally, we hope that a design framework will be developed in the future. By developing this design framework, we could create new ambient media more easily.

References

1. Aaron, S., Blackwell, A.F.: From sonic Pi to overtone: creative musical experiences with domain-specific and functional languages. In: Proceedings of the First ACM SIGPLAN Workshop on Functional Art, Music, Modeling Design, pp. 35–46. ACM, New York (2013)
2. An, P., Bakker, S., Ordanovski, S., Taconis, R., Eggen, B.: ClassBeacons: designing distributed visualization of teachers' physical proximity in the classroom. In: Proceedings of the Twelfth International Conference on Tangible, Embedded, and Embodied Interaction, pp. 357–367. ACM, New York (2018)
3. Atzori, L., Iera, A., Morabito, G.: The Internet of Things: a survey. Comput. Netw. **54**(15), 2787–2805 (2010)
4. Cadiz, J., Venolia, G., Jancke, G., Gupta, A.: Designing and deploying an information awareness interface. In: Proceedings of the 2002 ACM Conference on Computer Supported Cooperative Work, pp. 314–323. ACM, New York (2002)
5. Dahley, A., Wisneski, C., Ishii, H.: Water lamp and pinwheels: ambient projection of digital information into architectural space. In: Proceedings of CHI 1998 Conference Summary on Human Factors in Computing Systems, pp. 269–270, ACM, New York (1998)
6. Ishii, H., Ullmer, B.: Tangible bits: towards seamless interfaces between people, bits and atoms. In: Proceedings of the ACM SIGCHI Conference on Human Factors in Computing Systems (CHI 1997), pp. 234–241. ACM, New York (1997)
7. Ishii, H., et al.: ambientROOM: integrating ambient media with architectural space. In: Proceedings of CHI 1998 Conference Summary on Human Factors in Computing Systems, pp. 173–174. ACM, New York (1998)
8. Mao, C.-C., Liu, K.-H., Chiu, W.-C., Lin, C.-L., Chen, C.-H.: Citycell: an interactive OLED lighting system in public space. In: Proceedings of the 3rd International Conference on Communication and Information Processing, pp. 490–494. ACM, New York (2017)

9. Odom, W., Zimmerman, J., Davidoff, S., Forlizzi, J., Dey, A.K., Lee, M.K.: A fieldwork of the future with user enactments. In: Proceedings of the Designing Interactive Systems Conference, pp. 338–347. ACM, New York (2012)
10. Prante, T., Stenzel, R., Rocker, C., Streitz, N., Magerkurth, C.: Ambient agoras: InfoRiver, SIAM, Hello.Wall. In: Proceedings of CHI 2004 Extended Abstracts on Human Factors in Computing Systems, pp. 763–764. ACM, New York (2004)
11. van Dantzich, M., Robbins, D., Horvitz, E., Czerwinski, M.: Scope: providing awareness of multiple notifications at a glance. In: Proceedings of the Working Conference on Advanced Visual Interfaces, pp. 267–281 (2002)

Cognitive Dissonance in a Multi-mind Automated Decision System

Monte Hancock[1(✉)], Antoinette Hadgis[2(✉)], Katy Hancock[3(✉)],
Benjamin Bowles[2(✉)], Payton Brown[2], and Tyler Higgins[2(✉)]

[1] 4Digital, Los Angeles, CA, USA
practicaldatamining@gmail.com
[2] Sirius19, Melbourne, FL, USA
twomagpies115@gmail.com, benbowles2@gmail.com,
sirius19conf@gmail.com, ttjh61@yahoo.com
[3] Murray State University, Murray, USA

Abstract. We address the application of machine learning to the modeling of a specific cognitive behavior that brings two fundamental elements of human cognition into direct conflict: the purely emotional (referred to here as EMOS), and the purely rational (referred to here as NOOS).

An extensive development of the machine learning algorithmics for Knowledge-Based Systems (KBS) used for this work is presented.

Psychologists refer to the mental experience associated with this conflict as cognitive dissonance.

"Cognitive dissonance refers to a situation involving conflicting attitudes, beliefs or behaviors. This produces a feeling of mental discomfort leading to an alteration in one of the attitudes, beliefs or behaviors to reduce the discomfort and restore balance."–Saul Mcleod, University of Manchester, Division of Neuroscience & Experimental Psychology [1] (Fig. 1).

Early development efforts undertaken by AI researchers were necessarily focused on modeling very limited domains (e.g., Blocks World, Board Games, etc.) However, the range of possible behaviors in these domains was so narrow that the models developed were viewed as "toy problems" having little relation to the modeling of actual human behavior [2].

Modern learning machines have reached a level of maturity allowing the development of relatively high-fidelity models of group human behavior (e.g., social media traffic, collaborative filtering). This success is facilitated by the fact that group behavior models are actually models of the sampling distribution of the population of behaviors, rather than the behaviors of individuals.

The next natural step in the development of high-fidelity models of human behavior is to decrease the size of the groups being modeled, while simultaneously maintaining the scope and complexity of the tasks being performed.

As groups become smaller, the mental states and psychology of individual members play a greater role in the process of adjudicating group behavior.

A "double-minded" rule-based system was developed consisting of the two components EMOS and NOOS above. These two system components use the same decision-making algorithm, but different heuristics: to assess and combine facts, EMOS applies "soft emotional" factors, while NOOS applies "rigid

D. D. Schmorrow and C. M. Fidopiastis (Eds.): HCII 2019, LNAI 11580, pp. 41–57, 2019.
https://doi.org/10.1007/978-3-030-22419-6_4

principled" factors. Not surprisingly, the decisions they produce, given identical inputs, can be quite different.

Because the machines' preferences are numerically specified, the level of cognitive dissonance that arises in the double-minded machine as a whole can be quantified using the difference between the components' "commitments" to their separate conclusions. Further, for each preference, each machine has a numerical "psychological inertia" quantifying its reluctance to compromise on each of its preferences.

We selected a jury in a criminal trial as the domain for experiments with this system, the task to be performed is to consider facts-in-evidence and apply the rule bases to produce verdicts of "guilty", "not guilty", or "deadlocked". This scenario offers a fundamentally binary decision problem that can be easily understood without special knowledge. To make interpretation of the machine's decision processes transparent, EMOS and NOOS were developed with a conclusion justification capability by which each can express (in natural language) how the pieces of evidence affected their decisions.

Results and interpretation of the results of 255 "jury trials" are described.

An architecture for incorporating temporal awareness into the machine is given.

Keywords: Machine psychology · Cognitive dissonance · AI emotion

1 Emotional Machines

1.1 Why Design Emotion into Automated Systems?

Man-made systems exhibiting emotional behaviors intuitively similar to those exhibited by humans occur frequently in popular literature [3]. Such automata provide a convenient literary device for portraying bald human behaviors that are uninformed by human experience, and therefore uncomplicated by the checks and balances that humans acquire as they mature. However, to the extent fiction is successful in distilling behaviors in this way, its characterizations becomes less authentic avatars for humans.

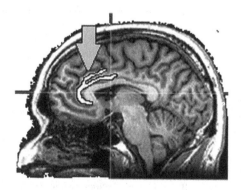

Fig. 1. In humans, perceptions of cognitive dissonance are correlated with activation of the Anterior Cingulate Cortex (indicated by arrow).

Automated systems are not endowed with what most people would regard as "emotions" unless these are essential to system effectiveness. Using a human voice in the interface to a query system [4] might not be essential to its proper functioning, but adding the appearance of emotion is probably a great help to gaining acceptance among a tech-skittish user base. Adding human-like "emotions" to a launch-control system makes an interesting movie concept [5], but no engineering sense.

Humans think of emotion as being somewhat spontaneous, presumably because of its unpredictability. Unpredictability can be emulated in an automated system by the introduction of random elements into its processes. Interestingly enough, such unpredictability is an essential aspect of "human intelligence". A system that always answers the same questions in exactly the same way will certainly not pass the Turing Test [6].

Must "emotions" in automated systems necessarily be of the intuitive "human-like" variety? Certainly not. To further refine what "emotion" in an automated system is and does, we consider in the next section the relevant system engineering principles.

1.2   A Model for Emotion in Automated Systems: Functional and Performance Requirements

System Engineers design systems to satisfy specified user requirements. These requirements are of two types:

- Functional Requirements: WHAT the system must do
- Performance Requirements: HOW the system must do it

Under a behaviorist model, functional requirements can be thought of as stimulus-response pairs linking system states to system responses: "When the engine gets hot, cool it down" might become the pair ($H \rightarrow C$), which is a rule informing the action of the system: when "Engine Hot" is the value of a particular system state variable, set a particular state variable's value to "Activate Engine Cooling System".

Such a rule needs more than just vague concepts like "hot" and "cool" for effective control (but, consider Fuzzy Logic [7]). Therefore, performance requirements are introduced to quantify, refine, and condition functional requirements. In the ($H \rightarrow C$) scenario, there would likely be associated performance requirements that call out specific temperature thresholds, and the specific type and aggressiveness of cooling the system is to perform.

These two types of requirements provide the basis for our understanding of "emotion" in automated systems. Hard principles will be likened to functional requirements specifying unquantified and unrestricted cause-and-effect operational rules. Emotional principles will be likened to performance requirements, which amend functional requirements by appending conditions which must be satisfied when the rule is applied.

Performance requirements can also be thought of as a means of establishing secondary system goals. For example, an autopilot system has a primary goal of getting an airplane from A to B, and a secondary goal of maintaining passenger comfort. The "emotion" here, such as it is, appears in the form of derived requirements like, "Do not subject passengers to excessive G-forces". The system is effective in getting passengers to their destination in a "kind" way (Fig. 2).

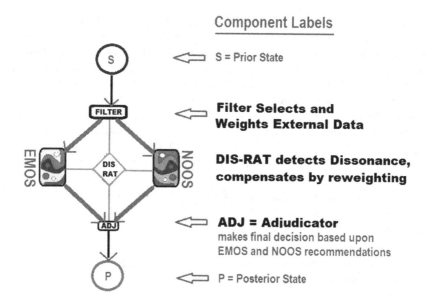

Component Labels

S = Prior State

Filter Selects and Weights External Data

DIS-RAT detects Dissonance, compensates by reweighting

ADJ = Adjudicator
makes final decision based upon
EMOS and NOOS recommendations

P = Posterior State

EMOS: "Emotional" Knowledge-Based System
NOOS: "Principled" Knowledge-Based System

Fig. 2. A single EMOS/NOOS decision system. It scans the data state S, applies the EMOS and NOOS Rule Bases independently, then adjudicates the two recommendations. If cognitive dissonance is present, DISRAT is invoked to negotiate an agreement to produce a final recommendation to update P.

2 Experiment Concept

A "double-minded" decision system was developed having two Knowledge-Based System components: the EMOS and NOOS Knowledge-Based Systems described above. These two system components use the same decision-making algorithm, but different heuristics: to assess and combine facts, EMOS applies "soft emotional" factors, while NOOS applies "rigid principled" factors. Not surprisingly, the decisions they produce, given identical inputs, can be quite different.

Because the machines' preferences are numerically specified, the level of cognitive dissonance that arises in the double-minded machine as a whole can be quantified using the difference between the components' "commitments" to their separate conclusions. Further, for each preference, each machine has a numerical "psychological inertia" quantifying its reluctance to compromise on each of its preferences.

We selected a jury in a criminal trial as the domain for experiments with this system, the task to be performed is to consider facts-in-evidence and apply the rule bases to produce verdicts of "guilty", "not guilty", or "deadlocked". This scenario offers a fundamentally binary decision problem that can be easily understood without special knowledge. To make interpretation of the machine's decision processes transparent, EMOS and NOOS were developed with a conclusion justification capability by which each can express (in natural language) how the pieces of evidence affected their decisions.

Once the double-minded machine produced a verdict, reinforcement learning was applied to the machines' preferences to reduce the measured cognitive dissonance to force a compromise. This was done subject to the condition that the psychological cost (given by the psychological inertia terms) would be kept small.

Because the case being tried involves 8 possible evidentiary facts, each of which could be present or absent, there are $2^8 = 256$ possible evidence suites a trial could present. All 256 cases were presented to the system, and the results tabulated.

3 Knowledge-Based Systems

3.1 Using Rules to Accrue Evidence

For simplicity and definiteness, the reasoning problem will be described here as the use of evidence to select one or more possible conclusions from a closed, finite list that has been specified a priori (the "Classifier Problem").

Expert reasoning is based upon facts (colloquially, "interpretations of the collected data"). Facts function both as indicators and contra-indicators for conclusions. Positive facts are those that increase our belief in specified conclusions. Negative facts are those that increase our disbelief in specified conclusions. Negative facts can also be thought of as being exculpatory: they impose constraints upon the space of conclusions, militating against those unlikely to be correct. Facts are salient to the extent that they increase belief in the "truth", and/or increase disbelief in "untruth".

Pieces of evidence are quantified by how they alter beliefs, independent of other pieces of evidence. That is, by the belief held when that piece of evidence is the only one known.

A rule is an operator that uses facts to update beliefs by applying biases. In software, rules are often represented as structured constructs such as IF-THEN-ELSE, CASE, or SWITCH statements. We use the IF-THEN-ELSE in what follows since they correspond to familiar colloquial processes.

Rules consist of an antecedent and a multi-part body. The antecedent evaluates a BOOLEAN expression; depending upon the truth-value of the antecedent, different parts of the rule body are executed.

The following is a notional example of a rule for classifying animals based upon their various attributes (features). It is intended to mimic the amount by which a human expert would alter her beliefs about an unknown animal should she determine whether or not it is a land-dwelling omnivore:

```
If (habitat = land) AND (diet = omnivorous) THEN
        INCREASE BELIEF(primates, bugs, birds)
        INCREASE DISBELIEF(bacteria, fishes)
ELSE
        INCREASE DISBELIEF(primates, bugs, birds)
        INCREASE BELIEF(bacteria, fishes)
End Rule
```

Using an INCREASE BELIEF function, and a DECREASE BELIEF function ("aggregation functions", called AGG in Fig. 3 below), many such rules can be efficiently implemented in a looping structure. Such functions can be thought of as "update methods", since they specify how existing beliefs are to be updated when new evidence is obtained.

Tj(**Fi**): truth-value of predicate j applied to fact **Fi**
bias(k,j,1): belief to accrue in conclusion k when predicate j true
bias(k,j,2): disbelief to accrue in conclusion k when predicate j is true
bias(k,j,3): belief to accrue in conclusion k when predicate j false
bias(k,j,4): disbelief to accrue in conclusion k when predicate j is false

Pseudo-Code	Explanatory Comments
IF Tj(**F**)=1 THEN	'if predicate j true for **Fi**...
FOR k=1 TO K	'for conclusion k:
Belief(v,k)=AGG(Belief(v,k), bias(k,j,1))	'true: accrue belief bias(k,j,1)
Disbelief(v,k)=AGG(Disbelief(v,k), bias(k,j,2))	'true: accrue disbelief bias(k,j,2)
NEXT k	
ELSE	'else if predicate j false for **Fi**...
FOR k=1 TO K	'for conclusion k:
Belief(v,k)=AGG(Belief(v,k), bias(k,j,3))	'false: accrue belief bias(k,j,3)
Disbelief (v,k)=AGG(Disbelief(v,k), bias(k,j,4))	'false: accrue disbelief bias(k,j,4)
NEXT k	
END IF	

Fig. 3. Multiple rule execution loop to accumulate beliefs and disbeliefs for feature vector v

More generally, define, as demonstrated in Fig. 3:

The process depicted in Fig. 3 creates for each feature vector v an ordered- tuple of *positive* class-membership beliefs Belief(k,v) = (b(v,1), b(v,2), ..., b(v,K)), and an ordered-tuple of *negative* class-membership beliefs Disbelief(k,v) = (d(v,1), d(v,2), ..., d(v,K)). These two vectors are then combined ("adjudicated") for each feature vector to determine a classification decision. For example, they can be adjudicated by simple differencing to create the vector of class-wise beliefs, **B**:

$$\boldsymbol{B}(v) = (Belief(v,1) - Disbelief(v,1), \ldots, Belief(v,K) - Disbelief(v,K))$$

This ordered-tuple of beliefs is the <u>belief vector</u> for the input feature vector v.

A final classification decision can be chosen by selecting the class having the largest belief. Confidence factors for the classification in several ways (e.g., the difference between the two largest beliefs).

Clearly, the inferential power here is not in the rule structure, but in the "knowledge" held numerically as biases. As is typical with heuristic reasoners, BBR allows the complete separation of knowledge from the inferencing process. This means that the structure can be retrained, even repurposed to another problem domain, by modifying only data; the inference engine need not be changed. An additional benefit of this separability is that the engine can be maintained openly apart from sensitive data.

3.2 Combining Accrued Evidence: Aggregation Methods

We propose a simple aggregation method that has many desirable properties; it can be modified to handle monotonicity or reasoning under uncertainty. Because it has a principled basis in probability theory, it can give excellent results as a classifier, and can be extended in a number of useful ways.

Proper aggregation of belief is essential. In particular, belief is not naively additive. For example, if I have 20 pieces of very weak evidence, each giving me 5% confidence in some conclusion, it would be foolish to assert that I am 100% certain of this conclusion just because $20 \times 5\% = 100\%$.

Important in all that follows, it is required that biases be in [0,1].

Illustrative Example of Belief Aggregation. Two rules, r_1 and r_2, having positive biases b_1 and b_2, respectively, are applied in sequence:

Belief(v,1) = 0.0 'the belief vector is initialized to the zero vector
Rule 1:
Belief (v,1) = AGG(Belief(v,1), b_1)) 'accrue belief bias = b_1
Rule 2:
Belief (v,1) = AGG(Belief(v,1), b_2)) 'accrue belief bias = b_2

What will the aggregate belief be after both rules are run? We define the simple aggregation rule for this two-rule system as the "probability AND". The aggregate belief from combining two biases, b_1 and b_2, is:

$$rules\, r_1 \text{ and } r_2 \text{ both run}, \ AGG = b_1 + b_2(1 - b_1) = b_1 + b_2 - b_1 b_2$$
$$= 1 - (1 - b_1)(1 - b_2)$$

If b_1 and b_2 separately give me 50% belief in a conclusion, after aggregation using this rule my belief in this conclusion is:

$$1 - (1 - 0.5)(1 - 0.5) = 0.75 = 75\%$$

(If this rule is applied for the case of twenty, 5% beliefs, we arrive at an aggregate belief of about 64% far from certainty.)

This simple aggregation rule says that we accrue additional belief as a proportion of the "unused" belief.

If a third rule r_3 with belief b_3 fires, we find:

aggregate belief$((r_1 \text{ and } r_2) \text{ and } r_3) =$
$(b_1 + b_2(1 - b_1)) + b_3(1 - (b_1 + b_2(1 - b_1))) = b_1 + b_2 + b_3 - b_1 b_2 - b_1 b_3 - b_2 b_3 + b_1 b_2 b_3$
$= 1 - (1 - b_1)(1 - b_2)(1 - b_3)$

In general, firing J rules r_j having isolated beliefs b_j gives aggregate belief:

$$aggregate\ belief\ (r_1\ AND\ r_2\ AND \ldots AND\ r_J) = 1 - \Pi_j \{1 - b_j\}$$

The aggregate belief can be accumulated by application of the simple aggregation rule as rules fire. For, if $J-1$ rules have fired, giving a current belief of $\underline{b}$, and another rule r_J having isolated belief b_J fires, the simple aggregation rule gives a new belief of $\underline{b} + b_J(1-\underline{b})$, which is easily shown to be in agreement with the above.

The simple aggregation rule is clearly independent of the order of rule firings, assumes values in [0,1], and has partial derivatives of all orders in the biases b_j. In fact, because...

$$\partial(1 - \Pi_m \{1 - b_n\})/\partial b_n = (\Pi_m \{1 - b_m\})/(1 - b_n)$$

... all partials having multi-indices with repeated terms are zero.

Important: biases accrued must be in [0,1]. The aggregation rule defined here will not work if negative biases are accrued. This is why we accrue positive belief and positive disbelief, then difference them.

Given a set of data having ground truth tags, an iterative cycle using this (or a similar) update rule can be used to <u>learn</u> the beliefs and disbeliefs that will cause the heuristics to give correct answers on the training set (Fig. 4):

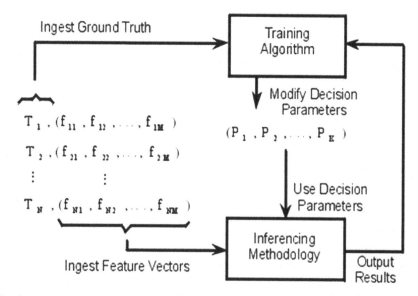

Fig. 4. Learning Loop: how trainable systems learn from data tagged with "ground truth."

Note that if a belief $b_j = 1$ is accrued, the aggregate belief becomes and remains 1, no matter what other beliefs might be accrued. Similarly, accruing a belief $b_j = 0$ has no effect on the aggregate belief. These are both consistent with intuitive expectations.

4 Experimental Data

To establish psychological "engrams" for EMOS and NOOS, researchers filled out a standard form specifying how each mind-set, EMOS and NOOS, might approach jury membership (Priors favoring guilt or innocence), and how reluctant they would be to change these to achieve a compromise verdict (Fig. 5).

KBS' Prior Beliefs in Conclusion 1 for an unknown Defendant before any evidence is seen

Conclusion 1: The Defendant is guilty.

0.4=EMOS prior probability that conclusion 1 is true
(EMOS belief that conclusion 1 is true before any evidence has been seen)

```
IF EMOS>=0.5 THEN
    EMOS Reason = "Suspicious people make our streets unsafe
for law-abiding people."
ELSEIF EMOS<0.5 THEN
    EMOS Reason = "The court system is racially biased and un-
fair."
END IF
```

RESISTANCE TO INCREASING EMOS' Conclusion 1 PRIOR PROBABILITY TO RESOLVE DISSONANCE:
3=EMOS Reluctance to increase prior confidence in conclusion 1
RESISTANCE TO DECREASING EMOS' Conclusion 1 PRIOR PROBABILITY TO RESOLVE DISSONANCE:
8=EMOS Reluctance to decrease prior confidence in conclusion 1

0.6=NOOS prior probability that conclusion 1 is true
(NOOS belief that conclusion 1 is true before any evidence has been seen)

```
IF NOOS>=0.5 THEN
    NOOS Reason = "Most people who are indicted for crimes are
guilty."
ELSEIF NOOS<0.5 THEN
    NOOS Reason = "Some laws are unjust and should not be en-
forced."
END IF
```

RESISTANCE TO INCREASING NOOS' Conclusion 1 PRIOR PROBABILITY TO RESOLVE DISSONANCE:
8=NOOS-up 0-10 Reluctance to increase prior confidence in conclusion 1
RESISTANCE TO DECREASING NOOS' Conclusion 1 PRIOR PROBABILITY TO RESOLVE DISSONANCE:
4=NOOS-down 0-10 Reluctance to decrease prior confidence in conclusion 1

Fig. 5. Human experimenter can specify the prior beliefs of EMOS and NOOS, and the level of their resistance to adjusting them during negotiations to reduce cognitive dissonance.

In essence, these priors quantify the minds' assessment of the law enforcement process. Those with relatively higher priors favoring a guilty verdict would be more confident that defendants arrested, charged, and brought to trial are actually likely to be guilty (Fig. 5).

To further fill out the psychological "engrams" for EMOS and NOOS, researchers filled out a standard form specifying how each mind-set, EMOS and NOOS, might assess the relative importance and evidentiary significance of particular facts in evidence (Fig. 6).

For this work, eight possible facts in evidence are used. Each fact can be either true or false. Because of this, it was felt that the term "fact in evidence" could be confusing, so the term "contingency" was selected to designate each piece of evidence. These were expressed as statements, known to be either true or false. The eight contingencies are:

1. TRUE or FALSE: The Defendant has a criminal history.
2. TRUE or FALSE: The Defendant has been identified as the perpetrator by eye witnesses.
3. TRUE or FALSE: There is forensic evidence that ties the Defendant to the crime.
4. TRUE or FALSE: The Defendant had a motive for the crime.
5. TRUE or FALSE: The Defendant is a member of a minority racial or religious group.

```
Contingency 6: The Defendant is under 26 years of age.

EMOS Rule:
6=EMOS 0-10 relative importance of this Contingency (piece of evidence)

If EMOS ≥ 5 then
    Young people deserve second chances!
Elseif EMOS < 5 then
    If people don't learn to obey the law when young they never will.
End if
4=EMOS up 0-10 Reluctance to increase default confidence in contingency
1=EMOS down 0-10 Reluctance to decrease default confidence in contingency


NOOS Rule:
5=NOOS 0-10 relative importance of this Contingency (piece of evidence)

If EMOS ≥ 5 then
    The young often act out of foolishness rather than criminal intent.
Else if NOOS < 5 then
    Ignorance is no excuse for bad behavior.
End if


8=NOOS up 0-10 Reluctance to increase default confidence in contingency
4=NOOS down 0-10 Reluctance to decrease default confidence in contingency
```

Fig. 6. Establish the relative evidentiary importance of Contingency 6; set up the justification statements to be reported by the KBS when considering Contingency 6; and determine the resistance to changing the Contingency 6 biases in order to reduce cognitive dissonance. (Experiments are based upon what brains are given as evidence; admissibility is not relevant).

6. TRUE or FALSE: The Defendant is under 26 years of age.
7. TRUE or FALSE: The Defendant has a history of gang membership.
8. TRUE or FALSE: The Defendant dropped out of high school.

Finally, the amounts by which EMOS and NOOS individually adjust their beliefs in light of the truth and falsity 9the "biases) of the eight contingencies was specified by researchers. This had to be done for each brain, for each contingency, for each conclusion. Many different specifications were run as experiments, primarily to assess the sensitivity of the brains to the various factors. To avoid inconsistency, ("schizophrenic" machines), each set of parameters for an experiment was prepared by one researcher (Fig. 7).

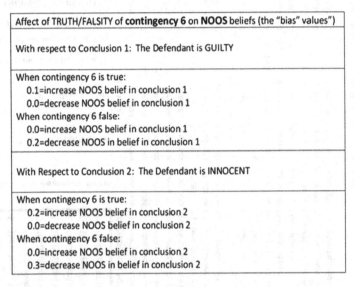

Fig. 7. Establish the "biases" that NOOS will apply when considering Contingency 6.

5 Details for a Specific Case: #155

Figures 8 and 9 below summarize the results for a particular case run: Case #155. The facts in evidence for this case are:

TRUE: The Defendant has a criminal history.
FALSE: The Defendant has been identified as the perpetrator by eye witnesses.
FALSE: There is forensic evidence that ties the Defendant to the crime.
TRUE: The defendant had a motive for the crime.
TRUE: The Defendant is a member of a minority racial or religious group.
FALSE: The Defendant is under 26 years of age.
TRUE: The Defendant has a history of gang membership.
TRUE: The Defendant dropped out of high school.

Apply EMOS rules to the evidence. Each fact (A – H at right) can be TRUE or FALSE (values in column 1).

The next-to-last column shows new Aggregate Beliefs as facts are applied.

The last column is the message EMOS will append to its explanation report.

Evidence legend

A — The Defendant has a criminal history.
B — The Defendant has been identified as the perpetrator by eye witnesses.
C — There is forensic evidence that ties the Defendant to the crime.
D — The Defendant had a motive for the crime.
E — The Defendant is a member of a minority racial or religious group.
F — The Defendant is under 26 years of age.
G — The Defendant has a history of gang membership.
H — The Defendant dropped out of high school.

EMOS ADJUST GUILTY BELIEF (Prior is 0.40)

	Evidence	If TRUE increase BEL-ief by	If TRUE increase DIS-belief by	If FALSE increase BEL-ief by	If FALSE increase DIS-belief by	Resulting BEL-ief	Resulting DIS-belief	Resulting Aggregate Belief	Justification for Adjustment
TRUE	A - HISTORY	0.004	0	0	0	0.4024	0.6	-0.1976	A person's past actions are predictors of their future actions.
FALSE	B- WITNESS	0	0	0	0.02	0.4024	0.608	-0.2056	Eye witnesses can make mistakes.
FALSE	C-FORENSICS	0	0	0	0.024	0.4024	0.617408	-0.215008	Evidence can be faked.
TRUE	D-MOTIVE	0.014	0	0	0	0.4107664	0.617408	-0.2066416	They wanted to do it.
TRUE	E-MINORITY	0.01	0	0	0	0.4166588	0.617408	-0.2007493	Minorities deserve a break in the courts.
FALSE	F-UNDER 26	0	0	0	0.024	0.4166588	0.6265903	-0.2099315	If people don't learn to obey the law when young they never will.
TRUE	G-GANG	0.016	0	0	0	0.4259922	0.6265903	-0.200598	Gangs encourage even demand criminal behavior.
TRUE	H-DROPOUT	0.01	0	0	0	0.4317323	0.6265903	-0.194858	The Defendant didn't have good employment opportunities.

EMOS ADJUST INNOCENT BELIEF (Prior is 0.60)

	Evidence	If TRUE increase BEL-ief by	If TRUE increase DIS-belief by	If FALSE increase BEL-ief by	If FALSE increase DIS-belief by	Resulting BEL-ief	Resulting DIS-belief	Resulting Aggregate Belief	Justification for Adjustment
TRUE	A - HISTORY	0.008	0	0	0	0.6032	0.4	0.2032	A person's past actions are predictors of their future actions.
FALSE	B- WITNESS	0	0	0	0.03	0.6032	0.418	0.1852	Eye witnesses can make mistakes.
FALSE	C-FORENSICS	0	0	0	0.036	0.6032	0.438952	0.164248	Evidence can be faked.
TRUE	D-MOTIVE	0.028	0	0	0	0.6143104	0.438952	0.1753584	They wanted to do it.
TRUE	E-MINORITY	0.02	0	0	0	0.6220242	0.438952	0.1830722	Minorities deserve a break in the courts.
FALSE	F-UNDER 26	0	0	0	0.036	0.6220242	0.4591497	0.1628745	If people don't learn to obey the law when young they never will.
TRUE	G-GANG	0.032	0	0	0	0.6341195	0.4591497	0.1749697	Gangs encourage even demand criminal behavior.
TRUE	H-DROPOUT	0.02	0	0	0	0.6414371	0.4591497	0.1822873	The Defendant didn't have good employment opportunities.

ADJUDICATE EMOS

GUILTY belief: 0.4317323   GUILTY Disbelief: 0.6265903   EMOS GUILTY AGG: -0.194858

INNOCENT belief: 0.6414371   INNOCENT Disbelief: 0.4591497   EMOS INNOCENT AGG: 0.1822873

EMOS DECISION: The Defendant is innocent.   Confidence = $|0.1822873 - (-0.194858)| = 0.3773453$

Fig. 8. Execution trace for Case #155 showing how EMOS used the evidence to arrive at its belief that the defendant in this case is innocent.

Apply NOOS rules to the evidence. Each fact (A - H at right) can be TRUE or FALSE (values in column 1).

The next-to-last column shows new Aggregate Beliefs as facts are applied.

The last column is the message NOOS will append to its explanation report

NOOS ADJUST GUILTY BELIEF	EVIDENCE	If TRUE increase BEL-ief by	If TRUE increase DIS-belief by	If FALSE increase BEL-ief by	If FALSE increase DIS-belief by	Resulting BEL-ief	Resulting DIS-belief	Resulting Aggregate Belief	JUSTIFICATION FOR ADJUSTMENT:
						Prior = 0.60			
TRUE	A - HISTORY	0.014	0	0	0	0.6056	0.4591497	0.2056	If they broke the law before they are likely to break it again.
FALSE	B - WITNESS	0	0	0	0.008	0.6056	0.4048	0.2008	Eye witnesses say what they are told to say.
FALSE	C - FORENSICS	0	0	0	0.036	0.6056	0.4262272	0.1793728	Police labs are known to be careless and unreliable.
TRUE	D - MOTIVE	0.012	0	0	0	0.6103328	0.4262272	0.1841056	Criminals don't need a good reason to commit a crime.
TRUE	E - MINORITY	0.004	0	0	0	0.6118914	0.4262272	0.1856643	The Defendant is a product of their environment.
FALSE	F - UNDER 26	0	0	0	0.012	0.6118914	0.4331124	0.178779	Ignorance is no excuse for bad behavior.
TRUE	G - GANG	0.006	0	0	0	0.6142201	0.4331124	0.1811076	Those who choose to join gangs are choosing criminality.
TRUE	H - DROPOUT	0.004	0	0	0	0.6157632	0.4331124	0.1826507	Laziness and poor work ethic are signs of bad character.
NOOS ADJUST INNOCENT BELIEF						Prior = 0.40			
TRUE	A - HISTORY	0.028	0	0	0	0.4168	0.4331124	-0.1832	If they broke the law before they are likely to break it again.
FALSE	B - WITNESS	0	0	0	0.012	0.4168	0.6048	-0.1880001	Eye witnesses say what they are told to say.
FALSE	C - FORENSICS	0	0	0	0.054	0.4168	0.6261408	-0.2093408	Police labs are known to be careless and unreliable.
TRUE	D - MOTIVE	0.024	0	0	0	0.4307968	0.6261408	-0.1953344	Criminals don't need a good reason to commit a crime.
TRUE	E - MINORITY	0.008	0	0	0	0.4353504	0.6261408	-0.1907904	The Defendant is a product of their environment.
FALSE	F - UNDER 26	0	0	0	0.018	0.4353504	0.6328703	-0.1975199	Ignorance is no excuse for bad behavior.
TRUE	G - GANG	0.012	0	0	0	0.4421262	0.6328703	-0.1907441	Those who choose to join gangs are choosing criminality.
TRUE	H - DROPOUT	0.008	0	0	0	0.4465892	0.6328703	-0.1862811	Laziness and poor work ethic are signs of bad character.
ADJUDICATE NOOS									
GUILTY belief: 0.6157632	GUILTY Disbelief: 0.4331124					NOOS GUILTY AGG: 0.1826507			
INNOCENT belief: 0.4465892	INNOCENT Disbelief: 0.6328703					NOOS INNOCENT AGG: -0.1862811			
NOOS DECISION:	The Defendant is guilty.					Confidence = |0.1826507 - (-0.1862811)| = 0.3689318			

A The Defendant has a criminal history.
B The Defendant has been identified as the perpetrator by eye witnesses.
C There is forensic evidence that ties the Defendant to the crime.
D The Defendant had a motive for the crime.
E The Defendant is member of a minority racial or religious group.
F The Defendant is under 26 years of age.
G The Defendant has a history of gang membership.
H The Defendant dropped out of high school.

Fig. 9. Execution trace for Case #155 showing how NOOS used the evidence to arrive at its belief that the defendant in this case is guilty.

EMOS believes that this defendant is innocent, while NOOS believes this defendant is guilty. After adjudication, this defendant was found innocent (Fig. 10).

ADJUDICATE				
EMOS	GUILTY belief: 0.4317323	GUILTY Disbelief: 0.6265903	EMOS GUILTY AGG: -0.194858	
EMOS	INNOCENT belief: 0.6414371	INNOCENT Disbelief: 0.4591497	EMOS INNOCENT AGG: 0.1822873	
EMOS		The Defendant is innocent.	Confidence = |0.1822873 -(-0.194858)| =	37.73%
NOOS	GUILTY belief: 0.6157632	GUILTY Disbelief: 0.4331124	NOOS GUILTY AGG: 0.1826507	
NOOS	INNOCENT belief: 0.4465892	INNOCENT Disbelief: 0.6328703	NOOS INNOCENT AGG: -0.1862811	
NOOS		The Defendant is guilty.	Confidence = |0.1826507 - (-0.1862811)| =	36.89%
VERDICT				
Dissonance:	0.00036341			
	The Defendant is innocent. (Dissonance is less than Threshhold (0.2) so high-confidence prevails			

Fig. 10. EMOS and NOOS disagree, so the verdict is decided if favor of the most confident brain.

When EMOS and NOOS are adjudicated, the defendant is found innocent (Fig. 10):

Each case represents a unique combination of evidence features. As an experiment, each of these 255 combinations of evidence was tried 30 times, and the proportion of guilty and innocent verdicts rendered by the double-minded BOT were tabulated. Data were then sorted (ascending) in the proportion of GUILTY verdicts rendered for that combination. The proportions of GUILTY and INNOCENT verdicts are plotted above in this sorted order (Fig. 11).

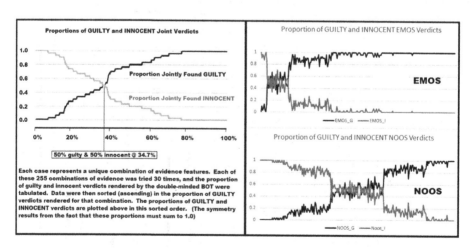

Fig. 11. Plots are shown sorted in ascending order of proportion of joint GUILTY verdicts. Plots show results of running each of the 255 combinations of evidence 30 times. On the left are plotted the proportions of verdicts rendered by EMOS and NOOS jointly, after adjudication. Proportions of verdicts rendered by EMOS, NOOS before adjudication for 30 runs of the 255 combinations of cases on right.

The same plot for EMOS and NOOS separately are on the right. With the settings defining EMOS for this experiment, the emotional reaction to the evidence resulting in a larger proportion of GUILTY verdicts, while NOOS seems relatively balanced. Together, EMOS AND NOOS moderated to rendering verdicts of guilty about 1/3 of the time.

6   Conclusions

This investigation has demonstrated that specifics aspect of unstructured mentation can be modeled as an efficient KBS. Further, the interplay between these aspects can be adjusted, and the affects observed.

Additional experiments (not detailed in this paper for the sake of brevity) used the psychological inertia terms to negotiate compromises between EMOS and NOOS in such a way that the total psychological cost (sum of inertia terms) could be minimized. It is an interesting and somewhat surprising fact that the approach taken to resolve cognitive dissonance is both conceptually similar to how humans might behave and is also an NP-Complete problem. In fact, minimizing the psychological cost during negotiation by the EMOS and NOOS brains is an instance of the Knapsack Problem [8].

The sensitivity of the EMOS – NOOS combination to small changes in bias settings was not as significant as had been anticipated, though some care was required in assigning biases to factors depending upon their specific manner of use.

The software implementation of the EMOS-NOOS KBS was very computationally efficient, requiring an average of only 13 microseconds to hear a case, adjudicate the results, and write the findings to disc, (single core INTEL i3 processor). This efficiency facilitated the execution of a large number of experiments.

The modeling approach here can be made trainable. Given a set of desired adjudicated verdicts, machine learning can be used to establish biases and priors that will, to the extent possible while maintaining model consistency, return those verdicts (and we have done this).

This also suggests a forensic application for a KBS model of a multi-component decision system. The internal hidden variables leading to a particular set of decision outcomes could be estimated by:

– holding known KBS parameters fixed
– training the KBS components so that the target decision are rendered

In this way, plausible estimates for hidden variables are obtained.

7   Future Work

The larger concept is to allow the decision system to move through time (Fig. 11). The upper track represents state variables sampled periodically from the external world and coded as the state sequence Pk. The lower track represents the decisions of the system coded as the state sequence Sk. The decision system also has available to it (through the diagonal connections) a history of prior input data and state estimates. This is to facilitate adjustment of the EMOS and NOOS rules bases, based upon a domain ontology (Fig. 12).

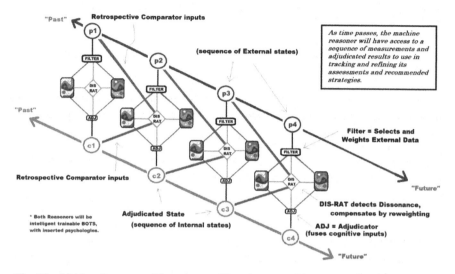

Fig. 12. Making the system Time-Aware: Time increases from upper-left to lower-right.

Acknowledgements. The authors would like to acknowledge the support of both the Sirius Project, and the INTJ Forum for their support of this work.

References

1. https://www.simplypsychology.org/cognitive-dissonance.html
2. Schoppek, W., Kluge, A., Osman, M., Funke, J.: Editorial: complex problem solving beyond the psychometric approach. Frontiers Psychol (2018)
3. Examples are too numerous to list, but the refinement of these works over time parallels the development of technology, at least from "Frankenstein" by Mary Shelley (1797–1851) to "Star Wars" by Steven Spielberg (1946)
4. For example, Apple, Inc.'s Siri. Introduced October 12, 2011
5. A.I. Artificial Intelligence, Steven Spielberg, Director (2001)
6. Oppy, G., Dowe, D.: The turing test. In: Zalta, E.N. (ed.) The Stanford Encyclopedia of Philosophy, Spring 2019 Edn. (2019, forthcoming). https://plato.stanford.edu/archives/spr2019/entries/turing-test/
7. Kosko, B.: Fuzzy Thinking: The New Science of Fuzzy Logic (1993). ISBN-13: 978-0786880218
8. Martello, S., Toth, P.: Knapsack Problems: Algorithms and Computer Implementations. Wiley, 14 December 1990

Bibliography

1. Minn. L. Rev. 455 (1948–1949) Jurimetrics–The Next Step Forward; Loevinger, Lee
2. Loevinger, L.: Jurimetrics: The methodology of legal inquiry. Law and Contemporary Problems, pp. 5–35 (1963)
3. About Jurimetrics: About the Journal. American Bar Association Section of Science & Technology Law and the Center for Law, Science & Innovation, n.d. Web. 26 April 2014. http://www.law.asu.edu/jurimetrics/JurimetricsJournal/AbouttheJournal.aspx
4. Bem, D.J.: Self-perception: an alternative interpretation of cognitive dissonance phenomena (PDF). Psychol. Rev. **74**(3), 183–200 (1967). https://doi.org/10.1037/h0024835. PMID 5342882
5. Zanna, M., Cooper, J.: Dissonance and the pill: an attribution approach to studying the arousal properties of dissonance. J. Pers. Soc. Psychol. **29**(5), 703–709 (1974). https://doi.org/10.1037/h0036651. PMID 4833431
6. Kiesler, C.A., Pallak, M.S.: Arousal properties of dissonance manipulations. Psychol. Bull. **83**(6), 1014–1025 (1976). https://doi.org/10.1037/0033-2909.83.6.1014. PMID 996211

A Hierarchical Characterization of Knowledge for Cognition

Monte Hancock[1,3], Jared Stiers[2,3], Tyler Higgins[2,3], Fiona Swarr[2,3], Michael Shrider[2,3], and Suraj Sood[2,3(✉)]

[1] Digital, Los Angeles, USA
[2] Sirius19, Melbourne, USA
surajsoodx@gmail.com
[3] University of West Georgia, Carrollton, GA, USA

Abstract. A modestly formal systematic representation for "knowledge" is presented in the context of structured cognition. This representation places knowledge artifacts ("data", "facts", "rules", etc. to be defined later) into a hierarchy. This hierarchy aligns naturally with "stages" or "levels" often observed in intelligent biological and mechanical systems engaged in structured cognition.

Each level within the knowledge hierarchy consists of knowledge artifacts that arise as specific relations on the level prior to it. This establishes a recursive relational algebra by which knowledge at all levels can be specified in terms of percepts ("data") at the bottom. This is quintessential essentialism as a ground for cognition [1].

res est forma eius
The thing is its form."

"

For the purposes of this work, an object of thought is regarded as equal to the assemblage of attributes it manifests. From the standpoint of cognition, nothing is sacrificed here, since percepts arising from this assemblage constitute the entirety of material available for structured cognition [2].

This formalizes a structured context for the analysis of cognition and the knowledge artifacts it uses. The U.S. Intelligence Community and the Department of Defense use similar but looser formalisms to support data fusion processes [3]. These are briefly described.

A brief case study is presented applying this knowledge representation to the analysis of an important pattern processing problem: the classification of distant military vehicles from their Doppler RADAR phase history.

Keywords: Knowledge model · Knowledge hierarchy · Cognitive model

1 Goals and Assumptions

Proposed here is an intuitive formulation, validation, and extension of a mathematical system within which high-fidelity models of a wide range of decision-support problems can be developed, assessed, and optimized. This optimization can be carried out on segments of a problem both individually, and in aggregate. By subjecting models

D. D. Schmorrow and C. M. Fidopiastis (Eds.): HCII 2019, LNAI 11580, pp. 58–73, 2019.
https://doi.org/10.1007/978-3-030-22419-6_5

derived within this formalism to mathematical analysis (e.g., visualization, modeling) rational solutions to real problems can be inferred in a principled, systematic way.

The essence of such a program will be a description of the fundamentals of the formalization to be created. This must include descriptions of representation of domain entities, methods for automated data preparation/processing, and a mechanism for automated reasoning. All of these must flow from the formal theory in a natural way.

2 Elements of the Approach

The customary approach to the analytic process is to employ analytic means to represent and prepare data, and discrete algebraic means for inferencing. This leads to a number of well-known difficulties:

The frame problem, reasoning that is inherently monotonic, super-polynomial inferencing time, and discrete inferencing that handles uncertainty as a "tacked on" afterthought.

We suggest that the representation be algebraic; the processing be geometric/topological; and, the inferencing be analytic/computational. This explicit choice of modalities allows principled formalization of the analytic process, and provides efficient satisficing solutions to these problems.

Representational foundation: an algebraic structure is proposed as the framework within which disparate domain entities are represented. This representation is applied so that

1. It makes extensive use of existing formalisms; we are not "starting from scratch".
2. Flexible abstraction is facilitated; the number and type of entity attributes made explicitly visible in the representation is adjustable.
3. The scope of entity definitions is labile; the representation allows an entity to be a signal sample, or a city.
4. It does not inherently limit the types of reasoning that can be applied to domain entities.
5. The representation is natural for machine implementation.
6. "Information" is an emergent property that arises through interpretation of entity relationships, rather than as context-free "magical contents" of individual, isolated entities. This means that relational structure is the basis of all domain knowledge, and geometric reasoning is the principal method by which it is derived.

Geometric toolkit: An interoperable collection of topological methods implemented in software is proposed for information extraction and refinement. These support

1. Parametric methods and unsupervised learning for the characterization of latent patterns.
2. Feature extraction, enhancement, evaluation, and winnowing.
3. User-centric, interactive high-dimensional visualization.

Inferencing mechanism: An analytic method is proposed as the means by which inferencing is performed. This allows

1. Linear-time non-monotonic reasoning, rather than the usual NP-Hard process required by graph-theoretic methods such as belief networks.
2. Declarative inferencing, that is, reasoning by direct computation. This makes knowledge "numeric", thereby facilitating the implementation of machine learning, as well as providing tractable approaches to the Frame Problem. Numeric knowledge is quantifiable, portable, and learnable by analytic methods (e.g., regression).
3. Natural representation, computation, and tracking of uncertainty, data pedigree, and conclusion confidence.

3   What Is Knowledge? Meta-Meta-Knowledge Models

We begin by defining a representation schema that is adequate to handle "knowledge" at every level of abstraction. The natural way to do this is to define knowledge according to a recursive method, so that, at the lowest level of abstraction, "knowledge" is nothing but data: fundamental percepts that are experienced and measured in the world. At higher levels of abstraction, "knowledge" posits relationships among entities that are at a lower level.

In this schema, "knowledge" exists at different formal levels. Advancement to the next level is recursive (from "knowledge" to "meta-knowledge" to "meta-meta...").

The simple relational algebra is best understood by considering an intuitive example:

Each level in the hierarchy will have a finite list of relation operators.

Level 0:
Relations: IS-A, ISNT-A (associates a percept with an attribute)

An element of knowledge at level = 0 would be a pair consisting of the null symbol and an attribute arising from some sensation or other measurement, such as a designation of time, place, or condition.

Example 1: 64 degrees.
Example 2: Tampa, Florida
Example 3: 2:00 p.m. today

At level 0, these attributes are not predicated of anything... they can be regarded as abstract properties.

Level 1:
Relations: all lower level relations, and some relational comparators and logical connectives (e.g., >, logical AND)

An element of knowledge at this level is a relation on data, such as the "ISA" relation (this datum ISA that datum), or any *wff* ("well-formed formula") in some predicate calculus.

Example 1: The temperature in Tampa, Florida today at 2:00 p.m. is 64 degrees.
Example 2: The temperature in Orlando, Florida today at 2:00 p.m. is 84 degrees.
Example 3: Power consumption in Florida increases with increasing temperature.
Level 2
Relations: all lower level relations, and Logical IMPLICATION.

An element of knowledge at this level is a relation on level 1 elements (such as a proof, or "argument"). For example

P1: The temperature in Tampa, Florida today at 2:00 p.m. is 64 degrees.
P2: The temperature in Orlando, Florida today at 2:00 p.m. is 84 degrees.
P3: Power consumption in Florida increases with increasing temperature.

Therefore,

C: The power consumption in Orlando, Florida is higher than the power consumption in Tampa, Florida at 2:00 p.m. today.
Level 3
Relations: all lower level relations, possibly others

This consists of what we normally call meta-mathematics, that is, the theory of formal systems.

Level k: an element of knowledge at this level is a relation on elements in levels 0 – $(k-1)$

These relations are formal associations among elements. If they are reflexive, symmetric and transitive, they are called equivalence relations. If they also preserve an arithmetic, they are called congruences. These can be represented as directed graphs and/or association matrices.

At levels above 0, it will probably be necessary to define operators that carry out transformations on elements (e.g., closures, proofs, resolution).

An expert system, for example, is just a collection of level 2 objects (rules) organized and executed by a level 3 object (an inference engine). In this way, we explicitly define what it means for a system to "know what it is doing": it is composed of level k objects, and has a level $k+1$ component to support learning.

General machine learning techniques would work naturally within this schema. There are learning paradigms that are appropriate to different levels. Manual learning can occur easily at levels 0 and 1, but automated techniques are probably required at higher levels. However, in a pure sense, each level is a formal relation, so all can be analyzed and represented using graph-theoretic methods.

As will be seen, black-box regression methods are just sophisticated ways of building what we will call level 1 ontologies.

4 The Representation Scheme as an Analogy

There is no such thing as formalism apart from representation. Formal reasoning in any domain requires an appropriate underlying symbology supported by inferencing machinery. Establishing a scheme capable of representing domain entities and their interactions is logically prior to everything else. This is where we begin [4].

The United States Intelligence Community (IC) has shown the way by its prescient attempts to create frameworks for such a scheme.

Two of these have gained wide acceptance in the IC. One, the NSA Reference Model, views the analytic process from a data-centric perspective. The other, the JDL Fusion Model, views the analytic process from a functional perspective. Placed side-by-side, they are seen to be similar attempts to represent the analytic activity in terms of a discrete hierarchy.

These hierarchies depict the systematic elevation of processing artifacts from mere phenomena (pure syntax) to cognitive artifacts (pure semantics). In this way, the analytic process winnows and refines sense experience, concentrating its latent value to produce actionable, decision-ready input for the human user [5].

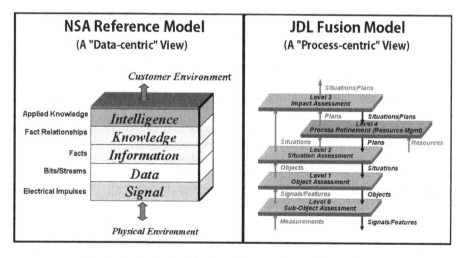

Fig. 1. Analysis viz. "data" and "process" according to the IC

These frameworks are notional depictions of what analysis does, but they are not formalizations. There is, however, one very important ingredient they share that makes them indispensable to any credible effort to formalize the analytic process: they were created by humans to describe what analysis conducted by humans is. A formalization congruent to these schemes will be comprehensible to humans.

5 Relation Towers

In Fig. 2 is a notional pictorial representation of a knowledge hierarchy.

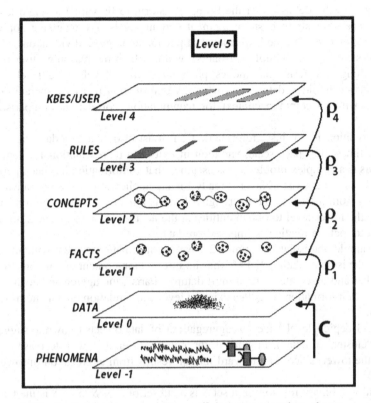

Fig. 2. Relation tower provides sufficient representation scheme.

We now describe the formalism for the representation scheme; it is depicted in Fig. 2.

Figure 2 shows a "tower" consisting of multiple layers. At layer -1 is the phenomenology of the real world. This is where waveforms, text, noumenal actors, and the phenomenology of reality reside.

At the next level above level 0, is *data*. Data consist of isolated measurements with no imposed context. They are depicted here as a collection of disconnected dots (a "sandbox"!). The next layer above that, level I, is the layer of *facts*. The next layer above that, layer 2, is the *concepts* layer. The next higher layer, layer 3, is the *rules* layer. Layer 4 is the reasoning layer.

The mathematical model proposed is based upon an established mathematical formalism.

The "lifting transform" that elevates entities through the successive layers of the tower is now described. It must not only accomplish the identifications with data, facts, etc., it must do this in the same way for each layer transition. That is, the transform used to elevate entities from the data level to the fact the level should be mathematically equivalent to the transform used to elevate facts to concepts, then from concepts to rules, and finally from rules reasoning (KBES).

This allows the definition of the lifting transform to be varied a single layer at a time, perhaps with slight customization at certain levels. By recursive application, entities are elevated from the lowest level of phenomenology (individual data items) to layer 4 (system having multiple reasoners), within which the reasoning user resides.

The lifting transform that moves phenomena from level -1 to level 0, C, is conventional-technology data conformation. Conformation recasts disparate data and places them into a common context for manipulation, correlation, comparison, and analysis.

Conformation is not just "reformatting"; it must also accommodate variations in precision, timescale, space-scale, variation in units and representational schemes, and so on. This is a complex problem. It is assumed that conformation has been performed at layer -1, and data exist at layer 0 ready for the application of the formalism.

Moving from a level to the next higher level is a matter of imposing a relation on entities in the lower level to obtain entities at the next higher level. That is, in moving from data to facts, a relation is imposed on data.

For example, the isolated datum "72°" has no information content without context. Similarly, the isolated datum "2 PM", has no information-content out of context. When tied together and associated with a third datum, "Paris", the aggregate becomes a fact bearing latent information. This fact exists at level 1 as a relation on data items residing at level 0.

Figure 1 depicts level 1 facts as aggregations of data objects (drawn to suggest that they are "clusters of data"). In general, the number of entities will decrease at higher levels in the tower, since each upward motion results from aggregation of lower level entities [5].

Formally, a binary relation on a set S is a subset of $S \times S$ ($S \times S$ is the Cartesian product of S with itself).

Each ordered pair depicts an association from the first item to the second item. This can be depicted graphically as showing the two items as vertices in a directed graph: small icons connected by an arrow starting at the first element and terminating at the second.

Moving from level 1 to level 2 in the relation tower is done by aggregating facts, that is, by imposing a binary relation on layer 1. These digraphs are referred to as concepts. Layer 2 is the concept layer, where collections of facts about entities in the domain of discourse reside.

Moving next from level 2 level 3 is accomplished in the same way: by imposing a binary relation on the concept layer. Here there is an opportunity to inject some logical formalism by assigning types to the edges in the digraphs. If the associations are logical connectives (conjunction, disjunction, and implication), the lifting transform can generate well-formed formulas in propositional logic. A digraph of ANDed and OR'ed antecedent concepts followed by an implication is a *rule*.

Finally, at level 4, collections of rules are aggregated to obtain rule clusters. An appropriately selected collection of rule clusters constitutes the raw material for a *reasoner.* Such systems are capable of inferencing within a domain, and when supported by the following

1. An appropriate non-monotonic formalism
2. A means of handling the frame problem
3. An adjudication logic

are able to perform automated cognition in a principled way.

By this process of imposing relations on lower-level objects to elevate entities to a higher level, the relation tower organizes entities of varying logical complexity within the domain of discourse in a principled way. Further, the same lifting transform is used at each level. After four applications of this lifting process, we have moved from the layer of uninterpreted phenomenology at level 0 to automated cognition, adaptive systems, and trainable software at level 4.

The relation tower does not describe <u>how</u> inferencing is conducted; rather, it provides a representation scheme within which inferencing will occur. There are many ways in which inferencing can occur in a structure of this sort. Because any mapping can be represented as a relation, the lifting transform is theoretically sufficient to support any kind of aggregation or connection one might want to impose on lower-level objects to obtain higher-level constructs.

5.1 Unification

The relation tower supports several important implementation considerations

1. Separate relation towers can be created for different problem spaces, areas-of-interest, or other decompositions of the analytic problem. This allows the creation of customized methods for specific problems; parallelization; and distributed computing.
2. More importantly, perhaps, is the fact that multiple towers can be <u>unified</u> at any desired level as a means to implement multi-level fusion.

Figure 3 is a notional diagram depicting the use of a relational tower to unify (fuse) two other towers at an intermediate level of processing

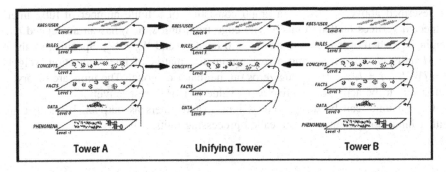

Fig. 3. Fusion can be implemented without developing additional algorithmics.

It is hypothesized that animal brains engage in parallel pattern association, matching percepts with sets of extracted image features previously annealed into plastic brain structures. Images are perceived to be members of the category which developed the associations giving the best match.

These plastic structures are emulated here by a collection of associative memories (AM's), each corresponding to a particular image category. Incoming percepts are matched against each AM, giving a matrix of responses. These responses are passed to a reasoning component (described below) responsible for adjudicating them to produce a classification to category for the image.

Markov Random Fields (MRF's) are used to implement the AM's. MRF's (certain implementations are referred to as Boltzmann Machines) are regressions that estimate binary system state variables from binary observables (e.g., condition true/false, present/absent). MRF's represent domain knowledge as a weight matrix $W = \{W_{ij}\}$ which quantifies the joint-likelihoods that binary state variables s_i and s_j are simultaneously "true". We use this matrix to impose a "soft" relation on the image features space to generate image responses and cross-responses for the image categories.

If we insist on exact matches in reasoning as above, the resulting inferencing scheme will be brittle. MRF's provide a natural remedy by replacing the crisp relation $r \sim s$ with a real-valued symmetric association weight matrix $W = \{W_{rs}\}$. The association weight W_{rs} will have a positive value when r and s usually occur together (are usually consistent), and will have a negative value when s and r usually do not occur together (are usually inconsistent). More precisely, when both $p(s|r)$ and $p(r|s)$ are large, W_{rs} will be positive; when both $p(not\ s|r)$ and $p(not\ r|s)$ are large, W_{rs} will be negative. Under this new definition, concepts are arbitrary sets of symbols whose elements generally have pairwise positive association weights, but might also include some element pairs having negative association weights. In keeping with the notion of crisp concepts above, it is customary to require W to be symmetric.

Unlike a simple cooccurrence matrix, W captures the relative strength of the associations of state variables in the context of the values assumed by all other state variables: it retains "context".

6 Constructing Markov Random Fields from Percepts

As described above, a MRF is a graph consisting of a finite number of fully inter-connected vertices (called "units"). The edge from unit s_i to s_j has an associated real weight W_{ij}, where it is required that $S_{ij} = 0$ for all i, and $W_{ij} = W_{ji}$. Specifically, there are M units $s_1, s_2, ..., s_M$, and a real, symmetric matrix of interconnection weights $W = \{W_{ij}\}$, $i, j = 1, ..., M$, having zeros on the major diagonal. The units are binary; a unit is said to be active or inactive as its value is 1 or 0, respectively.

Certain units can be designated as inputs, and others as outputs. Units that are neither input nor output units are called processing units.

Inferencing in MRF's proceeds according to the following update rule

1. Assign the (binary) values of the input units ("input clamping")
2. Randomly select a non-input unit s_i. Sum the connection weights of the active units connected to s_i, and make s_i active if the sum exceeds a preselected threshold value t_i, else make it inactive.
3. Repeat step 2 until the unit values stop changing
4. Read off the (binary) values of the output units

In this scheme, positive weights between units are excitatory, and negative weights inhibitory: an active unit tends to activate units with which is has a positive connection weight, and deactivate units with which it has a negative connection weight.

Formal mathematical analysis of the inferencing procedure above is facilitated by the introduction of an energy function (due to John Hopfield): Let V be a vector of binary inputs. The energy of the MRF at $V = (b_1, b_2, ..., b_L)$ is

$$E_W(V) = -(1/2) \sum_{i \neq j} s_i s_j w_{ij} + \sum_{i=1}^{M} S_i t_i$$

With this definition of network energy, Hopfield's updating rule is equivalent to the following: randomly select an unclamped unit; if toggling its state will lower the network energy, do so. Repeat until no single state change can lower the energy, then read the states of the output units.

The values of the input nodes are fixed, but the changes in the states of the processing and output units under updating cause changes in the energy of the network. This suggests that a supervised reinforcement learning algorithm called "simulated annealing" (first applied to Hopfield nets by Hinton and Sejnowski) can be used to train the network to associate desired outputs with given inputs

1. Assign the (binary) values of the input units ("input clamping")
2. Assign the desired (binary) values of the output units ("output clamping")
3. Randomly select a processing unit s_j
4. Set the state of s_i to active with a certain one.
5. Repeat step 3 for many epochs, allowing the network to relax to a local energy minimum
6. Once an energy minimum has been reached, step around the network, incrementing the weights between pairs of units that are simultaneously active (encourage future simultaneity!), and decrementing the weights between pairs having just one active unit (discourage future simultaneity!)

To avoid saturation of the weights, an additional updating sequence having both the processing and output units unclamped is often included.

Here T is a parameter (the *temperature*) that is gradually reduced as training proceeds. At low temperatures, the weight matrix is easily modified; as the temperature is reduced, the weights "lock" into their final values.

Simulated annealing develops a weight matrix that gives local minima in energy when the input pattern is "consistent" with the output pattern: the lower the energy, the stronger the consistency. The interconnect weights, then, codify constraints on outputs the machine is likely to produce when the inputs units are clamped and updating is applied. Notice that the machine can give different outputs for the same input owing to the random element in the update rule. It can be shown that the probability of "incorrect" outputs can be made arbitrarily small (Fig. 4).

- Imagine a ball rolling on some bumpy surface
- The position (state, height) of the ball at any instant represents the activity of the nodes in the network
- Minimum energy states give the element settings that are the networks' output

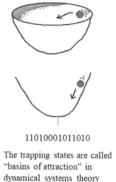

11010001011010

The trapping states are called "basins of attraction" in dynamical systems theory

Fig. 4. The Hopfield Network adjust the strengths (both positive and negative) between related entities seeking a set of connection weights that willo enable the machine to match news patterns of activated elements with previsously learned patterns of activation. It does this by "falling into" a minimum energy configuration when learned patterns are present.

Parameters associated with training include the number of input, processing, and output units; the initial and final temperatures; how the temperature is to be reduced (annealing schedule); the number of random updates applied during each relaxation epoch; the number of relaxation epochs; the values of the thresholds; and the amount by which to increment/decrement the weights after relaxation is complete. As with most trainable systems, there are no universally applicable heuristics for these assignments.

7   Reasoning with Knowledge: An Executable Ontology

The term ontology here refers to information structures which enumerate the entities, concepts, and relationships that obtain in a domain. We introduce the notion of an executable ontology, within which these relations are made dynamic and actionable by being embedded in adaptive algorithms. A natural way to do this is to bind domain relationships in parameterized rulesets.

By adjusting the parameters based upon experience, the executable component of the ontology can be made adaptive and trainable. This overcomes the difficult problems associated with developing conventional knowledge-based systems, Bayesian belief networks, and the like.

Human experts typically do a fairly good job of assessing the impact, relevance, and quality of individual heuristics in isolation. However, when knowledge-based systems are placed into operation, it is the entire of mix of heuristics in context that determines the effectiveness of the system.

Even humans having deep domain expertise have difficulty understanding the complex interactions of a large number of rules. For this reason, the optimization of existing rules is best done by automation of information theoretic techniques. A gradient-based rule optimization strategy can be used to calibrate existing rulesets to perform reasoning tasks. This is described in the following.

7.1   Hopfield Networks

The Hopfield Network Architecture was proposed at Caltech in 1982 by physicist John Hopfield. A Hopfield Net is an artificial neural network that is able to store certain memories or patterns in a manner that is functionally similar to animal brains.

The nodes in the network are vast simplifications of real neurons - they can only exist in one of two possible states - *firing* or *not firing* (there are no input or output neurons).

Every node is connected to every other node with some strength (or weight) but not with itself; connections are symmetrical "association weights". At any instant of time a node will change its state (i.e. start or stop firing) depending on the inputs it receives from the other nodes.

There exists a rather simple way of setting up the connections between nodes in such a way that any desired set of patterns can be made a stable firing pattern – minimize its energy. Thus, any desired set of "memories" can be burned into the network at the beginning using an "annealing" process.

8   Case Study: Learn Patterns in the Phase Structure of Doppler RADAR Reflections from Military Vehicles

Doppler RADAR operates by computing the way in which the motion of a reflecting surface affects a stream of incident radio waves. If the earliest wave sent out indicates the same distance to the surface as subsequent waves, that surface must be stationary with respect to the RADAR. However, if later waves return earlier or later than earlier waves, the surface must be respectively, approaching the RADAR, or receding from it. The amount of this change gives a direct measure of the relative speed of surface and RADAR.

This is the principle by which Doppler Weather RADAR detects complex atmospheric movements: winds moving in different directions and at different speeds hasten or delay RADAR waves in a way that allows the detection of tornadic vortices, distance storms, wind-shear, and micro-bursts.

Suppose now that the RADAR wave are incident upon the body of a military vehicle. Such vehicles usually have parts that themselves move as the vehicle moves: wheels, tracks, antennas, guns, hatches, etc. These moving objects add their own

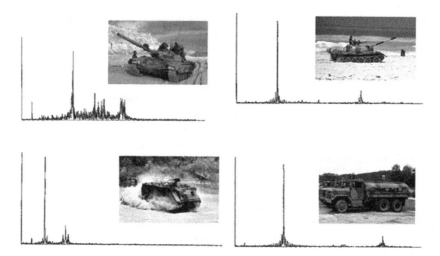

Fig. 5. The truck in the lower right-hand-corner is a low-value target... generally not of great concern, and usually not worth deploying expensive munitions to engage. The other three armored vehicles are high-value targets.

changes into a Doppler RADAR measurement, resulting in complex modulation of the radio waves incident on the vehicle.

If this modulation varies from vehicle to vehicle in a consistent way, it might be possible to infer the vehicle model from its Doppler RADAR modulation pattern. Distinguishing a convoy of trucks and a column of tanks from miles away offers clear tactical benefits.

When humans look at ragged waveforms, such as those in Fig. 5, they aren't drawn to the random-looking, disorganized stubble that constitutes most of the data; they naturally look for "clumps" that, taken together, tell a story. Humans can make sense of a few clumps of variable size and arrangement; they rest is inscrutable.

The difficulty with the vehicle classification problem, though, is that the "clumps" are not where the discriminating information lies.

The two vehicles on the left are both in the class "high-value targets"; yet their Doppler returns look nothing alike. The two vehicles on the right are in different classes; one high-value and the other low-value. Yet, their Doppler returns are virtually identical, "clumpologically" speaking.

The discriminating information, if it is present at all, is scattered across the random-looking stubble. It is distributed in a way, having cross-correlations of a sort that no human will ever be able to determine by visual inspection. What is needed is a method to determine which pieces of the Doppler RADAR trace being present/absent in combination (clumpy or not) that together indicate the vehicle model.

In other words, what is sought is the clique of energy peaks that is characteristic of each vehicle class.

This is exactly a relation. At level 0, there are bins for the various RADAR frequencies. Real numbers exist at level 0.

At Level 1, we will have a set of "facts" that link each RADAR bin with real number giving its relative energy when a vehicle of a given class is scanned.

At Level 2, we would like to have "rules" similar to:

{bin7 energy > 1%, bin12 energy > 1%, bin16 energy < 1%, bin25 energy > 1%} → HIGH-VALUE-TARGET

This takes us away from manual clumpology into the realm of trainable MRF. It is not necessary (in fact, not possible) to enter vehicle modulation information into a data base, because it isn't known. But an MRF can infer these patterns using simulated annealing.

A reference set of vehicles is assembled, and scanned with the RADAR, and their modulation patterns sampled over many trials. Using the relaxation training method described in section xxx, and MRF has these patterns annealed into it. When a test pattern is introduced at the machine's input elements, the output element corresponding to the vehicle model will become active.

Feature reduction begins by creating a bit string equal to the number of bins into which frequency has been quantized. Bit positions corresponding to energy levels above a threshold value are set to "1"; others are set to "0". This encodes the Doppler RADAR modulation as a vector of zeros and ones.

To enter this data into the Hopfield network, the input nodes corresponding to bit values of "1" are switched ON; nodes corresponding to bit values of "0" are switched OFF. The update cycle is then applied until the output bits stop changing, and the total energy is low. The output node active indicates the networks classification decision (Fig. 6).

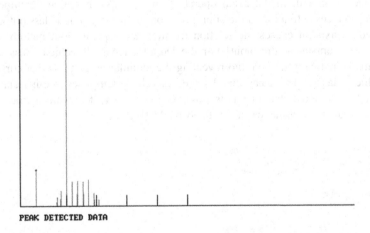

PEAK DETECTED DATA

Fig. 6. If the energy of a bin exceeds a threshold value, the corresponding input bit is set to "1"; otherwise it is set to "0".

When a Hopfield Network having 32 input nodes, 6 input nodes, and 2 output nodes is trained and run on blind test data, accuracy of classification to a High-value/Low-value vehicle was 10-percentage points higher than the competing regression classifier was realized. When a larger topology having 12 output nodes (the number of specific vehicle models) was trained on a large test set, accuracy improved again, and the machine was able to classify vehicle to specific model type about 60% of the time (Fig. 7).

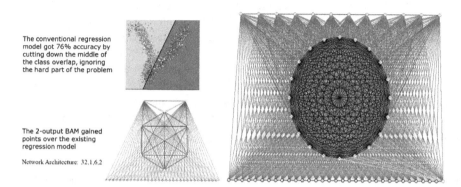

The conventional regression model got 76% accuracy by cutting down the middle of the class overlap, ignoring the hard part of the problem

The 2-output BAM gained points over the existing regression model

Network Architecture: 32,1,6,2

Fig. 7. Final Engine: lower-left is the high-low value MRF classifier architecture; the 32 input nodes are across the bottom, and the two output nodes are at the top. On the right is the larger 12-class MRF. In both of these architectures, every node is connected to every other node, and the input nodes are "clamped" to the input bit-string values.

Finally, since our particular MRF is a Boltzmann Machine, it is a bi-directional associative memory (BAM). Unlike most classifiers, trained BAMs can be run both forward and backward. In the usual operation mode, input nodes are clamped, and relaxation produces activation of the appropriate output node. But it a class output node is clamped, relaxation causes the relation defining the typical input pattern for that output class to appear at the input! For the Doppler RADAR vehicle classifier, the BAM is mathematically tricked into revealing the modulation patterns that correspond to the vehicle classes. These can then be used as vehicle templates in other classifiers. Our team has demonstrated that it is possible in this way to create a rule set that achieves the same performance of a trained BAM (Fig. 8).

```
NUMBER OF INPUT, PROCESSING, AND OUTPUT NEURONS: 32  6  2
NUMBER OF TRAINING PASSES: 200
LEARNING RATE: 2.999999932944775E-002
ANNEALING CONSTANT: 1.004999995231628
NEURAL ACTIVATION THRESHHOLD: .5
FINAL TEMPERATURE: .300000011920929

   CLASS     PROTOTYPICAL FEATURE VECTOR
  ---------  ------------------------------------------------------------

   TRACKS    4 5 3 8 9 4 6 10 21 9 5 6 2 6 2 6 4 8 5 7 7 6 6 7 4 6 1 3 4 2 7 6

   WHEELS    16 18 9 3 5 16 7 0 0 3 3 5 13 6 5 7 7 5 3 5 3 3 6 8 5 9 6 3 3 6 4 3

   ?
```

Fig. 8. Running the BAM is reverse produces the relative energy levels in each frequency bin that are the prototypical modulation patterns for both high-value and low-value vehicles.

References

1. Boyen, X., Koller, D.: Tractable inference for complex stochastic processes. In: Proceedings of the 14th Conference on Uncertainty in Artificial Intelligence, pp. 33–42, Madison (1998)
2. Delmater, R., Hancock, M.: Data Mining Explained. Digital Press (2001)
3. Doucet, A., de Freitas, N., Gordon, N. (eds.): Particle Filters Sequential Monte Carlo Methods in Practice. Springer, New York (2001)
4. Friedman, N., Koller, D., Pfeffer, A.: Structured representation of complex stochastic systems. In: Proceedings of the 15th National Conference on Artificial Intelligence, pp. 157–164, Madison (1998)
5. Hancock, M.: Practical Data Mining. CRC Press, Boca Raton (2011)
6. Battaglia, F., Borensztajn, G.: Structured cognition and neural systems: from rats to language. Neurosci. Biobehav. Rev. **36**(7), 1626–1639 (2012)
7. Essentialism. https://www.britannica.com/topic/essentialism-philosophy. Accessed 15 Feb 2019

A Study on the Development of the Psychological Assessment a Using Eye-Tracking: Focused on Eye Gaze Processing of Literacy Text

Joon Hyun Jeon[1], Gyoung Kim[2], and Jeong Ae Kim[1(✉)]

[1] Konkuk University, Seoul, Republic of Korea
{Naturaljeon, esgell3}@konkuk.ac.kr
[2] Media, Interface, and Network Design (M.I.N.D) Lab, Syracuse University, Syracuse, USA
gmkim@syr.edu

Abstract. The purpose of this study is to examine the possibility of improving a self-epic diagnosis tool by using a gaze processing method in terms of literacy therapy. In this study, we used an eye-tracker measuring eye-position and eye-movement to read subjects' reactions to the critical region in a context. With this technology, we focused on the fixation and saccade to see what vocabulary subjects focused and how their gaze was shifted. We also explored to find a relationship between their eye-movement and their degree of sympathy by examining their physiological data. This pilot study showed a potential use of a physiological data in the epic of self-test in literacy therapy.

Keywords: Psychological assessment · Gaze · Eye-tracking · Literacy · Psychological diagnosis through processing texts

1 Introduction

This study aims to examine the empathy on literary text through the way of gaze processing, and to explore the possibility of psychological diagnosis using literary text. The possibility that literary text can be utilized as a measure of psychological diagnosis has already been already proposed in the field of "literary therapeutics" and is called the "epic of self-test". It is based on the premise that literature itself and reaction to literary works are all holistic reflections of human psychology.

The problem with conducting psychological diagnoses using literary text is that it is difficult to ascertain whether the degree of response and empathy toward a text consisting of a long narrative is the result of understanding and response to the entire narrative context, or the result of excessive interpretation or distortion of parts of the narrative. Even if one understands the whole context and shows an empathetic response to the text, how to present and evaluate the level of empathy is difficult. Therefore, this study seeks to find a way to visually measure the human ability to empathize with literary text. For achieving this goal, this study employs the use of eye-tracker devices.

© Springer Nature Switzerland AG 2019
D. D. Schmorrow and C. M. Fidopiastis (Eds.): HCII 2019, LNAI 11580, pp. 74–83, 2019.
https://doi.org/10.1007/978-3-030-22419-6_6

With the recent development of a variety of eye movement control models (Engbert et al. 2002; Engbert et al. 2005; Morrison 1984; O'Regan 1990; Pollatsek et al. 2006; Reichle et al. 1998), research on reading through eye movement tracking is becoming more diverse. In addition to dealing with issues such as perceptual range, saccade-related phenomena, and saccade response times from one fixation, effects of language characteristics on fixation and saccade in real time has also come to be studied.

In accordance with this prior research, this study focuses on the area of interest (AOIs) that contains the core of the text and attempts to understand how much the subject's eyes are fixed and where they move in the key vocabulary and syllables that convey the main theme. This study also aims to discover whether there is a difference in sentences or words between groups with different empathetic abilities. This will help confirm the subject's level of empathy for the core of the literary text. If the subject's eyes deviate from the core, it can be said that the degree of empathy is low, and in the inverse case, that the degree of empathy is high.

Thus, this study attempts to ascertain whether the reading area differs from group to group depending on their empathic abilities (EQ) whether the response to literary text relates to the whole text or to some part of the text, and this correlates with the subject's empathic abilities (EQ).

2 Method

To determine the range of perception, the phenomenon related to saccade, and the time of the saccade response from one fixation based on the level of empathic ability of the viewer, twenty-two college students with literacy skills were recruited. All participants in the experiment have a corrected visual ability of 0.6 or higher and could read the text on the monitor easily. Using the eye-tracker, the subjects' eye movements were recorded as they read 16 narrative stimuli.

After dividing the students into two groups with high EQ scores and low EQ scores, The data was analyzed through fixation, saccade and the heat-map that appeared as they read the text. And factorial extraction and regression analysis of the narrative text survey, which consisted of a five-point Likert scale, was conducted to determine the correlation between 16 narrative texts and EQ scores.

2.1 Hypotheses

The concept of empathy, which is related to sympathy, is important not only in the field of literature, but also in the fields of psychoanalysis and psychological counseling.

Empathy is one feels when one experiences other people's positions or situations and fully understands them, and many studies have shown that people with good sympathy have a better understanding and feeling of narrative texts. However, there are few devices that can scientifically prove where empathy is achieved. In this study, subjects are expected to be easily aware of and empathize with familiar words or sentences.

Therefore, it is predicted that among familiar words or paragraphs, the areas of eye fixation or saccade will appear differently between people with higher and lower degrees of empathy. The following assumptions were derived:

H1: the cognitive process of reading leads eye movement: Focus on frequency of use and prediction (linguistic properties of words).

H2: The comprehension of sentences will vary depending on the level of empathy.

H3: Depending on the level of empathy, the level of interpretation of a particular word or paragraph will vary.

To test the above hypothesis, 16 narrative stimuli and eye-tracker, the epic of self test were used.

2.2   Stimuli

A 2D flat computer display with Full-HD resolution (1920x1090 pixels) was used for which participants would read the text. A single literary text was set up to be displayed on one screen. entirely. After participants finished reading the text on the screen, they proceeded by pressing the ESC button and writing a questionnaire. Psychological tests were read by participants to conduct 'emotional-response-type epic of self-tests'. The emotional-response-type epic of self-test consists of asking through five Likert scale, ranging from 'repulsion' to 'touching'. In other words, the emotional-response-type

---

The first question: <The Liver of Son and Stepmother>

> (1) A man raised a son with his wife, but when she died, he got a new wife.
>
> (2) When the new wife gave birth to a son, she said that she was ill falsely, and she must eat the liver of his ex-wife's son. (3) To save his new wife, the man ordered a butcher to kill his ex-wife's son and bring his liver. (4) When the butcher's wife asked the butcher to catch the dog instead of the child, he let go of the child and brought the dog's liver. (5) The ex-wife's son grew up well and became a high official. (6) The ex-wife's son punished his stepmother, and lived well with his father and the butcher and his wife.

■ Please check your feelings after reading to the above story.

① repulsion

② uninterested

③ nothing special

④ interested

⑤ touching

■ Please write down the reason why you feel that way.

Fig. 1. An Example of questions: the first question of "The Liver of Son and Stepmother"

epic of self-tests is designed to measure human beings' degree of empathy by measuring the degree of empathy for literary text.

The example is as follows (Fig. 1).

Sixteen questions were created by processing and summarizing the oral folktales from 'The Collection of Korean Oral Literature'.

In folktales, the process from the problem situation to the solution appears compressive (compressed). Thus, the subjects can identify the contents in a short period of time and have an immediate response. The 16 questions were divided into four areas according to the positions of children, the positions of men and women, the positions of husband and wife, and the positions of parents, centering on human relationships as a circular form. These four areas are called the realm of son and daughter, the realm of man and woman, the realm of husband and wife, and the realm of father and mother. They are designed according to the four levels and patterns of human relations. In other words, four levels of stories are arranged in order in each of the four areas. If the subject answers 16 questions, the result will be between 16 and 80 (Table 1).

Table 1. Configuration of clauses and sentences in questions

Question	1	2	3	4	5	6	7	8	9	10	11	12	13	14	15	16
Sentence	6	6	6	6	6	6	6	6	8	12	8	15	11	10	13	10
Clause	82	131	121	81	75	114	184	124	186	223	145	245	182	176	207	164

2.3 Tool

In this experiment, the eye tracker device (Gazepoint GP3 eye tracker) was used. The equipment has a collection rate of 60 Hz and a resolution of 0.5 to 1 degree of visual angle accuracy.

2.4 Procedure

The experiment was conducted individually and without distinction between groups. After preparing a consent form for participation in the experiment, the EQ test was conducted for group propensity analysis. To carry out the experiment in earnest, we explained the experiment to the participants, and nine points were presented on the screen in order to calibrate the subjects' eyes.

The 16 story questions in the emotional-response-type epic of self-test are given to participants in 16 projects. After the participant has finished reading the literary text, press the ESC button and select one of the five options in the questionnaire: repulsion, uninterested, nothing special, interested, or touching. Once the selection is complete, check that the line of sight is fixed in the center, and put the text of the second question on the screen so that the subject can re-execute the process that he or she performed in Question 1. This process is to be carried out through Question 16.

Scores that are converted through options selected by the subject will be used to analyze the differences in gaze processing between the group with a high level of empathy and that with a low level of empathy during the analysis phase.

3 Results and Discussion

3.1 Correlation Between Factors Analysis of Literacy Text and Empathic Abilities

Five factors were extracted from the experiment. They are "son and daughter epic", "father and mother epic", "man and woman epic", "husband and wife epic", and "personal epic".

The correlation between each factor was not significant and the "son and daughter epic" had a slight effect on the EQ score (Table 2).

Table 2. Correlation between factors analysis of literacy text and empathic abilities

Rotated Component Matrix

	Component						
	1	2	3	4	5	6	7
pa02	.824	.033	.038	−.036	.023	.315	.065
C01	.768	.150	.094	.129	.304	−.125	−.071
C02	.544	−.284	−.473	.025	−.100	−.412	−.154
C03	.373	.790	−.123	.168	−.061	.005	−.132
G02	−.157	.779	.212	.049	−.200	−.157	.035
cu02	−.071	.521	−.277	−.434	.130	.423	.323
pa03	.315	.472	.024	.281	.464	.267	.324
G03	.107	−.034	.906	−.085	.199	−.010	−.140
G01	−.073	.069	.648	.186	−.277	.103	.469
pa01	.278	.184	−.140	.803	−.086	.005	−.006
cu01	−.107	.091	.227	.742	.077	.329	.007
C04	−.221	−.250	−.359	.568	.313	−.092	−.387
pa04	.151	−.104	.227	.043	.867	.149	−.011
G04	.055	−.167	−.274	−.077	.655	−.419	.172
cu04	.128	−.121	.036	.173	−.009	.858	.005
cu03	−.039	−.034	−.019	−.092	.128	−.017	.951

Extraction Method: Principal Component Analysis.
Rotation Method: Varimax with Kaiser Normalization.

a. Rotation converged in 19 iterations.

Table 3. Correlation between factors analysis of "Son and Daughter epic"

Correlations

		T1	T2	T3	t4	t5	EQscore
T1	Pearson Correlation	1	.177	-.081	.101	.260	.306
	Sig. (2-tailed)		.430	.720	.654	.242	.167
	N	22	22	22	22	22	22
T2	Pearson Correlation	.177	1	.004	.050	.000	.009
	Sig. (2-tailed)	.430		.986	.824	1.000	.967
	N	22	22	22	22	22	22
T3	Pearson Correlation	-.081	.004	1	-.069	-.030	.018
	Sig. (2-tailed)	.720	.986		.760	.896	.935
	N	22	22	22	22	22	22
t4	Pearson Correlation	.101	.050	-.069	1	.064	.247
	Sig. (2-tailed)	.654	.824	.760		.777	.267
	N	22	22	22	22	22	22
t5	Pearson Correlation	.260	.000	-.030	.064	1	.089
	Sig. (2-tailed)	.242	1.000	.896	.777		.693
	N	22	22	22	22	22	22
EQscore	Pearson Correlation	.306	.009	.018	.247	.089	1
	Sig. (2-tailed)	.167	.967	.935	.267	.693	
	N	22	22	22	22	22	22

Model summary[b]

Model	R	R square	Adjusted R square	Std. error of the estimate	Change statistics					Durbin-Watson
					R square change	F change	df1	df2	Sig. F change	
1	.474[a]	.224	.042	9.29339	.224	1.229	4	17	.335	1.817

a. Predictors: (Constant), pa, G, cu, C
b. Dependent Variable: EQscore

The experiment results above show that the case was insufficient to explain the correlation between the text stimulant and the EQ. However, if sufficient cases are secured through the pilot test, there is a possibility that there will be a link between "son and daughter epic" and the EQ (Table 3).

3.2   Experiment Results of Eye-Tracking

As previously explained, the narrative stimulus that demonstrated a correlation with EQ was the "son and daughter epic" or the story that revolves around the relationship between a parent and child. To test the hypothesis that the level of empathy will change the response and understanding of a text, the upper and lower EQ score groups were established, and the frequency of eye fixation from each group was examined. In order to help visualize this, a heat map was implemented.

Here, the differences between the two groups in relation to the "son and daughter epic" are presented, and the survey contents between the two groups are compared.

In the second "son and daughter epic" question, the groups separated by the EQ score consisted of three participants each, and the frequency of eye fixation was checked (Table 4).

The results show that the upper group was focused on the word "up" from the sentence 'up to the top of the tree in the yard', and the lower group was concentrated on the words "child", and "left children at home" from the sentence 'mother left little brother and sister and a child at home and went to work'.

The lower group is focused on the first half of the story. The upper group is focused on the second half of the story. Furthermore, the lower group is focused on the characters, while the upper group is focused on the characters' actions. The point when the brother and sister went up to the sky, a symbol of their independence from their parents' world, is the sentence that determines the ending of the story.

Therefore, the fact that the upper group was focused on the sentence 'up to the top of the tree in the yard' means that the eyes were fixed longer and repeatedly at a story's point. One of the upper group participants chose 'interesting' for this question and responded by saying, "the sheer force of the young children, suspicious that the tiger is different from the mother's, is funny." Meanwhile, one of the lower group participants responded, "the content of a tiger eating a baby was horrifying", and chose 'repulsion'.

The higher the score on 5-point scale, the more attention is paid to the process of children's self-reliance and overcoming the conflict with the tiger. On the other hand, the lower the score, the more attention was paid to the element of tiger cruelty.

Table 4. The second question of "Son and Daughter epic"

This is to be confirmed through the acquisition and analysis of more participants (Table 5).

In the case of the third "son and daughter epic" question, the groups separated by the EQ score consisted of five participants each, and the frequency of eye fixation was checked.

The results show that the upper group was focused on the word "up" from a sentence 'up to the top of the tree in the yard', and the lower group is concentrated on the words "child", and "left children at home" from the sentence 'mother left little brother and sister and a child at home and went to work'.

As a result, the upper group is focused on the word 'her good fortune' from the sentence 'the third daughter said she lives in her good fortune', and 'to a charcoal dealer' from the sentence 'father sent the third daughter to a charcoal dealer.' Alternatively, the lower group was focused on the phrase 'proper price' from the sentence 'she asked him to sell the stones at a proper price'. Here we can see that while the upper group was focused on the conflict between the father and daughter, the lower group was focused on how the daughter became rich. In other words, the upper group is paying attention to problem situations triggered by relationships, while the lower group

Table 5. The third question of "Son and Daughter epic"

	Q02_Score high	Q02_Score low
fixation map		
Heat map		

is paying attention to wealth. This means that the upper group is focusing on context, while the lower group is focusing on motif.

One of the upper group participants chose 'touching' for this question and responded by saying, "I'm impressed by the part where the third daughter serves her parents who drove her away hard and became beggars." One of the lower group participants responded, "I wonder why the smart daughter lives with people who don't even have a parent's qualifications", and chose 'repulsion'.

The higher the score on the 5-point scale, the causal context of a rich daughter having a father is understood, and on the other hand, the lower the score, more subjective feelings are projected. This is to be confirmed through the acquisition and analysis of more participants.

Acknowledgments. I (Dr. Joon Hyun Jeon) am immensely grateful to Gyoung Kim, lab manager in the Media, Interface, and Network Design (M.I.N.D. Lab) at Syracuse University, for his comments on an earlier version of the manuscript, although any errors are my own and should not tarnish the reputations of his esteemed professional.

References

Engbert, R., Longtin, A., Kliegl, R.: A dynamical model of saccade generation in reading based on spatially distributed lexical processing. Vis. Res. **42**(5), 621–636 (2002)

Engbert, R., Nuthmann, A., Richter, E., Kliegl, R.: SWIFT: a dynamical model of saccade generation during reading. Psychol. Rev. **112**(4), 777–813 (2005)

Morrison, R.E.: Manipulation of stimulus onset delay in reading: evidence for parallel programming of saccades. J. Exp. Psychol. Hum. Percept. Perform. **10**, 667–682 (1984)

O'Regan, J.K.: Eye movements and their role in visual and cognitive processes Eye movements and reading. In: Kowler, E. (ed.) pp. 395–453. Elsevier (1990)

Reichle, E., Pollatsek, A., Fisher, D., Rayner, K.: Toward a model of eye movement control in reading. Psychol. Rev. **105**, 125–157 (1998)

Pollatsek, A., Reichle, E.D., Rayner, K.: Tests of the E-Z reader model: exploring the interface between cognition and eye-movement control. Psychology **52**, 1–56 (2006)

Zhu, Z., Ji, Q.: Eye and gaze tracking for interactive graphic display. Mach. Vis. Appl. **15**, 139–148 (2004). https://doi.org/10.1007/s00138-004-0139-4

Laura, A.G., Joachims, T., Gay, G.: Eye-tracking analysis of user behavior in WWW-search. In: Laura A.G. (ed.) Research Gate (2004). https://doi.org/10.1145/1008992.1009079

Salvucci, D.D., Goldberg, J.H.: Identifying fixations and saccades in eye-tracking protocols. In: Proceedings of the 2000 Symposium on Eye Tracking Research & Applications. ACM, pp. 71–78 (2000)

Impedances of Memorable Passphrase Design on Augmented Cognition

Lila A. Loos[✉], Michael-Brian Ogawa[✉], and Martha E. Crosby[✉]

University of Hawai'i at Mānoa, Honolulu, HI 96822, USA
{lila7194, ogawam, crosby}@hawaii.edu

Abstract. The quest for optimum computer authentication continues to stimulate resolution toward memorable password design. As human cognitive load processes the transfer of information from working memory to long-term memory, intrinsic and extraneous complexities present impedances to cognitive load. This study applies human cognition to the use of passphrases characterized as assembled words that are secure and easy to recall [23]. "Passwords and passphrases, when married with psychology and psycholinguistics, yield an authentication scheme that is revocable, memorable, and secure" [21, p. 2]. Subsequently, this study examines augmenting cognitive load of passphrase design to enhance memorability and effectuate authentication performance.

Keywords: Cognitive load · Passphrase · Memorability · Augmented cognition

1 Introduction

Present day authentication schemes are prone to security attacks and are difficult to use. In response to memorability struggles, users make slight modifications to their existing password when prompted to change their password. Consequently, security is not automatically enhanced by this requirement; therefore, a study proposes imposed passwords suggesting security and memorability are central to its design [9]. Although various findings consist of multi-factor verification and multi-modal biometric forms, cognitive demands remain for the user [1]. Additionally, passphrase or multi-word combination resolutions seek to improve human computer authentication in security and usability inquiry [5, 9]. Moreover, reducing cognitive load during authentication suggests to reduce input errors, improve performance and enhance the authentication experience [19].

Augmented cognition inquiries involve identifying support systems to current password and proposed passphrase structures. A study including assistance from mechanisms like a system shared secret [21] and a four-word system assigned passphrase suggests to increase memorability using reinforcement [10]. Similarly, passphrases using mnemonics implies improvement to security and recall [30]. Accommodating longer passphrase lengths combined with a validation system that authenticates common typing mistakes alludes to improve security and aids memory to ultimately minimize input error [17]. The selection and encoding of meaningful words as random entries suggests to initiate deep processing and increase passphrase length

© Springer Nature Switzerland AG 2019
D. D. Schmorrow and C. M. Fidopiastis (Eds.): HCII 2019, LNAI 11580, pp. 84–92, 2019.
https://doi.org/10.1007/978-3-030-22419-6_7

and security [31]. Likewise, personalizing passphrases selected from a random word set alludes to memorability [4]. Although existing studies contribute to relationships among password construction, recall and meaningfulness, research investigating cognitive psychology and password selection research are scarce [21].

As usability and access to computer systems remain problematic [4], further passphrase design inquiries, driven by cognitive theories, are expected to address memorability concerns [31]. Contributions that detail user-centric authentication design principles include adapting to user behavior and preferences to minimize interruptions and reduce cognitive load for successful authentication [19]. Inclusive to this study of advancing memorable passphrase investigation and addressing access to computer systems is the evaluation of personality traits and memory association influences on imposed and user created passphrases [14].

2   Theoretical Background

2.1   Cognitive Load

Cognitive load is the amount of mental effort expended by a learner interacting with instruction and is influenced by and interacts with novel learning instructions kept in short-term memory [24]. If instructions facilitate learning, cognitive load is germane otherwise cognitive load is extraneous [11]. A third type of load, intrinsic load, is distinguished as information difficulty [8]. The combination of germane, extraneous, and intrinsic cognitive loads suggests to establish learning within the constraints of working memory [2, 8, 11, 20, 24, 25]. Furthermore, it is implied that germane and extraneous load are influenced by the design of the learning model as opposed to the intrinsic nature of the information [20].

Multiple sensory channels are required to increase working memory capacity [8]. Information is processed through working memory prior to storage in long-term memory where it can be retrieved without temporary constraints. Impedances to cognitive load increase with disruptions that interfere with learning and ultimately with the transfer process to long-term memory. Complex word or instruction variability as well as difficulty associating meanings to words may necessitate significant exertion while cognitive workload could disturb the transport of learning to long-term memory [25]. Learning and recalling computer passphrases depend on the processing of information for sustained memorability.

Long-term memory enables recall and more importantly provides proficiency in high level processes such as problem solving [24]. In this study passphrase recall is used to measure cognitive load with the goal of discovering design outcomes to improve passphrase learning processes that transform to long-term memory. Unlike long-term storage of vast assembled information, working memory processes up to three interrelating components. These components rely on defined procedures to process interacting elements that expand its capacity [20]. Study results reported no significant difference in entropy between 3-word and 4-word passphrases noting that 3-word passphrases resulted in fewer errors [23]. Additionally, the number of characters in a passphrase rather than the words negatively affects usability while a sentence like structure that can be visualized positively affects usability. Considering the limitations of short-term memory, clear and concise passphrase instruction is an important design requirement.

2.2 Memory Implications

A study of visually presented words suggests participants tend to recall words from the beginning and end of the list consisting of four to six words [29]. User created passwords are discoverable and system generated passwords are difficult to remember resulting in preference for passphrases over long passwords comprised of characters. The study randomly selected short words from a large dictionary which provides greater security than characters in a password. Implications suggest passphrases are more memorable than a password composed of characters [9].

Although "it is not known what makes a password memorable" [7, p. 221], findings include password repetition and importance of the user's account supports memorability [7]. Therefore, frequent logins into a computer encourages passphrase memorability. Another viewpoint considers the critical management of remembering multitudes of passwords competing with each other to evoke the associated personal or professional account [32]. Findings include creating passwords based on a defined structure associated with the account elicits unique passwords and improves recall. Furthermore, a working memory study showed that interference such as competing representations associated with a passphrase interferes with the encoding process and decays retention unless it is refreshed by rehearsal [18]. Avoiding processing load and information loss during passphrase design are valid considerations.

Achieving mental efficiency is a result of reducing cognitive load and extending working memory. Likewise, "cognitive load always needs to be related to performance" [15, p. 6]. Discovering collaborative insights between memory and effort of the login task will further our examination of passphrase development.

3 Initial Cognitive Pilot Study

A repeated measures pilot study was conducted to determine memorability with imposed and user created passphrases. The imposed passphrases contained a series of three random words that were visually displayed prior to recall. The participants were allowed to freely select three words of their choice for their created passphrase. In both scenarios, a distractor was applied to clear working memory throughout the recall task.

Additionally, the study tested for personality traits using Rotter's [22] locus of control personality scale of internal and external measurement. "Applying psychological variables of locus of control to technology is expected to increase understanding of personality influences on the selection and construction of computer passwords and contribute to the design of memorable passwords" [14, p.8]. Internal control is a personality trait characterized as inhabiting self-reliant behavior whereas external control attributes circumstances upon encompassing surroundings [22]. Participants responded to a twenty nine question survey to determine their locus of control.

Moreover, the study consisted of timed memory dynamics tests using Ekstrom's et al. [6] associations for cognitive aptitude to determine its effect on passphrase recall.

Participants were presented with a list of objects and numbers as well as first and last names. After providing time for working memory to encode the information, they were evaluated on their ability to match objects with a number and first names with last names.

The participants were undergraduate university students considered to have prior knowledge of password construction as passwords were required to login to their university account. Information was provided using Qualtrics' anonymized online survey platform conducted at a university classroom.

3.1 Cognitive Pilot Study Results

Although the majority of the participants were recognized as internally controlled (Fig. 1), results show these participants were found in both the passphrase imposed and passphrase created groups.

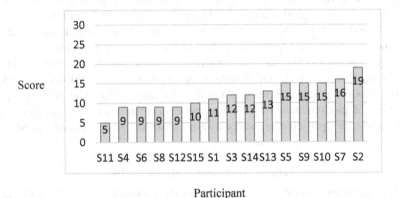

Fig. 1. Locus of control internal external personality traits. Maximum score = 23

Memory associations categorized by object number and first name last name recall tests resulted in the majority of participants scoring high when asked to recall the imposed and created passphrases (Figs. 2 and 3).

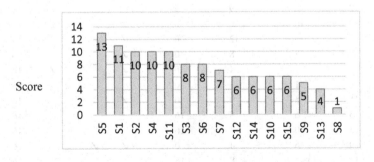

Fig. 2. Memory associative object number test. Maximum score = 15

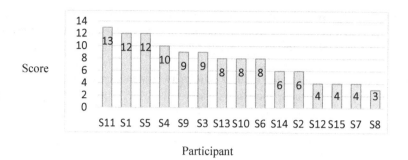

Fig. 3. Memory associative first name last name test. Maximum score = 15

The summary of outcomes listed by participant are presented in Table 1. A participant with a lower locus of control score is considered to possess the internal personality trait whereas a participant with a higher locus of control score is considered to possess the external personality trait. Results for the imposed passphrase recall test were bimodal. Therefore, approximately half of the participant population successfully recalled the imposed passphrase. The majority of these participants possess internal locus of control and are sufficient at memory associations. Most subjects who failed to recall the imposed passphrase selected the first and third words. This finding suggests recollection of random material produces common results among participants [16]. Additionally, all participants successfully recalled their created passphrases.

Table 1. Pilot study cognition results

	Personality Locus of control	Memory associative object number	Memory associative First name Last name	Passphrase recall imposed	Recall			Passphrase recall user created
Participant n = 15	Maximum 23	Maximum 15	Maximum 15	Maximum 16	First word	Second word	Third word	Maximum 16
P1	11	11	12	16				16
P2	19	10	6	16				16
P3	12	8	9	16				16
P4	9	10	10	0	✓		✓	16
P5	15	13	12	0	✓		✓	16
P6	9	8	8	0	✓		✓	16
P7	16	7	4	0	✓		✓	16
P8	9	1	3	0				16
P9	15	5	9	0	✓			16
P10	15	6	8	16				16
P11	5	10	13	16				16
P12	9	6	4	0	✓	✓		16
P13	13	4	8	16				16
P14	12	6	6	16				16
P15	10	6	4	0	✓		✓	16

Approximately twenty five percent of these students were considered externally influenced by the locus of control personality trait.

Distractors were presented during the recall of the imposed and user created passphrase tests. This procedure allowed processing in working memory for repeated recall [4, 24]. Unlike results from the imposed passphrase recall task, findings from the participant created passphrase task revealed that distractors provided no interference on working memory as all participants successfully recalled their created passphrase. Therefore, the role of distractors did not control participant behavior and was considered irrelevant to cognitive load [13].

4 Cognitive Pilot Study Modifications

The goal of the pilot study is to gather and analyze preliminary data, test the treatment and prepare the methods and procedures for a successful main study. It is intended to guide the planning and optimization of a large-scale investigation [27]. Furthermore, a pilot study improves validity and reliability of survey instruments [28] and allows the modification of questions and procedures that did not produce expected answers. Reporting the pilot study results will contribute to researchers considering similar inquiries.

The cognitive pilot study identified undesirable data produced by the participant created passphrase instrument. Although a distractor was presented during the recall task, all subjects repeatedly remembered their created password throughout the task. The hundred percent participant recall of user created passphrases is attributed to the design construction of passphrases. Students were instructed to freely select three random words of their choice. The new instrument will be modified to prescribe rule sets governing the revised procedure. The addition of rule sets are expected to increase intrinsic cognitive load and provide variability for recall. Since additive working memory is anticipated to affect memorability, the predicted results are anticipated to create variance toward a normal statistic distribution resembling a bell curve.

Furthermore, the results from the participant imposed passphrase instrument yielded bimodal data. Therefore, fifty percent of the sample successfully recalled the three random word passphrase. Modifications to the new instrument will include multiple schemes corresponding to the participant created tasks to probe passphrase recall. The new rule sets governing the imposed passphrases are expected to provide improved results guiding statistical analysis.

"Physiological techniques are based on the assumption that changes in cognitive functioning are reflected by physiological variables" [25, p. 22]. Future research involving physiological responses to cognitive load is encouraged to determine inferences of working memory on recall [13]. Additionally, study findings validate physiological instruments for cognitive load measurement [26]. To strengthen our projected outcomes, this study will conduct multiple physiological quantification techniques to examine cognitive load during recall of imposed and created passphrases. Study enhancement will include the following physiological responses to cognition: electroencephalography (EEG) changes in neural activity, electrodermal (EDA) changes to skin conductance, electrocardiogram (EKG) variability in heart rate, and electromyography (EMG) activity produced by the corrugator muscle [3].

These physiological factors are expected to enrich cognitive load measurements [2, 25] and determine implications to the phenomenon of recall abilities for the design of memorable passphrases. Similar to Szulewski's [26] results, this study's outcomes and patterns are anticipated to uncover correlation between passphrase recall, personality factors, and memory associations utilizing multiple physiological instruments. Measurements relative to study variables are projected to be explainable within and across participant groups.

Since physiological research is essential to advance techniques and their potential to measure cognitive load [25], we will apply modifications to the study design dynamics and measure recall using valid and reliable physiology, personality and memory associative instruments for the main study. These enhancements will focus on examining changes in cognitive discovery, correlation among the instrument outcomes and understanding of passphrase development and construction of memorable authentication.

5   Method

This study will be based on the descriptive quantitative approaches [12] designed with repeated measures. Therefore, participants will be measured once with surveys, memory associative tests, and physiological measures. Using a sample population, cognitive discovery will be made between the independent variables and the outcome dependent variables.

The dependent variables are the recall measurements from the imposed passphrase and user created passphrase treatments. These outcomes are the actual recall variables for both passphrase tasks that will be used in the broad scope of authentication development. The study results are expected to contribute to the behavioral understanding of passphrase selection and memorability.

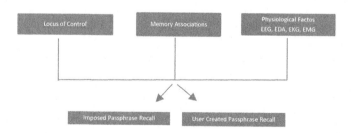

Fig. 4. Independent and dependent constructs

The independent variables measure extent of the locus of control personality test [22], memory associative test [6], and multiple physiology responses [3]. These constructs represent factors for identifying participant behavior. Measurements will be examined to discover performance magnitude of passphrase memorability.

Results are expected to establish associations and convergence between variables rather than causality. Therefore, results are presumed to produce a range of concepts explaining the study's measurement of cognitive load on passphrase recall. The research study is designed using validated personality and cognitive instruments as well as physiological measurements and can be replicated given its validity.

References

1. Al Abdulwahid, A., Clarke, N., Furnell, S., Stengel, I., Reich, C.: The current use of authentication technologies: an investigative review. In: 2015 International Conference on Cloud Computing (ICCC), pp. 1–8. IEEE, April 2015
2. Antonenko, P.D., Niederhauser, D.S.: The influence of leads on cognitive load and learning in a hypertext environment. Comput. Hum. Behav. **26**(2), 140–150 (2010)
3. Berntson, G. G., Cacioppo, J. T., Tassinary, L.G. (eds.): Handbook of Psychophysiology. Cambridge University Press (2017)
4. Blanchard, N.K., Malaingre, C., Selker, T.: Improving security and usability of passphrases with guided word choice. In: Proceedings of the 34th Annual Computer Security Applications Conference, pp. 723–732. ACM, December 2018
5. Bonneau, J.: Guessing human-chosen secrets (No. UCAM-CL-TR-819). University of Cambridge, Computer Laboratory (2012)
6. Ekstrom, R.B., Dermen, D., Harman, H.H.: Manual for Kit of Factor-Referenced Cognitive Tests, vol. 102. Educational Testing Service, Princeton (1976)
7. Gao, X., Yang, Y., Liu, C., Mitropoulos, C., Lindqvist, J., Oulasvirta, A.: Forgetting of passwords: ecological theory and data. In: 27th {USENIX} Security Symposium ({USENIX} Security 18), pp. 221–238 (2018)
8. Hollender, N., Hofmann, C., Deneke, M., Schmitz, B.: Integrating cognitive load theory and concepts of human–computer interaction. Comput. Hum. Behav. **26**(6), 1278–1288 (2010)
9. Jones, M.: Closing the Gap Between Memorable and Secure Passwords (2018)
10. Joudaki, Z., Thorpe, J., Martin, M.V.: Reinforcing system-assigned passphrases through implicit learning. In: Proceedings of the 2018 ACM SIGSAC Conference on Computer and Communications Security, pp. 1533–1548. ACM, October 2018
11. Kirschner, P.A., Ayres, P., Chandler, P.: Contemporary cognitive load theory research: the good, the bad and the ugly. Comput. Hum. Behav. **27**(1), 99–105 (2011)
12. Labaree, R.V.: Research Guides: Organizing Your Social Sciences Research Paper: Types of Research Designs (2009)
13. Lavie, N.: Attention, distraction, and cognitive control under load. Curr. Dir. Psychol. Sci. **19**(3), 143–148 (2010)
14. Loos, L.A., Crosby, M.E.: Cognition and predictors of password selection and usability. In: Schmorrow, Dylan D., Fidopiastis, Cali M. (eds.) AC 2018. LNCS (LNAI), vol. 10916, pp. 117–132. Springer, Cham (2018). https://doi.org/10.1007/978-3-319-91467-1_10
15. Mavilidi, M.F., Zhong, L.: Exploring the development and research focus of cognitive load theory, as described by its founders: interviewing John Sweller, Fred Paas, and Jeroen van Merriënboer. Educ. Psychol. Rev. 1–10 (2019)
16. Naim, M., Katkov, M., Tsodyks, M.: Fundamental Law of Memory Recall. bioRxiv, 510750 (2019)
17. Nielsen, G., Vedel, M., Jensen, C.D.: Improving usability of passphrase authentication. In: 2014 Twelfth Annual International Conference on Privacy, Security and Trust, pp. 189–198. IEEE, July 2014

18. Norris, D., Hall, J., Butterfield, S., Page, M.P.: The effect of processing load on loss of information from short-term memory. Memory **27**(2), 192–197 (2019)
19. Oviatt, S.: Human-centered design meets cognitive load theory: designing interfaces that help people think. In: Proceedings of the 14th ACM international conference on Multimedia, pp. 871–880. ACM, October 2006
20. Paas, F., Renkl, A., Sweller, J.: Cognitive load theory and instructional design: recent developments. Educ. Psychol. **38**(1), 1–4 (2003)
21. Pilson, C.S.: Tightly-Held and Ephemeral Psychometrics: Password and Passphrase Authentication Utilizing User-Supplied Constructs of Self. arXiv preprint arXiv:1509.01662 (2015)
22. Rotter, J.B.: Generalized expectancies for internal versus external control of reinforcement. Psychol. Monogr. Gen. Appl. **80**(1), 1 (1966)
23. Shay, R., et al.: Correct horse battery staple: exploring the usability of system-assigned passphrases. In: Proceedings of the Eighth Symposium on Usable Privacy and Security, p. 7. ACM, July 2012
24. Sweller, J.: Cognitive Load Theory. In: Psychology of Learning and Motivation, vol. 55, pp. 37–76. Academic Press (2011)
25. Sweller, J., van Merriënboer, J.J., Paas, F.: Cognitive architecture and instructional design: 20 years later. Educ. Psychol. Rev. 1–32 (2019)
26. Szulewski, A., Gegenfurtner, A., Howes, D.W., Sivilotti, M.L., van Merriënboer, J.J.: Measuring physician cognitive load: validity evidence for a physiologic and a psychometric tool. Adv. Health Sci. Educ. **22**(4), 951–968 (2017)
27. Thabane, L., et al.: A tutorial on pilot studies: the what, why and how. BMC Med. Res. Methodol. **10**(1), 1 (2010)
28. Van Teijlingen, E.R., Hundley, V.: The importance of pilot studies. Nurs. Stand. **16**, 33–36 (2002)
29. Ward, G., Tan, L.: Control processes in short-term storage: retrieval strategies in immediate recall depend upon the number of words to be recalled. Mem. Cogn. **47**, 1–25 (2019)
30. Woo, S.S., Mirkovic, J.: Memorablity and security of different passphrase generation methods. 정보보호학회지, **28**(1), 29–35 (2018)
31. Zhang, X., Clark, J.: Matrix Passwords: A Proposed Methodology of Password Authentication (2012)
32. Zhang, J., Luo, X., Akkaladevi, S., Ziegelmayer, J.: Improving multiple-password recall: an empirical study. Eur. J. Inf. Syst. **18**(2), 165–176 (2009)

Tokens of Interaction: Psychophysiological Signals, a Potential Source of Evidence of Digital Incidents

Nancy Mogire[1]([✉]), Randall K. Minas[2], and Martha E. Crosby[1]

[1] Information and Computer Sciences, University of Hawaii at Manoa,
POST 317 1680 East-West Road, Honolulu, HI 96822, USA
nmogire@hawaii.edu
[2] Shidler College of Business, University of Hawaii at Manoa,
2404 Maile Way Suite E601f, Honolulu, HI 96822, USA
rminas@hawaii.edu

Abstract. The human factor is a key component of any computing network just as are other tools and devices within it. At the same time, human emotion is highly responsive to the environment and this manifests in psychophysiological changes even when no physical reaction is observable. Therefore, a digital record of the state of body and mind can to one degree or another reflect the state of other components in a given network while the person is a part of it. Meanwhile, as the digital and physical worlds continue to converge cybersecurity is increasingly a day-to-day concern. Many crimes are now committed, mediated or witnessed through a digital device, and many operational artifacts of computing systems have later proved useful as evidence in digital investigations. Psychophysiological signals though unharnessed in this regard, could be a rich resource—in detecting occurrence, timing and duration of adverse incidents— owing to high human emotional responsiveness to the environment. Further, psychophysiological signals are hard to manipulate and so they are likely to provide a truer reflection of events. This is not only promising for investigations but as a potential feedback channel for monitoring safety and security in digital spaces, independent of human decision-making. This paper proceeds a dissertation study investigating psychophysiological signals for markers of digital incidents. Understanding and harnessing psychophysiological markers of digital incidents can enable designing of safer computing spaces through triggering appropriate controls to adaptively manage threats—such as cyberbullying and insiders threats.

Keywords: Psychophysiological markers · Cybersecurity ·
Digital investigations · Digital evidence · Threat management

1 Introduction and Related Work

In the digital age, many crimes occur within or around a digital device. These crimes are witnessed by the digital device. Sometimes incidents are caught on camera, but many times they are only decipherable by studying the artefacts left behind. In the

© Springer Nature Switzerland AG 2019
D. D. Schmorrow and C. M. Fidopiastis (Eds.): HCII 2019, LNAI 11580, pp. 93–110, 2019.
https://doi.org/10.1007/978-3-030-22419-6_8

digital space, various artefacts have been utilized as forensic evidence. These include computer system data such as login credentials, network traffic data and software usage metadata [1].

Meanwhile, human-computer interaction is fast moving from being an interaction between two disconnected entities into a more merged and symbiotic computational connection. More smart devices are being created every day to couple tightly with the human body. While this approach may increase convenience, the impact of adverse digital incidents is also likely to increase. This calls for improved methodologies geared towards safer digital spaces.

Broadly, there are two aspects in which safer digital spaces could be created: one would involve methods of real-time monitoring and adaptive response to threats, and the other is availability of digital metadata with evidential value. Such metadata can assist in solving open questions in forensic investigations to support justice and law enforcement functions.

There are many forms of forensic methodology, each with its own foci and applicability spectrum but none has proved perfect for exclusive use [2]. Hence, digital investigations benefit more with every additional form of acceptable evidence that is available.

In the past psychophysiological data was not typically available outside of clinical or specialized settings. Perhaps this explains why this form of data has yet to be widely explored as a form of digital evidence. Today however, many end user tools are capable of collecting and storing such data within their regular context of usage. Examples include fitness monitors, smart eyeglasses, smart clothing and even gaming headsets among numerous other smart body items. Even smartphones and other non-wearable devices are now able to collect human-generated artifacts such as motion, pressure and eye gaze data.

In light of that change, it is no longer far-fetched to study applications of human psychophysiological signals as a potential source of markers of digital incidents. Finding and harnessing such markers would yield an additional form of digital evidence of incidents, as well as a potential feedback channel revealing threats in the digital environment and triggering appropriate responses in a timely manner. Such threats could include cyberbullying and insider attacks.

There has been some work aimed at securing digital spaces. As an example, Mondal and Bours [3] define a continuous identification model based on hand swiping movements to continually verify that the authorized user is the one using the touch screen of a mobile device. They envision adding a forensic component if the model is used within a closed system that allows it to attempt to identify any intruder.

From a forensic perspective, an example usage of psychophysiological data involved the application eye gaze tracking to determine if witnesses recognized evidence that was in front of them [11].

2 Problem and Summary of Research Goal

Despite the potential to be a rich source of evidence data in digital investigations, psychophysiological signals have remained largely unexplored in this regard.

The human is a key component of a computing network with high emotional responsiveness to the environment. This responsiveness manifests as psychophysiological changes occurring even when no physical emotional reaction is observable. Further, psychophysiological signals are hard to manipulate and so they are likely to provide a true reflection of digital incidents. Therefore, analyzing recorded signals for structural changes, could reveal information regarding the state of other components in the computing network during the same time period. Identifying and harnessing this information resource can yield evidence towards digital investigations, as well as enable innovations that support a safer and more secure computing environment.

The goal of this research work is to investigate how various psychophysiological signals are affected by response-evoking properties of stimuli—such as novelty, salience, aversiveness, and the element of surprise. This involves investigating the timing of onset of interaction with a digital event, duration of interaction and timing of the end of such interaction.

3   Theoretical Background

Recording psychophysiological data allows for circumventing conscious decision-making through probing involuntary feedback channels [12]. Psychophysiological change is a manifestation of emotional reaction to events [4]. There are various psychophysiological feedback channels of emotional response. Examples include electrodermal activity - measuring skin conductance responses, electromyographic activity - facial muscle movements, electrocardiogram activity - heart pumping activity and electroencephalography - brain electrical activity.

3.1   Emotion in HCI

Emotion is a consequence of human appraisal of a situation and is a reflection of the resulting affect [4]. There can be various response eliciting properties in a stimuli. These include novelty, significance, salience, surprise, intensity, arousal [60].

Emotion-based studies have been done under various analysis approaches. One major dichotomy is between the **Discrete vs Dimensional approaches.** Discrete emotion [5] theories analyze emotion through its manifestation in physical expressions and functions drawn from a discrete and limited set e.g. interpretation of facial or hand expressions by function. Dimensional theories [6] on the other hand do not evaluate emotions as discrete and limited but rather as a large range of emotional states within a defined dimensional space (e.g. a one dimensional model defining motivation towards action such as approach-avoidance model).

For this study, we apply the dimensional evaluation model and are concerned with intensity of responses triggered by those emotions rather than categorization of the specific type of arousal or valence. This would be ideal for analyzing signal when the source person is not available to be observed or questioned—as would be common in an investigation context.

In turn, there are various dimensional approaches in studies involving psychophysiological response. These include models such as Valence-Arousal [6],

Approach-Avoidance [7], and Threat-Challenge [8]. For this work, we apply the Valence-Arousal model. In the **Valence - Arousal model,** emotion is considered in terms of its location within the valence-arousal dimensions quadrant i.e. high arousal positive, high arousal negative, low arousal positive or low arousal negative e.g. anger may be evaluated as highly negative and highly aroused. Valence-Arousal can be calibrated using pictures from the International Affective Picture System (IAPS) which are standardized and pre-rated for valence-arousal [9]. These dimensions have been found to show considerable consistency across cultures [10].

3.2   Psychophysiological Feedback Channels: Basis of Markers of Response to Stimuli

Electromyography in Corrugator Supercilii

Overview of Mechanism
Electromyography (EMG) measures activity of muscles. Muscles are the bodily tissue that generates and transmits force [13]. The electromyogram (EMG) is an electrical signal generated following muscle contraction. This signal reflects the electrical and not the mechanical, events of the contraction [14]. Cardiac and skeletal muscle groups are striated i.e. they are composed of bundles of thin fibers known as fibrils [15]. Each striated muscle is innervated by a motor nerve through which neural signals are delivered. There neural signals—muscle action potentials (MAPS)—are responsible for all actions of striated muscles [16].

EMG voltage changes as a result of multiple muscle action potentials (MAPs) across many muscle fibres within several motor units rather than a direct measure of muscle tension, contraction or movement. Hence, the signal measured using surface electrodes is attributed to muscle activity in a given muscle region or site rather contraction of a specific muscle [13]. Specific location of muscle contraction is difficult to determine due to the close proximity in the arrangements of striated muscles and the non-specificity of surface electrodes [13, 17]. EMG is measured using surface electrodes due to their non-invasive nature and because psychophysiological research questions are concerned with muscle sets rather than motor units within muscles. Surface EMG measurement detects ongoing muscular contraction in situations where simple observation by eye is too imprecise [13].

EMG Markers of Response to Stimuli
Subtle psychological processes often cause EMG activation without any accompanying visually perceptible actions or visceral changes [18, 19]. For example, while muscle activation will accompany facial expressions, muscle activation can also occur without the occurrence of any overt facial distortions, such as when activation is weak or fleeting or suppressed [20].

Negative sensory stimuli and mild negative imagery cause increased activation over the brow region also known as corrugator supercilii. This upper face site is the muscle region that draws brows inward and downward, sometimes forming vertical wrinkles [13]. Activity over the brow region (corrugator supercilii) varies inversely as a function of affective valence of stimulus i.e. negative affect such as disgust leads to increased activation over the brow region [13]. Several studies have shown increased EMG activity in the corrugator supercilii region when negative imagery or sensory stimuli is experienced [21]. As an example, when participants with depression were asked to imagine unpleasant experiences they displayed increased EMG activity in the corrugator supercilii region [24]. Non-depressed participants showed similar patterns although at a lower scale. Other negative emotions that have been found to increase EMG activity at the brow region include anger, fear, sadness, surprise [22], and disgust [23]. These studies suggest that the corrugator supercilii may be a suitable EMG feedback site when testing for high valence and negative affect.

Decrease in activity of the corrugator stimuli has been observed following positive stimuli such as presenting participants with pictures of smiling faces [26]. Some studies indicate that corrugator EMG may even suffer from interference when some non-negative emotion is faked.

In a deception study—using EMG and facial action coding—of individuals who pleaded for return of their missing relatives in televised recordings, deceptive participants showed reduced contraction of the corrugator supercilii compared to honest participants [25]. Faces of these individuals also showed masked smiles while they tried to put on sad looks. Half the individuals in the study had been eventually convicted of murdering the persons prior to pleading with them to return home. This may be an indication that general EMG activity can be interfered with by feigning an emotion. or simply that EMG is not a good indicator for deception. However, there is not much literature studying deception with EMG.

The baseline is used to determine the onset of stimuli-induced affect in EMG measurements. The true physiological baseline of EMG is zero [13]. In this ideal case, the lowest empirical baseline would be the level of noise in the recording equipment. However, muscles are unlikely to be completely at rest, especially in a laboratory setting, as the participant is never completely relaxed. Therefore, the baseline is considered to be the EMG activity that exists in the absence of experimental stimuli [13]. Once the baseline has been determined, changes in signal frequency can be interpreted to be signalling the ongoing response to stimuli.

Signal amplitude is commonly used as the dependent variable in psychophysiological experimentation [13]. Counting or averaging the EMG peaks in amplitude or tallying directional changes or signal crossings can be used to gauge EMG activity provided a high sampling rate is used [27]. However, some researchers consider Integrated EMG signal—the total energy of an EMG at a given time—to be a more meaningful way of measuring of overall muscle contraction than by counting or averaging amplitude peaks [28–30]. There are various techniques used in deriving the integrated EMG e.g. computing the arithmetic mean of a rectified and smoothed EMG [13].

Electrocardiography (ECG)

Overview of Mechanism

ECG is a measure of heart rate variability assessed using various metrics in the time or frequency domains. The time and frequency domain methods are complementary ways of characterizing the same sets of variances [31]. Time domain methods include measures of variance of heart periods and of their distributions as well as geometric methods based on heart period distributions [32]. Measures include standard deviation of the normal beat-to-beat intervals. Frequency domain methods decompose the heart period variance into frequency bands [32, 33]. One example method that can be used to approximate any periodic time-varying waveform is the fast fourier transform, by which the summation of a finite set of pure sinusoids of differing amplitudes can be computed [31–33].

High frequency heart rate variability is associated with variations in parasympathetic control in respiration [34, 35]. This variability is widely used as an index of vagal heart control [34–38]. The basal heart rate variation (Respiratory Sinus Arrhythmia - RSA) can be influenced by factors such as age, activity and posture. Hence RSA variability across subjects may not offer valid comparison in vagal control without controlling for those factors. The within-subject variability may be more valid as a vagal control measure provided properties such as posture and activity are taken into account [31]. Low frequency variability is associated with both autonomic branches and hence not regarded as a pure index of either [31], even though low frequency variability can be useful in measuring baroreflex action and cognitive workload [39, 40]. This low frequency bands have also been applied in indexing of autonomic balance which is evaluated along the sympathetic-parasympathetic spectrum [31, 41].

Cardiovascular measures have been used in study of arousal, stress, emotion and cognitive processes [31]. For example, high frequency heart rate variability has been found to indicate attentional capacity and performance [45]. Heart rate variability has also been found to decrease with increased workload [31, 40, 42]. Heart rate changes have also been found to distinguish between tasks involving external stimuli such as listening to noise, and internal stimuli such as attention to information processing [46]. RSA/RSA reactivity can account for a third of the variability in a between-subjects psychomotor vigilance task [31, 50]. Pre-task RSA has been found to predict performance in cognitive tasks requiring short-term memory [31, 51]. Baroreflex measures—reflecting the blood pressure control activity of baroreceptors arising in the arteries—have also been found to be highly responsive to psychological events such as mental effort and stress [42–44]. Stress has been found to reduce baroreflex gain.

ECG Markers of Response to Stimuli
Heart rate variability is highly responsive to increased arousal, workload and mental effort. Reduction in HRV is associated with increased arousal [31], increased workload [40, 42], and increased mental effort [45]. High frequency heart rate variability can predict the level of attentional capacity and performance in a person [45].

The "heart rate" refers to the number of heart beats per minute (bpm). Activity of the heart occurs in cardiac cycles with each consisting of the events between one heartbeat and another. In each cardiac cycle, there is a period when the heart does not pump blood, and one when the heart pumps; These are referred to as the diastole and systole respectively. The diastole and systole represent the blood pumping activity of the heart. This pumping action helps maintains the flow of oxygenated blood into the lungs and the rest of the body [31].

Blood flow is regulated by intrinsic mechanisms arising locally within cardiac tissue as well as extrinsic ones arising from hormonal or autonomic effects. blood flow can be altered by local mechanisms which adjust tissue structure to meet the need (e.g. when a cancerous tumor occurs, blood vessels will increase to meet the increasing need for additional blood flow). Interactions with extrinsic mechanisms such as the actions of autonomic neurons are also able to cause activations that affect aspects of cardiac function [52–56]. These activations can alter the interval between one heartbeat and the next and hence change the heart rate. The parasympathetic stimulation is more predominant than sympathetic stimulation in control of heart rate [34]. Heart rate has a generally linear relationship with parasympathetic activity while it has some non-linear relationship with sympathetic activity [31].

The QT interval of the ECG represents the time from ventricular excitation until the return to resting state ranging between 200–500 ms. The intervals are shorter with higher heart rates. The QRS peak which lies between the Q and T points corresponds in particular to the timing of the invasion of the myocardium which is the peak response to the electrical activation. This segment lasts about 100 ms unless there is a block in the branches within the conduction system resulting in a prolonged interval [31]. Hence a change in measured ECG signal, would be observed within 500 ms from the onset of the stimulus that caused it.

Heart period is the time in msec between adjacent heart beats, measured between successive R spikes in the ECG and it is a value reciprocal to heart rate. The two values can therefore be converted into each other and neither is dominant as the primary metric for cardiac measurement. In spite of that, heart period has been recommended for circumstances where a strictly linear relationship with autonomic inputs is desired [31]. A change in activation in either autonomic branch will lead to about the same change in heart period regardless of the baseline [57, 58]. Due to this mostly linear relationship, the use of heart period has been recommended as opposed to heart rate as a metric when changes in heart period are anticipated from autonomic responses as in psychophysiology experimentation [31, 57].

Heart rate has a non-linear relationship with autonomic events and its measurement may often suffer the effects of accentuated antagonism between the two autonomics branches [31, 59]. Heart period appears to be less disturbed by these interactions and hence is also recommended for studies where cardiac function may vary widely such as under varied experimental manipulations [31, 57].

Heart period is often converted to and represented as heart rate, although it is sometimes used on its own [31]. Change in heart period reflects the autonomic effect of response to a stimulus. This indicates that the disturbance in heart period would continue to persist while the stimulus continues to elicit responses that may correspond to the duration of the stimulus.

Electrodermal Activity (EDA)

Overview of Mechanism
EDA measurement reflects the level of electrical resistance generated by the skin. Resistance has been long known to decrease in response to sensory stimuli [61]. This means that the skin becomes a better electrical conductor in the moment after receiving stimuli. EDA can be measured endosomatically or exosomatically—internal or external measurement respectively [60]. In this section we reference exosomatic EDA measurement which is more suitable for this work. Using this method, the change in resistance is observed by passing a small current onto surface electrodes on the skin during the presentation of the stimuli [60].

The mechanism underlying the changes in skin surface resistance is the eccrine sweat glands whose primary function is thermoregulation of the body by evaporative cooling [60]. However, eccrine glands have been found to be responsive to physiologically significant stimuli. The eccrine glands on plantar and palmar surfaces in particular, have been thought to be more responsive to physiological stimuli that to heat, possibly due to the high gland density in those regions [60, 62, 63].

Innervation in the sweat glands arises predominantly from fibers in the sympathetic chain [63]. As such, specific conductance response (SCR) has been found to correlate highly with activation in the sympathetic nervous system [64]. The amygdala in particular has been found to show high levels of activation when stimuli elicited SCR [65, 66].

As sweat level increases in a given sweat duct—columnar components of the gland that open onto the surface of the skin and act as variable resistors—resistance in the sweat duct decreases. Hence, the change in sweat levels and the resulting changes in resistance lead to the changes in the measured EDA. The measure commonly used for controlled experiments is the amplitude of skin conductance response (SCR) which is the amount of increase in conductance from the onset of the response to the peak [60]. This measure is a part of the skin conductance level (SCL) which is the overall tonic level of conductivity of the skin. In some cases, SCL is a more suitable measure than SCR. An example is SCL usage in studies of continuous situations without specific stimuli for which a SCR can be measured [60].

When experimental stimuli is repeated, an average SCR size can be computed—either magnitude or amplitude—to represent the average response value across trials. A magnitude average includes trials that returned no responses, while an amplitude average includes only non-zero trials. The former measure is commonly used but both have applications where they are suitable [60, 67].

EDA has been found to respond to many types of tasks and it is argued that SCRs on their own are hard to link to a specific psychological response to stimuli such as anxiety or anger [68]. However, knowledge of the stimulus conditions coupled with carefully controlled experimental paradigms enables such inferences to be done. Another way that SCR to stimuli property relationships have emerged is through consistencies occurring between concurrently observed brain and skin activations [60].

There are a number of other disadvantages with EDA. First, EDA measures often suffer interference in the way of superimposed responses [67, 69] (i.e. the size of a response is a function of time since the previous response and size of that response if superimposed). Hence, EDA studies typically require long interstimulus intervals of at least 20–60 s. Another problem is inter-individual variance due to extraneous differences between individuals [60]. Range correction was initially proposed as a solution, where each individuals range is computed separately and their response quantified within that range [70]. This method is however unsuitable for some situations e.g. comparing individuals in different ranges [71]. A different solution to this problem is the use of within-subject standardized scores to adjust for individual differences.

Another problem with SCR-based studies are that SCRs can be impacted by habituation as the stimulus becomes more familiar. This causes decline and gradual disappearance of the SCRs. Various measures of habituation can be computed including trials-to-habituation count, decline of SCR across trials as assessed in interaction effect in analysis of variance, or regression of SCR magnitude on the log of trial number [60, 72, 73].

Due to the limitations of EDA, it is recommended to use it as one of multiple measures in experiments. That approach enables the researcher to tap into EDA's general utility as an indicator of arousal and attention, while also being able to get specific about psychophysiological state via a better differentiator, such as heart rate [60].

EDA Markers of Response to Stimuli

Increased SCRs have been observed when parts of the brain are involved in effortful activity [60]. As an example, SCR has been found to increase at the decision-making stage prior to risky or bad decisions in gambling tasks [74]. Thermal pain stimuli also showed increase in SCR [75] matching with increased responses in the neural regions that respond to pain including the thalamus and ACC. During rest, SCRs increased in line with brain activation in the ACC—anterior cingulate cortex which is associated with consciousness [76].

Other examples of SCR eliciting properties of stimuli include: novelty, unexpectedness, aversiveness, salience and emotional significance [60]. For instance, a discrete stimuli that elicits a response to the significance (salience) property in stimuli is the guilty knowledge test also referred to as Concealed information Test [77].

SCRs that are elicited by any non-aversive stimuli are initially considered to be an orienting response (OR) [72]. A minimum amplitude threshold is used to determine when SCRs can be linked to the properties of the specific stimuli. This minimum is commonly set between 0.01 and 0.05 µS [60]. Further, a minimum latency window is also set to prevent counting of spontaneous responses—non-specific skin conductance responses (NS-SCRs) e.g. as might result from bodily movements—as responses to experimental stimuli. Typically, SCRs beginning inside 1–3 or 1–4 s windows from the stimuli onset are considered as elicited by that stimuli [60].

Elicited response is determined against the baseline—set between 0.01 and 0.05. The size of elicited SCR often ranges between 0.1 and 1.0 microvolts with variations arising from environment [60], methodology and individual differences [78].

Where stimuli presentation is not discrete such as video games or in situations continuing over long periods, SCL and frequency of NS-SCR measures are preferred over SCR because the tonically varying levels of arousal. Both measures will show change between resting level to anticipation level and then to action level for almost any task [60].

In continuous tasks, SCL has been found to increase in response to both external stimuli such as loud sound, and internal stimuli such as information processing tasks [46]. SCL and frequency of NS-SCR have been found to be responsive to various continuous psychological stimuli situations including: anticipation and performance of any task [46], task switching and video gaming [60]. This observation indicates that tonic EDA is applicable for indexing processes related to energy regulation and it has been interpreted that the EDA responses are caused by effort of allocating information resources [60] which increases autonomic activation.

These measures also show an increase during non-task related continuous stimuli such as fear, anger and suppression of facial emotion display during viewing of a movie. Social stimulation through emotions such as stress and anxiety also caused an increased SCL and frequency of NS-SCR [60].

Electroencephalography (EEG)

Overview of Mechanism

Nunez and Katznelson [89] discuss that EEG signals occur and are recordable without the need for any deliberate stimuli application. However, changes in EEG signal, occurring due to response to a specific stimuli are referred to as evoked potentials.

Brain responses to stimuli are quantified by measurement of EEG amplitude or energy changes [90]. EEG signals can respond to various types of stimuli including visual stimuli, somatosensory stimuli and motor imagery [91]. Examples of media used to present visual stimulation include the Snodgrass & Vanderwart picture set [90] which has been found to induce highly synchronized neural activity in the gamma band [92]. Lists of acronyms have also been used as stimuli for Visual Evoked Potentials [91].

EEG has been utilized for brain-computer interfaces (BCI) in medical and non-medical settings. Several applications have been derived including: automated diagnosis of epileptic EEG using entropies [93]; automated drowsiness detection using wavelet packet analysis [94]; EEG-based mild depression detection using feature selection methods and classifiers [95]; neuro-signal based lie detection [96]; authentication [90] and continuous authentication.

EEG/ERP Markers of Response to Stimuli

ERPs represent EEG signals resulting from exposure and response to a particular stimuli. They run from 0 to several hundred milliseconds [79]. These waveforms are characterized by sets of positive and negative peaks labelled P and N respectively. These labels indicate the direction of polarity following a stimuli exposure and usually have a number specifying the ordinal position or the latency of the peak [79].

ERP Peaks are different from ERP components which are the changes that reflects a specific neural or psychological process [80]. ERP components arise from electrical activations in the brain and are then conducted through the skull and onto the scalp where they are measured using scalp electrodes [81]. The peaks in an ERP waveform do not directly translate to an underlying ERP component, although early peaks reflect sensory responses while later peaks represent motor and cognitive responses. To obtain ERP components, various statistical procedures can be used. Two major methods used to isolate ERP components are principal component analysis (PCA) and independent component analysis (ICA). PCA finds the components that individually account for the largest variation in activation while ICA finds those components that are maximally independent [79].

Various ERP components can be obtained from EEG depending on the nature of stimuli. These include the P3 component, Mismatch Negativity (MMN), N2 posterior-contralateral (N2pc) component and several others. P3 amplitude depends on the probability of occurrence of the target stimulus—as defined by the task—amongst the more frequently occurring non-target stimuli [82]. The P3 component can be found by subtracting the amplitude of the rare stimuli from the frequent one. The onset time of the resulting difference wave corresponds to the reaction time needed to perceive and categorize a stimulus [80].

When participants ignore the stimuli, no P3 wave is elicited by the unexpected [80]. However, surprising sounds typically elicit a negative potential, from 150–250 ms, even while the stimuli is ignored. This negative potential trend is known as mismatch negativity (MMN) [79, 83]. MMN is known as pre attentive or automatic potential due its arising in task-unrelated stimuli [79]. This potential can be calculated as a difference wave in a similar manner as the P3 component. This potential has been applied for assessing processing in locked-in individuals, such as preverbal infants [84].

N2 posterior-contralateral (N2pc) component occurs during a participant's focused attention to a target while multiple stimuli are presented laterally on each trial. It appears at the posterior electrode 200–300 ms after onset of each stimuli array [85]. Deflection at the left hemisphere electrode reflects when the target is located in the right hemisphere and vice versa. [79]. N2pc can be used to infer that an object has elicited a shift in attention [86]. The timing of N2pc can help to compare how fast attention shifts for different types of targets [87] and among different participants or groups [88].

4 Research Study Design

4.1 Task Description

Participants will play an on-screen game during a session in which their psy-chophysiological activity will be recorded. Various sensors will used to measure signals of this activity. The specific measurements will include electrodermal activity – to record the skin conductance responses, electromyographic activity to record the facial muscle movements, electrocardiogram activity to record the heart pumping activity and electroencephalography to record the brain electrical activity.

The stimuli of interest will consist of unexpected interruptions that we introduce into the participant's session. These interruptions are selected for their computer security implications—participants will be debriefed with full information after the task. We will be using plain non-security events as a control.

The purpose of this study is to assess if the signals collected by the sensors during the session contain any information that can be useful as markers of the events that occured. If these events repeatedly create significant structural changes in physiological signal, then such signal could be examined for specifics such as onset, duration and cessation of the event.

This would then open the way to studying how best to harness such markers towards various applications e.g. as a viable source of digital forensic evidence.

4.2 Selection of Psychophysiological Measures

In selecting physiological measures to base the study on, considerations included ease of obtaining a signal reading, non-intrusiveness, cost, and the reliability of signal to reflect the changes in psychophysiological state. In addition, the significance of psy-chophysiological states have been found to draw from activity across different feedback channels rather than to reflect a discrete response domain [31, 47–49].

For this work, ECG, EDA, EMG, EEG were selected. All can be measured with relative ease and measurements are digitized automatically. A combination rather than a single one of these methods is often better at closely tracking both sympathetic nervous system (SNS) and parasympathetic nervous system (PNS) responses.

4.3   Selection of Stimuli

In order to correlate physiological measures to specific emotional states, it is necessary to select stimuli that create conditions for the participant to be able to reach the necessary emotional state during the task e.g. to unambiguously elicit disgust or amusement or anger. At the same time, the stimuli should not cause or allow for conditions that may hinder response e.g. by eliciting shame when the study is not concerned with shame or by allowing unusual levels of boredom.

For this study, participant sessions will be interrupted by various unwanted events, which will constitute the stimuli of interest. It is intended that response to negative interruptions will mimic the affect caused by events typical as part of cyber incidents.

Between the interruptions, participants will play an on-screen game. Therefore they will perform a cognitive task as a distractor and in between stimuli. The purpose of the cognitive task is to ensure that the participant is mentally engaged with the digital device environment. They study task will require them to interact with the event, and that will enable us to link physiological responses occurring at the same time to these events.

However, we are also interested in recording the timing of their response in order to determine where to find the corresponding markers if any within the recorded signal. Timing will also help us infer when and how psychophysiological signal reflects response to the stimuli by comparing the onset of the physiological response markers, mouse or keyboard response markers and onset of exposure to stimuli.

Therefore, it is important to reduce the amount of extraneous delay between the onset of the participant's mouse or keyboard response and the other two pieces of onset timing information. Engaging the participant in a moderate cognitive task within the stimuli environment ensures that they are alert and ready to begin interacting with the stimuli as soon as is natural for them.

The gaming task in particular is an ideal cognitive task because no memorization is needed. Each gaming component is resolved while it is visible, and nothing is lost during the interruptions. Hence as soon as the stimuli exposure begins, the participant's cognitive resources are released to the stimuli.

4.4   Validating Responses

Even when the stimuli are valid for the task, the responses and their intensities will differ by participant. In addition, different responses may be combined. Verification and calibration methods can include coding of facial behavior by automated systems or cultural informers as well as self-reporting by the participants.

In this study, the task requires a mouse or keyboard response to stimuli, allowing us to estimate the timing when responses began. This is coupled with a repeated measure strategy to allow for comparison of signal patterns. Each stimuli will be presented multiple times during the course of the study. Further, we do not classify the responses in discrete emotion terms—e.g. anger, frustration, surprise—but rather as a measure of intensity of valence or arousal affect.

For comparison, there will be stimuli that will tend towards neutral in nature e.g. a call to rate the game or sign up to a newsletter about the game and there will be interruptions of a security nature with implications unknown to the participant during the task—they will be debriefed with full information at the end of the session. The study will assess the signals collected at the sensors during the session, for any structural information corresponding to the interrupting events that occurred during the session. If there is such information, and responses to neutral events present structural information that differs from that of responses to non-neutral events, then the latter can be regarded as psychophysiological markers of those types of events.

These markers can then be utilized to examine for specifics of when and how long such events occured.

5 Summary and Conclusion

The human factor is a key component of the computing network, and human emotion is highly responsive to its environment. Hence, psychophysiological signal activity can hold a lot of information about an individual's experiences while using a computing device. This form of metadata while largely unexplored for this purpose, is particularly promising as it is preconsciously controlled. This makes it less susceptible to human decision-making errors and deliberate tampering than other forms of computing metadata.

With cybersecurity becoming a concern in many contexts, psychophysiological markers of digital incidents can be useful as forensic evidence. Such markers could also be used in designing of tools that help create safe digital spaces, by triggering appropriate controls to protect individuals. Further these markers could be used to manage insider threats also by triggering appropriate controls to secure digital resources that are in use by a potentially rogue insider.

This work describes the theoretical background and the study aimed to examine the relationship between digital events with high valence or affect properties, and the structural properties of recorded psychophysiological signals occurroccurringing simultaneously.

We seek to determine if the signals recorded during these events contain any information that can be useful as markers of the events that occured. If these events repeatedly create significant structural changes in psychophysiological signal, then such signal could be further examined for specifics such as onset, duration and cessation of the event. This would in turn lead to studying how best to harness such markers towards various applications (e.g. as a viable source of digital forensic evidence).

References

1. Resendez, I., Martinez, P., Abraham, J.: An Introduction to Digital Forensics (2017)
2. Dessimoz, D., Champod, C.: Linkages between biometrics and forensic science. In: Jain, A. K., Flynn, P., Ross, A.A. (eds.) Handbook of Biometrics, pp. 425–459. Springer, Boston (2008). https://doi.org/10.1007/978-0-387-71041-9_21
3. Mondal, S., Bours, P.: Continuous authentication and identification for mobile devices: Combining security and forensics. In: 2015 IEEE International Workshop on Information Forensics and Security (WIFS) (2015)
4. Moors, A., Ellsworth, P.C., Scherer, K.R., Frijda, N.H.: Appraisal theories of emotion: State of the art and future development. Emot. Rev. 5(2), 119–124 (2013)
5. Mauss, I.B., Cook, C.L., Gross, J.J.: Automatic emotion regulation during anger provocation. J. Exp. Soc. Psychol. 43(5), 698–711 (2007)
6. Osgood, C.E., Suci, G., Tannenbaum, P.H.: The Measurement of Meaning, p. 335. University of Illinois Press, Urbana (1957)
7. Carver, C.S., Harmon-Jones, E.: Anger is an approach-related affect: evidence and implications. Psychol. Bull. 135(2), 183 (2009)
8. Tomaka, J., Blascovich, J., Kelsey, R.M., Leitten, C.L.: Subjective, physiological, and behavioral effects of threat and challenge appraisal. J. Pers. Soc. Psychol. 65(2), 248 (1993)
9. Lang P, Bradley, M.M.: The International Affective Picture System (IAPS) in the study of emotion and attention. In: Handbook of emotion elicitation and assessment, p. 29, 19 April 2007
10. Osgood, C.E.: Semantic differential technique in the comparative study of cultures. Am. Anthropol. 66(3), 171–200 (1964)
11. Watalingam, R.D., Richetelli, N., Pelz, J.B., Speir, J.A.: Eye tracking to evaluate evidence recognition in crime scene investigations. Forensic Sci. Int. 280, 64–80 (2017)
12. Mcdonough, B.E., Don, N.S., Warren, C.A.: Differential event-related potentials to targets and decoys in a guessing task. J. Sci. Explor. 16(2), 187–206 (2002)
13. Tassinary, L.G., Cacioppo, J.T., Vanman, E.J.: The somatic system. In: Cacioppo, J.T., Tassinary, L.G., Berntson, G.G. (eds.) Handbook of Psychophysiology. Cambridge Handbooks in Psychology, 4th edn, pp. 151–182. Cambridge University Press, Cambridge (2016)
14. Roberts, T.J., Gabaldón, A.M.: Interpreting muscle function from EMG: lessons learned from direct measurements of muscle force. Integr. Comp. Biol. 48(2), 312–320 (2008)
15. Schmidt-Nielsen, K.: Animal Physiology: Adaptation and Environment. Cambridge University Press, Cambridge (1997)
16. Sherrington, C.S.: The integrative action of the nervous system. J. Nerv. Ment. Dis. 57(6), 589 (1923)
17. David Kahn, S., Bloodworth, D.S., Woods, R.H.: Comparative advantages of bipolar abraded skin surface electrodes over bipolar intramuscular electrodes for single motor unit recording in psychophysiological research. Psychophysiology 8(5), 635–647 (1971)
18. Graham, J.L.: A new system for measuring nonverbal responses to marketing appeals. In: 1980 AMA Educator's Conference Proceedings, vol. 46, pp. 340–343 (1980)
19. Rajecki, D.W.: Animal aggression: implications for human aggression. Aggress. Theor. Empir. Rev. 1, 189–211 (1983)
20. Cacioppo, J.T., Bush, L.K., Tassinary, L.G.: Microexpressive facial actions as a function of affective stimuli: replication and extension. Pers. Soc. Psychol. Bull. 18(5), 515–526 (1992)

21. Larsen, J.T., Norris, C.J., Cacioppo, J.T.: Effects of positive and negative affect on electromyographic activity over zygomaticus major and corrugator supercilii. Psychophysiology 40(5), 776–785 (2003)
22. Chen, Y., Yang, Z., Wang, J.: Eyebrow emotional expression recognition using surface EMG signals. Neurocomputing 30(168), 871–879 (2015)
23. Kreibig, S.D., Samson, A.C., Gross, J.J.: The psychophysiology of mixed emotional states. Psychophysiology 50(8), 799–811 (2013)
24. Schwartz, G.E., Fair, P.L., Salt, P., Mandel, M.R., Klerman, G.L.: Facial muscle patterning to affective imagery in depressed and nondepressed subjects. Science 192(4238), 489–491 (1976)
25. Porter, S., Ten Brinke, L., Wallace, B.: Secrets and lies: involuntary leakage in deceptive facial expressions as a function of emotional intensity. J. Nonverbal Behav. 36(1), 23–37 (2012)
26. Dimberg, U., Lundquist, L.O.: Gender differences in facial reactions to facial expressions. Biol. Psychol. 30(2), 151–159 (1990)
27. Loeb, G.E., Loeb, G., Gans, C.: Electromyography for Experimentalists. University of Chicago Press, Chicago (1986)
28. Gartside, I.B., Lippold, O.C.: The production of persistent changes in the level of neuronal activity by brief local cooling of the cerebral cortex of the rat. J. Physiol. 189(3), 475–487 (1967)
29. Basmajian, J., De Luca, C.: Muscles Alive: Their Functions Revealed by Electromyography, 5th edn. Williams & Wilkins, Baltimore (1985)
30. Goldstein, I.B.: Electromyography: a measure of skeletal muscle response. In: Handbook of Psychophysiology, pp. 329–365 (1972)
31. Berntson, G.G., Quigley, K.S., Norman, G.J., Lozano, D.L.: Cardiovascular Psychophysiology. In: Cacioppo, J.T., Tassinary, L.G., Berntson, G.G. (eds.) Handbook of Psychophysiology. Cambridge Handbooks in Psychology, 4th edn, pp. 183–216. Cambridge University Press, Cambridge (2016)
32. Camm, A.J., et al.: Heart rate variability: standards of measurement, physiological interpretation and clinical use. Task force of the European Society of Cardiology and the North American Society of pacing and electrophysiology. Circulation 93(5), 1043–1065 (1996)
33. Berntson, G.G., et al.: Heart rate variability: origins, methods, and interpretive caveats. Psychophysiology 34(6), 623–648 (1997)
34. Berntson, G.G., Cacioppo, J.T., Quigley, K.S.: Cardiac psychophysiology and autonomic space in humans: empirical perspectives and conceptual implications. Psychol. Bull. 114(2), 296 (1993)
35. Eckberg, D.L.: Physiological basis for human autonomic rhythms. Ann. Med. 32(5), 341–349 (2000)
36. Berntson, G.G.: Autonomic cardiac control III. Psychological stress and cardiac response in autonomic space as revealed by pharmacological blockades. Psychophysiology 31(6), 599–608 (1994)
37. Cacioppo, J.T.: Social neuroscience: autonomic, neuroendocrine, and immune responses to stress. Psychophysiology 31(2), 113–128 (1994)
38. Grossman, P., Kollai, M.: Respiratory sinus arrhythmia, cardiac vagal tone, and respiration: within-and between-individual relations. Psychophysiology 30(5), 486–495 (1993)
39. Stuiver, A., Mulder, B.: Cardiovascular state changes in simulated work environments. Front. Neurosci. 8, 399 (2014)
40. Van Roon, A.M., Mulder, L.J., Althaus, M., Mulder, G.: Introducing a baroreflex model for studying cardiovascular effects of mental workload. Psychophysiology 41(6), 961–981 (2004)

41. Malliani, A.: The pattern of sympathovagal balance explored in the frequency domain. Physiology **14**(3), 111–117 (1999)
42. Stuiver, A., De Waard, D., Brookhuis, K.A., Dijksterhuis, C., Lewis-Evans, B., Mulder, L.J.: Short-term cardiovascular responses to changing task demands. Int. J. Psychophysiol. **85**(2), 153–160 (2012)
43. del Paso, G.A., González, I., Hernández, J.A.: Baroreceptor sensitivity and effectiveness varies differentially as a function of cognitive-attentional demands. Biol. Psychol. **67**(3), 385–395 (2004)
44. Steptoe, A., Sawada, Y.: Assessment of baroreceptor reflex function during mental stress and relaxation. Psychophysiology **26**(2), 140–147 (1989)
45. Porges, S.W.: Autonomic regulation and attention. Attention and information processing in infants and adults, pp. 201–223 (1992)
46. Lacey, J.I., Kagan, J., Lacey, B.C., Moss, H., Black, P.: Expression of the Emotions in Man. International Unlv. Press, New York (1963). (edited by Knapp, PH)
47. Norman, G.J., Berntson, G.G., Cacioppo, J.T.: Emotion, somatovisceral afference, and autonomic regulation. Emot. Rev. **6**(2), 113–123 (2014)
48. Lang, P.J.: Emotion's response patterns: the brain and the autonomic nervous system. Emot. Rev. **6**(2), 93–99 (2014)
49. Levenson, R.W.: Emotion and the autonomic nervous system: Introduction to the special section. Emot. Rev. **6**(2), 91–92 (2014)
50. Henelius, A., Sallinen, M., Huotilainen, M., Müller, K., Virkkala, J., Puolamäki, K.: Heart rate variability for evaluating vigilant attention in partial chronic sleep restriction. Sleep **37** (7), 1257–1267 (2014)
51. Capuana, L.J., Dywan, J., Tays, W.J., Elmers, J.L., Witherspoon, R., Segalowitz, S.J.: Factors influencing the role of cardiac autonomic regulation in the service of cognitive control. Biol. Psychol. **102**, 88–97 (2014)
52. Hall, J.E.: Guyton and Hall Textbook of Medical Physiology e-Book. Elsevier Health Sciences, Philadelphia (2010)
53. Armour, J.A.: Potential clinical relevance of the 'little brain' on the mammalian heart. Exp. Physiol. **93**(2), 165–176 (2008)
54. Brack, K.E.: The heart's 'little brain' controlling cardiac function in the rabbit. Exp. Physiol. **100**(4), 348–353 (2015)
55. Randall, W.C., Wurster, R.D., Randall, D.C., Xi-Moy, S.: From cardioaccelerator and inhibitory nerves to a heart brain: an evolution of concepts. In: Shepherd, J.T., Vatner, S.F. (eds.) Nervous Control of the Heart, pp. 173–199. HarwoodAcademic Publishers, Amsterdam, January 1996
56. Richardson, R.J., Grkovic, I., Anderson, C.R.: Immunohisto chemical analysis of intracardiac ganglia of the rat heart. Cell Tissue Res. **314**(3), 337–350 (2003)
57. Berntson, G.G., Cacioppo, J.T., Quigley, K.S.: The metrics of cardiac chronotropism: biometric perspectives. Psychophysiology **32**(2), 162–171 (1995)
58. Parker, P., Celler, B.G., Potter, E.K., McCloskey, D.I.: Vagal stimulation and cardiac slowing. Auton. Neurosci. Basic Clin. **11**(2), 226–231 (1984)
59. Quigley, K.S., Berntson, G.G.: Autonomic interactions and chronotropic control of the heart: heart period versus heart rate. Psychophysiology **33**(5), 605–611 (1996)
60. Dawson, M.E., Schell, A.M., Filion, D.L.: The Electrodermal System. In: Cacioppo, J.T., Tassinary, L.G., Berntson, G.G. (eds.) Handbook of Psychophysiology. Cambridge Handbooks in Psychology, pp. 217–243. Cambridge University Press, Cambridge (2016)
61. Fere, C.: Note on changes in electrical resistance under the effect of sensory stimulation and emotion. Comptes rendus des seances de la societé de biologie. **5**, 217–219 (1888)

62. Edelberg, R.: Electrical activity of the skin: its measurement and uses in psychophysiology. In: Handbook of Psychophysiology, pp. 367–418 (1972)

63. Shields, S.A., MacDowell, K.A., Fairchild, S.B., Campbell, M.L.: Is mediation of sweating cholinergic, adrenergic, or both? A comment on the literature. Psychophysiology **24**(3), 312–319 (1987)

64. Wallin, B.G.: Sympathetic nerve activity underlying electrodermal and cardiovascular reactions in man. Psychophysiology **18**(4), 470–476 (1981)

65. Cheng, D.T., Knight, D.C., Smith, C.N., Helmstetter, F.J.: Human amygdala activity during the expression of fear responses. Behav. Neurosci. **120**(6), 1187 (2006)

66. Phelps, E.A., Delgado, M.R., Nearing, K.I., LeDoux, J.E.: Extinction learning in humans: role of the amygdala and vmPFC. Neuron **43**(6), 897–905 (2004)

67. Venables, P.H., Christie, M.J.: Electrodermal activity. In: Martin, I., Venables, P.H. (eds.) Techniques in Psychophysiology, vol. 54. Wiley, Chichester (1980)

68. Landis, C.: Psychology and the psychogalvanic reflex. Psychol. Rev. **37**(5), 381 (1930)

69. Grings, W.W., Schell, A.M.: Magnitude of electrodermal response to a standard stimulus as a function of intensity and proximity of a prior stimulus. J. Comp. Physiol. Psychol. **67**(1), 77 (1969)

70. Lykken, D.T., Rose, R., Luther, B., Maley, M.: Correcting psychophysiological measures for individual differences in range. Psychol. Bull. **66**(6), 481 (1966)

71. Lykken, D.T., Venables, P.H.: Direct measurement of skin conductance: a proposal for standardization. Psychophysiology **8**(5), 656–672 (1971)

72. Siddle, D., Stephenson, D., Spinks, J.A.: Elicitation and habituation of the orienting response. In: Siddle, D.(ed.) Orienting and Habituation: Perspectives in Human Research, pp. 109–182. John Wiley, Chichester (1983)

73. Lader, M.H., Wing, L.: Physiological Measures, Sedative Drugs, and Morbid Anxiety. Oxford University Press, London (1966)

74. Bechara, A., Damasio, H., Damasio, A.R., Lee, G.P.: Different contributions of the human amygdala and ventromedial prefrontal cortex to decision-making. J. Neurosci. **19**(13), 5473–5481 (1999)

75. Dube, A.A., Duquette, M., Roy, M., Lepore, F., Duncan, G., Rainville, P.: Brain activity associated with the electrodermal reactivity to acute heat pain. Neuroimage **45**(1), 169–180 (2009)

76. Fan, J., et al.: Spontaneous brain activity relates to autonomic arousal. J. Neurosci. **32**(33), 11176–11186 (2012)

77. Verschuere, B., Ben-Shakhar, G., Meijer, E. (eds.): Memory Detection: Theory and Application of the Concealed Information Test. Cambridge University Press, Cambridge (2011)

78. Boucsein, W.: Electrodermal Activity. Springer, Boston (2012). https://doi.org/10.1007/978-1-4614-1126-0

79. Luck, S.J., Kappenman, E.S.: Electroencephalography and event-related brain potentials. In: Cacioppo, J.T., Tassinary, L.G., Berntson, G.G. (eds.) Handbook of Psychophysiology. Cambridge Handbooks in Psychology, 4th edn, pp. 74–100. Cambridge University Press, Cambridge (2016)

80. Kappenman, E.S., Luck, S.J.: ERP components: the ups and downs of brainwave recordings. In: Luck, S.J., Kappenman, E.S. (eds.) Oxford Handbook of Event-Related Potential Components. Oxford University Press, New York (2012)

81. Nunez, P.L., Srinivasan, R.: Electric Fields of the Brain: The Neurophysics of EEG. Oxford University Press, Oxford (2006)

82. Squires, K.C., Donchin, E., Herning, R.I., McCarthy, G.: On the influence of task relevance and stimulus probability on event-related-potential components. Electroencephalogr. Clin. Neurophysiol. **42**(1), 1–4 (1977)
83. Woldorff, M.G., Hackley, S.A., Hillyard, S.A.: The effects of channel-selective attention on the mismatch negativity wave elicited by deviant tones. Psychophysiology **28**(1), 30–42 (1991)
84. Trainor, L., McFadden, M., Hodgson, L., Darragh, L., Barlow, J., Matsos, L., Sonnadara, R.: Changes in auditory cortex and the development of mismatch negativity between 2 and 6 months of age. Int. J. Psychophysiol. **51**(1), 5–15 (2003)
85. Luck, S.J.: Electrophysiological correlates of the focusing of attention within complex visual scenes: N2pc and related ERP components. In: Kappenman, E.S., TLuck, S.J. (ed.) The Oxford Handbook of Event-Related Potential Components, pp. 329–360. Oxford University Press, Oxford (2012)
86. Sawaki, R., Luck, S.J.: Capture versus suppression of attention by salient singletons: electrophysiological evidence for an automatic attend-to-me signal. Atten. Percept. Psychophys. **72**(6), 1455–1470 (2010)
87. Kappenman, E.S., MacNamara, A., Proudfit, G.H.: Electrocortical evidence for rapid allocation of attention to threat in the dot-probe task. Soc. Cogn. Affect. Neurosci. **10**(4), 577–583 (2014)
88. Lorenzo-López, L., Amenedo, E., Cadaveira, F.: Feature processing during visual search in normal aging: electrophysiological evidence. Neurobiol. Aging **29**(7), 1101–1110 (2008)
89. Nunez, P., Katznelson, R.: Electric Fields of the Brain. Oxford University Press, New York (1981)
90. Zuquete, A., Quintela, B., Cunha, J.: Biometric Authentication using electroencephalograms: a practical study using visual evoked potentials. Electronica E´ Telecomunicacoes, vol. 5, no. 2 (2010)
91. Palaniappan, R.: Electroencephalogram-based brain-computer interface: an introduction. In: Miranda, E., Castet, J. (eds.) Guide to Brain-Computer Music Interfacing, pp. 29–41. Springer, London (2014). https://doi.org/10.1007/978-1-4471-6584-2_2
92. Snodgrass, J., Vanderwart, M.: A standardized set of 260 pictures: norms for name agreement, image agreement, familiarity, and visual complexity. J. Exp. Psychol.: Hum. Learn. Mem. **6**(2), 174–215 (1980)
93. Acharya, U., Molinari, F., Sree, S., Chattopadhyay, S., Ng, K., Suri, J.: Automated diagnosis of epileptic EEG using entropies. Biomed. Signal Process. Control **7**(4), 401–408 (2012)
94. da Silveira, T., Kozakevicius, A., Rodrigues, C.: Automated drowsiness detection through wavelet packet analysis of a single EEG channel. Expert Syst. Appl. **55**, 559–565 (2016)
95. Li, X., Hu, B., Sun, S., Cai, H.: EEG-based mild depressive detection using feature selection methods and classifiers. Comput. Methods Programs Biomed. **136**, 151–161 (2016)
96. Cakmak, R., Zeki, A.: Neuro signal based lie detection. In: 2015 IEEE International Symposium on Robotics and Intelligent Sensors (IRIS) (2015)

The Artificial Facilitator: Guiding Participants in Developing Causal Maps Using Voice-Activated Technologies

Thrishma Reddy[1], Philippe J. Giabbanelli[2(✉)], and Vijay K. Mago[1]

[1] Department of Computer Science, Lakehead University, Thunder Bay, ON, Canada
{tshivara,vmago}@lakeheadu.ca
[2] Computer Science Department, Furman University, Greenville, SC, USA
giabbanelli@gmail.com

Abstract. Complex problems often require coordinated actions from stakeholders. Agreeing on a course of action can be challenging as stakeholders have different views or 'mental models' of how a problem is shaped by many interacting causes. Participatory modeling allows to externalize mental models in forms such as causal maps. Participants can be guided by a trained facilitator (with limitations of costs and availability) or use a free software (with limited guidance). Neither solution easily copes with large causal maps, for instance by preventing redundant concepts. In this paper, we leveraged voice-activated virtual assistants to create causal models at any time, without costs, and by avoiding redundant concepts. Our three case studies demonstrated that our artificial facilitator could create causal maps similar to previous studies. However, it is limited by current technologies to identify concepts when the user speaks (i.e. entities), and its design had to follow pre-specified rules in the absence of sufficient data to generate rules by discriminative machine-learned methods.

Keywords: Amazon Alexa · Causal maps · Mental models · Participatory modeling · Virtual assistant

1 Introduction

Problems as diverse as ecological management or obesity are often called *complex*, or 'wicked'. While the complexity sciences provide many definitions and tools to measure complexity[1], complex problems often share at least two traits which are central to this paper. First, they are *multifactorial*. The traditional reductionist approach trying to fix the 'root' cause does not lend itself well to a complex problem [1], and may even cause harm through unintended consequences [2]. Rather, the emphasis is often on mapping [3] and navigating [4] the

Research funded by MITACS Globalink Research Award, Canada.

[1] For an overview of the complexity sciences, see the map at http://www.art-sciencefactory.com/complexity-map_feb09.html.

© Springer Nature Switzerland AG 2019
D. D. Schmorrow and C. M. Fidopiastis (Eds.): HCII 2019, LNAI 11580, pp. 111–129, 2019.
https://doi.org/10.1007/978-3-030-22419-6_9

complex system of interactions between factors that contribute to, and/or are impacted by, a problem of interest. Second, dissemination and implementation research emphasizes that solutions to complex problems often require *coordinated actions between stakeholders* from multiple sectors (i.e., a multiactor view [5]). For instance, actions regarding population obesity involve sectors as varied as food production, the built environment (e.g., to promote walkable cities and access to fresh food), mental and physical well-being [6]. Coordinated actions should produce a coherent policy, which implies that stakeholders work together at least by sharing a mission [7].

It can be challenging to assess whether stakeholders share a mission when operating in a complex system of interactions. They may have different views or 'mental models' on how the factors interact, which may lead to very different takes on interventions. In the case of ecological management, one stakeholder may ignore the pressure of fishing and instead focus on the environment (e.g., enough nutrition for the fish, not too many predatory birds) while another may acknowledge that fishing reduces the fish population but downplay its importance [8]. Stakeholders may also have the same views but express them differently, for example by naming factors in different ways depending on their fields, which can create a communication gap [9,10]. Consequently, complex problems involving multiple stakeholders often involve *participatory modeling*, which allows to externalize [11] and hence compare [8] the mental models of stakeholders. There are various approaches to participatory modeling, depending on whether the objective is to be able to *simulate* a system [12,13] (e.g., to quantitatively assess *how much* effect an intervention would have) or only capture its *structure* [14] (e.g., to qualitatively assess *what* an intervention would affect). In the example of obesity, qualitative approaches may be realized by systems dynamics or agent-based modelling [15] while qualitative approaches may generate 'systems maps' or 'diagrams' [16]. The creation of systems maps is particularly important either as an endpoint (for qualitative analysis of stakeholders' mental models), or as a step toward the creation of quantitative models [14] (e.g., starting with a Causal Loop Diagram to produce a Systems Dynamics model). Causal maps are a widely used form of systems maps, in which concepts are represented as nodes and their causal connections are captured through directed edges (Fig. 1).

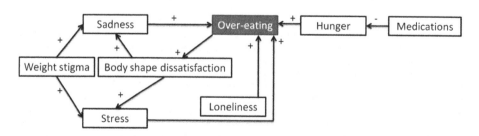

Fig. 1. Sample causal map where "over-eating" is the problem of interest [17].

Participants interested in developing causal models have often done it with the support of a trained facilitator, who elicits concepts and causal relations [18–22]. Alternatively, tech-savvy participants may receive training and independently develop causal models using software such as cMap (common in education research), MentalModeler (most used in socio-ecological systems), or Vensim (typical in health and systems engineering). However, both approaches have limitations. A trained facilitator can provide ample guidance, but may be costly or unavailable. A software may be free and available anytime, but it does not guide the participant through the process of building a causal map. In addition, both approaches rely on a visual inspection of the map as it is built, which does not easily scale as participants start to have many concepts and/or interrelationships. For example, a participant may add a concept that is actually a synonym of a concept already present. To notice this redundant concept, the facilitator and/or participant would need to manually look at all other concepts, which becomes prohibitive as the number of concepts increases.

There is thus a need for an approach to causal model building that can be available at any time, without costs, and scales easily. In this paper, we address this need by leveraging voice-activated virtual assistants (Amazon Alexa) to design and implement a *virtual facilitator*. Our solution guides participants in developing a model through a conversation (like a human facilitator), but is available at anytime without cost (as a software) and continuously examines the map to avoid typical issues such as synonymy of concepts.

The remainder of this paper is organized as follows. In Sect. 2, we provide background information on the process to create a causal map, and we briefly discuss recent uses of conversational agents built on Amazon Alexa, Microsoft's Cortana, and Apple Siri. In Sect. 3, we present the process that our artificial facilitator follows, and we cover its implementation in Sect. 4. Several examples are offered in Sect. 5, where a participant interacts with our technology to develop a model. Videos of the interaction are provided as supplementary online material. Finally, Sect. 5 contextualizes the implications of this work for the development of causal models and participatory modeling in general.

2 Background

2.1 Why Do We Create Causal Maps?

A causal map is a *conceptual model*. In Modeling and Simulation (M&S), conceptual models are the first stage of model development before quantifying nodes and relationships (mathematical model [23]) and possibly implementing the model as code (computational model). Conceptual models serve multiple objectives such as identifying key elements and aspects (thus delineating the boundaries of a system) or externalizing hypotheses through a transparent list of expected relations [14]. These objectives may be sufficient to warrant the development of a conceptual model as a final product. In this case, the conceptual map is often analyzed using network theory[2]. A common type of analysis is the identification of clusters or communities to divide a complex system into broad themes, as exemplified by the Foresight Obesity Map [25,26], maps for the Provincial Health Services Authority [27], or the recent work of Allender, McGlashan and colleagues [28,29]. Other analyses may include the centrality, to identify leverage points in a system [30,31]; an inventory of loops, to better characterize and possibly change the dynamics of the system [32–34]; an exploration of disjoint paths between factors, to capture how a policy impacts an outcome in multiple ways [4,33]; or a comparison of maps, to understand how different are the mental models of participants [10,35].

Map-liked artifacts may be constructed solely from data, for instance as Structured Equation Models (SEM) or Fuzzy Cognitive Maps (FCMs) [36]. Alternatively, traces produced by an analyst in exploring the data can be structured in a map [37,38], or the literature on a topic can be synthesized into a map [39]. It would be overly reductive to categorize such data-driven maps as 'objective' compared to participant-driven maps being deemed 'subjective'. Data can also have "biases, ambiguities, and inaccuracies" [40] and the inference process to build a map may not be perfect. Our focus is on participatory modeling (PM), in which participants drive the development of causal maps. Participatory modeling serves a different (and sometimes complementary) purpose than data-driven modeling. As detailed elsewhere [17], data-driven modeling may strive for accuracy with respect to the data whereas PM aims to be transparent and representative of the participants' mental models. PM can thus be employed in 'soft' situations that lack data and rely on human expertise [41], to support decision-making processes [42], or to understand what actions would be acceptable to various stakeholders [43].

The elicitation process consists of externalizing the mental model of a participant or group into a map. The elicitation process is first and foremost a *facilitation* process: we want to support participants in expressing their perspectives, rather than judge whether what they think is 'right' given our own ideas. Research in cognitive sciences has long been concerned with how humans store mental models, or their "conceptualization of the world" [44]. This storage takes place in *semantic memory*, which provides functional *relationships* between objects. As we previously summarized, "if mental models are published and shared in the form of maps, it owes to the fact that we seek to capture semantic memory whose structure is network-based" [8]. On one extreme, freeform approaches such as Rich Pictures pose no constraints on the creation of maps [45], which simplifies the process for participants but limits the analytical possibilities. At the other extremes, concept maps and mind maps have a very structured process that lists concepts (e.g., via brainstorming), group them, link them, and

---

[2] A conceptual model is an aggregate model in which *factors* or *concepts* are connected. This is different from a 'social network', which is an individual model in which nodes represent individuals rather than factors. Although the methods are often similar (e.g., centrality, community detection), the application of network science to social networks is often presented as 'social network analysis' [24].

label the links. However, this process precludes the presence of some structures (e.g., mind maps are trees so they cannot contain cycles) which are important to characterize the dynamics of a system. Causal maps occupy an intermediate position: the development process is more guided than rich pictures, less restricted than concept maps and mind maps, and any network structure can be produced by participants[3].

2.2   How Do We Create Good Causal Maps?

The process to produce a map as shown in Fig. 1 is relatively simple: participants create concept nodes, and link them by indicating the causal relationship to be an increase ('+') or a decrease ('−') [48,49]. However, at least three issues may arise if the facilitator does not provide further guidance[4]. First, participants need to choose node labels that have an unambiguous quantification: having 'more' or 'less' of this concept should be a straightforward notion. For instance, labeling a concept as 'weather' does not work, since having more or less weather is undefined. However, having more or less rain would be defined. A facilitator thus regularly ensures that labels are quantifiable, or prompts for clarifications that would change the label. Second, users may forget about concepts that they already have, and add one with a similar name. Facilitators thus continuously monitor the maps to either avoid creating a redundant concept, or merge them once they are discovered. Given the tremendous potential for (subtle) variations in language, discovering equivalent concepts is a difficult problem, particularly as the number of concepts increases [9,10]. Third, case studies have shown that cognitive limitations make it difficult for participants to think of structures such as loops and disjoint paths [50,51]. In particular, Ross observed how peculiar it was that "those who set policy think only acyclically, especially since the cyclical nature of causal chains in the real world has been amply demonstrated" [52]. Without paying particular attention to loops, participants may produce star diagrams with the one central problem at the core, and every other factor directly connecting to it. Facilitators may thus prompt participants extensively for relationships, to minimize the risk of missing loops or additional paths [27,33].

---

[3] There are at least two limitations to this representation. First, networks or graphs only represent binary relationships. However, participants may think of non-binary relationships, for instance when three concepts are directly involved together. While we have long been aware that cognitive structures could generally be represented by relations between any number of concept (e.g. using a hypergraph), it has been common practice to limit the structure to a graph [46]. Second, the network is only used to represent what is true (i.e. the existence of a causal connection between two factors) rather than what is false. As noted by Johnson-Laird, mental models also include counterexamples, which are important in decision-making processes [47].

[4] Some of these issues are also addressed in our tutorials at https://www.youtube.com/watch?v=OdKJW8tNDcM and https://www.youtube.com/watch?v=D-2Q2IHclo4.

2.3 Smart Conversational Agents

The term 'conversational agent' may be used loosely for any system that can carry on a conversation with a human. However, there are significant differences across systems. Unlike chatbots, smart conversational agents are not limited to performing simple conversations. And unlike embodied conversational agents, they do not provide computer-generated characters to mimic the movements or facial expressions of a virtual interlocutor. Smart conversational agents are at the confluence of speech processing, natural language processing (NLP), and artificial intelligence (AI). As detailed by Williams and colleagues [53], voice-activated devices such as Amazon Alexa or Apple Siri start by converting what a user said (i.e., an audio utterance) into text using automatic speech recognition. Words are then processed through spoken language understanding (SLU) and passed onto a dialog state tracker (DST), which results in identifying an appropriate response. The words in the response are prepared by natural language generation (NLG), and turned into audio by text-to-speech (TTS).

Smart conversational agents can be designed in many ways, as shown in the recent review by Laranjo *et al.* applied to healthcare [54]. A conversation may not be oriented toward the completion of a specific *task*, but takes place for its own sake. The flow of the discussion may be *controlled* by the system and/or the user. *Interactions* can be via spoken language and/or written language. Finally, the *dialogue management* may take the user through a sequence of pre-determined steps (i.e., a finite-state system), elicit an input and parse it using a template to decide the dialogue-flow (i.e., a frame-based system), or take an agent-based approach to focus on beliefs and desires. In the specific healthcare context reviewed by Laranjo *et al.*, agent-based approaches were uncommon (1 study) while finite (6 studies) and frame-based systems (7 studies) were equally common [54]. However, when interactions rely on voice and a task has to be accomplished, then the frame-based design is so common that the system may be presented as a *slot-based dialog system* [55].

3 Process in an Artificial Facilitation

As described in Sect. 2.2, the process needs to (i) obtain concept labels that are quantifiable and distinct from labels already used, and (ii) help participants provide relationships to minimize the risk of missing essential structures such as loops. To help participants track relationships, a map building process can be conceptualized as a *graph traversal*: we want to elicit/visit all of the concepts (i.e. nodes) that pertain to the user's mental model, and we move from a concept to another using a relationship. Unlike a *graph exploration* in which we typically come back to the first node, the map building process ends on an arbitrary node.

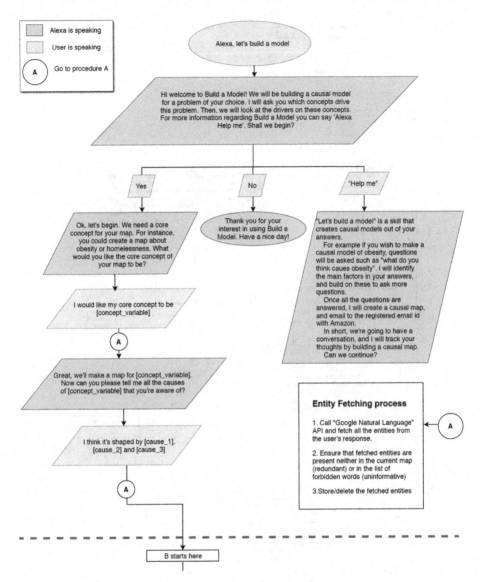

Fig. 2. Process to start a model and provide the first causes.

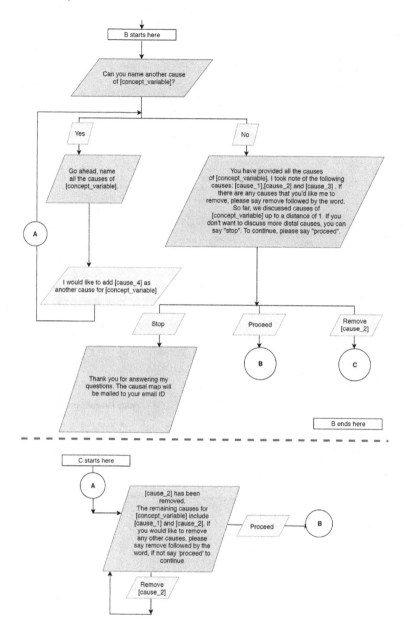

Fig. 3. Continuation of the process, showing how to get additional causes, get another layer of causes, or removing a causal edge.

Box 1. Sample conversation between a user and our facilitator using a depth-first search.

"Let's discuss obesity. What is one cause of obesity?
- Obesity is caused by unhealthy foods.
- What do you think would be one reason for having unhealthy foods?
- They're cheaper than healthy ones.
- And why are they cheaper?
- Because of government subsidies for food production.
- What's one driver of agricultural subsidies?
- I think that's going off topic.
- Ok. Let's go back. You said unhealthy foods are cheaper than healthy ones, because of agricultural subsidies. Any other reason?
- Not that I can think of.
- That's fine. Let's go back. You said people have unhealthy foods because of prices. Any other cause?"

Two typical approaches to a graph traversal are a depth-first search (DFS) and a breadth-first search (BFS). Starting from a root, a DFS follows *one* unexplored node, and from there visits another unexplored node, thus going as far as possible. When it cannot go further, it backtracks until it can branch in a new direction. This approach potentially undesirable in a facilitated process for at least three reasons. First, it can take participants on tangents, quickly going away from the main topic until they realize that factors are no longer relevant to the problem space. As a result, the map may be imbalanced, and a high cognitive load is placed on the individual who needs to frequently think of the problem's boundaries. Second, often going *back* to a node may feel less natural than going *forward*, possibly coming across as 'jumping' between ideas. Third, a DFS requires that the user only provides *one* new concept each time, and may thus ask many times about the same node. This is more cumbersome than providing all known concepts at once, at moving on. These points are illustrated through an example of a DFS-based conversation in Box 1.

Starting from a root, a BFS asks for *all* connected nodes. Intuitively, it acquires the complete layer of connected concepts at distance 1 from the starting one. Then, it goes through all of these concepts and acquires all of their neighbors, thus completing the layer at distance 2. By going through entire layers at a time, it avoids taking participants on tangents. By asking whether participants want to continue when an entire layer is done, it asks for a conscious monitoring of the problem boundary at specific moments instead of offloading this responsibility onto the user at every question. By going through layers, it only goes forward (i.e. uses a queue) instead of backward (i.e. using a stack in the DFS). Finally, by asking for all connected concepts at once, users have the natural opportunity to share all of their thoughts instead of restricting themselves to a single new concept. For these reasons, our artificial facilitator uses a breadth-first search. The functioning of a BFS is illustrated via a conversation in Box 2.

Box 2. Sample conversation between a user and our facilitator using a breadth-first search.

"Let's discuss obesity. Can you share all of the contributors to obesity?
- Eating unhealthy foods, lack of physical activity.
- Alright, these are the direct drivers. Do you want to go further and share what do you think drives these?
- Sure.
- What do you think would be the reasons for eating unhealthy foods?
- Could be a coping mechanism for stress or depression, an affordability issue because they're cheap, or a personal taste preference.
- And what about physical activity?
- Could be obesity itself, because it creates some barriers. Or a lack of access to facilities, or a fear of engaging in physical activity.
- So we've now looked at indirect drivers. Do you think it'd be relevant to discuss their causes?"

Note that, while the BFS is meant to cover more concepts, the appearance of previous concepts can create loops. As illustrated in Box 2, we have a loop from obesity to a lack of physical activity, which itself contributes to obesity.

As shown in Figs. 2 and 3, our process utilizes the layer-by-layer approach of the BFS. It also closely monitors the names of concepts, as shown in Fig. 2 (inset A). We actively prevent the creation of similar concepts, informing the user that they are already present in the map under a possibly different name. We also attempt to avoid the use of concepts that cannot be quantified, thus promoting more operational definitions of concepts. The technology used to realize these objectives is detailed in the next section.

4   Implementation of Our Artificial Facilitator

Our implementation is task-oriented as we seek to guide a participant in externalizing their mental model. The virtual facilitator controls the flow of the conversation by asking questions. Interactions in the deployed version are exclusively vocal, but developers in Amazon Alexa also have access to a console that takes written input (for testing only). Dialogue-management uses a frame-based system. All of these technical choices were briefly discussed in Sect. 2.3.

Our code is provided at https://github.com/datalab-science/causalMapBuilder. Our implementation involves several technologies, shown in a high-level view in Fig. 4 and detailed in Table 1. We use Amazon Alexa as it provides automatic speech recognition and text-to-speech, in addition to working on three out of four smart speakers [56]. We interact with the Alexa Skills Kit (ASK) through

Table 1. List of technologies and versions.

Technology	Version
Amazon web services	Accessed october 2018
Alexa skills kit (ASK)	
Dynamo DB	
Amazon S3	
Google natural language API	1.2 GA release
Python	3.6
NetworkX	2.2
NLTK	3.3
Ubuntu	18.04.1 LTS

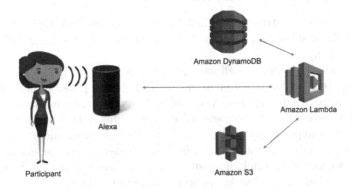

Fig. 4. High-level view of the prototype.

a program written in the Python language, stored on Amazon S3 (Amazon Simple Storage Service), which is invoked by Amazon lambda functions when objects are created or when intents are triggered through user interactions. The complete conversation log generated during a session with a user is stored in Amazon Dynamo DB, which is Amazon's fully-managed solution for NoSQL databases. The NetworkX library for Python serves to store and visualize the map. When the discussion ends, the visualization is emailed to the user together with a file containing a list of edges.

Google Natural Language API is queried extensively to find *entities*. Consider that the artificial facilitator asks "what causes obesity?" and the user responds "I believe that obesity is caused by an excess in eating and not enough exercise". Google Natural Language API will extract the entities 'obesity', 'excess', 'eating', and 'exercise'. Since an answer often includes a repetition of the subject, we automatically ignore user-provided entities that were part of the question. In this example, 'obesity' would be ignored, thus there are only three new concepts: 'eating', 'excess', and 'exercise'. As detailed in Sect. 3, we must ensure that the concepts are not already used. When a new concept node is created, we use

WordNet (accessed via the NLTK library in Python) to retrieve all cognitive synonyms (i.e., *synsets*). If the user later mentions an entity that belongs to these synsets, the artificial facilitator points out that it already exists under a different name.

A causal map is not supposed to have unquantifiable concepts, but users may lose track of this requirement. If Google Natural Language API identifies an entity which is unquantifiable, then our application can use it in nonsensical questions. For instance, 'excess' was identified as an entity although it is unquantifiable. The application may continue by asking "what causes excess?". We tested the application with 8 subjects over two months to identify such problematic entities. Since we cannot manually identify *all* such entities, we use the ones we identified as seeds to automatically fetch all similar entities, thus constituting a large dictionary of entities to ignore. The creation of this dictionary takes three steps performed using WordNet:

(1) We have a set of entities, identified during testing as both (i) fetched by the Google Natural Language API and (ii) unquantifiable. For instance, consider {lack, bunch}.
(2) For each word, we retrieve all its *hypernyms*, which are words with a broader meaning (e.g., color is a hypernym of red). Here, {lack, bunch} is transformed into {need, agglomeration, collection, cluster, gathering}.
(3) For each hypernym, we retrieve all its hyponyms, which are more specific words (e.g., hyponyms of color would include red, blue, and green). In this example, {need, agglomeration, collection, cluster} would be expanded into a large set including {lack, necessity, urge, ..., bunch, pair, trio, hive, crowd, agglomeration, batch, block, ensemble, ..., population}.

Amazon Alexa development features were altered during the development of the artificial facilitator. Our initial implementation relied extensively on an intent (i.e., a template) known as *AMAZON.LITERAL*, which allowed for free-form speech input instead of a defined list of possible values. This slot was deprecated on October 22, 2018. Consequently, the implementation presented here relies on custom slots.

5 Case Study: Creating Obesity-Related Maps

We used three case studies to test our system. In the first two case studies, we verified whether a participant could (re)create a previously developed causal map when using our artificial facilitator (Fig. 5). Leveraging the broad variety of languages and accents supported by Alexa, we set the device to Indian English for these two cases, as it is the language spoken by our participant. In the third case, the device was set to American English, and we tested additional features such as detecting redundant concepts or allowing the user to correct the map. All case studies were performed using an Amazon Echo Dot Device version 618571720. We recorded the discussion and the resulting map

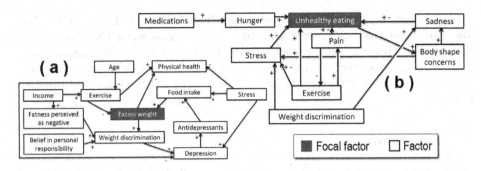

Fig. 5. Two previously published causal maps from health behaviors research [17]. Each map is centered on a different problem or 'focal factor'. The original study added an intervention to these maps as part of a virtual trial.

that our artificial facilitator emailed to the participant. To provide full disclosure, our three recordings can be viewed at https://www.youtube.com/playlist? list=PL7UTR3EL44zrkwrcDkiSwV-7kL0Nv6fQ5

Our first two case studies demonstrated that the *structure* of the maps could correctly be created using our artificial facilitator. We observed three issues due to the automatic detection of entities. First, it can lead to significantly shorter concept labels (https://www.youtube.com/watch?v=57tq0w4OEPw&t=324s). The original map stated that weight discrimination was driven by excess weight, fatness perceived as negative, and a belief in personal responsibility. Our automatic process resulted in weight discrimination being driven by weight, fatness, and responsibility. This loses some nuances: it is not fatness in itself that leads to discrimination, but the societal belief that fatness is an undesirable trait. The problem is aggravated when concepts that should be different are shortened such that they are indistinguishable. For instance, 'cardiovascular diseases' and 'metabolic diseases' are very different medical situations. However, entity recognition sees both as 'diseases' and thus conflates them, which results in structural errors for the map. Second, entity recognition is a bottleneck of the application in terms of time: users can have to silently wait for several seconds before entities have been processed. These awkward silences disrupt the flow of the discussion. Finally, accents can lead to very different performances in terms of entity recognition. Results are not only different between Indian and American participants, but also among Americans (e.g., from the South or the Midwest). As noted by Rachael Tatman, the training dataset for smart speakers results in working "best for white, highly educated, upper-middle-class Americans, probably from the West Coast, because that's the group that's had access to the technology from the very beginning" [57].

The third case study demonstrated that additional features of our artificial facilitator worked as specified. For instance, the participant stated that over-eating was caused by over-indulgence, but these two concepts are considered interchangeable per WordNet. Consequently, the artificial facilitator informed the user (https://www.youtube.com/watch?v=U2mYkSLE9NE&t=40s). We also confirmed that users were able to remove causes when they have been incorrectly captured (https://www.youtube.com/watch?v=U2mYkSLE9NE&t=213s). Finally, we verified that the virtual facilitator did repeat questions when prompted by the user (https://www.youtube.com/watch?v=U2mYkSLE9NE&t=95s).

6 Discussion

In collaborative modeling, participants externalize their mental models into various artifacts such as causal maps. This externalization can be guided by a trained facilitator, but there may be associated costs, and availability is limited. Alternatively, free software can be used at any time to create causal maps but they do not guide participants. In addition, neither facilitators nor current software can easily cope with larger causal maps, for instance, to avoid the creation of redundant concepts. To address these limitations, we designed an artificial facilitator that leverages voice-activated technologies. We implemented the prototype via Amazon Alexa, and demonstrated its features through three case studies.

As our system constitutes the first use of voice-activated technologies to build causal maps in participatory modeling, we are at the early stage of a multi-year process. There are several opportunities to improve the system or address additional research questions in the short- and medium-term. In the short-term, our prototype faces two limitations. First, we used hand-crafted rules, which is more in line with early spoken dialog systems than with current ones. Other approaches use generative methods (e.g., Bayesian networks) which often involve hand-crafted parameters, or discriminative methods where parameters are inferred by machine learning from the data. As stated by Henderson, "discriminative machine-learned methods are now the state-of-the-art in dialog state tracking" [55]. However, machine learning requires data to learn from. There is currently no corpus of model building involving a facilitator and one participant. Such sessions are often conducted with *many* participants, and the recordings are not released as the consent forms generally include an anonymity clause. Designing a better artificial facilitator will thus start by assembling a large set of recordings between a facilitator and a participant, for instance by modeling a system in which participants would be comfortable in publicly sharing their perspectives.

Second, our approach extensively relies on Alexa followed by Google Natural Language API to identify entities. Our prototype struggled with creating causal maps with specialized terms (e.g., from the medical domain) as Alexa could not identify them in speech and/or the API would not see them as relevant entities. The API may improve over time, and it may also be assisted with ontologies to identify (i) which specialized terms may be used, and (ii) which term is likely to be used following another one. Similarly, improvements in the API would reduce the processing time which currently results in many awkward seconds of silence. We note that improvements in the API or in Alexa Skill Kit will automatically benefit the quality of our application, without changes in our design or implementation.

In the medium-term, research may explore how an artificial facilitator can provide guidance in aspects that are necessary yet challenging for trained facilitators. The structure of causal maps is normally analyzed *after they have been built*, for instance by identifying leverage points via centrality [30,31] or inventorying loops that drive the dynamics of the map [32–34]. However, a large map of a complex system that contains no loops may already be identified as problematic, suggesting that some causal edges are potentially missing. Consequently, the artificial facilitator can leverage network algorithms to analyze the structure of the map as it is built, thus informing participants of potential issues and approaches to address them. The artificial facilitator can also build on natural language processing in many ways that go beyond the identification of entities. Causal maps sometimes start with a brainstorming process, in which many concepts are generated and then *grouped*. Our artificial facilitator can use the semantic relatedness of concepts to inform the user about potential themes, which may result in combining several overly-detailed concepts into a more abstract category.

7   Conclusion

We successfully used Alexa to develop a voice-activated assistant that guides a user in creating a causal map. We addressed the challenge of finding appropriate concept names. In future work, we will automatically inform the user when concepts related to a theme may be used instead of narrowly defined concepts, and we will monitor the structure of the map as it is being built to support users in identifying loops.

Supplementary Material

Our code is available at https://github.com/datalab-science/causalMapBuilder. Our three case studies (Sect. 5) as well as a video overview can be accessed at https://www.youtube.com/playlist?list=PL7UTR3EL44zrkwrcDkiSwV-7kL 0Nv6fQ5

Acknowledgments. The authors are indebted to Mitacs Canada for providing the financial support which allowed TR to perform this research at Furman University, while mentored by PJG (local advisor) and VKM (home advisor). Publication costs are supported by an NSERC Discovery Grant for VKM.

Contributions. TR implemented the software and produced the videos demonstrating its functioning. PJG wrote the manuscript and designed the workflow of the virtual facilitator. TR was advised by PJG and VKM, who jointly initiated the study. All authors read and approved of this manuscript.

References

1. Heitman, K.: Reductionism at the dawn of population health. In: El-Sayed, A.M., Galea, S. (eds.) Systems Science and Population Health, chap. 2, pp. 9–24. Oxford University Press (2017)

2. Fink, D.S., Keyes, K.M.: Wrong answers: when simple interpretations create complex problems. In: El-Sayed, A.M., Galea, S. (eds.) Systems Science and Population Health, chap. 3, pp. 25–36. Oxford University Press (2017)

3. Giabbanelli, P.J.: Analyzing the complexity of behavioural factors influencing weight in adults. In: Giabbanelli, P.J., Mago, V.K., Papageorgiou, E.I. (eds.) Advanced Data Analytics in Health. SIST, vol. 93, pp. 163–181. Springer, Cham (2018). https://doi.org/10.1007/978-3-319-77911-9_10

4. Giabbanelli, P.J., Baniukiewicz, M.: Navigating complex systems for policymaking using simple software tools. In: Giabbanelli, P.J., Mago, V.K., Papageorgiou, E.I. (eds.) Advanced Data Analytics in Health. SIST, vol. 93, pp. 21–40. Springer, Cham (2018). https://doi.org/10.1007/978-3-319-77911-9_2

5. Riley, B., et al.: Systems thinking and dissemination and implementation research. In: Brownson, R.C., Colditz, G.A., Proctor, E.K. (eds.) Dissemination and Implementation Research in Health: Translating Science to Practice, chap. 9, p. 143. Oxford University Press (2017)

6. Verigin, T., Giabbanelli, P.J., Davidsen, P.I.: Supporting a systems approach to healthy weight interventions in British Columbia by modeling weight and wellbeing. In: Proceedings of the 49th Annual Simulation Symposium, Society for Computer Simulation International, p. 9 (2016)

7. Dubé, L., et al.: From policy coherence to 21st century convergence: a whole-of-society paradigm of human and economic development. Ann. N. Y. Acad. Sci. **1331**(1), 201–215 (2014)

8. Lavin, E.A., et al.: Should we simulate mental models to assess whether they agree? In: Proceedings of the Annual Simulation Symposium, Society for Computer Simulation International, p. 6 (2018)

9. Gupta, V.K., Giabbanelli, P.J., Tawfik, A.A.: An online environment to compare students' and expert solutions to ill-structured problems. In: Zaphiris, P., Ioannou, A. (eds.) LCT 2018. LNCS, vol. 10925, pp. 286–307. Springer, Cham (2018). https://doi.org/10.1007/978-3-319-91152-6_23

10. Giabbanelli, P.J., Tawfik, A.A.: Overcoming the PBL assessment challenge: design and development of the incremental thesaurus for assessing causal maps (ITACM). Technol. Knowl. Learn. **24**, 161–168 (2019)

11. Voinov, A., et al.: Tools and methods in participatory modeling: selecting the right tool for the job. Environ. Model. Softw. **109**, 232–255 (2018)

12. Giabbanelli, P.J., Crutzen, R.: Using agent-based models to develop public policy about food behaviours: future directions and recommendations. Comput. Math. Methods Med. **2017** (2017). https://www.hindawi.com/journals/cmmm/2017/5742629/abs/

13. Maglio, P.P., Mabry, P.L.: Agent-based models and systems science approaches to public health. Am. J. Prev. Med. **40**(3), 392–394 (2011)

14. Giabbanelli, P.J., Jackson, P.J.: Using visual analytics to support the integration of expert knowledge in the design of medical models and simulations. Procedia Comput. Sci. **51**, 755–764 (2015)

15. Xue, H., et al.: Applications of systems modelling in obesity research. Obes. Rev. **19**(9), 1293–1308 (2018)

16. de Pinho, H.: Generation of systems maps: mapping complex systems of population health. In: El-Sayed, A.M., Galea, S. (eds.) Systems Science and Population Health, chap. 6, pp. 61–76. Oxford University Press (2017)
17. Giabbanelli, P.J., Crutzen, R.: Creating groups with similar expected behavioural response in randomized controlled trials: a fuzzy cognitive map approach. BMC Med. Res. Methodol. **14**(1), 130 (2014)
18. Lich, K.H., et al.: Extending systems thinking in planning and evaluation using group concept mapping and system dynamics to tackle complex problems. Eval. Program Plan. **60**, 254–264 (2017)
19. Frerichs, L., et al.: Mind maps and network analysis to evaluate conceptualization of complex issues: a case example evaluating systems science workshops for childhood obesity prevention. Eval. Program Plan. **68**, 135–147 (2018)
20. Gray, S., et al.: The structure and function of angler mental models about fish population ecology: the influence of specialization and target species. J. Outdoor Recreat. Tour. **12**, 1–13 (2015)
21. Befus, D.R., et al.: A qualitative, systems thinking approach to study self-management in women with migraine. Nurs. Res. **67**(5), 395–403 (2018)
22. Yourkavitch, J., et al.: Interactions among poverty, gender, and health systems affect womens participation in services to prevent HIV transmission from mother to child: a causal loop analysis. PloS One **13**(5), e0197239 (2018)
23. Giabbanelli, P.J., Torsney-Weir, T., Mago, V.K.: A fuzzy cognitive map of the psychosocial determinants of obesity. Appl. Soft Comput. **12**(12), 3711–3724 (2012)
24. Zhang, S., de la Haye, K., Ji, M., An, R.: Applications of social network analysis to obesity: a systematic review. Obes. Rev. **19**(7), 976–988 (2018)
25. Finegood, D.T., Merth, T.D., Rutter, H.: Implications of the foresight obesity system map for solutions to childhood obesity. Obesity **18**(S1), S13–S16 (2010)
26. Jebb, S., Kopelman, P., Butland, B.: Executive summary: foresight tackling obesities: future choices project. Obesity Reviews **8**, vi–ix (2007)
27. Drasic, L., Giabbanelli, P.J.: Exploring the interactions between physical well-being, and obesity. Can. J. Diabetes **39**, S12–S13 (2015)
28. Allender, S., et al.: A community based systems diagram of obesity causes. PLoS One **10**(7), e0129683 (2015)
29. McGlashan, J., et al.: Comparing complex perspectives on obesity drivers: action-driven communities and evidence-oriented experts. Obes. Sci. Pract. **4**, 575–581 (2018)
30. McGlashan, J., et al.: Quantifying a systems map: network analysis of a childhood obesity causal loop diagram. PloS One **11**(10), e0165459 (2016)
31. Knapp, E.A., et al.: A network approach to understanding obesogenic environments for children in pennsylvania. Connections (02261766) **38**(1) (2018). https://doi.org/10.21307/connections-2018-001
32. Owen, B., et al.: Understanding a successful obesity prevention initiative in children under 5 from a systems perspective. PloS One **13**(3), e0195141 (2018)
33. Giabbanelli, P., et al.: developing technology to support policymakers in taking a systems science approach to obesity and well-being. Obes. Rev. **17**, 194–195 (2016)
34. Rwashana, A.S., et al.: Advancing the application of systems thinking in health: understanding the dynamics of neonatal mortality in uganda. Health Res. Policy Syst. **12**(1), 36 (2014)

35. Giabbanelli, P.J., Tawfik, A.A., Gupta, V.K.: Learning analytics to support teachers' assessment of problem solving: a novel application for machine learning and graph algorithms. In: Ifenthaler, D., Mah, D.-K., Yau, J.Y.-K. (eds.) Utilizing Learning Analytics to Support Study Success, pp. 175–199. Springer, Cham (2019). https://doi.org/10.1007/978-3-319-64792-0_11

36. Singh, M., et al.: Building a cardiovascular disease predictive model using structural equation model & fuzzy cognitive map. In: 2016 IEEE International Conference on Fuzzy Systems (FUZZ-IEEE), pp. 1377–1382. IEEE (2016)

37. Pillutla, V.S., Giabbanelli, P.J.: Iterative generation of insight from text collections through mutually reinforcing visualizations and fuzzy cognitive maps. Appl. Soft Comput. **76**, 459–472 (2019)

38. Pratt, S.F., et al.: Detecting unfolding crises with visual analytics and conceptual maps emerging phenomena and big data. In: 2013 IEEE International Conference on Intelligence and Security Informatics (ISI), pp. 200–205. IEEE (2013)

39. Ozawa, S., Paina, L., Qiu, M.: Exploring pathways for building trust in vaccination and strengthening health system resilience. BMC Health Serv. Res. **16**(7), 639 (2016)

40. Lukoianova, T., Rubin, V.L.: Veracity roadmap: is big data objective, truthful and credible? Adv. Classif. Res. Online **24**, 4 (2014)

41. Jetter, A.J.: Fuzzy cognitive maps for engineering and technology management: what works in practice? In: PICMET (2006)

42. Jordan, R., et al.: Twelve questions for the participatory modeling community. Earth's Future **6**(8), 1046–1057 (2018)

43. Giabbanelli, P.J., Adams, J., Pillutla, V.S.: Feasibility and framing of interventions based on public support: leveraging text analytics for policymakers. In: Meiselwitz, G. (ed.) SCSM 2016. LNCS, vol. 9742, pp. 188–200. Springer, Cham (2016). https://doi.org/10.1007/978-3-319-39910-2_18

44. Binder, J.R., Desai, R.H.: The neurobiology of semantic memory. Trends Cogn. Sci. **15**(11), 527–536 (2011)

45. Lewis, P.: Rich picture building in the soft systems methodology. Eur. J. Inf. Syst. **1**(5), 351–360 (1992)

46. Sommerfeld, E., Sobik, F.: Operations on cognitive structures - their modeling on the basis of graph theory. In: Albert, D. (ed.) Knowledge Structures, pp. 151–196. Springer, Heidelberg (1994). https://doi.org/10.1007/978-3-642-52064-8_5

47. Johnson-Laird, P.N.: Mental models and human reasoning. Proc. Natl. Acad. Sci. **107**(43), 18243–18250 (2010)

48. Mago, V.K., et al.: Analyzing the impact of social factors on homelessness: a fuzzy cognitive map approach. BMC Med. Inform. Decis. Mak. **13**(1), 94 (2013)

49. Mago, V.K., et al.: Fuzzy cognitive maps and cellular automata: an evolutionary approach for social systems modelling. Appl. Soft Comput. **12**(12), 3771–3784 (2012)

50. Axelrod, R.: Decision for neoimperialism: the deliberations of the British Eastern Committee in 1918. In: The Cognitive Maps of Political Elites, Structure of Decisions, chap. 4, pp. 77–95 (1976)

51. Axelrod, R.: Results. In: The Cognitive Maps of Political Elites, Structure of Decisions, chap. 9, pp. 221–250 (1976)

52. Ross, S.: Complexity and the presidency. In: The Cognitive Maps of Political Elites, Structure of Decisions, chap. 5, pp. 96–112 (1976)

53. Williams, J., Raux, A., Henderson, M.: The dialog state tracking challenge series: a review. Dialogue Discourse **7**(3), 4–33 (2016)

54. Laranjo, L., et al.: Conversational agents in healthcare: a systematic review. J. Am. Med. Inform. Assoc. **25**(9), 1248–1258 (2018)
55. Henderson, M.: Machine learning for dialog state tracking: a review. In: Proceedings of The First International Workshop on Machine Learning in Spoken Language Processing (2015)
56. Canalys: Amazon reclaims top spot in smart speaker market in Q3 2018 (2018)
57. Harwell, D.: The accent gap. The Washington Post (2018)

Human Cognition and Behavior in Complex Tasks and Environments

Augmented Cognition for Socio-Technical Systems

Scott David and Barbara Endicott-Popovsky[✉]

University of Washington CIAC, Seattle, WA 98105, USA
{sldavid, endicott}@uw.edu

Abstract. Research into situated and embodied cognition and related investigations have suggested that "cognition" takes place not only in the human brain, but also in externalities such as the environment (situated cognition) and the body (embodied and morphological computing). This article explores two propositions and their implications for "augmented cognition" resulting from these expanded, systemic views of cognition.

Keywords: Augmented cognition · Socio-technical

1 Introduction

Two propositions result from an expanded systems view of cognition:

First proposition, that the externalization of cognition involves recruitment of both external technology AND external people/institutions into individual processes of "cognition." The latter human/social externalities invite consideration of new strategies for "cognitive augmentation." We need socio-technical augmentation for socio-technical cognition.

Second proposition, that augmentation can be viewed as part of a spectrum of "cognition support" that ranges from leveraging processes for improvement and enhancement at one end to applying de-risking (e.g., noise reduction, trusted sources of data, etc.) at the other end. In other words, "augmentation" can be achieved through both enhancement of function and also through the mitigation of the diminishment of function – both strategies yield additional cognitive resources.

Some strategies and tactics are effective all along that spectrum. To the extent of that relationship, and if the first proposition above (i.e., that we need socio-technical augmentation for socio-technical cognition) is tractable, then future socio-technical forms of cognitive augmentation might be derived from existing strategies of socio-technical behavioral de-risking that have applied historically to the behaviors of people and institutions. Those de-risking strategies emerge in myriad business, legal, technical and social (BLTS) realms. Existing strategies for supporting reliable socio-technical systems can help to both augment and de-risk socio-technical cognition.

D. D. Schmorrow and C. M. Fidopiastis (Eds.): HCII 2019, LNAI 11580, pp. 133–142, 2019.
https://doi.org/10.1007/978-3-030-22419-6_10

1.1   Introduction of Analysis

The notions of cognition have always reflected a combination of individual and social elements. As the digital revolution has enhanced the ability to capture and measure ever-more-granular attributes of interactions, the scales have continued to tilt toward perceiving cognition as migrating from what was an intimate and individual process toward a more complex system of "socio-technical" processes. As our understanding of cognition changes, it invites modification in our strategies and operations aimed at augmenting cognition.

As our ability to clarify the concept and locus of "cognition" continues to migrate from individual human brains onto complex hybrid socio-technical systems, our strategies and tactics for augmenting such expanded "cognition" will also change. Future augmentations of cognition will be most effective if designed, developed and deployed with intentionality and awareness of the systemic nature and mechanisms of post-Internet networked cognition.

We suggest that "augmentation" and "de-risking" of systems share a common root of performance reliability at a given level of performance. From that starting point, we conclude that future strategies of cognitive augmentation can be gainfully informed by historical models of socio-technical behavioral de-risking. This statement is supported by an expanded view of cognition, which sees "situated cognition" as being possible in both external inanimate objects AND also in other people and in external institutional interactions. Applying this approach this paper also seeks to anticipate some of the potential future vectors of augmentation of these expanded systems with reference to historical structures of de-risking performance and behaviors of existing socio-technical systems [1].

1.2   Where Does Cognition Take Place?

Thought and cognition have traditionally been understood to reside in the brain of individual humans, and many long-standing and foundational rules, laws, and customs (such as culpability for actions under law) are based on that understanding. Increasingly, however, there is evidence that cognition and consciousness is more subtle, and extended, than that conception. Cognition is amenable to "systems" analysis [2].

Cognition is increasingly recognized, and fruitfully analyzed, as taking place within and outside human brains in systems of scaled perception and meaning making. In fact, in the last several decades, the concept of "cognition" has migrated into the external world, where it is increasingly viewed as being "situated" and/or "embodied" in external inert physicality. From this perspective, the brain is a virtual "antenna" that is tuned by formal learning and informal experience into language and culture that foster the interactions from which the "mind" emerges.

What are the possible analytical and operational insights that might be derived if the processes of cognition are not just informed by physical environments (e.g., in perceptual consciousness), but also take place in those environments? As an initial matter, consider that even those who embrace the notion that cognition takes place in the brain recognize that those patterns of cognition are formed from external inputs. Education, indoctrination, training, etc. are examples of the formalization and institutionalization of that externality. There are also myriad informal external sources of cognitive patterning, for example learned social norms, touching a hot stove, etc.

Consider, for example, an adopted infant from one culture who is raised in a foreign culture – the grown-up adult's language, cultural preferences, attitudes and cognitive patterns will be a product entirely of the adoptive land. In fact, it is relevant that from this perspective, education and social learning are forms of "augmentation" of cognition, although the augmentations precede the acts of cognition in preparing individual infant "minds" to apply patterns of meaning to later perceptions and cognition. Education is cognitive augmentation.

That point may initially seem rhetorical and academic, but it forms the basis for our later assertion that standard education and training in standard policies and rules is a form of "meaning security" that is necessary for the sustainable reliability of information networks. Since situated cognition of all sorts takes place on those networks, future strategies of "augmented cognition" will include heavy measures of education (and re-education) into rules and norms from which reliable socio-technical systems are built. There will be policy-based constraints on individual liberty associated with these standards, just as red lights and shoplifting rules constrain liberty at present. These forms of augmentations will have costs like those that arise in the social contract or the "golden rule" – both of which constrain liberties in exchange for enhanced de-risking and leverage.

2   Part 1 – The Wandering Concept of "Cognition'

2.1   Cognition in the Brain – Augmentation in the External Environment

Cognition is commonly thought to reside in the mind which in turn resides in the brain. In the 1960's, notions of augmented cognition started from the presumption that cognition is a process that takes place in the human brain. In this early paradigm, augmented cognition is typically understood to refer to augmentation of human (brain) cognition by external technologies.

For example, Douglas Engelbart defined augmented human intelligence in similar terms as "increasing the capability of a man to approach a complex problem situation, to gain comprehension to suit his particular needs, and to derive solutions to problems [3]." From the beginning, the source of augmentation was considered an externality to human cognition. In fact, DARPAs augmented intelligence program originally anticipated human machine dyad pairs [4].

It is notable, however, that these earlier conceptions were based on some significant assumptions. The Oxford English dictionary defines "cognition" as: "The action or faculty of knowing, knowledge, consciousness; acquaintance with a subject [5]." The OED defines "Augment" as: "To make greater in size, number, amount, degree, etc. To increase, enlarge extend [6]." Notably, neither term is bound to a specific system, and that contextual ambiguity is typically tolerable in casual conversation about augmented cognition; however, where the intention is to analyze and enhance specific systems of "augmented cognition" for future applications, disambiguation is a necessary first step.

2.2   Extended Minds –Social Cognition

In the 1960's, sociologists such as Erving Goffman explored the relationship of the "self" to interactions, from both expressive and perceptual angles [7]. The result was a recursive house of mirrors of meaning, with social and technical elements intermixing to inform frameworks of cognition. Kuhn observed other social components to individual cognitive function [8]. In both of these cases, the individual's cognition and consciousness (sense of self) are observed to be rectified and valorized by interactions with other individuals (rather than inert objects) in social interactions.

In both of these analyses, there is no suggestion that cognition resides outside of the mind of the individual, but rather that the individual's cognitive patterns (existential and paradigmatic) are heavily influenced by people and institutions in the external environment. This has direct impact on the search for future strategies for augmenting cognition. We can best augment them if we first know what they are.

2.3   Extended Minds –Situated Cognition in Inert Physicality

Various theories of situated cognition and embodied cognition embrace the extension of the concept of cognition to external environmental elements.[1] Unlike traditional notions of "augmentation," these non-brain sources of cognitive support are not seen as mere external sources of "augmentation," but rather as foundational elements of the act of cognition itself.

At some level, this is an issue of semantics; however, it is not a mere academic exercise because a more holistic view of the apparatus of cognition (brain, language, culture, rules, other brains, etc.) combined with consideration of the information processes that are advanced through cognition/intelligence (e.g., data plus meaning equals information) suggests increased solution phase space for nagging challenges of de-risking broadly distributed and scaled information networks. We cannot fully and effectively augment cognition if we misapprehend where and how it occurs.

2.4   Socio-Technical Solutions for Socio-Technical Problems

The heterogeneous nature of augmentation in socio-technical systems is such that it introduces new classes of variables into the analysis of augmentation and performance integrity of cognitive systems. In fact, this article will observe that cognition results from hybrid socio-technical systems that are composed of humans in their respective environments and myriad inter-dependent elements from business, legal, technical and social (BLTS) domains.

By parsing the concepts, we can expose additional degrees of freedom in networked information/cognition systems, enabling more highly refined development of such systems, with downstream benefits of enhanced reliability and predictability of

---

[1] This article will not seek to disambiguate these multiple meanings, but will apply a definition that locates the source of augmentation external to the system of cognition. This assumes that if the augmentation is internal to that system of cognition, it would not be referred to as augmented cognition, but rather just "cognition."

experience for information seekers and data subjects alike. In fact, the advances in reliability and predictability of integrity of information systems due to advances in "meaning integrity" will help dissipate many of the current concerns voiced under the general categories of "security," "privacy," and "liability mitigation" online. This is because those concerns are just symptoms of underdeveloped integrity in information channels, all of which can be improved with enhanced "meaning" systems. This relationship of augmentation and de-risking is further elaborated in the section on meaning integrity below.

2.5 What Insights Are Derived from These Alternative Conceptions of Cognition?

The concept of "cognition" has migrated from the brain to physical externality. Concepts such as "social cognition," "embedded cognition," "morphological cognition/computing," "situated cognition," etc. all share the characteristic of analyzing the systems (including systems of systems) that can be said to be involved in "cognition."

The language and the concepts of these extended mind paradigms are fluid and frequently overlap due to the intrinsically ambiguous nature of the concepts of cognition and consciousness. While some part of these notions may initially seem like exercises in semantics,[i] brief reflection reveals that the parsing of these concepts can also lead to additional potential phase space for cognitive augmentation solutions.

As an initial matter, consider that their shared recognition of the relevance (and even necessity) of the external environment in the processes of cognition invites consideration of what additional strategies of "augmentation" associated with such external environments might be explored in these forms of extended minds and their extended systems of cognition. The question takes on added importance with the rise of socio-technical systems, where hybrid influences of three types of entities – people, institutions and things on cognition. What analytical construction could help tame the complexity of these myriad inputs?

3 Part 2 - The "Cognitive Unit" as Reaction Vessel for Converting Data into Information by Adding Meaning

3.1 What Is a "Cognitive Unit?"

In this paper we assert that - Information can be produced, and cognition can take place, in any system, at any scale, in which meaning is applied to data to yield information. We will refer to these virtual information reaction vessels as "cognitive units." Following this definition, systems of "augmentation" of these cognitive units are those which can increase the information yield of a given set of data and/or perceptions that are introduced to those cognitive units.

That sounds very theoretical, but application of a simple algorithm can help to make it operational: Data plus meaning equal information (consistent with Claude Shannon [9]). Therefore, any structure of meaning that can yield more information from a given set of data offers a potential indication of the presence of a cognitive unit.

In turn, the identification of the attributes of that "cognitive unit" help to reveal paths to its cognitive augmentations.

3.2   Socio-Technical Systems that Are Cognitive Units Can Leverage Data to Enhance Information Yields

A car and a driver are together an example of a socio-technical system. Even a perfectly tuned car cannot de-risk a reckless (or intentionally criminal) driver. Where the driver and the car both perform in accordance with expectations (set by technical specification for the car and rules and norms for the driver), the socio-technical system is de-risked.

Networked information systems are also socio-technical systems. Even a perfectly engineered Internet cannot de-risk a reckless (or intentionally criminal) online user. Most cybersecurity work has been focused on engineering the technology and protecting the data as a way of de-risking the internet. That strategy is necessary but insufficient to achieve de-risking. In fact, the term "users" is misleading in the case of socio-technical systems such as the Internet, because people and institutions are not just external users of a tool, but themselves critical components of the sociotechnical information system of the Internet. Markets are another example. The internet (and markets) would be inert if there were no users (participants). They are both examples of socio-technical systems.

Just as millions of untrained and unaware drivers would cause roadways to be so dangerous as to be un-usable, so too are information networks negatively affected in the absence of rules. The car and driver should also be treated as a unit in certain levels of risk mitigation analysis.

3.2.1   How Does the Analytical Construction of a Socio-Technical System Aid in the Identification of a Cognitive Unit?

Narratives and paradigms are transferred between and among people and institutions in socio-technical structures. Examples are students learning in schools, people behaving in conformity with laws, employees following company policies, companies following supply chain contracts and industry standards, etc. By spreading "meaning" these organizations "de-risk" future interactions at cumulatively massive scales.

Humans and human institutions regularly rely on these networks to situate their cognition, with the result that the augmentations of such situated cognition already involve systems outside the human brain. Those mental systems that are outside the human rain invite consideration of separate augmentation pathways that more traditional "augmented" cognition.

What is new is the simultaneous mutual situated cognition of information reaction vessels in one another. These are the mechanisms that dynamically create and perpetuate meaning, much like the waves of synaptic activity (not the synapses themselves) are thought to give rise to traditional cognition in the human brain. In sociotechnical cognitive systems, many of the components are not inert embodiments of technology, but rather interacting meaning reactors. Data flows through and among these reactors, feeding the meaning mechanisms. There are many meaning mechanisms around us. Brains/minds, institutions, markets. AI is an emerging meaning mechanism

– hence its existential threat to existing meaning making systems of humans and institutions.

This article uses the concept of augmented cognition (frequently described as external technical boosts for activities in the human brain) as a starting point to suggest that a fruitful line of analysis can be supported by revisiting the notions of both augmentation and cognition with a more systems-oriented approach. This approach is supported by many lines of inquiry over thousands of years but has been ignored under the paradigm of individual as the thinking unit. Bottom line is that cognition might be said to reside in those systems that apply meaning to data thereby creating "information". These are the reaction vessels of information and therefore cognition.

To de-risk a socio-technical system, the hybrid behaviors of the social and technical elements must be made more reliable and predictable. Many of those same strategies might prove useful in augmenting the forms of cognition that take place in those socio-technical cognitive units.

Consider, for example, that in both cases, the quest for reliability and predictability can be seen to simultaneously de-risk and leverage (augment) the processes of those socio-technical systems. To the extent that "cognition" is viewed as taking place in these systems, that cognition can be said to be enhanced by these strategies.

Security from Reliability
Security is performance or operation in accordance with expectations. From this definition, security (and privacy and liability) can be pursued through the path of reliability and predictability. The reliability of socio-technical systems requires reliability of technology and reliability of people/institutions. Technology is made reliable by conformity to specifications. People/institutions are made reliable by conformity to rules and laws.

What is needed for people and institutional reliability? Not specs – but rules, incentives and penalties. This applies to any entity with discretion, whether human or organizational. Since, their "reliability" is less predictable as a result of their potential for the exercise of discretion, need to have penalties, and incentives to draw that discretion in a more reliable (and secure and private) direction.

4   The Missing Piece – Meaning Standards for Cognitive Integrity and SI System Integrity

4.1   Data Interoperability Versus Meaning Interoperability

If these mechanisms of supporting technical and data interactions are broadly present, why aren't the resulting information systems more interoperable? Data flows across technical systems are largely interoperable, since data, in its purest form is relatively inert, and data and technical interoperability have the advantage of being based on the laws of physics which are, for all present purposes, uniform across the globe.

4.2   Meaning Security" Has Yet to Be Explicitly Recognized and Developed

HIPAA, GLB, GDPR apply "data" security to the challenge of information security. That is necessary, but insufficient for information security. Data plus meaning equals information. So, information security requires both data security AND meaning security; however, the term "meaning security" is unfamiliar. In addition, what does it have to do with augmented cognition?

This paper asserts that the challenge of the current and future period is no longer data interoperability, but "meaning" interoperability. To accomplish "information security" a system must apply strategies for both "data security" AND "meaning security." The ultimate failing of HIPAA, GLB, GDPR and other Fair Information Practice Principles (FIPPs)- based "privacy" rules is that they emphasize "data" security, ignore "meaning security," and treat all data as equally information laden (i.e, equally "surprising" in Shannon terms). This relieves plaintiffs of proving harm. Since data is dual use technology, this is too blunt an instrument.

4.3   Secrecy Is Dead – Manage Risk (and Augment Cognition) with "Meaning" Reliability/Security Not Data Security

The foregoing should not be interpreted as advocacy for the death of security [10]. Instead, it is based on a realistic assessment of the realities of seeking to maintain security controls over exponentially increasing interaction systems (and the exponentially increasing risks that they create) [11]. From this perspective, data security is a losing battle [12]. It is still important to raise the costs of unauthorized access (and the consequent dissipation of information arbitrage), but is ultimately doomed to fail. That failure is caused by the billions of people and their trillions of "thumb swipes" performed every day. Those swipes reflect information seeking behavior. The collective pressure of that behavior will ultimately doom efforts to provide data that feeds the insights being sought.

4.4   Augmented Cognition and de-Risking Are Linked Because the Both Depend on the Presence of Meaning to Convert Data into Useful Information

Data without meaning cannot inform a party. If the party is not informed, cognition cannot be said to have occurred (or been augmented!), future risks cannot be avoided.

4.5   Pathways to Meaning Augmentation/Security

The meaning making mechanisms of individuals are informed by education, narratives, priors, etc. The meaning making mechanisms of institutions (business, government, civil society) are more specifically programmed – set forth in foundational documents and regulations such as articles, bylaws, contracts, constitutions, etc. Shared meaning across a population enhances the likelihood that the population will be similarly informed by a given set of data/perceptions. Similar information supports similar behavioral responses, i.e., red means stop ample. The light does not stop traffic, it is the agreement that red signals the engagement of stopping behavior that stops traffic.

4.6 Augmentation and de-Risking Are Linked Because Both Preserve Cognitive Resources

The link is made clearer through recognition that security, privacy and liability mitigation are all symptoms of the underlying illness of a lack of "integrity" (as variously measured) of the input and output channels of information that form the myriad feedback loops (operating at multiple levels) from which cognition emerges. As cognition migrates from the brain to hybrid socio-technical systems, the connections of these channels become more extended and thereby more vulnerable to intrusions on integrity. New strategies that augment cognition will include those that can mitigate the threats and vulnerabilities of these extended channels, and thereby decrease the resources (including cognitive load) that are associated with maintaining the reliability of those channels. Those resources can then be re-directed toward more directly cognitive tasks. This is roughly akin to trying to write a novel in a quiet room versus a noisy room.

5 Conclusions

In this paper, we observe that prior notions of cognition and therefore its augmentation, should be revisited to account for the emergence of broadly scaled "situated cognition" in online information networks, and the increasing hybridization of socio-technical information networks. We also propose that a variety of new vectors are made available for augmentation as cognition migrates (or is outsourced?) to massively networked and distributed socio-technical systems, where we each are integrated together with our institutions as users, and also "augmenters" of an emerging "cognitive commons."

6 Recommendations

The interaction trends in these disintermediated, distributed information networks has challenged traditional notions of cognition as a social phenomenon, and also institutional meaning making mechanisms. This undermines institutional power by moving traditional communications to channels that are not subject to normal regulatory channels. We need to do a better job of rendering the humans in the system more reliable. This is a case for developing compensating controls that will meet the security objectives of the system that remain unmet by existing tools. This includes the business and legal constraints (rules) that will de-risk systems from a cybersecurity perspective.

7 Future Work

It is the intent of the authors to pursue research into these alternate controls, discover patterns of behavior and activity that will render the human aspects of systems more reliable and evolve standards for these practices to disseminate to others.

References

1. David, S.: The Atlas of Risk Maps. CIAC, Seattle (2016)
2. Senge, P.: The Fifth Discipline. Doubleday, New York (2006)
3. Englebart, D.: Augmenting human intellect: a conceptual framework. SRI Summary Report AFOSR-3223 Prepared for: Director of Information Sciences, Air Force Office of Scientific Research, Washington DC, October 1962
4. Insert reference to DARPA Dyad pairs
5. Oxford English dictionary defines "cognition"
6. OED defines "Augment" as: "
7. Goffman, E.: The presentation of Self in Everyday Life. Monograph No. 2, Social Sciences Research Centre, University of Edinburgh (1956)
8. Kuhn, T.: The Structure of Scientific Revolutions, 2nd edn. University of Chicago Press, Chicago (1970)
9. Shannon, C.: A mathematical theory of communication. Bell Syst. Tech. J. **27**, 379–423 (1948)
10. David, S., Endicott-Popovsky, B.: Security beyond secrecy: practical strategies to address emerging cybersecurity paradoxes through professional and stakeholder education and co-management architectures designed to cultivate community-situated, non-technical structures of group synthetic intelligence (aka "Neighborhood Watch"). In: 19th HCI International, Vancouver, Canada (2017)
11. Endicott-Popovsky, B., Frincke, D.: Adding the fourth 'R': a systems approach to solving the hacker's arms race. Paper presented at Hawaii International Conference on System Sciences (HICSS) 39 Symposium: Skilled Human-intelligent Agent Performance: Measurement, Application and Symposium. Kauai, Hawaii, January 2006. http://www.itl.nist.gov/iaui/vvrg/hicss39/4_r_s_rev_3_HICSS_2006.doc
12. Endicott-Popovsky, B., Endicott-Popovsky, B.: The probability of 1. J. Cyber Secur. Inf. Syst. **3**(1), 18–19 (2015)

Using Eye Tracking to Assess the Navigation Efficacy of a Medical Proxy Decision Tool

Soussan Djamasbi[1](✉), Bengisu Tulu[1], Javad Norouzi Nia[1],
Andrew Aberdale[1], Christopher Lee[2], and Susanne Muehlschlegel[2]

[1] Worcester Polytechnic Institute, Worcester, USA
{djamasbi,bengisu,jnorouzinia,aaberdale}@wpi.edu
[2] University of Massachusetts Medical School, Worcester, USA
Christopher.Lee@umassmed.edu,
Susanne.Muehlschlegel@umassmemorial.org

Abstract. Making a life-or-death decision about the best course of action for a loved one is emotionally taxing and cognitively complex. Decision aids can help reduced the burden by providing relevant important information that facilitates decision making while allowing the decision maker to process the information at their own pace. Given the intense cognitive effort required to process complex medical alternatives, it is crucial to provide a decision aid that is easy to understand and easy to use. We designed a digital decision aid based on a validated paper-based decision aid targeting proxies making survival or comfort care decisions for their loved ones suffering traumatic brain injury. In this study, we focused on the effects of navigation design, which plays an essential role in helping users consume the provided content. We used eye tracking to study users' information processing and navigation behavior throughout the entire decision making process. Our results showed that one of our navigation designs reduced the feeling of "lostness" and improved the overall perception of the usability of the tool. The same navigation design improved the information processing behavior by increasing engagement with the system and helped decision makers spend more time processing content rather than searching for the provided information.

Keywords: User experience · Decision making · Decision support tool · Eye tracking · Navigation

1 Introduction

Traumatic Brain Injury (TBI) is one of the leading causes of death and disability in adults in the U.S. [1]. Since severe TBI patients are at best minimally responsive, surrogate decision-makers – as decision makers of the patient – routinely face the life-or-death decision about continuation of care (survival) or comfort measures only (comfort). A loved one of the patient typically serves as the surrogate, and the surrogate is typically under a great deal of emotional stress. Not only is the surrogate tasked with making life-changing and/or life-ending decisions quickly on behalf of a patient who

© Springer Nature Switzerland AG 2019
D. D. Schmorrow and C. M. Fidopiastis (Eds.): HCII 2019, LNAI 11580, pp. 143–152, 2019.
https://doi.org/10.1007/978-3-030-22419-6_11

has been suddenly incapacitated due to TBI, the surrogate may very well lack the medical knowledge to make informed decisions.

To alleviate some of the psychological pain and burden in making such critical decisions on someone else's behalf, surrogates are typically provided with important information during a meeting with the physicians who are responsible for the care of the TBI patient. Given the fast paced healthcare environment, the surrogates do not usually have enough time to absorb the information provided and ask questions that are important for them as they make this critical decision. To help reduce pressure on surrogates and prepare them on their own pace for the meeting with the physician, paper-based decision aids have been developed. [2] These decision-aids provide relevant information to assist surrogates in making an informed decision in cooperation with a physician.

In a recent project, a paper-based decision aid was developed for addressing the needs of surrogates of patients suffering from moderate to severe TBI [3]. We translated this paper-based decision-aid into an web-based digital decision aid to address the needs of surrogates who would like to access the decision aid online and share the decision aid with others who are involved in the decision making process. As part of the design process, we decided to focus on the navigation of the web-based tool to take better of advantage of its interactive ability. Although there are plenty of other factors that impact usability, navigation was chosen for this study because of its control over the content and testing would deliver the greatest impact on utility and presentation in the shortest period of time. The objectives of this project were to determine the navigation design that can: (a) minimize the strain placed on the surrogate decision maker and (b) improve surrogate's ability to find information and to comprehend the information presented in the tool. To achieve these objectives, an experimental eye tracking study was used to gain insight about the possible impact of navigation design on information processing behavior.

2 Methodology

The experiment required the development of two decision aid tools. One was a control site, which was the original online tool. A second alternate site was designed using the same content, but with a hypothetical improvement in the navigation method. Multiple navigation formats were considered for implementation for the alternate site. Macro-Navigation, Static Navigation, and Sub-Menus Navigation were the three draft formats considered. Examples of these formats are shown in Fig. 1. As shown in Fig. 2, the original online tool utilized horizontal, top-justified, static navigation. Ultimately vertical, left-justified, static navigation was selected for the final design of the alternate tool. This design was selected because of its simplicity and because we wanted to encourage users to view the pages in the order they were listed on the navigation bar. Previous eye-tracking research suggested that left justified navigation may help to achieve this design goal [4].

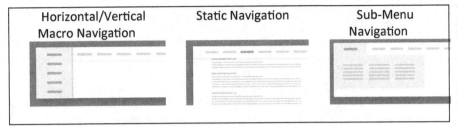

| Horizontal/Vertical Macro Navigation | Static Navigation | Sub-Menu Navigation |

Images Source: https://agentestudio.com/blog/website-navigation-design

Fig. 1. Selected navigation methods

The next step was to design the content of the navigation bar on the alternate tool. This required numerous version improvements by trial and error. The first of these versions derived from the original site, utilizing the same titling of pages and quantity of links. The second version focused improvement on formatting adjustments and the hierarchy of pages. As mentioned before, content of pages was not changed from the original, nor was the order in which pages were presented. Only the way pages were presented within the navigation bar was changed. For example, in the original version of the tool, subsections of a page were not represented on the navigation bar. In the new prototype, each subsection was listed as a separate entry on the navigation bar (Fig. 2). The third and final version expanded on the number of links available, including links to each of the 18 available pages to reduce the number of clicks it takes to switch pages and ultimately maximize the usefulness to the user. Standardizing the format for each link, and keeping a simple navigation bar layout was expected to minimize cognitive strain. Additionally, after moving away from each page a checkmark was displayed next to the item as a visual cue to remind users that the pages with checkmarks were viewed. As mentioned above, a major navigation design goal in this study was to provide a navigation overview for users and to encourage them to view the provided material in the order they were listed on the left navigation bar.

3   Experiment

Sixteen graduate and undergraduate students from Worcester Polytechnic Institute participated in this study. Participants were randomly assigned to the two groups: *control group* (the existing online decision tool with a top navigation bar) and *experimental group* (the alternate online decision tool with the new navigation design). All participants completed the same task, which required them to first read a scenario and then based on that scenario act as a medical proxy for a "loved one" using their randomly assigned tool. The example scenario is displayed in Article A: Test Scenario of the Appendix. Following completion of their assigned tasks users were interviewed to gather information about their experience.

This same task has been used previously to test the efficacy of the online tool.

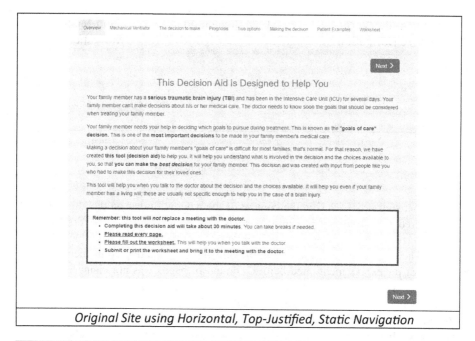

Original Site using Horizontal, Top-Justified, Static Navigation

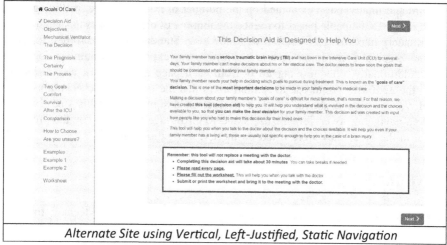

Alternate Site using Vertical, Left-Justified, Static Navigation

Fig. 2. Implemented navigation comparison

4   Results

We used self-reported measures to compare the navigation experience between the two groups in our study [5, 6]. These surveys measured the degree to which participants described feeling "lost" when using the online decision tool ("lostness"). The users' overall reactions to the usability of the tool was measured using a System Usability Scale (SUS).

The result of a two-tailed t-test showed significant differences in "lostness" scores between the two groups (Mean$_{left\ Navbar}$ = 6.5, Mean$_{top\ Navbar}$ = 5.2, t = 4.10, df = 15, p = 0.005, g = 1.9). The effect size for this phenomenon was large (>0.8) as shown by the value of Hedge's g (1.9). These results, which are displayed in Fig. 3, show that participants rated the navigability of the new design significantly better.

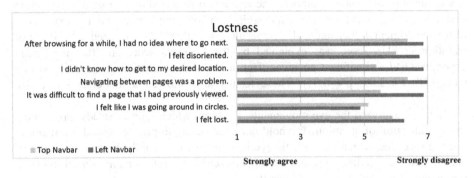

Fig. 3. Scores for lostness items.

SUS uses 10 items to measure a user's overall experience with a technology. These items then are typically converted to a single score ranging from zero to 100. SUS scores between 71–85 represent a good overall subjective rating for usability, scores between 85–90 represent an excellent rating, and scores above 90 represent best imaginable rating [7]. The single SUS score is widely used in industry research to assess the subjective reactions of users to a technological product or service [8].

The results of a t-test did not show significant differences in the single SUS scores between the two groups in our study (Mean$_{left\ Navbar}$ = 80.95, Mean$_{top\ Navbar}$ = 77.50, t = 0.48, df = 15, p = 0.64). Similarly, we did not find significant differences between individual SUS items between the two groups. Figure 4 displays the scores for individual SUS items.

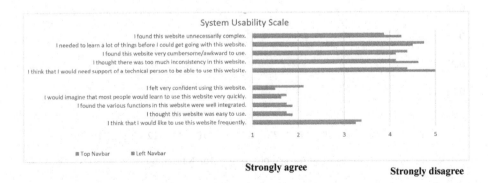

Fig. 4. Scores for SUS

To assess the impact of the navigation on overall usability of the decision tool, we used the following regression model:

$$SUS = b0 + b1 * Lostness \qquad (1)$$

Where SUS represents the single SUS scores and "Lostness" the average of individual items of the lostness survey. The regression results showed that the navigation design had a significant positive relationship with how people rated the user experience of the experimental decision tool. The results show that 31% (adjusted $R^2 = 0.31$) of variation in SUS score was explained by the Lostness score and that navigation experience had a significant positive effect on the overall reaction to the decision tool (B = 9.50, t = 2.89, p = 0.01). These results also showed a moderate effect size ($f^2 = 0.15$).

Next we analyzed the eye movements. Fixation, which refer to steady gazes, are a strong indication of attention. We hold our gaze steady to process visual information. We use saccades, which are ballistic eye movements, to abruptly change the center of our attention. Saccades, represent attempts to search for information. Visual information is not processed during saccades [9].

The analysis of the eye movement data showed that average fixation duration on each page was different between the two groups (Fig. 5). The results of t-test showed that participants in the experimental group (left navigation bar) on average had significantly longer glances (fixation durations) on each page ($Mean_{left\ Navbar} = 0.24$, $Mean_{top\ Navbar} = 0.23$, t = 2.23, df = 34, p = 0.03, g = 0.7). The effect size for this phenomenon was almost large as shown by the value of Hedge's g (0.7).

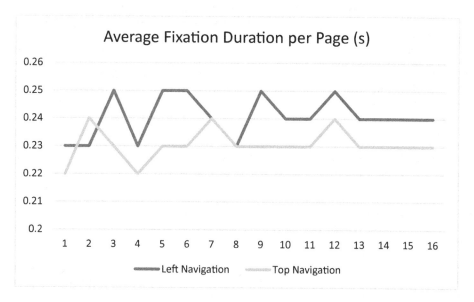

Fig. 5. Comparing average fixation duration for each page on the decision tool

We also looked at the ratio of fixation duration to visit duration on each page (Fig. 6). Fixation duration in this analysis refers to the total duration of fixations on a page. Visit duration refers to the total duration of both fixations and saccades on a page. This metric (fixation-to-visit duration) represents the ratio of information processing to search on each page [9, 10].

The analysis of the eye movement data showed that the fixation-to-visit ratio on each page was different between the two groups. The results of t-test showed that participants in the experimental group (left navigation bar) on average had significantly larger ratios on each page (Mean$_{left\ Navbar}$ = 0.89, Mean$_{top\ Navbar}$ = 0.88, t = 3.26, df = 34, p = 0.0003, d = 0.9). The effect size for this phenomenon was large (Cohen's d = 0.9) indicating that people in the left navigation bar group spent significantly less time searching and more time processing the provided information. Figure 6 visualizes this difference between the two groups for each page in the decision tool.

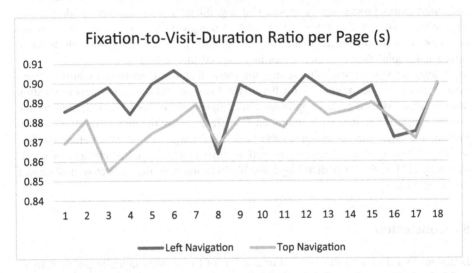

Fig. 6. Comparing the ratio of total fixation total visit duration for each page on the decision tool

4.1 Discussion

In this paper we examined the impact navigation design on perceived lostness and usability of a medical decision tool. The results showed significantly more favorable navigability ratings for the new navigation design. These results showed that the new navigation design significantly improved navigation experience. While we did not find significant differences in SUS scores between the two treatment groups, the results of a regression analysis showed that navigability played a significant positive role in the subjective ratings of the overall usability of the decision tool. This finding is consistent with prior work that shows a strong positive link between navigability and system usability scores [11].

Eye movement supported the above self-report analysis. T-tests showed significant differences between the two treatments in average fixation duration per page and the ratio of information processing to search per page. These results along with the significantly more favorable results of navigability, indicate that people in the new navigation design group were significantly more engaged with the material and spent more time absorbing the information rather than searching for it.

4.2   Limitations and Future Research

As in any experiment, the results of this study are limited by the setting, population, and sample size. This experiment was conducted in a laboratory setting using a student population. Future research can extend the generalizability of these results by conducting field studies using different populations. Similarly, increasing the sample size is likely to provide more insight about the impact of navigation design on user reactions and information processing behavior. Our significant results showed moderate and large effect sizes for the impact of navigation design on user reactions and viewing behavior. This in turn suggests that increasing sample sizes is likely to yield to more significant implications for the factors investigated in our study.

While the system usability scores indicated that the perceived usability of the system was in a good range (78 for the original design and 81 for the new navigation design), they also suggested opportunity for improving the tool. As indicated by the results, subjective ratings were improved from 78 to 81 by the new navigation design, making the SUS ratings closer to the "excellent range" of 85–90. Future studies, refining the navigation experience as well as other changes such as changes in visual hierarchy [12] of the individual pages are likely to improve the subjective usability of the decision tool.

5   Conclusion

Decisions that deal with the survival or comfort of loved ones naturally put their users under significant cognitive and emotional strain. It is then of utmost importance to make these tools as useable as possible to help proxies make critical decisions.

This paper addresses a first step toward improving an online msTBI which helps medical proxies to decide the best course of action for their loved ones. To achieve this goal, we focused on improving the navigation experience of the online decision tool and tested its impact on perceived navigability, and usability as well as information processing behavior. Our results show that the new navigation design had a significant positive impact on subjective reactions as well as objective measures of attention and engagement captured through eye movements. These results confirm the efficacy of the new navigation design and suggest that this design may be particularly effective in developing medical decision tools for proxies. Our results also show the value of eye tracking in user experience design in general and in navigation design in particular. Eye tracking provides a continuous measure of information processing behavior and as such is an invaluable tool for providing a more comprehensive view of user experience of a decision tool.

Appendix

Article A: Test Scenario

Instructions for usability testing of the decision aid

"Helping families make "goals of care" decisions for their family members with traumatic brain injury (TBI)"

Imagine that your loved one is sick in a neurologic intensive care unit (neuroICU) after falling down the stairs and suffering a traumatic brain injury. Your loved one has been on a life support machine (a mechanical ventilator) for over 10 days—much longer than average. He/She is 62 years old and remains very sleepy from the brain injury.

Your loved one is so sick that he/she can't make decisions for him/herself. He/She needs you to help make a decision about what to do next.

The computer program you are about to use is called a **decision aid**. It helps people think through a difficult decision. Please use the decision aid to make your decision for your loved one. We will ask you to think aloud while you are using the decision aid. After you make your decision, we will ask you a few questions about your experience with the decision aid that you used.

---

*Remember, this is an **imaginary situation**. It does not relate directly to any particular patient. We do not record any personal information. This is confidential!

*Remember, we are testing the decision aid, not you. There are no right or wrong decisions or answers. As long as you do your best to read and understand the material to make a decision, your input will be invaluable to us.

Thanks for your time! Your feedback helps us to make better decision aids to help people in very difficult, stressful situations.

References

1. Traumatic Brain Injury: Hope Through Research (2018). Fri, 2018-09-07 15:51; National Institute of Neurological Disorders and Stroke. https://www.ninds.nih.gov/Disorders/Patient-Caregiver-Education/Hope-Through-Research/Traumatic-Brain-Injury-Hope-Through
2. Khan, M.W., Muehlschlegel, S.: Shared decision making in neurocritical care. Neurol. Clin. **35**(4), 825–834 (2017)
3. Muehlschlegel, S., et al.: Derivation of a pilot decision aid for goals-of-care discussions in critically-ill moderate-severe traumatic brain injury patients and interim feasibility trial results. In: Annals of Neurology. Wiley, Hoboken (2018). 07030-5774
4. Djamasbi, Soussan, Siegel, Marisa, Tullis, Tom: Visual hierarchy and viewing behavior: an eye tracking study. In: Jacko, Julie A. (ed.) HCI 2011. LNCS, vol. 6761, pp. 331–340. Springer, Heidelberg (2011). https://doi.org/10.1007/978-3-642-21602-2_36
5. Webster, J., Ahuja, J.: Enhancing the design of web navigation systems: the influence of user disorientation on engagement and performance. MIS Q. **30**(3), 661–678 (2006)

6. Brooke, J.: SUS: a "quick and dirty" usability scale. In: Jordan, P.W., Thomas, B., Weerdmeester, B.A., McClelland, I.L. (eds.) Usability Evaluation in Industry, pp. 189–194 (1996)
7. Bangor, A., Kortum, P., Miller, J.: Determining what individual SUS scores mean: adding an adjective rating scale. J. Usability Stud. **4**(3), 114–123 (2009)
8. Tullis, T.: Measuring the user experience collecting, analyzing, and presenting usability metrics. In: Albert, B. (ed.) Interactive Technologies, 2nd edn. Elsevier, Amsterdam (2013)
9. Djamasbi, S.: Eye tracking and web experience. AIS Trans. Hum.-Comput. Interact. **6**(2), 37–54 (2014)
10. Shojaeizadeh, M., et al.: Detecting task demand via an eye tracking machine learning system. Decis. Support Syst. **116**, 91–101 (2019)
11. Djamasbi, S., et al.: Web Experience and Growth. In: AMCIS, pp. 1–7 (2015)
12. Djamasbi, S., Hall-Phillips, A.: Eye tracking in user experience design. In: Schall, A.J., et al. (eds.) Eye Tracking in User Experience Design. Morgan Kaufmann, Waltham (2014)

Considerations for Human-Machine Teaming in Cybersecurity

Steven R. Gomez, Vincent Mancuso, and Diane Staheli[(✉)]

MIT Lincoln Laboratory, Lexington, MA 02421, USA
{steven.gomez,vincent.mancuso,diane.staheli}@ll.mit.edu

Abstract. Understanding cybersecurity in an environment is uniquely challenging due to highly dynamic and potentially-adversarial activity. At the same time, the stakes are high for performance during these tasks: failures to reason about the environment and make decisions can let attacks go unnoticed or worsen the effects of attacks. Opportunities exist to address these challenges by more tightly integrating computer agents with human operators. In this paper, we consider implications for this integration during three stages that contribute to cyber analysts developing insights and conclusions about their environment: data organization and interaction, toolsmithing and analytic interaction, and human-centered assessment that leads to insights and conclusions. In each area, we discuss current challenges and opportunities for improved human-machine teaming. Finally, we present a roadmap of research goals for advanced human-machine teaming in cybersecurity operations.

Keywords: Cybersecurity · Cyber · HCI · Teaming · Interaction · Sensemaking · Situational awareness · Artificial intelligence

1 Introduction

With ever-increasing reliance on networked information systems, cybersecurity is a critical component of almost every military, government, and private-sector organization. As organizations deploy new technologies for their respective missions, and adversarial capabilities advance, it is clear that the goal of cybersecurity must be to *maintain* a strong defensive posture and effectively resolve incidents, rather than *achieve* some level of security and move onto the next goal. In general, this maintenance process involves mitigating vulnerable systems (including tools, people, and workflows), as well as continually observing

DISTRIBUTION STATEMENT A. Approved for public release. Distribution is unlimited.

This material is based upon work supported under Air Force Contract No. FA8702-15-D-0001. Any opinions, findings, conclusions or recommendations expressed in this material are those of the author(s) and do not necessarily reflect the views of the U.S. Air Force.

D. D. Schmorrow and C. M. Fidopiastis (Eds.): HCII 2019, LNAI 11580, pp. 153–168, 2019.
https://doi.org/10.1007/978-3-030-22419-6_12

and analyzing the environment for activity that might enable—or be evidence of—exploitation of both known and unknown vulnerabilities.

In the event of a security incident, analysts must determine a benign cause or understand the extent of malicious activity, identifying any adversary and their goals, their capabilities, and the intended target effects [4]. The addition of an intangible, logical cyber environment creates new complexities for analysts compared to physical security domains. For example, traditional physical domains have ecological and contextual anchors that play a role in decision-making processes, helping with validation and verification of events and courses of actions [24]; however, cyber environments lack easily observable anchors. Environments can also change rapidly and significantly over time with few physical constraints, and change is typical even under normal conditions. To make matters worse, adversaries may take steps to hide evidence of their activities. As a result, it is difficult to design and deploy any "canary in the coal mine" for security analysts that is reliable, easy to observe and interpret, and suggests a clear follow-up response.

In fact, analysts' insights about a current security posture are primarily guided through interactions with data and mediated by computer systems, magnifying the importance of Human-Computer Interaction (HCI) challenges in this domain. Security analysts often are responsible for tasks that are very cognitively demanding: collecting, analyzing, and interpreting large, dynamic volumes of data to confirm the presence of a threat [11], while at the same time unable to prove an environment is safe with total certainty. Improved coordination between humans and machines is a promising approach for addressing these challenges but is not well understood in the cyber domain, where analyses are highly exploratory and failure to arrive at clear, justifiable conclusions is costly (e.g., failure to halt an ongoing attack).

In this paper, we explore ways to apply or enable Human-Machine Teaming (HMT), where analysts work alongside machines responsible for some duties or sub-tasks traditionally held by humans, for cyber defense. Specifically, we focus on analysis and monitoring practices in cyber defense, rather than the security of individual systems, which may highly specific to individual environments. For simplicity, we consider machine teammates in the form of software agents, or intelligent components within software applications, rather than physical devices (e.g., robots). Our goal is to understand how current challenges that analysts face could significantly benefit from using machines for Intelligence Augmentation (IA) of analysts, or as Artificial Intelligence (AI) systems that interact with analysts after performing tasks autonomously. We explore the following questions:

- What human-centered challenges exist when performing tasks with existing tools and analytics for cyber defense?
- What potential benefits can be gained from improved HMT and IA/AI during these activities?
- What are high-impact research directions that could enable these benefits?

While HMT has been studied in past systems—notably, where humans have supervisory control over unmanned vehicles [5,25,28]—it has not been studied extensively in the cyber domain. One reason could be that past HMT research assumes humans can act as supervisors who can verify and re-vector their machine teammates as needed; however, unlike vehicle control or similar applications, it is non-trivial for an analyst to supervise in a traditional sense—primarily through observation of another's activities—and verify the behavior of an analytic or agent in the cyber domain. As such, we imagine that successful HMT in cyber is as much about effective bilateral communication of complex findings as it is about task delegation or instruction.

In order to understand how HMT can improve cyber defense, and how to get there through novel HCI and security research, we contribute a set of top challenges that occur during different stages of a data-analysis pipeline for security; we outline opportunities for HMT in each stage; and we discuss implications and a research roadmap that will enable these HMT opportunities. We note that the challenges and opportunities identified are not exhaustive, but reflect key areas for improvement based on our observations of defensive cyber operations and analysis activities.

2   Model of Activities for Cybersecurity Sensemaking

In order to understand the state of security analysis challenges and where teaming can help, we consider a simple model of stages of human-initiated activities that support cyber sensemaking. Cybersecurity operations happen within a sociotechnical system, with strong interplay between humans, technology, and data. Roughly speaking, raw data must be collected and organized, then transformed by algorithms and user interfaces; then humans discover and synthesize knowledge and possible narratives that explain the data.

Teams or individuals performing these analysis activities may develop unique practices over time, but some models have been proposed to describe generally what steps are involved in cyber analysis. For example, D'Amico et al. [9] describe these security analysis activities with respect to three stages:

1. *threat detection*, where analysts collect and analyze primary sources of data;
2. *situation assessment*, where analysts bring in more data sources, and convert the analyses into actionable knowledge; and
3. *threat assessment*, where analysts look across incidents, correlating with intelligence, making predictions, and proposing mitigation strategies.

We note that while threat detection may begin with an alert generated automatically by an analytic in the environment, e.g., from an intrusion-detection system (IDS) like Bro/Zeek [1], these tasks are primarily driven by people. In some ways, these tasks mirror steps taken to operationalize data for a particular use case (here, cybersecurity): from raw data to information to knowledge, sometimes using analysis products like visualizations as inputs to later analysis stages [6].

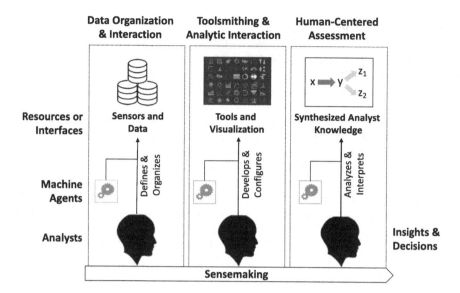

Fig. 1. Descriptive model of cyber sensemaking activities at three levels: data organization and interaction, toolsmithing and analytic interaction, and human-centered assessment.

In the remainder of this paper, we consider a model of cyber sensemaking based on the one by D'Amico et al., but generalized modestly to underscore the types of resources, technologies, and interfaces needed in each stage. Analysts sometimes must pivot across tasks and hypotheses, so have we separated out goals (which can be nested and held in parallel) from information and HCI affordances. In fact, threat detection and assessment activities are highly iterative, so analysts doing threat assessment might discover capabilities of an adversary that cause them to go back to the detection activity. As shown in Fig. 1, our model includes:

1. *data organization and interaction*, where analysts organize cyber data feeds and perform data-wrangling activities like filtering and cleaning;
2. *toolsmithing and analytic interaction*, where analysts use tools like visualizations to interpret information that has been transformed by algorithms or analytics; and
3. *human-centered assessment*, where people work with this information to construct and communicate high-level knowledge about threats or an environment.

At each stage of this model, there are critical human-centered activities that make use of machine interfaces and agents, ranging from graphical user interfaces to alerting tools that run without regular human guidance. We believe many activities can be improved beyond the current state of the art using machines that further augment analysts' performance, enable new analyses, or lighten analysts' workloads. In the following sections, we describe existing challenges and new opportunities in each of these stages.

3   Considerations for Data Organization and Interaction

Data organization is an early stage in the analysis pipeline that is critical for downstream activities, like analysts assessing normal or abnormal conditions in the cyber environment and responding. By "organization", we refer to analysts' ability to gather and structure data in a manner that is suitable for further analysis. This process might also include interaction with data in order to clean, filter, and otherwise prepare them for analysis. While some tools may be used to interact with raw data at this stage, we distinguish those from tools developed for analysts with the goal of extracting actionable information from the data, which is discussed in Sect. 4.

Typical examples of raw data include event feeds from Security Information and Event Management (SIEM) tools and databases; records of network flows; hardware and software inventories on endhosts; and directory services for users, among others. In order to support effective sensemaking about the security of the environment, the data must capture the most relevant information about potential threats, and the data feeds themselves must be protected from compromise that could poison downstream analysis.

Understanding what data must be collected and ensuring that feeds themselves are operational (sometimes using simple analytics or monitoring tools) are important ongoing security tasks for cyber defenders. Due to the sensitive nature of data collected, another consideration is that necessary data must be available to downstream analytics and analysts, and that proper data hygiene, like archiving and protecting confidentiality, is maintained.

3.1   Challenges

Collecting High-Integrity Data. Planning for effective collection and maintaining data feeds is cognitively intensive. Along with structured information, people form mental models of operations in the environment, as well as the needs or mission of the environment, in order to plan for and understand sensors. But this model might be incomplete or become inconsistent with changes over time. Unnecessary or incorrectly-configured sensors can result in an overwhelming amount of data that is costly to manage or impacts operations—for example, by disrupting users or reducing the performance of systems like networks or endhosts. At the same time, blind spots in the network might also form and result in incomplete analyses or, worse, lead to misguided conclusions (i.e., incorrectly "clearing" a potential attack vector that remains vulnerable or exploited).

Maintaining high awareness of operations in order to address sensor issues cannot easily be solved with off-the-shelf solutions. Resources to acquire (or develop) and deploy sensors that blanket the environment—and to manage the potential deluge of data from them—can be prohibitively expensive or result in low signal-to-noise for later analysis. Stakeholders for an environment must make choices about what to observe and when, using previous knowledge and intuition. This kind of task is important enough that it was featured in the 2016 VAST Challenge [32] and included determining which one of several data streams to enable mid-way through the exercise, given observations about the environment in the time leading up to the choice. It is not obvious how to construct automatic decision systems for these cyber choices.

Organizing Information. Cleaning and organizing raw data into useful information schemas is a critical part of the early stages of data analysis, and is both time consuming and demands advanced expertise. In some cases, this "data wrangling" can account for 80% or more of analysts' time [19] and cannot easily be outsourced, as it requires both domain knowledge and technical proficiency akin to programming [16]. Even aligning different time-series data, which might appear to be an easy task, can be challenging due to how different cyber data types are reported. For example, vulnerability reports are snapshots at a set point in time, and streaming data sources operate in real-time. Other activities require understanding the meaning and utility of data later in its lifecycle; for example, data might be removed due to resource requirements or hygiene practices, and one must decide what is safe to purge without impacting ongoing or future analyses of security incidents. Stakeholders for these data usually rely on automated approaches using simple heuristics (e.g., log rotation) due to the volume and rate at which new data are collected, even if there is some chance an old record might be needed at a later time. Furthermore, conventional ways of indexing records by time make it difficult to understand potential relationships of interest between aged-out data and preserved records.

Managing Access. Protecting confidentiality of data is challenging. Providing too much access to data can threaten security broadly, while providing too little access can prevent human defenders or analytics from observing and addressing potential security issues elsewhere in the environment. Even methods that aggregate data from multiple source might reveal sensitive information or allow it to be inferred. Current solutions to these problems usually involve both people, policy, and automated systems, where usability at the interface is a critical concern.

Ensuring that data is available only to those who are authorized and need it typically involves using access-control mechanisms that rely on curated rules that map user roles or attributes to needed resources. Methods for authoring rules that are usable by human operators have been studied in prior work [3,20], but less understood is how best to inform rule curators about access needs in the network as they evolve over time. Users requesting additional access may not understand the access-control system well enough to clearly state their needs. Others who no longer need access to resources may never proactively request removing this access if keeping it does not hinder their new objectives. Finally, in cases where access control is not enforced but sensors can assess risk or detect violations of confidentiality (e.g., identifying ongoing or completed exfiltration from data stores), presenting actionable and timely information can be difficult: alerts can overwhelm analysts or provide too little precision to be useful.

3.2 Opportunities for HMT

Automatic Assessment of the Data Platform. Machines are particularly well-suited for workloads that involve monitoring changes in data volume or velocity, so there may be opportunities to use machines to identify where additional sensing might be desired based on statistical patterns—even if understanding the nature or intent of these changes must be determined by human teammates. There is a need for visual analytics that present well-justified recommendations for data handling and provide what-if analysis capabilities for these recommendations. Implementing changes to sensors is another area where HMT is important, because the autonomy of a machine for deploying sensors could be limited by its physical interface. For example, installing new sensing applications into a network controller for a software-defined network is currently feasible by a software agent, but installing physical proximity card readers would require human assistance or robotics.

Regular testing of sensors and the data platform in the environment, which might otherwise be tedious or difficult to repeat without human error, could be performed routinely by a machine teammate. An example might be regularly initiating events in the sensed environment, like network flows, that are expected to be detected and recorded in data storage, then verifying these records as expected. The machine could escalate alerts about unexpected behaviors to stakeholders quickly, or maintain a digest of normal test outcomes in order to avoid alerting human teams without any required action. This process parallels current approaches for automated testing and building of software tools, among others; however, determining that an outcome is normal in the environment is likely to be more involved than running unit or integration tests in controlled test environments.

Shared Representations of Mental Models and Goals. Research toward developing comprehensive sets of structured data types, tasks, and goals for cyber analysis could facilitate closer interaction between machines and humans, for whom externalizing mentals models is typically very difficult. Heer notes that shared representations let both parties "contribute to, and adaptively learn from, solutions to shared problems" [16]. Enabling the analyst to more easily verify the representations a machine is working with also builds trust. We believe these representations could also help provide domain-specific ways in which data feeds might be organized—for example, to more automatically associate incoming observations or removed data with ongoing analysis cases.

Smart Data-Access Monitoring and Control. Another opportunity exists to leverage machines for fine-grained access-control maintenance. Regularly revisiting and verifying resource needs for users or agents (e.g., through interactive confirmation) in an environment can be tedious or error-prone for human operators if permissions are more fine-grained and change periodically, but machines could perform this ably. Furthermore, machine teammates could integrate analytics that enable these interactions to be more targeted, like cross-referencing a user's current access authorizations against their actual use based on observations in the environment. This use of HMT might free operators in charge of access control from performing maintenance tasks (e.g., or enable this practice) and providing more time for complex access-control strategies, like policy development or network management and segmentation.

4 Considerations for Toolsmithing and Analytic Interaction

The next level in our sensemaking pipeline involves people using interactive, analytic applications like visualizations to analyze the cyber data, as well as the developers who produce these tools. For some types of analysis, and for analysts with programming or scripting proficiency, the same person may assume both roles. In other cases, developers must understand enough about the available data and how analysts might use it to design effective tools.

Machines play a role here primarily as tools that transform and present data to human analysts. These tools act more as "teammates" in HMT as they perform tasks that go beyond what is precisely directed by their user. Some machines may initiate interactions with humans independently, for example by identifying patterns that would be difficult for an analyst to notice and escalating alerts for people to explore and verify. Other machines may be purely reactive, running analytic routines on data after analyst-driven interactions (e.g., visual analytics). In both cases, machines must be trusted by their teammates and communicate or display information effectively; otherwise, the added work or liability—if findings are not reliable—of using these machines threaten their long-term adoption by analysts.

4.1 Challenges

Trusting Integrity of Analytic Tools. Tools that operate with integrity in this stage transform and present information to the analyst in a way that does not cause a misleading interpretation about what is happening in the environment. Buggy implementations of tools threaten integrity, as do poor interface designs or visual encodings of data. Information displays must be legible so analysts can decode the information, interpret it, and integrate it into a larger narrative.

The ability to trust that a tool has integrity is critical, but it is often impossible or impractical to verify a tool's correctness based on its source code. Spot-checking that an analytic produces the expected output is difficult in practice (outside of testing) where ground truth is expensive to learn and might require additional tools that must be trusted themselves. As a result, analysts use tools mindful of the fact that they could be misleading. In fact, visualizations can be misleading, unintentionally or even deliberately, by using encoding methods that subvert an analyst's ability to draw conclusions from the data [29]. In cases where the encoded data has high volume or velocity, as in streaming analytics, an analyst can also be misled if the tool minimizes or hides relevant information before it is decoded by the analyst.

Anticipating Effective Tool Designs for Humans. Tool developers must understand enough about human capabilities to build tools that are legible and interpretable. This means designing tools that work effectively in consideration of perceptual and cognitive factors that could affect an analyst's ability to decode information from the interface, which could be text-based, a visualization, or use another output modality like sound.

Some design techniques have been developing as a work-around for perceptual limitations. For example, in many use cases of visual analytics, the number of events in a dataset exceeds the number of pixels available to encode the data. (In cybersecurity, sensors may collect data over months or years that could be relevant to a single incident, such as a sophisticated network intrusion.) Existing visualization approaches for handling this volume on-screen include Focus+Context techniques [7, 21], which combine an overview containing broad context with user-directed exploration for fine-grained information, and interactive views that support Shneiderman's mantra of "overview first, zoom and filter, then details-on-demand" [30]. In cases where a subset of data is presented, it is also important that a tool is clear about what data are excluded.

Model-driven visualization design—by modeling tasks, humans, and data, and making decisions based on simulations of user performance—is a compelling idea to account for human factors during toolsmithing. However, there are few time-tested human performance models that are mature enough to guide design decisions. Principles like Fitts' Law [22] and design-evaluation tools that utilize cognitive modeling (e.g., CogTool [18]) can guide simple UI design decisions (e.g., optimizing mark size or position), but generally modeling visualization effectiveness is notoriously difficult, especially for exploratory data analysis (EDA) tasks [13]. As a result, design practices often rely on gaining intuition about the application area (i.e., cybersecurity) and iterating with expert users to refine tools, which can be time consuming if done with proper rigor.

Designing for Partial Analysis and Knowledge Transfer. During security operations, analysts often must hand off findings to another person (e.g., during shift changes) for continued exploration and as context for future events. One obstacle is the difficulty of communicating one's mental model of a situation or environment. Tools are needed that go beyond exploratory data analysis (EDA) and help analysts compose narratives that include hypotheses and findings, estimates of uncertainty, and an accounting of what data was analyzed or not. Maintaining rich histories of these analysis records could pose technical challenges. Partial analysis products may need to be compressed or updated when later information is available.

4.2   Opportunities for HMT

Provenance Tracking in Visual Analytics. We believe an important step for HMT is designing mechanisms for analytic provenance in order to support trust in machine teammates and their products. Analytic provenance is traditionally about understanding through interactions with analysis tools how humans arrive at insights [26]. This is very important in order for analysts to establish trust in others findings. Machine teammates must also endeavor to provide evidence that their analyses have been executed in correct and justified ways, in order to get buy-in from human teammates or supervisors about conclusions or recommendations. Ways of building provenance tracking into tools and algorithms to use representations of provenance (i.e., usually large graphs) are rich areas to explore, especially since complex analytics might involve machine learning or other approaches that are difficult to explain on a step-by-step basis. Make machine learning "explainable" to the analysts who depend on them in HMT—not just model developers and toolsmiths, as recent work has focused on (e.g., [35])—is an important future goal. Part of communicating provenance also includes effectively describing uncertainty in the analysis [36], which is an ongoing research area in information visualization.

Living Notebook and Narrative Visualization. Building on provenance, there is an opportunity to use visual analytics that capture both human inputs and findings by machine analytics into a narrative that can adapt over time. This would support ongoing analyses and knowledge transfer between teammates with less context about previous events. New tools like "living notebooks" [8] that evolve over time are for potential method handling streaming data, which must be quickly integrated into existing cases or analyses. Annotations and feedback provided by humans could be used to refine the narrative, while machines could learn from this feedback to better handle future data.

User-Performance Modeling for Cyber Tool Design. Modeling how well a visualization or other tool might support an analyst's cyber task could supplement existing ways for designing effective tools in this domain, which include design studies (see [23] for examples) that are valuable but expensive to perform. As we mentioned earlier, modeling tools can be used to get fast, quantitative predictions on performance indicators like task speed. However, effectively modeling cyber tasks requires more research because they tend to encompass both routine interactions (like pulling up and searching logs) that are straight-forward to model, as well as exploratory or less-structured brainstorming tasks. Reusable modules that instrument user interfaces, both for downstream model fitting and other performance monitoring, would be useful for visual analytics and other HMT interfaces.

5   Considerations for Human-Centered Assessment

While advanced technologies may be responsible for collecting, reducing, and processing data during initial analyses, human analysts are the primary drivers

of cybersecurity understanding and sensemaking today. They interact with analytics and other analysts to produce information and knowledge for the purpose
of situational awareness and decision making in the organization. To be effective at this stage, analysts must be able to create linkages between information
about the network, the world, and their team [14]. This information must be
correlated with external information on emergent threats and threat actors, as
well as known attack signatures. Finally, analysts must also be able to fuse and
share this information with their local and broad organizational teams to create
a holistic picture of security across the organization [31].

5.1   Challenges

Multi-source Information Fusion. At the individual level, analysts must correlate information between multiple sources to produce knowledge, and communicate this knowledge to their superiors. As information and knowledge are passed
up through the organizational hierarchy, findings from multiple analysts must
be translated from discoveries to insights, and eventually into a broader picture. This process of information fusion lets analysts achieve improved accuracy
and understanding, compared to looking at an individual source of information.
The fusion occurs over five levels: data assessment, object assessment, situation
assessment, impact assessment, and process refinement [34]. At the analyst level,
this is often discussed as "hard" information fusion, in which the focus of data is
from hard sensors collecting objective information. As information is moved up
the organization, the fusion moves to "hard/soft" where the hard information is
fused with subjective information, which might be more uncertain, inaccurate,
or subjective [15]. At all levels, information fusion is a cognitively-demanding
task that requires memory, merging and conflict resolution, and de-confliction
to ensure that final conclusions are accurate and actionable.

Information Sharing Across Organizational Structure. As analysts process information and reveal incidents or other status indicators, they are responsible for
communicating this information up the chain for the purpose of awareness and
decision making. Before doing so, the analyst must make a judgement call of
whether or not a piece of information should be shared. If the analyst shares too
much, she may cause information overload to her superiors; on the other hand,
if she does not share enough, this degrades the situational awareness of the organization. The decision can be stressful or cognitively taxing. Research has shown
that humans are more likely to share commonly-known information, while high-
value, unique information they possess is not communicated [17,33]. At each
level of the organizational hierarchy, information and knowledge is further dis-
tilled, fused with other information, and summarized. Where an analyst may be
responsible for assessing an individual security event, his supervisor will have to
understand the interdependencies across multiple events, look for patterns, and
understand how to allocate resources.

Performance Measurement. Individual and organizational behavior requires monitoring and self-regulation of their actions, in order to adjust for emergent threats or to improve their overall performance. Regulation of behavior based on performance is a meta-cognitive task, in which an analyst must monitor her own cognitive behavior for task-specific knowledge, her understanding of that knowledge, and her affective responses to the activity [10]. Understanding one's performance and competencies in a particular area is a critical element in enabling trust and team dynamics [2], and is useful in assessing performance of individuals and teams. Additionally, without such information, supervisors cannot correctly allocate resources or balance the workload across their teams, which could help increase team performance [12].

5.2   Opportunities for HMT

Intelligent Information and Context Fusion. Information fusion requires pulling and aggregating findings from multiple sources, both hard (e.g., data) and soft (e.g., analyst reports) to form a summative understanding of the broader organizational picture. With tools that help create linkages between the information an analyst receives, the sources of the data and their trustworthiness and constraints, analysts can build more context around the information they are provided. Current workflows for building context like this can be ad hoc and use many tools. Unified interfaces that help synthesize and share knowledge and hypotheses between team members could lead to more systematic or streamlined analyses.

We previously discussed the difficulty in sharing information across analysts and the organizational hierarchy. Without context (like threat or analysis priorities that are communicated top-down by decision makers), it is difficult to know what information to share upwards; but without more information, it can be difficult to refine or understand some contexts. Natural language processing could help this issue by helping making it easier for analysts to construct context. As mentioned earlier, tracking sources of data and analytic provenance can help a person receiving synthesized information to learn how it was generated. This could help reduce potential data overload, allowing analysts to better understand and communicate their needs, and ensuring that information that needs to be communicated and shared.

Performance Monitoring Capabilities. Current research in neuroergonomics and physio-behavioral monitoring is making significant advances in developing metrics of fatigue, stress, and other state-based metrics that are linked to human interaction and performance while using technology [27]. Machines can use this information to augment a supervisor's intelligence and assist, or automate, tasks like load scheduling, resource allocation, and workload balancing across the team. Additionally, performance-measurement outcomes may be used to communicate information about analysts objectively up the chain; this can aid a supervisor in composing teams and allocating training. Similarly, there is potential to use HMT in situations where machines passively observe individual differences and strengths, and provide suggestions to leadership for how best to deploy teams.

Table 1. Research roadmap for improved HMT in cybersecurity

When	Research goal	D	T	HCA
Near	Methods for guiding data collection; curation assistance for those with little or no developer experience	✓	.	.
	Task modeling and representation for cyber defense operations	✓	✓	.
	Organizational Knowledge Management capabilities for intelligent information sharing	.	.	✓
	Explainable machine learning (ML) for analytic developers	.	✓	.
Mid	Improved visualization and analytics that provide distilled narratives of multi-dimensional change over time	.	✓	✓
	Accurate models for human performance in cyber defense	.	✓	✓
	Human cognitive and affective state detection	.	.	✓
	Tools for tracking and communicating analytic provenance	.	✓	.
	Natural language understanding for precise analysis tasks and wrangling data	✓	.	✓
	Support for externalizing and sharing mental models of an environment and analysis goals	✓	.	✓
Far	Explainable ML available for analysts using ML-based analytics	.	✓	✓
	Human cognitive augmentation for performance improvement	.	.	✓
	Tools for operations that adapt to individual needs and team composition	.	✓	.

6   Roadmap

The purpose of this paper was to present observations about the current state of cyber sensemaking activities, their associated challenges, and suggest opportunities for HMT in this domain. We believe the security and HCI communities can advance toward these opportunities by pursuing a research agenda at that intersection. In Table 1, we summarize some objectives in line with this agenda. This table is not meant to be complete or the product of a rigorous research-space analysis; rather, it describes some milestones related to the challenges in this paper that we think are achievable within the near (2–5 years), mid (5–10 years), and far (10–20 years) time frames. Each direction corresponds to one or more of the sensemaking stages discussed earlier: data organization and interaction (D), toolsmithing and analytic interaction (T), and human-centered assessment (HCA).

Future work in this area should not simply focus on the development of novel tools and technologies; instead we urge researchers to take a problem-based approach to addressing challenges in cyber sensemaking and analysis. Our intuition is that this will involve more closely-integrated HMT, so we can allow humans to focus on tasks that leverage their strengths and improve their decision making. New capabilities can help support provenance, correlation, and communication across the different layers of sensemaking—enabling effective and rapid pivoting from each phase and supporting the analysis missions for which security analysts are responsible: threat detection, situation assessment and threat assessment. As more of this roadmap is achieved, it is critical for researchers to maintain awareness of existing and emerging challenges, ensuring that we leverage technologies like AI/IA and HMT in an effective and strategic manner.

7   Conclusion

In this paper, we considered current challenges involved in human-centered aspects of cybersecurity operations, focusing primarily on difficulties in analyzing and communicating findings about complex cyber environments. Many of these challenges result from information management and sensemaking of highly dynamic, multi-dimensional data. These activities traditionally have been driven by humans in the cybersecurity domain, where verifiably-complete understanding of an environment or incident is difficult or impossible to achieve; as such, it is critical to have clear and justifiable partial findings, which is beyond existing capabilities of autonomous intelligent agents. Other challenges related to human factors arise due to the fast-changing and cognitively-demanding work of security analysts.

To address the challenges, we identified opportunities for improved interactions and teaming between security analysts and machines. These opportunities exist in each of three stages of a cybersecurity-analysis pipeline model, including: (1) data organization and interaction, (2) toolsmithing and analytic interaction and (3) human-centered assessment at the level of individuals up through groups and higher-level stakeholders in an organization. Many of these opportunities must be enabled by new research directions in the security and HCI fields. Based on this, we outlined several priorities for researchers.

References

1. The Zeek Network Security Monitor. https://www.zeek.org/
2. Austin, J.R.: Transactive memory in organizational groups: the effects of content, consensus, specialization, and accuracy on group performance. J. Appl. Psychol. **88**(5), 866–878 (2003)
3. Beckerle, M., Martucci, L.A.: Formal definitions for usable access control rule sets from goals to metrics. In: Proceedings of the Ninth Symposium on Usable Privacy and Security - SOUPS 2013, p. 1. ACM Press, Newcastle (2013). http://dl.acm.org/citation.cfm?doid=2501604.2501606
4. Caltagirone, S., Pendergast, A., Betz, C.: The diamond model of intrusion analysis. Technical report, Center For Cyber Intelligence Analysis And Threat Research Hanover MD, July 2013. https://apps.dtic.mil/docs/citations/ADA586960
5. Chen, J.Y.C., Barnes, M.J.: Human-agent teaming for multirobot control: a review of human factors issues. IEEE Trans. Hum.-Mach. Syst. **44**(1), 13–29 (2014)

6. Chen, M., et al.: Data, information, and knowledge in visualization. IEEE Comput. Graph. Appl. **29**(1), 12–19 (2009)
7. Cockburn, A., Karlson, A., Bederson, B.B.: A review of overview+detail, zooming, and focus+context interfaces. ACM Comput. Surv. **41**(1), 2:1–2:31 (2009). https:// doi.org/10.1145/1456650.1456652
8. Cook, K.A., Burtner, E.R., Kritzstein, B.P., Brisbois, B.R., Mitson, A.E.: Streaming visual analytics workshop report. Technical report PNNL-25266, 1417447, March 2016. http://www.osti.gov/servlets/purl/1417447/
9. D'Amico, A., Whitley, K., Tesone, D., O'Brien, B., Roth, E.: Achieving cyber defense situational awareness: a cognitive task analysis of information assurance analysts. In: Proceedings of the Human Factors and Ergonomics Society Annual Meeting 49(3), 229–233, September 2005. https://doi.org/10.1177/ 154193120504900304
10. Efklides, A.: Metacognition and affect: what can metacognitive experiences tell us about the learning process? Educ. Res. Rev. **1**(1), 3–14 (2006). http://www.sciencedirect.com/science/article/pii/S1747938X06000029
11. Fink, G.A., North, C.L., Endert, A., Rose, S.: Visualizing cyber security: usable workspaces. In: 2009 6th International Workshop on Visualization for Cyber Security, pp. 45–56, October 2009
12. Funke, G.J., Knott, B.A., Salas, E., Pavlas, D., Strang, A.J.: Conceptualization and measurement of team workload: a critical need. Hum. Factors **54**(1), 36–51 (2012). https://doi.org/10.1177/0018720811427901
13. Gomez, S., Laidlaw, D.: Modeling task performance for a crowd of users from interaction histories, pp. 2465–2468. ACM, May 2012. http://dl.acm.org/citation. cfm?id=2207676.2208412
14. Gutzwiller, R.S., Hunt, S.M., Lange, D.S.: A task analysis toward characterizing cyber-cognitive situation awareness (CCSA) in cyber defense analysts. In: 2016 IEEE International Multi-disciplinary Conference on Cognitive Methods in Situation Awareness and Decision Support (CogSIMA), pp. 14–20, March 2016
15. Hall, D.L., McNeese, M., Llinas, J., Mullen, T.: A framework for dynamic hard/soft fusion. In: 2008 11th International Conference on Information Fusion, pp. 1–8, June 2008
16. Heer, J.: Agency plus automation: designing artificial intelligence into interactive systems. Proc. Natl. Acad. Sci. **116**, 1844–1850 (2019). https://www.pnas.org/content/early/2019/01/29/1807184115
17. Jefferson, T., Ferzandi, L., McNeese, M.: Impact of hidden profiles on distributed cognition in spatially distributed decision-making teams. In: Proceedings of the Human Factors and Ergonomics Society Annual Meeting, vol. 48, no. 3, pp. 649–652, September 2004. https://doi.org/10.1177/154193120404800380
18. John, B.E., Prevas, K., Salvucci, D.D., Koedinger, K.: Predictive human performance modeling made easy. In: Proceedings of the SIGCHI Conference on Human Factors in Computing Systems, CHI 2004, pp. 455–462. ACM, New York (2004). https://doi.org/10.1145/985692.985750, Event-Place: Vienna, Austria
19. Kandel, S., Paepcke, A., Hellerstein, J.M., Heer, J.: Enterprise data analysis and visualization: an interview study. IEEE Trans. Vis. Comput. Graph. **18**(12), 2917–2926 (2012)
20. Krishnan, V., Tripunitara, M.V., Chik, K., Bergstrom, T.: Relating declarative semantics and usability in access control. In: Proceedings of the Eighth Symposium on Usable Privacy and Security - SOUPS 2012, p. 1. ACM Press, Washington, D.C. (2012). http://dl.acm.org/citation.cfm?doid=2335356.2335375

21. Lamping, J., Rao, R., Pirolli, P.: A focus+context technique based on hyperbolic geometry for visualizing large hierarchies. In: Proceedings of the SIGCHI Conference on Human Factors in Computing Systems, CHI 1995, pp. 401–408. ACM Press/Addison-Wesley Publishing Co., New York (1995). https://doi.org/10.1145/223904.223956

22. MacKenzie, I.S., Buxton, W.: Extending Fitts' law to two-dimensional tasks. In: Proceedings of the SIGCHI Conference on Human Factors in Computing Systems, CHI 1992, pp. 219–226. ACM, New York (1992). https://doi.org/10.1145/142750.142794, Event-Place: Monterey, California, USA

23. Mckenna, S., Staheli, D., Meyer, M.: Unlocking user-centered design methods for building cyber security visualizations. In: 2015 IEEE Symposium on Visualization for Cyber Security (VizSec), pp. 1–8, October 2015

24. McNeese, M.D.: How video informs cognitive systems engineering: making experience count. Cogn., Technol. Work. **6**(3), 186–196 (2004). https://doi.org/10.1007/s10111-004-0160-4

25. Mouloua, M., Gilson, R., Kring, J., Hancock, P.: Workload, situation awareness, and teaming issues for UAV/UCAV operations. In: Proceedings of the Human Factors and Ergonomics Society Annual Meeting, vol. 45, no. 2, pp. 162–165, October 2001. https://doi.org/10.1177/154193120104500235

26. North, C., et al.: Analytic provenance: process+interaction+insight. In: CHI 2011 Extended Abstracts on Human Factors in Computing Systems, CHI EA 2011, pp. 33–36. ACM, New York (2011). https://doi.org/10.1145/1979742.1979570

27. Parasuraman, R.: Neuroergonomics: research and practice. Theor. Issues Ergon. Sci. **4**(1–2), 5–20 (2003)

28. Parasuraman, R., Barnes, M., Cosenzo, K., Mulgund, S.: Adaptive automation for human-robot teaming in future command and control systems. Technical report, Army Research Lab, Human Research and Engineering Directorate, January 2007. https://apps.dtic.mil/docs/citations/ADA503770

29. Rogowitz, B.E., Treinish, L.A., Bryson, S.: How not to lie with visualization. Comput. Phys. **10**(3), 268–273 (1996). https://doi.org/10.1063/1.4822401

30. Shneiderman, B.: The eyes have it: a task by data type taxonomy for information visualizations. In: Proceedings of the 1996 IEEE Symposium on Visual Languages, VL 1996, p. 336. IEEE Computer Society, Washington, DC (1996). http://dl.acm.org/citation.cfm?id=832277.834354

31. Staheli, D., et al.: Collaborative data analysis and discovery for cyber security. In: Symposium on Usable Privacy and Security (SOUPS) Workshop on Security Information Workers (2016)

32. Staheli, D., et al.: VAST challenge 2016: streaming visual analytics. Technical report, MIT Lincoln Laboratory Lexington United States, October 2016. https://apps.dtic.mil/docs/citations/AD1033423

33. Stasser, G., Stewart, D.: Discovery of hidden profiles by decision-making groups: solving a problem versus making a judgment. J. Pers. Soc. Psychol. **63**(3), 426–434 (1992)

34. Steinberg, A.N., Bowman, C.L.: Revisions to the JDL data fusion model. In: Handbook of Multisensor Data Fusion, pp. 65–88. CRC Press (2008)

35. Wongsuphasawat, K., et al.: Visualizing dataflow graphs of deep learning models in TensorFlow. IEEE Trans. Vis. Comput. Graph. **24**(1), 1–12 (2018)

36. Xu, K., Attfield, S., Jankun-Kelly, T.J., Wheat, A., Nguyen, P.H., Selvaraj, N.: Analytic provenance for sensemaking: a research agenda. IEEE Comput. Graph. Appl. **35**(3), 56–64 (2015)

Do We Need "Teaming" to Team with a Machine?

Craig Haimson[(⊠)], Celeste Lyn Paul, Sarah Joseph, Randall Rohrer,
and Bohdan Nebesh

U.S. Department of Defense, Washington, D.C., USA
{crhaims, clpaul, skjosep, rmrohre,
banebes}@tycho.ncsc.mil

Abstract. What does it mean for humans and machines to work together effectively on complex analytic tasks? Is human teaming the right analogue for this kind of human-machine interaction? In this paper, we consider behaviors that would allow next-generation machine analytic assistants (MAAs) to provide context-sensitive, proactive support for human analytic work – e.g., awareness and understanding of a user's current goals and activities, the ability to generate flexible responses to abstractly-formulated needs, and the capacity to learn from and adapt to changing circumstances. We suggest these behaviors will require processes of coordination and communication that are similar to but at least partially distinguishable from those observed in human teams. We also caution against over-reliance on human teaming constructs and instead advocate for research that clarifies the functions these processes serve in enabling joint activity and determines the best way to execute them in specific contexts.

Keywords: Human-machine symbiosis and Human-machine interface ·
Human-machine teaming

1 Introduction

The more sophisticated a system's ability to perform complex tasks in coordination with its users, the more it seems to function as something more akin to a human work partner than a mere tool [1]. Such collaboration between humans and technology is often referred to as human-machine teaming (HMT) (e.g., [2, 3]) because of its resemblance to human teamwork. HMT has been heralded as the key to transforming automation-enabled work practices across a number of domains, many of which are of critical importance to national defense [e.g., 4]. But just how important are human teamwork behaviors to HMT? Is a more human-like machine teammate necessarily a better machine teammate? This paper focuses on the potential role of HMT in one domain – intelligence analysis – and explores the extent to which human teamwork is an appropriate model for these HMT use cases.

Intelligence analysis comprises a set of interrelated activities that generate evidence-based information products from collected information, often with the goal of answering critical questions about adversaries' attributes, associations, beliefs, intentions, and actions. Many of these activities depend upon a human analyst's ability to

D. D. Schmorrow and C. M. Fidopiastis (Eds.): HCII 2019, LNAI 11580, pp. 169–178, 2019.
https://doi.org/10.1007/978-3-030-22419-6_13

find and fuse information acquired across multiple heterogeneous datasets. With the volume, velocity, and variety of data constantly growing, intelligence professionals require increasingly sophisticated tools to enable them to keep pace with fast moving events and the signatures these events generate. Advances in machine learning are driving the development of new analytic technologies capable of recognizing and responding to meaningful patterns across these datasets; however, to maximize their utility to analysts, these technologies must be managed by intelligent software agents that can deploy analytics in a coordinated fashion on behalf of human analysts who constrain, shape, and consume the consolidated results of their activities. We refer to these software agents as machine analytic assistants (MAAs) and envision that they will facilitate intelligence analysis by collaborating with their human partners on shared analytic projects.

HMT research often draws on concepts from human teaming to inform and ground HMT principles [e.g., 1, 3] since there is a wealth of research exploring effective human teaming processes. We will discuss several such processes and suggest how they may facilitate joint analytic work by human analysts and MAAs. However, we will also identify some ways in which human-machine teams may differ from their all-human counterparts, at least for the intelligence analysis use cases with which our research is concerned. Based on this assessment, we suggest that HMT research should not set itself the task of fully emulating human teamwork but should instead just focus on determining the functions and interaction designs that maximize the overall efficiency and effectiveness of joint human-machine work.

2 Envisioned Characteristics of MAAs

Intelligence analysis typically entails multiple iterative tasks involving searching, filtering, evaluating, and fusing information contained within a large number of sources, driven initially by broad exploratory goals that evolve into more focused, hypothesis-driven objectives [5]. Analysts may need to use a variety of tools and methods to find relevant information, expose associations between entities of interest, and preserve key results for further analysis or reporting. Moreover, they must often perform many of these tasks manually, which can severely limit the quantity of data they can consider and amount of information they can extract and synthesize.

MAAs will support the intelligence process by coordinating activities of data analytics that help human analysts find and combine information in ways that satisfy intelligence requirements, similar to software agents that support other forms of exploratory data analysis [e.g., 6]. Machine analytics operating on text [e.g., 7], images [e.g., 8], or other media can classify and cluster data, identify important concepts and relationships, identify anomalies, and reveal and quantify key trends. MAAs will serve as intelligent gateways to these powerful capabilities, assembling and executing multi-analytic workflows to generate summarized findings that meet analyst needs, both by responding to analysts' explicitly expressed requests and also supplying additional recommendations based on knowledge of analysts' mission goals and analytic history. In these ways, MAAs will help analysts find and organize data for efficient and effective review and assessment. MAAs will expand their repertoire of behaviors by learning new workflows and conditions of use, either through passive observation of users' actions or active participation in demonstration/training sessions with users [e.g., as in 9, 10].

Note that although MAAs will utilize logical reasoning to support data-driven inference and other logical functions in pursuit of these activities, they are unlikely to possess a level of knowledge or cognitive sophistication required for more than rudimentary analysis and reporting tasks; moreover, the nuances of the legal policies that govern the conduct of intelligence analysts [e.g., 11–14] are too context-dependent, and the potentially disastrous consequences of automation failure are too severe to allow for anything beyond this in the foreseeable future, whatever the degree of artificial intelligence achieved. Thus, MAAs will not replace human analysts but will instead assume their more time-consuming and laborious information retrieval and manipulation tasks, freeing analysts to devote more time to interpretation (although MAAs could potentially also help structure the analysis process itself, as in [15]).

Achieving these modest yet still ambitious goals will require advances in software agent technologies that afford MAAs the ability to:

- Learn and understand users' evolving goals and maintain an awareness of available data and analytics that can satisfy them.
- Orchestrate and execute actions with flexibility, responding appropriately as requirements and results accumulate and change over time.
- Recognize when results reveal opportunities for useful follow-on analysis and then program and execute a new set of actions to exploit those opportunities in accordance with existing guidelines and constraints.
- Operate with a reasonable degree of independence to insulate users from an otherwise constant barrage of requests for input and validation while still making the most of available data and computational resources. This includes identifying conflicting goals and resolving lower level conflicts to avoid wasting time and computational resources on lower priority tasks.

Ideally, MAAs will have the ability to evaluate circumstances and consult policies that help them determine whether and when to request approval for a given task they will complete on their own Such an ability would minimize user involvement in routine tasks while ensuring a human is "in the loop" for more complicated, riskier decision making (e.g., about tasks that are resource intensive) and deeper, contextually-dependent analysis (e.g., about results that may have multiple interpretations or important national security implications). Johnson et al. [16] recommend a "combine and succeed" approach to allocating work between humans and automation, allowing for varying degrees of human involvement and multiple ways of achieving a given task based on circumstances. We believe this approach will be essential to the effective use of MAAs, which will vary in the level of autonomy with which they identify current information needs, decide which combination(s) of data and analytics are most likely to satisfy these needs, configure and execute analytics against appropriate data sources, and manipulate and interpret results (see [17] for a discussion of varying levels of autonomy across similar classes of tasks).

3   Human Teaming Behaviors and MAAs

The degree of interdependence envisioned between human and MAA tasks and the need for human-machine interactions that coordinate interdependent human-MAA work will create MAA HMT challenges. Research must address these challenges to ensure the successful application of MAAs to intelligence analysis. The complexity of the agents' behaviors will require that MAAs at times operate with different task sub-goals than their users currently hold while still working towards common overall objectives. In turn, the potential for MAAs to operate with different sub-goals than their users will create a need for users to ensure automation is aligned with their own situational understanding and mission priorities, in order to regulate use of computational resources, prevent adverse MAA activities, and maximize the fit and utility of machine contributions. Achieving these objectives will require the development of sophisticated teaming functions and human-machine interaction methods that allow both analysts and MAAs to coordinate their activities efficiently and effectively.

Human teaming seems a natural analogue for joint activity involving multiple autonomous-yet-interdependent actors. Decades of work in industrial and organizational psychology and management science have produced a rich literature on human team processes [e.g., 18–20], and there are many important lessons to be learned from this research about what makes human teams function effectively. We see a number of parallels between human-MAA teaming and human teaming, and we believe principles of human teaming should inspire and inform MAA HMT research and development; however, we do not feel that all principles of human teaming are equally relevant to MAA HMT and/or should necessarily be expressed in human-MAA interactions the same way they are in human teams.

Consider work by Salas and colleagues [18]. They conducted an extensive review and thematic analysis of two decades of human teaming research and identified five major factors that appear to affect the success of human teams: team leadership, mutual performance modeling, backup behavior, adaptability, and team orientation. These factors, along with the coordinating mechanisms of shared mental models, mutual trust, and closed-loop communication, all appear to be important for successful human teamwork involving either collaboration (team members work on a common task) or coordination (team members work on separate interdependent tasks contributing to a shared outcome). Tables 1 and 2 discuss Salas et al.'s human teamwork factors (Table 1) and coordinating mechanisms (Table 2), along with our observations regarding their applicability to MAA HMT.

As shown, many aspects of Salas et al.'s framework are highly applicable to MAA HMT use cases. There is solid evidence demonstrating the importance of these factors and coordinating mechanisms for human teaming, and it seems clear they will be important for MAA HMT as well. For example, functions that support mutual performance monitoring will allow analysts and MAAs to detect each other's errors, mitigating their impact. Similarly, capabilities that facilitate the development of shared mental models will enable analysts and MAAs to better understand each other's information needs, which should encourage proactive sharing and more efficient communication.

Table 1. Salas et al. [18] teamwork factors and applicability to MAA HMT

Factor	Description	Applicability to MAA HMT
Team leadership	Planning, assigning, coordinating, and facilitating team activities in accordance with knowledge of evolving objectives and conditions. Also involves evaluating, developing, and encouraging team personnel	User will direct MAA by providing goals, constraints, and feedback based on mission objectives and understanding of context. User will develop MAA by providing feedback on correctness/utility of its work, and teaching it new analytic procedures. MAA could plan/assign/coordinate some tasks and even instruct junior users ****Not applicable: User will not need to motivate MAA**
Mutual performance monitoring	Maintaining awareness of teammate performance to assess needs and identify errors	User and MAA will monitor each other's performance to infer teammate's goals, plans, and needs; identify disagreements in priorities or interpretation of data; and detect errors in teammate's decisions or actions
Backup behavior	Taking over some of a teammate's tasking to provide relief during periods of high workload. Also involves proactively offering information or support in anticipation of a teammate's needing it, or providing feedback when a teammate commits errors or has difficulty performing a task	User and MAA will provide corrective feedback when they identify errors in each other's performance, and they will proactively offer information in anticipation of each other's needs ****Not applicable: Workload rebalancing. User will never perform a task for MAA as long as MAA knows how to do it, and if MAA can execute a task for user, it will always do so**
Adaptability	Modifying team work plans and processes based on changing needs and circumstances	User and MAA will tailor analytic methods and MAA's level of autonomy according to changing complexity and uncertainty in requirements, data, and results
Team orientation	Considering teammates' perspectives and needs, effectively leveraging teammates' contributions to achieve one's own tasks, and valuing team's success over one's own self-interest	User will need to accept MAA's help and utilize it effectively, which could be complicated by fears that automation is "taking over" analysis process. ****Not applicable: MAA will not have interests or goals apart from those of user; thus, it is not clear either user or MAA would need to adopt the kind of collective outlook human teaming requires**

Table 2. Salas et al. [18] coordinating mechanisms and applicability to MAA HMT

Factor	Description	Applicability to MAA HMT
Shared mental models	Knowledge of a team's goals and tasks, as well as the dependencies that exist between them	Both user and MAA will need to understand how their tasks affect each other's work and contribute to joint goals; will enable task coordination and anticipation of each other's information needs
Mutual trust	Trusting that teammates will competently and conscientiously execute assigned tasks, accept and respect each other's contributions, and act in ways that benefit team	User will need to have sufficient trust in MAA's competence to allow it to work independently ****Not applicable: It is not clear that MAA's programmed acceptance of user's commands would constitute trust, and it is also not clear that user's trust in MAA would be the same as their trust in a human teammate. Trust in a human teammate includes believing the teammate will not act in ways that promote their self-interest at the team's expense; MAAs will not have self-interest, so user should not suspect MAA's "motives"**
Closed-loop communication	Communication in which communicants confirm they have received and understood each other's messages	User and MAA will need to engage in closed-loop communication to ensure messages are received and interpreted correctly. MAA will also need to communicate its inferences about user's goals and needs (user should not need to infer MAA's goals and needs, as MAA will always communicate these explicitly). Will require mechanisms for user and MAA to identify and correct miscommunications or misunderstandings

In contrast, the notion of workload-related backup seems less relevant to MAA HMT. This is partly because we assume MAAs and human analysts will generally perform different kinds of work, which would preclude their taking over each other's excess tasking (i.e., human-MAA teamwork will be more coordination-based than collaboration-based). However, we also expect MAAs and human analysts to be differentially affected by analytic workload. Machines do not have the same kinds of processing limitations as humans, so it is hard to imagine a situation in which a human would need to take over tasking from an MAA purely to lighten its workload, or in which an MAA would not perform a task for its user if it knows how to do it. It is possible that more collaborative HMT (e.g., humans and robots working together in urban search and rescue) would allow for workload-related backup, but it would seem to require that both human and machine teammates experience the same kinds of processing limitations (e.g., only being able to be in one physical location at a time). Thus, task rebalancing appears to expose one way in which human and machine teammates (and their associated teaming styles) may differ.

The more social aspects of team leadership, team orientation, and mutual trust highlight additional (and arguably more dramatic) differences that will exist between human and machine teammates. Software does not have an independent sense of its own interest; thus, factors like mutual trust, which helps ensure human teammates are willing to cooperate even when cooperation brings additional risks (e.g., the potential that teammates will act in ways that further their own interest at the expense of others'), seem inapplicable to machines. Moreover, while an analyst's willingness to work with and trust an MAA could be thought of as somewhat akin to team orientation and mutual trust, it is not clear the underlying constructs are the same. Although users may interact with machines in ways that resemble their interactions with humans, humans and machines are fundamentally different types of entities, and users fundamentally know this; thus, it not clear humans have the same underlying thoughts and feelings when they interact with machines as they do with humans. For example, humans need not concern themselves with the social costs of mistreating automation (other than ones they might experience if other humans observe them) and probably cannot truly empathize with or expect empathy from machines that do not experience life in the same way they do.

Our review of Salas et al.'s [18] framework suggests that MAA HMT will resemble human teaming but differ from it as well, especially when it comes to aspects of teaming that seem more dependent on teammates' core natures and the types of relationships they can form with each other. These kinds of differences need not prevent researchers from exploring the benefits of partnering humans with automation or from drawing inspiration from the human teaming literature in developing HMT capabilities; however, they do suggest the research community might think twice before assuming HMT need necessarily be the same as human teaming or that the interaction methods that support HMT need necessarily emulate human teammate interactions.

4   Discussion

We believe the future of intelligence analysis depends on the development of MAAs that can partner with analysts to improve the efficiency and effectiveness with which they exploit available data; without it, analysts will not realize the full benefits of analytics that have the potential to save them from extreme information overload. MAAs will be able to act with a great deal of independence and flexibility, thus eliminating much of the manual work currently associated with the use of analytic tools. However, commanding and controlling such technologies will pose challenges that must be addressed if intelligence analysis is to make the most of these technologies.

Many HMT researchers have based their recommendations on prior research in human teaming [e.g., 16, 21], and we are beginning to investigate ways in which lessons from the human teaming literature may be applied to HMT for MAAs. For example, we are exploring the kinds of dialogues analysts may need to engage in with automation to establish common ground and coordinate activity through negotiation. Note that these are among the top challenges Klein and colleagues [21] discuss with regard to HMT. We expect MAAs will participate in collaborative planning with their users as in [22] to establish joint high level goals and agree upon tasks to be performed separately in support of shared objectives. We also expect MAAs will communicate with their users about their interpretations of results, offering, defending, and/or challenging different explanations and hypotheses they or their users propose regarding the meaning or implications of a piece of data or analytic finding. Foundational work on dialogues between software agents [e.g., 23] will provide a basis for some of these dialogues about what actions to take and what results mean, but research from informal logic and human discourse analysis [e.g., 24] may also provide important background.

Although we have a strong interest in human teaming, we do not assume that all human teaming behaviors are relevant to HMT, or that the behaviors that are relevant need necessarily be expressed through styles of interaction that resemble human social interaction. On the contrary, we feel there is a critical need for research into under-standing what form these dialogues and the processes that underlie them need to be to maximize efficiency and effectiveness of joint work in a particular domain. Assuming the research community should strive to fully emulate human teams imposes a daunting set of research requirements that may be unnecessary or even counterproductive if the end goal is optimizing human-machine task performance versus creating synthetic equivalents of machines' human counterparts (see discussion by [25]). It may seem reasonable to assume humans will be most comfortable and effective engaging with machines in a manner that mimics their interactions with other humans. However, treating tools like people is a fairly recent phenomenon in our species' natural history, and it seems just as reasonable to assume humans will be most comfortable and effective interacting with machines in ways that satisfy the computational requirements of joint work while maintaining distinct roles for users and the things they use. These interactions may not constitute "true" teaming in the human sense, but users may not need true teaming to work with automation successfully: They may only need some-thing that accomplishes the teaming functions required for successful use of the technology.

We do not exclude the possibility that there will be circumstances in which effective human-machine coordination requires that humans are able to interact with machines in ways that closely resemble how they would interact with human team-mates. However, rather than starting with the question of how to emulate human-like teaming behaviors in human-machine teams, we propose the research community instead begin by asking what functions are necessary for coordinating the tasks humans and machines will be performing in support of common goals. It can then explore the efficacy of different methods of instantiating these functions, clearly distinguishing computational goals from algorithms and means of implementation [26]. Some methods may have clear advantages, while others may be roughly equivalent in terms of efficiency and effectiveness measures, allowing choice of methods to be driven largely by development costs, user preference, and fit with operational systems and settings. We suspect the methods best suited to enabling HMT will vary with the types of tasks on which humans and machines are collaborating, and we suggest that understanding the features that cause different types of tasks to require different methods should be a research priority. Understanding these features will enable the community to generalize findings from one domain to another.

There may be scientific and practical benefits to more basic HMT research that seeks to replicate human teaming in human-machine teams in as direct and authentic a way as is possible, not the least of which would be gaining further insight into the nature of teamwork itself. However, at this time, we believe a more applied research agenda that treats HMT as a means to an end rather than an end in itself holds more promise for delivering solutions that best achieve successful coupling of humans and machines on a particular set of tasks. If a given HMT approach enables analysts to use MAAs to perform their work more efficiently and effectively, it will have achieved its purpose.

References

1. Lyons, J.B., Mahoney, S., Wynne, K.T., Roebke, M.A.: Viewing machines a teammates: a qualitative study. In: AAAI Spring Symposium Series, Palo Alto, CA, pp. 166–170 (2018)
2. McDermott, P.L., Walker, K.E., Dominguez, C.O., Nelson, A., Kasdaglis, N.: Quenching the thirst for human-machine teaming guidance: helping military systems acquisition leverage cognitive engineering research. In: Gore, J., Ward, P. (eds.) 13th International Conference on Naturalistic Decision Making, Bath, UK, pp. 236–240 (2017)
3. Parasuraman, R., Barnes, M., Cosenzo, K.: Adaptive automation for human-robot teaming in future command and control systems. Int. C2 J. 1(2), 43–68 (2007)
4. Ryan, M.: Human-machine teaming for future ground forces. Center for Strategic and Budgetary Assessments (2018)
5. Pirolli, P., Card, S.K.: The sensemaking process and leverage points for analyst technology. In: Proceedings of the 2005 International Conference on Intelligence Analysis (2005)
6. St. Amant, R., Cohen, P.: Intelligent support for exploratory data analysis. J. Comput. Graph. Stat. 7(4), 545–558 (1998)
7. Jurasfky, D., Martin, J.H.: Speech and Language Processing, 2nd edn. Prentice Hall, Upper Saddle River (2008)
8. Li, S.Z., Jain, A.: Handbook of Face Recognition, 2nd edn. Spring, London (2011). https://doi.org/10.1007/978-0-85729-932-1
9. Allen, J., et al.: PLOW: a collaborative task learning agent. In: Proceedings of the Twenty-Second Conference on Artificial Intelligence (AAAI 2007), pp. 1514–1519 (2007)
10. Azaria, A., Krishnamurthy, J., Mitchell, T.M.: Instructable intelligent personal agent. In: Proceedings of the Thirtieth Conference on Artificial Intelligence (AAAI 2016), pp. 2681–2689 (2016)
11. Reagan, R.W.: Executive Order 12333: United States Intelligence Activities, U.S. Federal Register (1981)
12. Bush, G.W.: Executive Order 13284: Amendment of Executive Orders, and Other Actions, in Connection With the Establishment of the Department of Homeland Security, Federal Register (2003

13. Bush, G.W.: Executive Order 13355: Strengthened Management of the Intelligence Community, Federal Register (2004)
14. Bush, G.W.: Executive Order 13470: Further Amendments to Executive Order 12333, United States Intelligence Activities, Federal Register (2008)
15. Tecuci, G., Marcu, D., Boicu, M., Schum, D.: COGENT: cognitive agent for cognitive analysis. In: AAAI Fall Symposium Series, Arlington, VA, pp. 58–65 (2015)
16. Johnson, M., Bradshaw, J.M., Hoffman, R.R., Feltovich, P.J., Woods, D.D.: Seven cardinal virtues of human-machine teamwork: examples from the DARPA robotic challenge. IEEE Intell. Syst. **29**, 74–80 (2014)
17. Parasuraman, R., Sheridan, T.B., Wickens, C.D.: A model for types and levels of human interaction with automation. IEEE Trans. Syst. Man Cybern. Part A: Syst. Hum. **30**(3), 286–297 (2000)
18. Salas, E., Sims, D.E., Burke, C.S.: Is there a "big five" in teamwork? Small Group Res. **36**(5), 555–599 (2005)
19. Marks, M.A., Mathieu, J.E., Zaccaro, S.J.: A temporally based framework and taxonomy of team processes. Acad. Manag. Rev. **26**, 356–376 (2000)
20. Tuckman, B.W.: Developmental sequence in small groups. Psychol. Bull. **63**, 384–399 (1965)
21. Klein, G., Woods, D.D., Bradshaw, J.M., Hoffman, R.R., Feltovich, P.J.: Ten challenges for making automation a "team player" in joint human-agent activity. IEEE Intell. Syst. **19**, 91–95 (2004)
22. Ferguson, G., Allen, J.: A cognitive model for collaborative agents. In: AAAI Fall Symposium Series, Arlington, VA, pp. 112–120 (2011)
23. Parsons, S., Wooldridge, M., Amgoud, L.: An analysis of formal interagent dialogues. In: Castelfranchi, C., Johnson, W.L. (eds.) Proceedings of the First International Joint Conference on Autonomous Agents and Multi-Agent Systems, New York, NY, pp. 394–401 (2002)
24. Walton, D., Macagno, F.: Profiles of dialogue for relevance. Informal Logic **36**(4), 523–562 (2016)
25. Groom, V., Nass, C.: Can robots be teammates? Benchmarks in human-robot teams. Interact. Stud. **8**, 483–500 (2007)
26. Marr, D.: Vision: A Computational Approach. Freeman & Co., San Francisco (1982)

Automating Crime Informatics to Inform Public Policy

Katy Hancock[1(✉)] and Monte Hancock[2]

[1] Murray State University, Murray, KY, USA
khancock11@murraystate.edu
[2] 4Digital, Los Angeles, CA, USA
Sirus19conf@gmail.com

Abstract. Violent crime is a critically important community issue. Government has attempted to address this problem in a variety of ways, with varied levels of success. However, there are only a certain number and type of factors that can be addressed by government action; which are most important?

In this paper we address this question by "reverse engineering" the crime prediction problem. Intuition suggests that the collection of factors most informative in predicting crime will include, as a subset, the primary causal factors of crime. If this is true, it makes sense to develop ways to identify and objectively quantify these most informative predictive factors.

We characterize the K-metric (loosely related to the F-Measure) for assessing the effectiveness of measured features for crime prediction. This metric is used to substantially reduce the number of factors needed to capture the total information of a many-feature dataset. Further, all features in the set can be rank ordered by their K-metric values, providing an automated means of identifying and objectively quantifying potentially causal factors for intervention services.

Keywords: Crime prediction · Informatics · Feature selection

1 Introduction

In the context of Augmented Cognition, this work is intended to show how specially designed information theoretic and visualization tools can provide insight into complex data sets, where the information binding is unknown, and the relative information content low. The base empirical problem selected to demonstrate the tools is the prediction of community crime rates using gross sociographic factors. As such, communities will be the example used in the background.

1.1 First Data Set

Two different data sets were used. The first data set used was a combination of two separate data sets: the Part I Uniform Crime Reports (UCR) and the Healthy People 2010 (HP) data set. The UCR are compiled annually by the U.S. Federal Bureau of Investigation (FBI) from crime reports from over 18,000 law enforcement agencies nationwide; Part I includes reports of criminal homicide, forcible rape, robbery,

© Springer Nature Switzerland AG 2019
D. D. Schmorrow and C. M. Fidopiastis (Eds.): HCII 2019, LNAI 11580, pp. 179–191, 2019.
https://doi.org/10.1007/978-3-030-22419-6_14

aggravated assault, theft, burglary, motor vehicle theft, and arson. For this experiment, crime reports from county law enforcement agencies were used. The HP data set is collected under the Centers for Disease Control and Prevention (CDC) as part of the U. S. Department of Health's Healthy People agenda. The HP data set is comprised of health information from over 3,000 counties across the United States. This information includes demographics, obesity rates, birth and death rates, preventative health care access, mental health measures, number of physicians, and a plethora of other health related information. The UCR and HP data sets were combined to create a data set with features describing community demographics, health, and crime. The FAE was used to predict crime using the demographic and health information.

1.2 Second Data Set

The second data set was the Communities and Crime (CC) data set created by Redmond (2009) by combining socioeconomic and law enforcement data from the 1990 U. S. Census and the 1990 Law Enforcement Management and Administrative Survey respectively. The CC is available on the University of California Irvine (UCI) Machine Learning Repository website. The CC data set includes community features, such as percent of the population considered urban, family income, percent of 2 parent families, unemployment rates, racial and ethnic distributions, number of illegal immigrants, number of police officers, and many more. The CC data set also contains per capita violent crime rates calculated by using population and counts of violent crimes in the United States (murder, rape, robbery, and assault).[1] Similar to the first data set, the FAE was used to predict violent crime using the community features. The tables and results reported in this paper are derived from the second data set, which was determined to contain more latent information for this study.

2   Background

2.1 Features and Feature Selection

Features are measurable attributes of entities of interest. For example, a community may be an entity of interest with features such as crime rate, the percent of the population that is unemployed or which has a high school diploma, the number of primary care physicians, and many others. Features are used to detect, characterize, and classify these entities-for example, a community with a crime rate above 1,000 per 100,000 population may be classified as "high crime".

Additionally, features are data elements that can be nominal or numeric. For instance, a community may be given the nominal measure of "highly educated" or have the numeric measure of a crime rate of 200 crimes per 100,000 persons. Nominal features are often coded as numeric values so that analytic methods can be applied to quantify and transform them. Thus, "poorly educated" might be coded as 1, "moderately educated" might be coded as 2, and "highly educated" might be 3.

---

[1] For a full description of the CC data set, see: https://archive.ics.uci.edu/ml/datasets/Communities +and+Crime.

As in the case above of the community, it is often the case that entities of interest will have multiple, sometimes many, features. This means that data modelers must determine which of these features to use, in what combinations, to create the "best" models for a given application. Indeed, whether for legal reasons, grant requirements, or organizational functioning, there is a wealth of information about community features that government, nonprofit, research, and other agencies collect with regard to communities. Some of this information might be helpful in solving problems, while other information might be superfluous or even harmful.

Not all features are created equal. Features can be thought of as symbolic evidence for or against some conclusion about the corresponding entities. Some features are more salient than others, that is, they provide more readily usable and reliable information for addressing the application at hand. Some features are "toxic" in the sense that including them in a feature set might mislead or dilute the information content of otherwise salient features. It is important, then, to determine which features, used together as a suite, provide the "best" view of the problem space.

For a simple example, consider policy makers in a community which has been dealing with an alarming number of drownings. The policy makers might see that their community's ice cream consumption is highly correlated with drowning deaths. Subsequently, this information might lead them to make policies aimed at reducing ice cream sales when, in fact, ice cream consumption is a misleading or "toxic" feature for the problem of reducing drownings. A tool that can identify "season" or "temperature" as highly salient and "ice cream consumption" as toxic would lead to better policy decisions, such as hiring more lifeguards in the summer.

Features can be selected using statistical, probabilistic, analytic, and information theoretic methods. While there are various methods available for selecting features[2], the authors will not be reviewing them here. The purpose of this paper is to consider the special characteristics of an information theoretic sampling methodology created to automate feature extraction from large feature sets. First, though, it is important to understand the two major reasons that feature extraction is difficult: (1) the enigmatic nature of the evidentiary power of features and (2) the complexity of feature selection.

The Enigmatic Nature of the Evidentiary Power of Features. While the toxicity of ice cream consumption as a feature might be obvious in the drowning example above, for many problems, the usefulness of the features is not so clear. In fact, the evidentiary power of features is rarely obvious; in such cases, special methods are required to quantify them. The next section explains why automation is essential to effective feature selection.

The Complexity of Feature Selection. Given N features, there are 2^N possible feature sets, making feature selection exponentially complex. This means a data set with a mere 10 features would have over 1,000 possible feature sets, while 20 features would result in over 1 million. Data sets in the real world frequently have far more features than this.

---

[2] <u>Data Mining Explained</u> (Hancock 2001) and <u>Practical Data Mining</u> (Delmater and Hancock 2011).

To further illustrate this complexity, one can compare two problems. First, select the best feature set for a classification problem, where: (1) the entire data set consists of 250 features for each object and (2) using a super-computer, one can evaluate 1 trillion feature sets per second. The goal is to determine the best subset of features by checking all subsets. Second, wear down an Earth-sized sphere of solid iron, where once every 10 thousand years, a tiny moth flies by and brushes the globe with its wing, knocking off a single atom. The goal is to wear away the entire Earth-sized iron sphere. The moth will have completely worn away eight of the Earth-sized spheres centuries before all the feature sets have been evaluated. Clearly, complex problem spaces can create an overwhelming number of solution sets-an efficient way of selecting the best feature set in a limited amount of time is necessary.

3   Method

3.1   Feature Analysis Engine

The authors have developed a Feature Analysis Engine (FAE) that compares the information content of features, in context, for a specific classification task. The strategy is similar to forming a sampling distribution for feature performance. The FAE uses uniform random samples of features to inform a classifier. When the classifier for a particular random selection of features is run, its performance is added to the bin for each feature in that sample. "Good" features will be those that have participated in many good outcomes when they are present, and lower quality outcomes when they are absent. In this way, over many feature suite samples, the feature selector indicates the average contribution each feature makes to the performance of random feature suites of which it is a member.

3.2   The Feature Evaluation Algorithm

Let the set of feature vectors consist of L vectors, each having N features (i.e., coordinates in each feature vector) in the data set. The analysis consists of sampling trials, with three initial steps:

- Select a proportion of vectors to be used for training, with the remainder to constitute the blind set.
- Determine a maximum sample size, $1 <= M <= N$, for the number of features to select on a trial.
- Segment the set of feature vectors into a training set and a blind set, having the same proportion of every ground truth class in both.

Each trial proceeds as follows:

1. Uniform randomly select from 1 to M features for use on this trial. (note M/L is the probability that any particular feature will be selected for a trial.)
2. Train the classifier on the training set to recognize the ground truth assignment of a vector having the selected features.
3. Apply the trained classifier to the blind set (which also uses the selected features).
4. Add the performance of the classifier on the blind set into the corresponding histogram bin for each feature in the sample selected for this trial.

Over many trials, this procedure empirically computes the expected value of the performance of a classifier for each feature, when used as part of randomly selected feature suite of various sizes.

In the work described in this paper, three performance measures were aggregated for each feature: The proportion of the blind vectors correctly classified (range [0–1]); the geometric mean of the class Precisions (range [0–1]); and, the geometric mean for the class Recalls (range [0–1]). These measures are the components of the important two-class statistical decision metric called the F-Measure. The corresponding performances for the "feature absent" cases are computed in the same way using all the bins but leaving out each feature in turn.

From the "aggregate" Blind Accuracy, Precision, and Recall for each feature (as described above) is formed what we call the K-metric. According to this metric, the relative value of feature j in the dataset is given by:

$$K(j) = (A(j, 1) - A(j, 2))(B(j, 1) - B(j, 2))(C(j, 1) - C(j, 2))$$

where, for features j = 1, 2, 3, ... N:

A(j, 1) = Aggregate Accuracy on blind set when feature j is used
A(j, 2) = Aggregate Accuracy on blind set when feature j is not used
B(j, 1) = Aggregate Precision on blind set when feature j is used
B(j, 2) = Aggregate Precision on blind set when feature j is not used
C(j, 1) = Aggregate Recall on blind set when feature j is used
C(j, 2) = Aggregate Recall on blind set when feature j is not used

Notice that for "noise" features, one or more of the factors of K could be negative. This will happen in instances for which the classifier is not able to detect and ignore "noise" features. This can be handled by setting very small factors in the expression for K to some minimum non-negative value. This situation is rare but does occur.

It was also seen that the convergence of the histogram bins is fairly stable, rapidly settling down after a few hundred trials (i.e., a "large sample"). Figure 1 presents two typical examples taken from a run of over 4,000 trials:

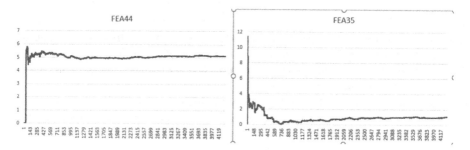

Fig. 1. Rapid convergence of two feature histograms to stable values in two typical runs

3.3   Prediction of Crime

The Feature Analysis Engine was applied to the datasets collected; the ground truth used was a measured crime rate value (#annual violent crimes/population). Initially the ground truth values were quantized into 8 integer values, conceptually representing extremely low, very low, low, low nominal, high nominal, high, very high, extremely high. Several experiments, statistical analysis, and visualization indicated that the data did not support estimation at this quantization level, and only a blind accuracy of 44% could be achieved. Figure 2 contains a confusion matrix showing that the transition from low crime rate to high crime rate covers several classes, making them impossible to distinguish in a manner that generalizes:

Confusion Matrix:

TRUE\MACH	1	2	3	4	5	6	7	8	Recalls
1	157	58	6	1	0	1	0	3	69.46903
2	51	65	19	16	1	1	2	1	41.66666
3	5	27	40	18	2	1	2	0	42.10526
4	2	17	18	13	1	1	4	2	22.41379
5	0	7	5	10	1	1	6	2	3.125
6	1	4	5	10	5	0	4	3	0
7	0	2	5	8	2	1	11	4	33.33334
8	0	2	3	6	3	1	7	8	26.66667
Precisions	72.7	35.7	39.6	15.9	6.7	0.0	30.6	34.8	

Fig. 2. A classification accuracy of 85% is achieved when there are only two quantization levels

For this reason, the ground truth was re-quantized into two classes: Lower and Higher. With two quantization levels, a classification accuracy of 85% was achieved (Fig. 2).

Using the two-level ground truth values, metric measurements were tabulated over 4,000 trials. The resulting K-metric histogram bin averages are plotted in Table 1, and graphically in Fig. 3.

Table 1. Empirical Assessment of Relative Feature Quality (K-Metric)

Feature File: uci.csv Norm: 1 Trials: 4255	**Empirical Assessment of Relative Feature Quality**									
	population	household	racepctbla	racePctWh	racePctAsi	racePctHis	agePct12t	agePct12t	agePct16t	agePctt
Metric	Fea 1	Fea 2	Fea 3	Fea 4	Fea 5	Fea 6	Fea 7	Fea 8	Fea 9	Fea 10
%ACC_when used	34.12289	33.31466	35.99184	36.34529	33.32308	34.43372	33.62272	33.60655	33.70614	34.108
Prec_when used	17.96521	17.75452	20.14341	20.3189	17.85457	17.91679	18.17315	17.48004	17.91845	18.228
Reca_when used	17.50904	17.5571	19.70927	20.08158	17.59873	17.60944	17.89521	17.22211	17.66631	18.035
%ACC_when NOT used	32.06586	32.11455	31.94712	31.94966	32.10565	32.03991	32.10135	32.09329	32.10043	32.064
Prec_when NOT used	16.53357	16.54718	16.39727	16.40803	16.53219	16.53197	16.52789	16.56026	16.54485	16.516
Reca_when NOT used	16.2313	16.22984	16.09407	16.09319	16.21782	16.22097	16.21537	16.24628	16.23104	16.198
K	3.76285	1.923121	54.77783	68.56337	2.223134	4.602772	4.204722	1.358225	3.165636	6.4269

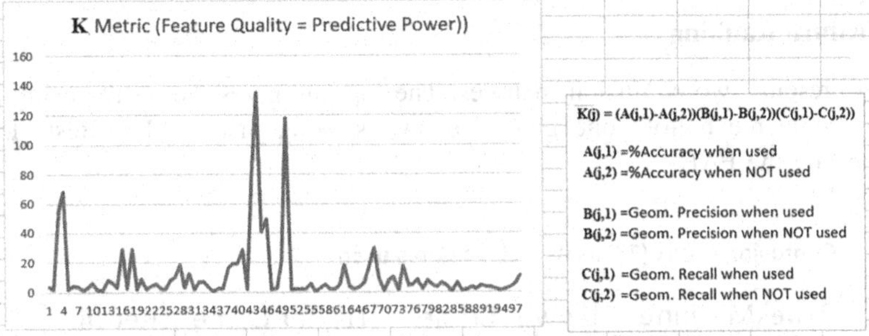

K Metric (Feature Quality = Predictive Power))

$K(j) = (A(j,1)-A(j,2))(B(j,1)-B(j,2))(C(j,1)-C(j,2))$

$A(j,1)$ =%Accuracy when used
$A(j,2)$ =%Accuracy when NOT used

$B(j,1)$ =Geom. Precision when used
$B(j,2)$ =Geom. Precision when NOT used

$C(j,1)$ =Geom. Recall when used
$C(j,2)$ =Geom. Recall when NOT used

Fig. 3. The taller peaks indicate the most informative features.

As stated above, the base empirical problem selected to test the FAE is the prediction of local crime rates using gross sociographic factors. Crime prediction, of course, is a problem that has been studied for a long time and is of importance to lawmakers, schools, businesses, non-governmental organizations (NGOs), and individuals. Indeed, a wide variety of sociodemographic, health, and economic factors have been tied to crime and delinquency. For example, mental health issues (Lipsey et al. 2010), male gender (Bayer and Pozen 2005), age (Moffitt 1994), and minority race (Hartney and Vuong 2009; Reisig et al. 2007) have all been found to predict higher rates of offending. In addition, the provision of adequate healthcare services has been tied to lower levels of reoffending after release from confinement (Hancock 2017; Kim et al. 1997). Furthermore, urbanicity, region of the country, temperature, and social class have been tied to crime in numerous studies (Siegel and Worrall 2013). In addition, family variables have been identified as risk factors for delinquency: these include having parents or siblings with criminal histories, coming from single parent homes, inconsistent parental discipline, having siblings, not being first-born, having abusive or neglectful parents, and separation from parents (Shader 2001). Moreover, a wide variety of school (i.e. harsh disciplinary policies) and neighborhood (i.e. high poverty, high crime) variables have been noted as risk factors for delinquency (Mack et al. 2015; Shader 2001; Taylor and Fritsch 2015). See Table 3.

Clearly there are a plethora of features that have been used to predict crime, yet there is also disagreement on whether and how these things might be related to crime. Thus, it is evident that the issue of crime is a complex problem space with less than obvious relationships between various features and the crime rate. It is important for policy makers to address the right features to most effectively and efficiently address the crime problem.

4   Experimental Results

4.1   Feature Ranking

Figure 4 presents two confusion matrices. The top one gives the results using all the features, while the bottom one gives the results using just the 10 "Best" features designated by the FAE.

Confusion Matrix (2 Class): ALL Features used

True\Machine	Lower Crime	Higher Crime	Recall
Lower Crime	1451.3	156.7	0.9025
Higher Crime	149.3	236.7	0.6132
Precision	0.9067	0.6017	

Confusion Matrix (2 Class): Using just the 10 "Best" Features

True\Machine	Lower Crime	Higher Crime	Recall
Lower Crime	1489.9	215.1	0.8738
Higher Crime	99	280	0.7388
Precision	0.9377	0.5655	

Fig. 4. Top table is the confusion matrix for the classifier when all features are used. Bottom table is the confusion matrix when only the 10 best features, as determined by the FAE, are used. This demonstrates that little information loss results when removing features identified as less informative are removed.

Table 2 on the next page presents the features sorted in descending order of their K-metric.

Table 2. Features ranked by how much they contribute to Blind Classification %Accuracy, Precision, and Recall.

Fea Rank	Log2(Q)	FeatureName
1	7.077954	PctKids2Par
2	6.885442	Pctllleg
3	6.397896	PctFam2Par
4	6.099366	racePctWhite
5	5.77552	racepctblack
6	5.635663	PctTeen2Par
7	5.35007	PctYoungKids2Par
8	4.906302	PctPersDenseHous
9	4.875083	pctWInvInc
10	4.862233	pctWPubAsst
11	4.824799	TotalPctDiv
12	4.369184	NumIlleg
13	4.28821	MalePctNevMarr
14	4.251834	FemalePctDiv
15	4.245187	PctPopUnderPov
16	4.159409	PctLargHouseFam
17	4.143998	PctVacantBoarded
18	4.045454	PctPersOwnOccup
19	3.998719	MalePctDivorce
20	3.67785	PctNotHSGrad
21	3.441019	PctUsePubTrans
22	3.359945	NumUnderPov
23	3.350506	PctHousOccup
24	3.259045	PctHousLess3BR
25	3.190718	medFamInc
26	3.042824	PctHousNoPhone
27	3.015528	medIncome
28	3.006192	HispPerCap
29	2.918522	OwnOccLowQuart
30	2.86857	HousVacant
31	2.815784	PctUnemployed
32	2.810102	PctEmploy
33	2.707228	MedRent
34	2.684139	agePct65up
35	2.65233	pctWWage
36	2.645663	RentLowQ
37	2.605876	PersPerRentOccHous
38	2.573814	blackPerCap
39	2.501965	PctImmigRec10
40	2.435991	MedYrHousBuilt
41	2.38216	PctRecImmig8
42	2.312566	PctNotSpeakEnglWell
43	2.297209	PopDens
44	2.202503	racePctHisp
45	2.199223	whitePerCap
46	2.13873	PctVacMore6Mos
47	2.107376	NumStreet
48	2.078594	OwnOccMedVal
49	2.07201	agePct12t21
50	1.972799	PctLargHouseOccup
51	1.911826	population
52	1.856538	RentMedian
53	1.846851	MedOwnCostPctIncNoMtg
54	1.835946	OwnOccHiQuart
55	1.802582	PctEmplManu
56	1.765658	PersPerOwnOccHous
57	1.662495	agePct16t24
58	1.652508	PctBornSameState
59	1.65047	pctWFarmSelf
60	1.647402	PctForeignBorn
61	1.599766	PctLess9thGrade
62	1.588268	LandArea
63	1.531958	PctOccupManu
64	1.508409	PctOccupMgmtProf
65	1.472902	PctWorkMom
66	1.429152	pctWSocSec
67	1.376209	PctRecImmig5
68	1.270759	indianPerCap
69	1.235236	NumInShelters
70	1.164698	AsianPerCap
71	1.152595	racePctAsian
72	1.145041	PctImmigRecent
73	1.112857	PctWorkMomYoungKids
74	1.102538	PctSameState85
75	1.053915	PersPerFam
76	0.979784	PctSpeakEnglOnly
77	0.971732	NumImmig
78	0.965095	PctRecImmig10
79	0.954633	PctImmigRec8
80	0.948322	PctImmigRec5
81	0.94345	householdsize
82	0.915945	perCapInc
83	0.86869	PctWOFullPlumb
84	0.865512	PctBSorMore
85	0.864056	PctSameHouse85
86	0.852064	pctWRetire
87	0.843909	PersPerOccupHous
88	0.770808	MedOwnCostPctInc
89	0.743902	numbUrban
90	0.650907	pctUrban
91	0.5425	PctHousOwnOcc
92	0.441722	agePct12t29
93	0.14965	PctEmplProfServ
94	-0.0309	MedNumBR
95	-0.05191	PctSameCity85
96	-0.06957	MedRentPctHousInc
97	-2.23086	RentHighQ
98	-8.68618	PctRecentImmig

Features Ranked by Q, a measure of how much they contribute to Blind Classification %Accuracy, Precision, and Recall

Table 3. Correlation Coefficients among the 10 most informative features (according to FAE) and Crime Rate. (Crime rate is in the bottom row (GT-Class). It has 7 levels, with 1 being lowest crime rate, and 7 being highest. Therefore, positive correlations in the bottom row are indicative of factors that, when larger, correlate with higher crime rates. Negative correlations in the bottom row are indicative of factors that, when larger, correlate with lower crime rates.

Top Crime rate Predictors	PctKids2Par	PctIlleg	PctFam2Par	racePctWhite	raceptblack	PctTeen2Par	PctYoungKids2Par	PctPersDenseHous	pctWInvInc	pctWPubAsst	GT_Class
PctKids2Par	1.000										
PctIlleg	-0.870	1.000									
PctFam2Par	0.985	-0.843	1.000								
racePctWhite	0.702	-0.803	0.638	1.000							
raceptblack	-0.736	0.811	-0.701	-0.794	1.000						
PctTeen2Par	0.908	-0.798	0.917	0.614	-0.690	1.000					
PctYoungKids2Par	0.931	-0.812	0.936	0.597	-0.656	0.837	1.000				
PctPersDenseHous	-0.383	0.428	-0.312	-0.640	0.158	-0.268	-0.304	1.000			
pctWInvInc	0.744	-0.635	0.703	0.596	-0.493	0.610	0.706	-0.568	1.000		
pctWPubAsst	-0.760	0.669	-0.737	-0.590	0.445	-0.630	-0.775	0.554	-0.751	1.000	
GT_Class	-0.740	0.732	-0.706	-0.684	0.619	-0.658	-0.669	0.467	-0.590	0.583	1.000

Features having both higher predictive power for high crime AND higher positive correlation with higher crime are highlighted.

A notable aspect of these data is that the top predictors of lower crime rate have to do with percentage of two-parent families (four of the top eight).
All four of the familial elements have more predictive power than any of the economic indicators in the data set.

5 Conclusions

The three most informative predictors of crime are the percent of children with two parents, the percent of illegal immigrants, and the percent of families with 2 parents. Indeed, 5 of the 10 top features involve family structure features. Thus, the FAE indicates that measures of family structure and stability are the most informative features in the set, superior to measures of educational attainment, economic and employment factors, age distributions, and racial and ethnic distributions. This information can be invaluable to policy makers and point to policies and programs aimed at strengthening families.

Beyond these specific implications, the results also suggest the benefit of using the FAE to inform the policy process. Indeed, the importance of family factors in the current study conforms to prior research and theory on crime and delinquency, underscoring the utility of the FAE for identifying important decision factors. Furthermore, the usefulness of the FAE could be extended to a broad range of important problems: expanding access to health care, improving community health, bolstering order maintenance in institutional settings, optimizing judicial and parole decision-making, facilitating the creation of case plans for troubled individuals and families, refining hospital efficiency, maintaining safety in schools, and many more. Identifying the most salient factors in these and other problems can improve response effectiveness, save money, and properly allocate resources to things that matter rather than to "toxic" factors.

6 Future Work

Given that this work suggests the relative superiority of family structure in the prediction of violent crime, future work will naturally evaluate the validity of this finding in other ways. This would include the use of datasets that express family structure in a variety of ways, and at different scales (e.g., multi-family neighborhoods). Further, the question of whether these factors are causal or merely correlative, and the precise characterization of any underlying mechanism(s) warrants investigation.

Another possible avenue for future work would be to approach the crime issue from another angle: rather than identifying factors that predict high crime rates, or "risk factors," the FAE could be used to identify factors which predict low crime rates, or "protective factors." Protective factors can lessen the impact of risk factors, effectively buffering individuals and communities from negative elements in their environment (Pollard et al. 1999). As with factors that predict crime, there is disagreement about what actually constitutes a salient protective factor-scholars tend to view these factors through the lenses of either being the absence of the risk factor (i.e. broken home versus unbroken home) or being factors which interact with risk factors (strong parental supervision mitigates poor attitudes towards rules) (Shader 2001). It may be that these lenses and the debate surrounding these factors have left important protective factors unidentified, as the focus is primarily on identified risk factors. The FAE may be a first step in identifying these-as such, policy makers can work to reduce the most salient risk factors while also trying to increase the presence of the most salient protective factors.

Indeed, prior research has indicated that addressing <u>both</u> risk and protective factors is necessary for crime reduction to be effective (Pollard et al. 1999). Future research can also investigate the use of the FAE with problem spaces outside of crime and delinquency, as outlined in the conclusion above.

Data Source Details and Citations for the Second Data Set

(this material is taken from the UC Irvine Data Mining Repository)

Related dataset used in Redmond and Baveja (2002) 'A data-driven software tool for enabling cooperative information sharing among police departments' in European Journal of Operational Research 141 (2002) 660–678

The authors downloaded the "second dataset" from the UCI Machine Learning Repository, which cited the following sources:

- U. S. Department of Commerce, Bureau of the Census, Census Of Population And Housing 1990 United States: Summary Tape File 1a & 3a (Computer Files)
- U.S. Department of Commerce, Bureau Of The Census Producer, Washington, DC and Inter-university Consortium for Political and Social Research Ann Arbor, Michigan (1992)
- U.S. Department of Justice, Bureau of Justice Statistics, Law Enforcement Management And Administrative Statistics (Computer File) U.S. Department of Commerce, Bureau of The Census Producer, Washington, DC and Inter-university Consortium for Political and Social Research Ann Arbor, Michigan. (1992)
- U.S. Department of Justice, Federal Bureau of Investigation, Crime in the United States (Computer File) (1995)
- Redmond, M. A. and A. Baveja: A Data-Driven Software Tool for Enabling Cooperative Information Sharing Among Police Departments. European Journal of Operational Research 141 (2002) 660–678.

Abstract: Communities within the United States. The data combines socio-economic data from the 1990 US Census, law enforcement data from the 1990 US LEMAS survey, and crime data from the 1995 FBI UCR.

References

Bayer, P., Pozen, D.E.: The effectiveness of juvenile correctional facilities: public versus private management. J. Law Econ. **48**(2), 549–589 (2005)

Hancock, K.: Facility operations and juvenile recidivism. J. Juvenile Justice **6**(1), 1–14 (2017)

Hartney, C., Vuong, L.: Racial and Ethnic Disparities in the US Criminal Justice System. National Council on Crime and Delinquency, Oakland (2009)

Kim, J.Y., et al.: Successful community follow-up and reduced recidivism in HIV positive women prisoners. J. Correct. Health Care **4**, 5–17 (1997)

Lipsey, M.W., Howell, J.C., Kelly, M.R., Chapman, G., Carver, D.: Improving the Effectiveness of Juvenile Justice Programs: A New Perspective on Evidence-Based Practice. Center for Juvenile Justice Reform, Georgetown Public Policy Institute (2010)

Mack, K.Y., Peck, J.H., Leiber, M.J.: The effects of family structure and family processes on externalizing and internalizing behaviors of male and female youth: a longitudinal examination. Deviant Behav. **36**(9), 740–764 (2015). https://doi.org/10.1080/01639625.2014.977117

Moffitt, T.: Natural histories of delinquency. In: Weitekamp, E., Kerner, H.J. (eds.) Cross-National Longitudinal Research on Human Development and Criminal Behavior, pp. 3–65. Kluwer, Dordrecht (1994)

Pollard, J.A., Hawkins, D., Arthur, M.W.: Risk and protective factors: are both necessary to understand diverse behavioral outcomes in adolescence? Soc. Work Res. **23**(3), 145–158 (1999)

Redmond, M.A., Baveja, A.: A data-driven software tool for enabling cooperative information sharing among police departments. Eur. J. Oper. Res. **141**, 660–678 (2002)

Reisig, M.D., Bales, W.D., Hay, C., Wang, X.: The effect of racial inequality on black male recidivism. Justice Q. **24**(3), 408–434 (2007). https://doi.org/10.1080/07418820701485387

Shader, M.: Risk Factors for Delinquency: An Overview. Office of Juvenile Justice and Delinquency Prevention, Washington, DC (2001)

Siegel, L.J., Worrall, J.L.: Essentials of Criminal Justice, 8th edn. Wadsworth, Belmont (2013)

Taylor, R.W., Fritsch, E.J.: Juvenile Justice: Policies, Programs, and Practices, 4th edn. McGraw Hill Education, New York (2015)

Delmater, R., Hancock, M.: Practical Data Mining. CRC Press, Boca Raton (2011)

Hancock, M.: Data Mining Explained. Digital Press, Newton (2001)

Redmond, M.: Communities and Crime Data Set. UCI Machine Learning Repository (2009). http://archive.ics.uci.edu/ml/datasets/communities+and+crime

Visualizing Parameter Spaces of Deep-Learning Machines

Monte Hancock[1](✉), Antoinette Hadgis[2](✉), Benjamin Bowles[2](✉),
Payton Brown[2](✉), Alexis Wahlid Ahmed[2](✉), Tyler Higgins[2](✉),
and Nikki Bernobic[2](✉)

[1] 4Digital, Los Angeles, CA, USA
practicaldatamining@gmail.com
[2] Sirius19, Melbourne, FL, USA
{twomagpies115@gmail.com, benbowles2@gmail.com,
alexis.wahlid.ahmed@gmail.com, kylebrown1997@me.com,
tjh361@yahoo.com

Abstract. This paper applies visualization techniques to the high-dimensional parameter spaces of a particular type of neural network, the Multi-Layer Perceptron (MLPs). MLPs are the archetypal "deep-learning" machines, digital imitations of biological brains. MLP parameter spaces are data structures holding the neuron interconnect-weights "learned" by these machines during training. These structures are abstractions of biological systems and are not generally amenable to analysis by conventional means.

The terms parameter space and (the more general) machine space will be used here interchangeably.

There are several things we would like to know about this machine as a solution to our problem, among them:

- Has our training algorithm converged "well"? (Is the trainer stuck, walking a ridge, etc.).
- Did we select the proper architecture for this classifier? (Indicated by the relative complexity of metrics in machine space).
- Did we select the proper training algorithm for this classifier? (Indicated by the regularity of the performance in machine space).
- Have we found a global optimum? (Are other optima/paths to optima visible?).
- How stable is our machine? (Are we sensitive to round-off, e.g., are we balancing on a needle?).

These questions are very difficult to answer analytically and are not very effectively addressed by sampling the *feature space* (e.g., blind testing). These are questions about the *machine space*.

Beginning with some background for those not familiar with deep-learning machines, we present a methodology for visualizing their parameter spaces, and describe how visual analysis of these spaces can inform developers.

Methods are shown for creating visual representations of performance manifolds, and how these representations assist in the characterization of certain model pathologies. In particular, performance manifolds for deep-learning machines (multi-layer perceptrons) are shown and interpreted.

© Springer Nature Switzerland AG 2019
D. D. Schmorrow and C. M. Fidopiastis (Eds.): HCII 2019, LNAI 11580, pp. 192–210, 2019.
https://doi.org/10.1007/978-3-030-22419-6_15

Keywords: Dual spaces · Visualization · Machine learning · Perceptron

1 Background

In what follows, vectors and vector-valued functions are indicated by underscoring, e.g., $\underline{V}$, $\underline{A}(x)$; the norm of a vector is denoted $|\underline{V}|$. The set of real numbers is denoted by R, and n-dimensional Euclidean space by R^N.

Parametric regression models are characterized by the specification of a model architecture and instantiated by the assignment of values to the model parameters. The set of all possible parameter assignments for a given model is its parameter space. For example, the parameter space for the linear functionals $L(R^K)$ is the dual space, $L * (R^K)$. Another example of a parameter space is R^{K+1} regarded as the space of vectors of coefficients for the real polynomials of order K in the real variable x. Parameter spaces can be defined for neural networks, kernel-based classifiers, and other parametric models by unambiguously assigning each model parameter (weight, coefficient, parameter, etc.) to a dimension in some Euclidean space R^D.

Adaptive learning problems are often formulated as regressions whose values are known on some training set. For these problems, an instance from a family of functions is sought that maps each training datum to a specified target value. The development of classifiers, estimators, and a wide range of decision support problems is a parameter space search that constitutes an instance of machine learning.

Assessment artifacts like correlation tables on the input side, and confusion matrices on the output side, provide secondary measures of model operation: they can be used to assess model inputs and outputs, but not the operational processes that tie these together. Obtaining primary measures of model operation requires analysis of the parameters determining how model outputs are developed from feature inputs; it is the parameters that contain information about model stability (round-off, convergence, range problems) and model pathology (e.g., over/under parameterization, multi-co-linearity).

However, while visualization is often used to analyze, conform, and regularize input and output data, it is almost never applied to the model parameters developed, owing to their high-dimensionality and general inscrutability.

The machine learning problem can be viewed as a search of some parameter space for settings that enable a regression function to model the outputs of a decision operation. This problem can be thought of in terms of the underlying geometry of these spaces, suggesting the use of ray tracing as a method for visualizing aspects of machine learning.

A family of functions selected as candidates for a parametric regression (and so having a common domain and range) is a Model Space.

The common domain of the functions in the Model Space is called the Feature Space; it contains the sets of features to be mapped to their respective target values. The common range of the functions in the Model Space is the Goal Space; it contains the target values for the regression.

A symbolic representation of an attribute is referred to as a feature (e.g., size, quality, age, color, etc.). Features can be numbers, symbols, or complex data objects.

Features are usually reduced to some simple form before modeling is performed; for the sake of definiteness, we will assume here that all features are represented by real numbers.

In the context of modeling, it is customary to identify an entity being modeled with the totality of its features (c.f., the philosophy of "Essentialism"). Once features for modeling have been selected (and coded as real numbers), a feature space can be defined by assigning each feature to a specific position in an ordered "tuple". Each instance of an entity having the given features is then represented by a point in n-dimensional Euclidean space: its feature vector. This Euclidean space, the feature space for the domain, has dimension equal to the number of features.

When known, the target value for a feature vector is often referred to as its "ground truth", and the set of all possible target values for a regression problem is its Goal Space. For continuous estimation problems (e.g., assigning a credit score), the Goal Space is usually a subset of R (e.g., [350, 850]). For classifiers, the Goal Space is usually an initial segment of the Natural Numbers (e.g., class identifiers 1, 2, ..., M).

For many problems, building a regression model requires a set of feature vectors, each having a known target value. Machine learning methods that use such a "set having ground truth" are referred to as supervised learning methods.

The set of all vectors of parameters which instantiate a particular function in the Model Space is the Parameter Space.

Example I. The diagram in Fig. 1 below depicts a binary (two class) classifier in a 3-dimensional Feature Space. As a two-class problem, the Goal Space is the two-point set $\{1, 2\}$.

The feature vectors of the two classes have been separated by a plane. Class 1 vectors are to the left and above the plane, and Class 2 vectors are to the right and below the plane. Such a separating structure in feature space is called a "Decision Surface".

All decision surfaces that are planes in this feature space are completely characterized by the values of the three coefficients A, B, and C; therefore, the Parameter Space for this parametric regression problem is R^3.

Example II. An estimator is desired for tomorrow's temperature at a particular location as a function of time-of-day. A parametric regression for this modeling problem might use the following:

- Feature Space (the empirical inputs to the model): [0, 24] in hours after midnight (time-of-day)
- Goal Space (the regression outputs): real interval [−100, 130] in degrees Fahrenheit (temperature)
- Model Space (the set of regression functions): Real polynomials of degree < 6 on [0, 24]
- Parameter Space: R^6 (ordered 6-tuples of coefficients that instantiate a polynomial regression)

Developing parametric regressions is an optimization problem, where some algebraic, iterative, or recursive technique is employed to find, for the set of points in

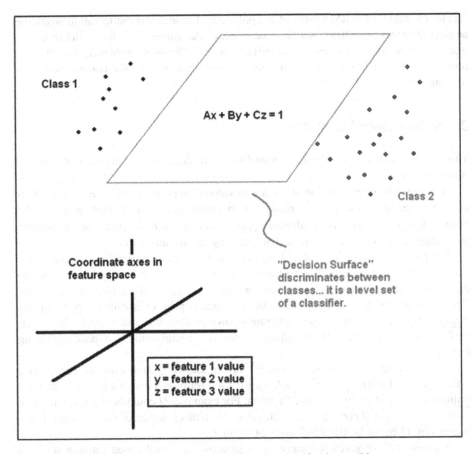

Fig. 1. A pictorial representation of a classification problem.

Feature Space, a vector of parameters in the Parameter Space that instantiates a function in Model Space mapping Feature Space inputs to "correct" Goal Space outputs.

Parameters are the (usually numeric) settings necessary to make model formalisms explicit. Examples of parameters for various learning machine regressions are:

- Polynomials: Coefficients (constant multipliers for powers of the variable(s))
- Fourier Series: Fourier Coefficients (amount of energy in each frequency)
- MLP: Interconnect weights (strengths of connections between neurons)
- BBNs: conditional probability tables (conditional probabilities connecting facts to conclusions)
- Kernel-Based-classifiers (SVM, RBF): regression coefficients, base-point coordinates
- Metric (K-means, RCE, matched filters): cluster centroids and weights

To facilitate optimization, an objective function is defined on the Parameter Space. This objective function is usually a real valued function having the property that its extrema correspond to vectors of parameters whose model instances are solutions of the corresponding regression problem. In the example just given, a reasonable objective function would be the RMS error of the polynomial model; the optimization would be an algorithm (e.g., Gradient Descent) that searches Parameter space for model instances that minimize the RMS regression error. For classification problems, an oft used objective function is the percent of total instances assigned to the correct class in a training set.

2 Performance Manifolds

This section defines *performance manifolds* and describes them for kernel-based regression models.

A parametric regression on R^N is a real-valued function $G : R^D \times R^N \Rightarrow R$. This notation is employed to make a distinction between the feature vectors at which the instantiated regression is evaluated ($\{(x_1, x_2, \ldots, x_N)\} \varepsilon R^N$), and the regression parameters ($\{(c_1, c_2, \ldots, c_D)\} \varepsilon R^D$) characterizing an instance of G.

Let f be a function in a Model Space, and let E be an objective function for a parametric regression problem on a training set. Define a scalar field Q by assigning to each vector in the Parameter Space the corresponding value of the objective function. The restriction of Q to a manifold in the parameter space Q is called a <u>performance manifold for</u> f. Each point of f will generate its own performance manifold. In this paper, performance manifolds will be affine spaces of co-dimension 1 (space-separating hyperplanes).

Let T be a set of feature vectors in R^N having known ground truth for each vector (viz. supervised learning). When such a set is used for model construction, it is called a <u>training set</u>. Let Q be an objective function. For example, Q could be the percentage of vectors of T assigned correct regression values from this instance of G; for a continuous regression, Q could be the RMS error of G on T.

Suppose we are given a parametric regression G, a well-posed training set T of feature vectors, and an appropriate objective function Q. Then each vector of parameters $\underline{p} \in R^D$ has associated with it a corresponding performance score Q on that training set. $Q : R^D \Rightarrow R$ is a scalar field and finding an optimum instance of G is a search of R^D for extremal values of this scalar field. Such an optimum resulting from a training activity will here be called the "Best" machine.

To facilitate visualization of the field Q, which is almost always high-dimensional, we might extract and display low-dimensional affine subsets of R^D colorized by the values assumed by Q. These manifolds, which are "slices" of the parameter space colorized using the values of Q, are performance manifolds. As will be seen, visualization of these manifolds gives insight into the performance of the underlying parametric regression models.

2.1 Performance Manifolds for Kernel-Based Regressions

Performance manifolds for kernel-based regressions are relatively simple, as the following analysis for $G{:}R \times R \Rightarrow R$ illustrates.

Let $x_1, x_2, \ldots, x_N \in R^1$ be a training set, T. Let the kernel functions be $F_i(x)$, $i = 1, 2, \ldots, D$.

Define the vector-valued function $F_k = (F_1(x_k), F2(x_k), ..., F_D(x_k))$, $k = 1, 2, ..., N$, and let $\delta = (\delta_0, \delta_1, ..., \delta_{0D-1} \in R^D$.

For $x \in R^1$, let $G_0(x) = \sum_{i=0,D-1} a_i F_i(x)$ be a regression model based on the kernel functions F_i, having coefficients a_i. Then $G_0(x)$ is a polynomial in x of degree $D - 1$. Suppose that this regression model has error 0 (i.e., it is a "Best" machine for this problem).

Let $G_\delta(x) = \sum_{i=0,D-1} (a_i + \delta i) F_i(x)$ be the corresponding model whose i^{th} coefficient a_i has been perturbed by δ_i.

The modeling error introduced by perturbing the coefficients of G at x_k by can be denoted:

$$
\begin{aligned}
E_k(\underline{\delta}) &= G_0(x_k) - G_{\underline{\delta}}(x_k), k = 1, 2, ..., N \\
&= \sum_{i=0,D-1} a_i F_i(x_k) - \sum_{i=0,D-1} (a_i + \delta_i) F_i(x_k) \\
&= -\sum_{i=0,D-1} \delta_i F_i(x_k) \\
&= -\underline{\delta} \cdot \underline{F}_k
\end{aligned}
\tag{1}
$$

The RMS error of $G_\delta(x_k)$ as an approximation of $G_0(x_k)$ over the set $x_1, x_2, ...,$ $x_N \in R^1$ is:

$$
\begin{aligned}
E(\underline{\delta}) &= \sqrt{\left[(1/N) \sum_{k=1,N} (E_k(\underline{\delta}))^2\right]} \\
&= (|\underline{\delta}|/\sqrt{N}) \sqrt{\left[(1/N) \sum_{k=1,N} \left\{|F_k|^2 \cos^2(\theta_{\underline{\delta},k})\right\}\right]}
\end{aligned}
\tag{2}
$$

where $\theta_{\delta,k}$ is the angle between δ and $\underline{F}_k$ in R^D. Therefore, the 2-dimensional performance manifolds of $Q = E(\delta)$ show a roughly radial symmetry, with highest performance at $\{a_0, a_1, ..., a_{D-1}\}$, dropping off at larger distances. In Fig. 2, the error is 0 in the center where $|\underline{\delta}| = 0$, and increases with increasing $|\underline{\delta}|$. Figure 3 shows an example of how problem complexity can affect performance manifolds.

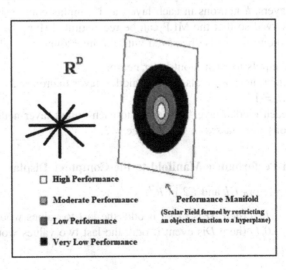

Fig. 2. The performance manifold for a kernel-based regression has radial symmetry

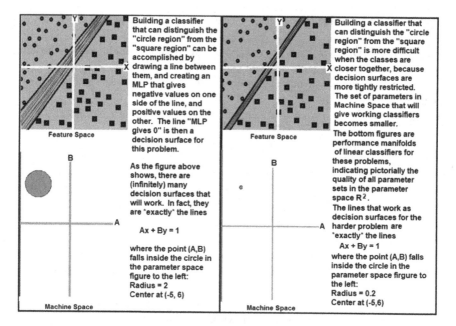

Fig. 3. Simple pictorial example showing feature space and parameter space for a binary classifier in the cartesian plane.

3 Creation of Performance Manifold Displays

In order to establish the parameter space R^D for visualizing performance manifolds for MLP's, the model parameters can be placed in some systematic order. Here we use input layer neuron weights, hidden layer neuron weights, finally output layer weights. The topology (#layers, # neurons in each layer, and # inputs coming into each layer) must also be preserved so that the MLP can be reconstituted (Fig. 4).

The MLP topology must also be saved with the interconnect weights:

features = # inputs to each input-layer neuron = 3
input-layer neurons = # inputs to each hidden layer neuron = 2
hidden layers = 1
neurons in each hidden layer = # inputs to each output layer neuron = 4
output neurons = # classes in Goal Space = 2.

3.1 Mapping a Performance Manifold to the Computer Display

Consider the two vectors *E1* and *E2* in R^D:

$\underline{E}_1$= *(1,0,1,0, ...,1,0)* (here *D* is even; if odd, the last to values would be 0,1)
$\underline{E}_2$= *(0,1,0,1, ...,0,1)* (here *D*is even; if odd, the last two values would be 1,0)

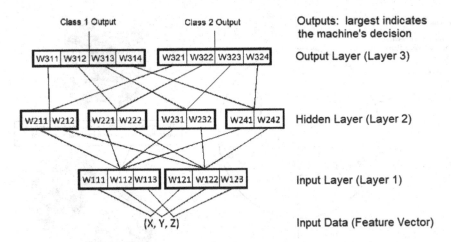

The 22 interconnect weights for this MLP instance can be represented as the parameter space vector:

P = (W111,W112,W113,W121,W122,W123,W211,W212,W221,W222,W231,W232,W241,W242,W311,W312,W313,W314,W321,W322,W323,W324)

Fig. 4. Labeling they components of a multi-layer perceptron

Given Q in R^D, we select the 2-dimensional plane that is the linear span of $\underline{E}_1$ and $\underline{E}_2$. The restriction of Q to this hyperplane is a performance manifold. Each point in this 2-manifold has an associated field value, given by the objective function. The "Best" machine is established as the origin of a coordinates for the display and is positioned in the center of the field of view. The line passing through the center and parallel to *E1* becomes the horizontal axis for the display; the line passing through the center and parallel to *E2* becomes the vertical axis for the display.

Motion within the manifold parallel to the vertical axis adds or subtracts a fixed amount from all of the odd-numbered parameters of the regression (because of how *E1* is defined); Motion within the manifold parallel to the horizontal axis adds or subtracts a fixed amount from all of the even-numbered parameters of the regression (because of how *E2* is defined). By virtue of this coordinatization of the performance manifold, this single two-dimensional manifold captures unbiased information about the effect of every parameter, no matter how many there are, and in an intuitive way.

Motion to the left and up in this presentation causes half of the parameters to increase, and half to decrease. These changes tend to cancel out for small changes, giving most performance manifolds for well-conditioned regressions a upper-left to lower-right symmetry; along such diagonals, performance is often roughly constant... experience has shown this to be a sign of good convergence. Examples are in Sect. 5 (Fig. 5).

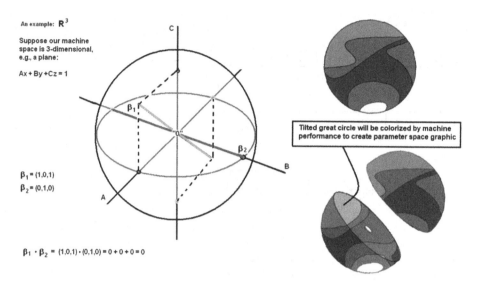

An example: **R**³

Suppose our machine
space is 3-dimensional,
e.g., a plane:

Ax + By +Cz = 1

β_1 = (1,0,1)
β_2 = (0,1,0)

$\beta_1 \cdot \beta_2$ = (1,0,1)·(0,1,0) = 0 + 0 + 0 = 0

Tilted great circle will be colorized by machine
performance to create parameter space graphic

Fig. 5. The slice along a "great circle" has the "best machine" at the center. This great circle disc is rendered as the "Flat" manifold.

3.2   Interpreting Performance Manifolds

Direct mathematical analysis of MLPs is difficult, but in extensive experiments on a variety of problems, the authors have made several empirical observations:

- Underdetermined MLPs (topology too simple to converge to a complete solution) tend to have
- Featureless performance manifolds and so, tiny gradients.
- Over-determined MLPs (topology more complex than necessary for convergence to a complete solution) tend to have highly-complex and isolated sub-optimal extrema.
- Local extrema appear in the manifold as isolated regions of locally high performance, "foothills" at which learning methods relying on gradient techniques (such as Back Propagation) tend to become trapped.

3.3   Streaming a Sequence of Manifolds as Movie Frames to Analyze Parameter Space Context

Even before being mapped to a 2-dimensional display, our "flat" performance manifolds are informed only by a slice of the parameter space, limiting their information content. These displays provide no information about what is happening away from the screen hyperplane. What is really needed is a means of visualizing the "depth" of the parameter space in a neighborhood of the "Best" machine. This requires multiple slices. This can be achieved by arranging a sequence of "flat" manifolds as the frames of an animation, showing how performance changes as the viewer approaches the "Best" machine from a distance in the parameter space.

The basic idea is very simple: instead of just a single $N-1$ slice of parameter space, colorized sequences of $N-1$ dimensional spherical surfaces converging to the training solution are visualized in a movie. Each frame is rendered as a ray traced frame. The frames are then sequenced and merged an animated gif.

One vexing technical conundrum is how to hold constant the area being visualized, since the field of view shrinks to zero as the viewer approaches the "Best" machine at the core of the manifold. Our solution is to move the screen plane closer to the center of the ball at the rate necessary to keep its apparent size constant (Fig. 6).

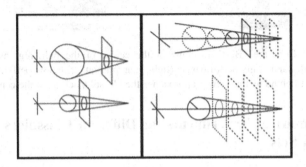

Fig. 6. Reduce the field-of-view to compensate for the decreasing image size

3.4 Animating the Approach Through Parameter Space to an "Optimal" Machine

The following example shows the view as a series of smaller and smaller nested spherical "shells" are penetrated on the approach to the "Best" machine at their center. This was produced by ray tracing the performance manifold of a three-layer MLP with nine feature inputs, five nodes in the hidden layer, and two nodes in the output layer.

The MLP was trained and the graphic was produced by processing a blind hold-back set. The performance of the machine is determined as the classification accuracy on the entire blind set; this value (in the range 0% to 100%) is linearly mapped to one of four gray values for the rendering. In the gray scale version of the performance manifold, the color gradients are in 25% increments from white to black, with light gray being the absolute best machine.

For this example, the Performance Manifold generated 250 frames of animation in 2 h on a core I7 Intel based computer with 6 gigabytes of DDR 3600 MHz memory with a Windows 7 64-bit Operating System.

Figure 7 below are frames from movies depicting a hyperspace approach to the optimized machine by ray-cast animation of the hyper-spherical performance manifolds with continuously decreasing radius.

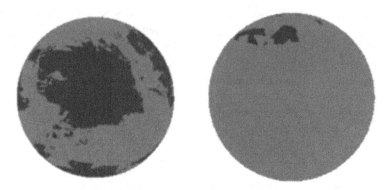

Fig. 7. Far-away from the "Best" machine at the center of the ball, regions of both low performance (dark) and high performance (light) are present. The low-performance regions shrink and are left behind as the viewer approaches the "Best" machine at the center of the ball.

4 Some Empirical Experiments for Different Classifiers and Data Sets

For one set of experiments, four data sets were used, each having a different level of classification difficulty (as determined by an ambiguity score).

The data sets were:

- The UC Irvine Breast Cancer Data Set (Very Easy)
- Synthesized clusters having moderate overlap (Moderately easy)
- Prediction of the number of valence electron of an element based upon several of its physical properties (melting point, etc.) (Difficult)
- The parity-9 problem (Very Difficult)

Four classifiers, each using a different paradigm, were applied to each data set. The classifiers were:

- Linear Discriminant (weak)
- Nearest neighbor (weak)
- Maximum Likelihood (strong)
- Multi-Layer Perceptron (very strong)

As expected, the performance manifolds for weaker classifiers present a less variegated visual scene. More powerful classifiers present a textural complexity (indicating complex interactions between model parameters), local extrema, etc. It is not possible to include samples of all these investigations here.

4.1 Additional Experiments

In Fig. 8, the same MLP topology was used to train a "Best" machine for four problems of varying complexity. The figure depicts performance manifolds at the same fixed distance from the "Best" instance of this MLP for each problem. White represents the highest performance, with successively darker colors indicating lower performance. As suggested earlier in Fig. 3, harder problems tend to have smaller regions of high performance, other factors being equal.

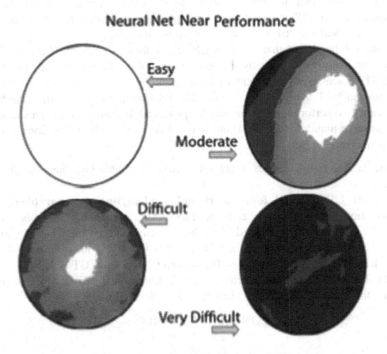

Fig. 8. Harder problems have smaller regions of high performance

5 Performance Manifolds and Modeling Problems

As with other regression formalisms, numerical adequacy, model stability, and model Error for an MLP show up in the regression parameters, and so, in performance manifolds. Certain visual artifacts in displays of performance manifold often indicate specific pathologies. Some of these pathologies are:

- Under-parameterization: the MLP architecture used lacks sufficient representational power to discriminate between the ground-truth classes. Either the number of layers or number of artificial neurons should be increased.
- Over-parameterization: the MLP architecture used has more parameters than needed to discriminate between the ground-truth classes. It is over-fitting and/or modeling random error in addition to data. Either the number of layers or number of artificial neurons should be decreased.
- There are multi-collinearities among the input features. That is, the features are not linearly independent, but related by some simple rule. Because MLP's use linear combinations during computation, features related in this way are redundant, introducing additional complexity without adding information.

5.1   Visual Analysis of Performance Manifolds for Trained MLPs

Once training is complete (takes about 3 min or so), the machine generates two performance manifolds: one colorized by the %-accuracy of the machines surrounding the "BEST" one found, and the other colorized by the precision metric of the machines surrounding the "BEST" one we found. Each pixel in the display represents an entire neural network near the "BEST" one found during training.

It is the left-hand plot that colors machines by their "Geometric Accuracy". The right-hand plot colors machines by the percentage of the data they classified correctly. The "BEST" machine is in the center of each plot.

The trained MLP is also applied to the Blind Set (data not used to train the MLP), and the same two performance manifolds are produced. If training has been successful, the performance manifolds for the Blind Set will be very similar to the corresponding plots for the Training Set.

Example A. 3 layers, 4 features, 3 layer-1 neurons, 7 hidden layer neurons, 4 output neurons.

File: AxeT_L3F4I3H7C4, **four linearly independent, linearly separable clusters**.

Observations: Feature data for consists of four uniform random, 25-member clumps of radius 0.25 in 3-space positioned at the basis vectors and origin. That is, Clump 1 has GT = 1, with centroid $\sim(0,0,0)$; Clump 2 has GT = 2, with centroid $\sim(1,0,0)$; Clump 3 has GT = 3, with centroid $\sim(0,1,0)$; and Clump 4 has GT = 4, with centroid $\sim(0,0,1)$. The classes can be separated using 2-dimensional planes as decision surfaces, and the clump centroids are virtually uncorrelated (Fig. 9).

CONFUSION MATRIX:

1|	25	0	0	0
2|	0	25	0	0
3|	0	0	25	0
4|	0	0	0	25

Fig. 9. Typical symmetry, connectedness, and simplicity for a well-posed classifier

Example B. 3 layers, 4 features, 3 layer-1 neurons, 7 hidden layer neurons, 4 output neurons.

Feature data for SQUARE.csv is a regular 17×34 grid (x, y) in the unit square. Its GT is: $1 + int(0.5 + 2 * (x + y))$.

Observations: The ground truth for this is a quadratic relationship between the two input features; this relationship can be seen in the warping of the performance manifold for precision on the left. It is also evident in the fine structure of the %-accuracy manifold on the right, though this is unlikely to be visible in a printed copy. Once again, the symmetry of the display is from upper-left to lower-right (Fig. 10).

CONFUSION MATRIX (TRAINING SET):

	20	0	0	0	0
1		20	0	0	0	0
2		0	149	0	0	0
3		0	0	253	0	0
4		0	0	0	140	0
5		0	0	0	1	15

Fig. 10. Quadratic interrelationship among features visible in the performance manifold

Example C. 3 layers, 4 features, 5 layer-1 neurons, 11 hidden layer neurons, 2 output neurons.

File: 2x2B_L3F4I5H11C2, **Multicollinearity present.**

Feature data uniform random (x, y, z, w) in unit 4-cube, with $x + y = z + w$. GT random binary {1, 2}.

Observations: This data set was created by intentionally introducing multi-collinearity into the feature vectors: the sum of features 1 and 2 equals the sum of features 3 and 4. This type of information redundancy has been observed to introduce many linear artifacts into the performance manifold, as the linear relationship in feature space cascades through the neurons of the MLP, showing up as lines of constant performance, but at varying slopes. The upper-left to lower-right symmetry has been destroyed (Fig. 11).

CONFUSION MATRIX
(TRAINING SET):

1		40	10
2		9	41

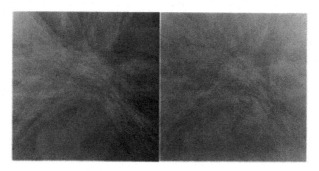

Fig. 11. Loss of symmetry and linear artifacts strong multicollinearity among features

Example D1. 2 layers, 4 input features, 4 layer-1 neurons, no hidden layer, 2 output neurons.

File: rndB_L2F4I4H0C2, **Under-Determined MLP Topology (too few layers and/or neurons).**

Observations: The two ground truth classes were assigned randomly. This was done so that any significant patterns that might appear in the performance manifolds would result from the machine's topology rather than patterns in the input feature vectors. Notice that the machine makes some headway in classifying the training data; but this disappears in the blind test, as it must for random ground truth.

Notice also that the strongest (brightest) visual artifacts in this display run from lower-left to upper-right, the reverse of the pattern occurring in well-conditioned architectures (Fig. 12).

Performance on Blind Set:
  Accuracy: 47%
  Class1 precision: 53.1%
  Class2 precision: 47.1%
  Class1 recall: 46%
  Class2 recall: 48%

CONFUSION MATRIX (TRAINING):

1		38	12
2		21	29

CONFUSION MATRIX (BLIND):

1		23	27
2		26	24

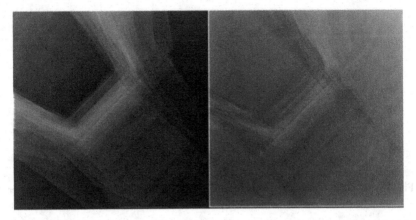

Fig. 12. Rigid, non-standard structure suggests MLP is under-determined

Example D2. 4 layers, 4 input features, 2 layer-1 neurons, 5 hidden layer neurons, 2 output neurons.

File: rndB_L4F4I2H5C2, **Well-determined MLP Topology (right number of layers and neurons).**

Observations: The performance manifold indicates that this architecture is well-conditioned for this problem. It begins to make headway on the training set, and even for the Blind data, it manages to produce elevated recall for Class2 vectors... better than expected for a random ground truth assignment.

Primary symmetry is upper-left to lower-right (Fig. 13).

Performance on Blind Set:
    Accuracy: 46%
    Class1 precision: 44.1%
    Class2 precision: 47.0%
    Class1 recall: 30%
    Class2 recall: 62%

CONFUSION MATRIX (TRAINING):
    1| 32   18
    2| 7   43

CONFUSION MATRIX (BLIND):
    1| 15   35
    2| 19   31

Fig. 13. Typical symmetry, connectedness, and simplicity for a well-posed classifier

Example D3. 5 layers, 4 input features, 6 layer-1 neurons, 13 hidden layer neurons, 2 output neurons.

File: rndB_L5F4I6H13C2,**Over-determined MLP Topology (too many layers and or neurons).**

Observations: Over-parameterization (too many layers, too many neurons) caused the performance manifold to bifurcate, with the upper branch having smooth gradient, and the lower riddled with a complex distribution of local extrema. A gradient technique would likely not be able to traverse the lower branch... and this is the branch containing the "Best" machine. As is often the case, this over-determined machine controls the training set well; but this is misleading, because this performance does not generalize to blind data. Primary symmetry is upper-left to lower-right (Fig. 14).

CONFUSION MATRIX (TRAINING SET):
```
1|    48    2
2|     2   48
```

CONFUSION MATRIX (BLIND SET):
```
1|    24   26
2|    24   26
```

Performance on Blind Set:
Accuracy: 50%
Class1 precision: 50%
Class2 precision: 50%
Class1 recall: 48%
Class2 recall: 52%

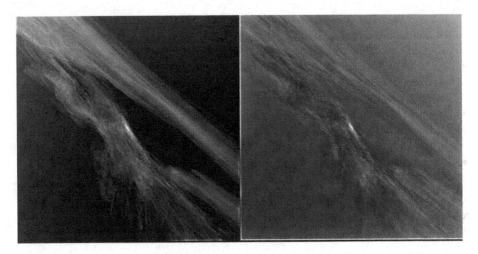

Fig. 14. Bifurcated into a stable band, and a chaotic band. Suggests MLP is over-determined.

6   Conclusions

- All modeling involves at least three spaces: feature space, where the empirical data resides; goal space, where the ground-truth resides; and machine space, where the regression parameters reside.
- We usually think of doing modeling in the feature space. But models themselves can be modeled (meta-modeling) by applying modeling techniques to their machine spaces.
- Valuable information about machine architectures, training, and performance can be gained by meta-modeling.

We usually think of modeling in the feature space. However, models themselves can be modeled (meta-modeling) by applying modeling techniques to the parameter spaces. Valuable information about machine architectures, training, and performance can be gained by forensic meta-modeling. In particular, the use of ray cast animation in N-dimensional space makes possible direct, intuitive forensic analysis of learning machine implementations. Important phenomena can be direct observed in this way, such as:

- The presence and distribution of local extrema of the objective function used for machine learning
- Adequacy of the learning architecture.

7   Future Work

- Build a user interface that facilitates direct interact with the parameter/machine space. Functionality would include zooming in and out and spin the sphere to discover the range of solutions for the machine. Allow the user to manually "game" the coloration: adjusting the color boundaries of the model by accuracy to see how accurate a machine can get before it is not efficient.

- Use described visual methods not just for forensics, but for design as well. In an interface similar to the one described above, a user could be given the ability to select a machine by mouse click.
- Optimize the rendering process for speed to support direct user interaction.
- Additional cross-paradigm experimentation using other reasoning paradigms, such as Self Organizing Maps.
- Use for analysis of known pathologies (such as local extrema).

Acknowledgements. The authors would like to thank the Sirius Project for facilitating this research. The Principal Author would like to thank Aneta Compton for her contribution to making this work possible.

References

1. Appel, A.: Some techniques for shading machine renderings of solids. In: Proceedings of AFIPS JSCC, vol. 32, pp. 37–45 (1968)
2. Arbib, M.A.: Handbook of Brain Theory and Neural Networks, pp. 157–160. Bradford Books, MIT Press, Cambridge (2002)
3. Cortes, C., Vapnik, V.: Support-vector networks. Mach. Learn. **20**(3), 273–297 (1995)
4. Delmater, R., Hancock, M.: Data Mining Explained, pp. 145–146. Digital Press, Newton (2001)
5. Hancock, M.: Near and long-term load prediction using radial basis function networks. In: Progress in Neural Processing, vol. 5, pp. 113–114. World Scientific (1995)
6. Haykin, S.: Neural Networks: A Comprehensive Foundation, 2nd edn. Prentice Hall, Upper Saddle River (1998)
7. Leondes, C.T.: Knowledge-Based Systems: Techniques and Applications, vol. 2, 1st edn. p. 436. Academic Press, San Diego (2000)
8. Pearl, J., Russell, S.: Bayesian Networks, Handbook of Brain Theory and Neural Networks. MIT Press, Cambridge (2001)
9. Winkler, K.P. (ed.): Locke, John, An Essay Concerning Human Understanding, pp. 33–36. Hackett Publishing Company, Indianapolis (1996)

Geometrically Intuitive Rendering
of High-Dimensional Data

Monte Hancock[1]([⊠]), Kristy Sproul[2]([⊠]), Jared Stiers[2]([⊠]),
Benjamin Bowles[2]([⊠]), Fiona Swarr[2]([⊠]), Jason Privette[2]([⊠]),
Michael Shrider[2]([⊠]), and Antoinette Hadgis[2]([⊠])

[1] 4Digital, Los Angeles, CA, USA
practicaldatamining@gmail.com
[2] Sirius19, Melbourne, FL, USA
kristy.sproul@gmail.com, StiersJD@gmail.com,
benbowles2@gmail.com, f.Swarr@gmail.com,
privetteJasonLee@gmail.com, twomagpies115@gmail.com,
m.shrider@yahoo.com, Sirus19conf@gmail.com

Abstract. Visualization is arguably the most powerful manual analytic method. However, as enterprise data have grown in size and complexity, supporting fast and intuitive data visualization has been become increasingly difficult. In particular, when the data being analyzed have many dimensions (e.g., feature vectors having many attributes), conventional methods, when they can be used at all, often make representational compromises that produce non-intuitive and misleading graphical products.

Three conventional methods for displaying sets of high-dimensional vectors on a flat computer display are described: pair-plots, feature-winnowing, and parallel coordinate plots. These are compared with two methods the authors have formulated, Recursive Stereographic Projection (RSP), and Tensor Projection (TP), for producing visual displays of high-dimensional discrete data that are fast, intuitive, navigable, and informative, and tend to preserve geometric characteristics of the original high-dimensional data such as, occlusion, proximity, intersection, relative scale, spatial extent, shape, and coherence.

Several real-world examples of intuitive 2-dimensional renderings of high-dimensional data are given. Suggestions for future research are made, and references provided.

Keywords: Data visualization · High-dimensional data · Data mining

1 Feature Vectors and Feature Spaces

Vectors and vector-valued functions will be indicated by underscoring, e.g., V, A(x); the norm of a vector is denoted |V|. The set of real numbers is denoted by R, and N-dimensional Euclidean space by RN.

A <u>feature</u> is a numeric value chosen to encode an attribute of an entity being modeled (e.g., size, quality, age, color, etc.). While features can be numbers, symbols, or complex data objects, they are usually coded in numeric form before modeling is performed.

© Springer Nature Switzerland AG 2019
D. D. Schmorrow and C. M. Fidopiastis (Eds.): HCII 2019, LNAI 11580, pp. 211–224, 2019.
https://doi.org/10.1007/978-3-030-22419-6_16

Once features have been designated, a feature space can be defined for a domain by placing the features of each entity into an ordered array in a systematic way.

Each instance of an entity having the given features is then represented by a single point in n-dimensional Euclidean space, called its feature vector. The space in which the feature vectors reside is the feature space.

1.1   The Particular Benefits of Visualization

Visual analysis has significant advantages over tabular and chart-based analysis:

1. Visualization allows the analyst to develop "intuition" about the intelligence domain, detection of known and emerging patterns, and characterization of entity behaviors. When an analyst is able to see data, she develops an intuitive understanding of it that can be obtained in no other way.
2. Visualization can lead to the direct discovery of domain "rules".
3. Visualization is an effective way to scan large amounts of data for "significant differences" between classes. If the icons representing interesting behaviors are well separated from uninteresting behaviors along some axis, this will be visually apparent. The feature that axis represents, then, accounts for some of the difference between interesting and uninteresting behaviors.
4. Visualization is a good technique for detecting poorly conditioned data. When data are plotted, errors in data preparation often show up as unexpected irregularities in the visual pattern of the data.
5. Visualization is a good technique for selecting a data model. For example, data that is normally distributed will have a characteristic shape. Visual analysis lets the analyst "see" which of several models might be most reasonable for a population being studied.
6. Visualization is a good technique for spotting outliers and missing data. When data are visualized, outliers (data not conforming to the general pattern of the population) are often easy to pick out. Missing records may show up as "holes" in a pattern, and missing fields within a record often break patterns in easily detectable ways.

1.2   Visualization Methods

The most effective pattern recognition equipment known to man is "wetware": the mass of neurons between the users' ears. Manual knowledge discovery is facilitated when data are presented in a form that allows the human mind to interact with them at an intuitive level. Visual data presentation provides a high-bandwidth, naturally intuitive gate to the human mind.

Data visualization is more than depiction of data in interesting plots. The goal of visualization is to help the analyst gain intuition about the data being observed. Therefore, visualization applications frequently assist the user in selecting display formats, viewer perspectives, and data representation schemas that foster deep intuitive understanding. Of particular importance is the fact that visualization allows fast mental correlation of many attributes at once: it makes the joint distribution of features accessible.

Several conventional techniques exist for the visualization of 2 and 3-dimensional data: scatter-plots of features by pairs, histograms, relationship trees, bar charts, pie charts, tables, etc. Modern visualization tools often include other capabilities (e.g., searching, sorting, scaling). Advanced functions might include:

- drill-down
- hyper-links to related data sources
- multiple simultaneous data views
- roam/pan/zoom
- select/count/analyze objects in display
- dynamic/active views (e.g., Java applets)
- using animation to depict a sequence or process
- creative use of color
- statistical summarization
- use of glyphs (icons indicating additional information about displayed objects)

When data have more than 3-dimensions, however, they become more difficult to conceptualize. Graphical techniques for representing high-dimensional data are often challenging to implement and use.

2 Classical Methods

- Many conventional methods have been developed for the representation of high-dimensional data. Most rely on some non-geometric way to retain some of the information lost when dimensions are suppressed:
- Using "glyphs": indicating two features as spatial x-y coordinates, and other features using aliases such as color, shape, size, etc. This simple type of display is often supported by conventional OLAP tools.
- Performing orthographic projection ("feature winnowing"): using spatial dimensions to plot two or three features, while ignoring the rest
- Using trees and graphs to encode dimensions in different regions of the display (e.g., quadtrees)
- Plotting data in parallel coordinates, so that dimensions are spread horizontally across the display

2.1 Pair Plots

Pair plots are an extreme instance of feature-winnowing: plot the data in just two of the available N features and ignore the rest. As such, pair plots are not actually a method for visualizing high-dimensional data; but they are the most common approach to the dimensionality problem in practice.

When this approach is taken, there are two common strategies to mitigate information loss. Strategy 1 is to select the two features that are the most informative when used together. Strategy 2 is to plot thumbnails of every one of the N-choose-2 possible pairs of features.

These strategies have their own inherent problems. For Strategy 1, users will often choose to plot the data using the two features most highly correlated with the data being sought. However, since the most informative features are often correlated with each other, this strategy amounts to plotting data using redundant features, and often yields sub-optimal results.

For Strategy 2, the number of pair plots grows quadratically in the number of features; with just N = 10 features, the number of pair plots is already up to 45. Plotting data in pairs hides the contextual links between groups of features, so all 45 plots must be mentally re-contextualized by the user... who is likely plotting the data to learn the context in the first place. Further, it is a nasty (but little known) statistical fact that N columns of data can be pairwise correlated in every pair, but jointly uncorrelated. Since it is the joint-correlation that is the repository of reliable information, the pair plot approach can result in both imaginary and overlooked discoveries. Figure 1 is a pair plot of 5,000 vectors having 5 features each. There are 100 classes, but much discriminating information has been lost.

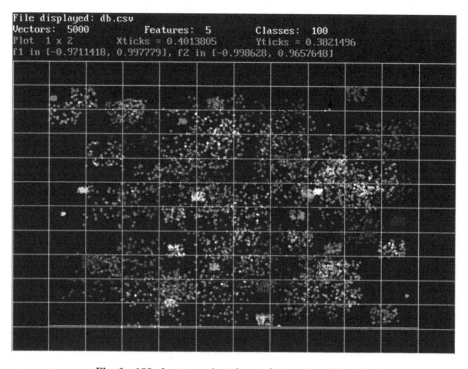

Fig. 1. 100 clusters projected onto features 1 and 2 only

2.2 The Bad News About Feature-Winnowing

Does N-dimensional visualization show me anything I can't see in lower dimensions? That is, how much damage is done by simply discarding a single feature, going from N-dimensions to $N-1$ dimensions? The following very real example shows that the loss of information can be substantial, even for very simple data sets.

For this example, we are given a data set of 4-dimensional feature vectors that is known to consist of five classes: A, B, C, D, and E, each containing over 1,000 vectors. The vectors of each class form tight and distinct clusters that do not intersect any of the others. If our graphics tools only support up to three dimensions, one of the features of our 4-dimensional space will have to be removed (Fig. 2).

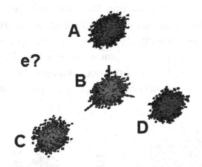

Fig. 2. Plot of feature vectors after feature-winnowing into 3-dimensional space (the coordinate axes can be seen extending out from the body of cluster B)

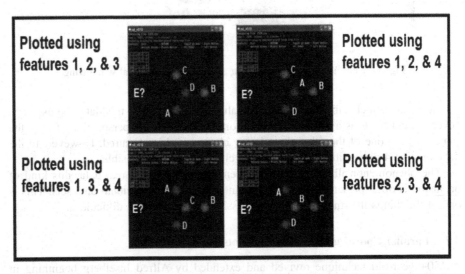

Fig. 3. Display the data in several projections to insure that something isn't "lost" by leaving out a "good" dimension.

To determine which feature to remove, each of the four possible projections is visualized in three dimensions; it doesn't seem to matter which is removed. However, cluster E is not visible in any of the renderings.

There are four 3-D feature-winnowing results:

1. Plot the data in features 1, 2, and 3 (Fig. 3, upper-left plot)
2. Plot the data in features 1, 2, and 4 (Fig. 3, upper-right plot)
3. Plot the data in features 1, 3, and 4 (Fig. 3, lower-left plot)
4. Plot the data in features 2, 3, and 4 (Fig. 3, lower-right plot)

No matter which of the four features is dropped, we obtain essentially the same view of the data, with the positions of the clusters permuted. But none of these views shows cluster E. Where is it?

If, instead of throwing away a feature, the data are viewed using one of the geometrically faithful N-dimensional visualizers in the data's native 4-dimensional space, the presence of cluster E is immediately revealed: a fifth, coherent 1,000-point cluster that cannot be seen in any 3D feature-winnowing. For this data set, removal of any of the four features causes all class E vectors to be located at the center of one of the other classes. Not only has it become invisible, its class identity has been lost (Fig. 4).

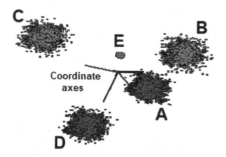

Fig. 4. Cluster E Cannot be seen in any lower-dimensional winnowing

No matter which subset of 1, 2, or 3 features is used to plot this data, the user will never see cluster E as a separate aggregation. From every 3D perspective, it is in the same place as one of the other three larger clusters and is obscured. However, In the data's native 4-dimensional space, all five clusters are easily visible.

The phenomenon illustrated here happens in practice with real data; this kind of occlusion is often present to some degree in large, high-dimensional data sets. This is one of the things that makes pattern processing in these spaces difficult.

2.3   Parallel Coordinates (Inselberg Plots)

A 150+ year-old technique revived and extended by Alfred Inselberg beginning in 1959, Parallel coordinates retain all dimensions. To produce graphical displays, a separate vertical axis is used for each dimension. A feature vector is no longer a point in space, but a trace of end-to-end line segments across the display.

It is customary for the leftmost vertical axis to receive the first feature, then moving to the right for features 2 through N. Highly coherent clusters show up as bands of traces across the display.

Parallel coordinate plots have proven to be quite useful in applications. If each class of vectors is colorized, or plotted on a different horizontal axis, these plots provide an excellent method of looking for highly discriminating features. In Fig. 5, for example, class 1 is on the top plot, and class 2 is on the bottom plot. It is readily seen that features 5, 6, and 8 might be used to distinguish class 1 from class 2.

Parallel coordinates, since they do not retain an intuitive analogy with the underlying data geometry require experience to interpret. The notions of cluster proximity, occlusion, shape, and spatial context are mostly lost.

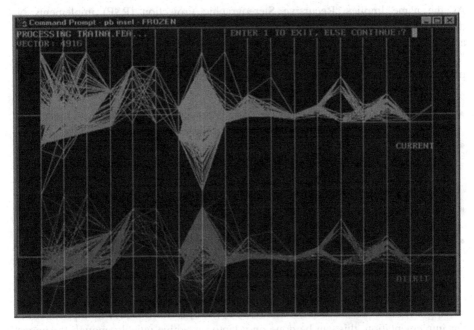

Fig. 5. Parallel Coordinate plot of classes (CURRENT, ATTRIT)

3 Geometrically Intuitive Methods

Two methods will now be presented that the authors have formulated and applied to many real-world problems involving the visualization of discrete data (sets of feature vectors) having from 3 to 100 dimensions: Recursive Stereographic Projection (RSP), Tensor Projection (TP).

These methods render colorized graphical displays of high-dimensional discrete data that:

- are informative while remaining geometrically intuitive
- support on-screen data annotation (e.g., lasso clusters, pseudo-color visual artifacts)
- TP is fast enough that desktop machines can smoothly navigate 100-dimensional data sets having 50,000 points. RSP can smoothly navigate 20-dimensional data sets having 5,000 points.
- can stream data frames to disc to create movies of high-dimensional data sets in their native spaces tend to preserve geometric characteristics of the original high-dimensional data such as, occlusion, proximity, intersection, relative scale, spatial extent, shape, and coherence

3.1   Recursive Stereographic Projection (RSP)

As the name implies, Recursive Stereographic Projection (RSP) implements N-dimensional visualization using a recursive stereographic projection, a mathematical analogy between a sphere and a plane missing one point.

A positive real value r is chosen for the radii of the spheres by means of which the projections are done. The viewer is positioned at (r, 0, 0, ..., 0).

Each down-projection follows the same pattern (Fig. 6):

- A hyperplane to receive the projected data is perpendicular to the position vector of the viewer, and at distance equal to the r on the far side of the origin.
- All features vectors are z-scored coordinate-by-coordinate. Each vector is then normalized to length r. These two steps project the entire data set onto the $N - 1$ dimensional sphere centered at (r/2.0.0.,...0) having radius r/2. This places the data at the center of the display, which is where the algorithm maps the origin.
- The recursion proceeds by stereographically projecting each data point from the viewer location at the topmost point of the sphere onto a target hyperplane.
- Each projection reduces the dimension of the data by 1. The recursion is repeated on each data point until a set of two-dimensional data are obtained. This pair is plotted on the display as the location of the original point.

Visual navigation is accomplished by moving the feature vectors. For example, rotating a feature vector about the line (r, 0, 0, ..., 0) is a fast matrix multiplication. In our implementation, this can be done on a laptop machine quickly enough to support smooth motion for 20 or fewer dimensions and 5,000 or fewer vectors. By repeating this operation using a small rotation angle, a "fly-around" survey of the data from various viewer perspectives is created.

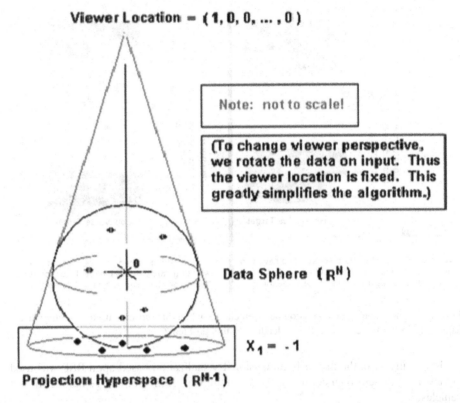

Fig. 6. Each projection to the next lower dimension is a projection from the normalized data on the surface of a ball, and the hyperplane beneath the ball. This is a one-point perspective projection, relative distances are scaled by how far they are from the viewer.

As an example of the utility of this visualizer, consider the following application.

Work was underway on an important system for detecting and classifying opposing military vehicles from a distance of several miles. A sensor was developed to collect "signatures" from the vehicles so that human analysts, and eventually intelligent software, would be able to create catalogs of distant military threats so that the evolving posture of adversaries could be assessed in real-time.

Vehicle signatures were first collected at a test range. Sometimes the signatures were easy to match to vehicles, but there were quite a few instances for which dangerous offensive vehicles looked like trucks. Sending light armor to engage a "line of trucks" which turns out to be a column of tanks is a fatal mistake (Fig. 7).

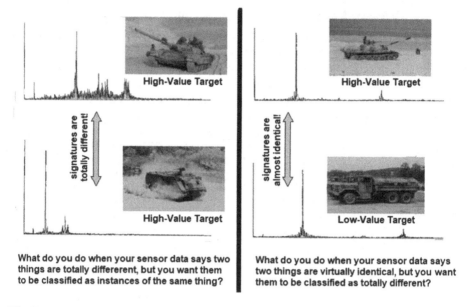

Fig. 7. Just how bad is the confusion between classes when this sensor is used? Exactly what is happening in the feature space that is leading to this problem?

In this instance, the use of N-dimensional visualization based upon RSP was used to view a 32-dimensional coding of the signature waveforms for a large number of test vehicles.

When visualized in its native 32-dimensional space, the signature data presents as three coherent, overlapping clusters. The "branch" at the upper left consists almost entirely of vectors of class1; the "branch" at the upper right consists almost entirely of vectors of class2. The clump at the bottom, consisting of about 25% of the data in even class proportions, is an unresolvable mixture (Fig. 8).

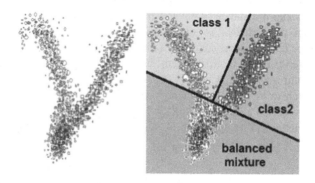

Fig. 8. In 32-dimensional space, it was easy to see what was happening

Using this knowledge, we construct a classifier by segmenting the data using two perpendicular hyperplanes and labeling the bounded regions as in Fig. 8.

This classifier will have a classification accuracy of about 75% from the class1 and class2 regions, which are heterogeneous by class. Randomly assigning a class decision to points from the balanced mixture will be correct about 12.5% of the time, giving an overall classification accuracy on the training data of 87.5%

It is unlikely that a classifier would be built this way, but the value here is that it establishes the fact that any classifier that does not have a classification accuracy of at least 87.5% on this training set is inferior. Further, we knew that it would not be necessary to redesign the sensor (a massive expense) but could obtain reasonable performance with modest software tweaking.

3.2 Tensor Projection (TP)

For clarity, we pose an illustrative example (Fig. 9). Other instances are analogous.

Consider the example of a 9-dimensional feature vector $V = (f_1, f_2, \ldots, f_9)$ in R^9.

Let $B = (b_1, b_2, b_3)$ and $A = (a_1, a_2, a_3)$ be vectors of real parameters,

Form a tensor product to induce a mapping from R^9 to R^2, where $V \to X = (x_1, x_2)$:

$$\begin{pmatrix} b_1 & 0 & b_3 \\ 0 & b_2 & 0 \end{pmatrix} * \begin{pmatrix} f_1 & f_2 & f_3 \\ f_4 & f_5 & f_6 \\ f_7 & f_8 & f_9 \end{pmatrix} * \begin{pmatrix} a_1 \\ a_2 \\ a_3 \end{pmatrix} = \begin{pmatrix} x_1 \\ x_2 \end{pmatrix}$$

Fig. 9. This tensor product provides a parametric linear mapping from the feature space to R^2

To render the vector V in R9 on a flat display, we plot (x1, x2).

In general, given a feature vector $W = (w_1, w_2, \ldots, w_m)$ in R^M, we form a tensor product as above:

- Begin with the N-by-N identity matrix I_N, where N is the smallest perfect square $\geq$ M. Set $K = \sqrt{N}$.
- load W into I_N in some standard order (we use row-major order).
- Select at parameter vector B in R^K and creating two rows as above with the top row having 0 in the even positions, and the bottom row having 0 in the odd positions, with the coordinates of B interleaved as above. (Exactly how this interleaving is chosen is somewhat arbitrary, of course).
- Select parameter vector A in R^K, place it into the column vector to the right of I_N.
- Plot the resulting tensor product (x_1, x_2) for W.

The coordinates of A and B provide parameters that the user can manipulate to modify the mapping for feedback in real time.

The TP high-dimensional visualization creates geometrically faithful views of data points as they actually reside in their high-dimensional space. Nearness, occlusion, and perspective are preserved.

We have implemented the GUI concept depicted in Fig. 10.

Initially, all the data are at the center of the display, because all parameters are set to 0.

The boxes at the upper left-hand corner of the large rectangular display box have a blue boundary. These are the transformation parameter controls (a_1, a_2, etc.). The white line is "zero"; values above the white line are positive, and values below the white line are negative.

The last (rightmost) bar is the "zoom" bar. It allows you to zoom in or out of the central region of the display. A negative zoom value will produce a mirror image of the corresponding positive zoom value.

Holding down the left mouse button, the user drags a transformation parameter color bar up and down. This sets the corresponding transformation parameter. Different parameter values give different projections of the feature data, changing the display accordingly, in real-time.

The boxes along the upper right-hand corner of the large rectangular display box have a red boundary. These are the navigation parameter controls (b_1, b_2, etc.). For all but the bottom bar, the white line is "zero"; values to the right of the white line are positive, and values to the left of the white line are negative.

This positive-negative paradigm does not apply to the bottom bar. The last (bottommost) bar is the "icon size" bar. It allows you to change the size of the icons used to represent data points. The values range from 0 (1 pixel) to a maximum positive radius.

Holding down the left mouse button, the user drags a navigation parameter color bar left and right. This sets the corresponding navigation parameter. Different parameter values effectively allow you to view the data from different locations, changing the display accordingly, in real-time.

Figure 10 is a screen shot of an implementation of a TP display after loading a file containing 100 clusters each consisting of 20 points in a 100-dimensional space. The clusters are plainly visible, and intuitive. What can't be seen on this screen-shot is the fact that TP allows the user to rotate and "flex" the data in various ways, zoom, change the icon size, and save lower dimensional representations of the data.

Dimension reduction for visualization produces successively lower-dimensional representations of the data, which can be saved for analysis or as feature winnowing products.

As an example of the utility of this visualizer, consider the following application.

Face-to-face interactions among groups of people follow certain conventions. These conventions provide an orderly structure for effective communication. This results in the formation of a variety of associations/relationships, such as "cliques", as

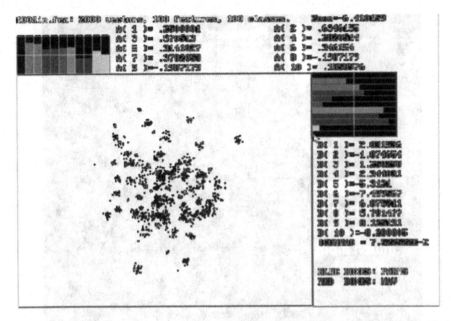

Fig. 10. TP allows on-screen scaling of coordinates so that clusters can be easily identified

people gather around and/or disperse driven by interests, demographics, and goals. These associations have been studied extensively, and analysis of the resulting social structures can be used to characterize the interests, goals, and even personalities of those involved.

Interactions among the participants on social media platforms also give rise to complex social structures, which are the basis of social engineering, forensic data mining, and behavior analysis.

When viewed graphically, certain types of social media interaction become immediately apparent: cliques can be analyzed visually, complex interactions that would otherwise defy detection can be seen, and forecasting and planning are enhanced. In particular, visualization of high-dimensional social media structures can be used to support profiling for marketing products, assessing and influencing public opinion, and estimating malware risk and susceptibility.

FACT: Human and BOT behaviors often look very different in this type of visualization.

A large collection of social media posts was numericized using text processing techniques. The resulting feature space was then visualized using the TP visualizer, showing a clear difference between BOT behaviors and human behaviors on this social media platform. This has phenomenon has since been observed visually in this way on other social media platforms (Fig. 11).

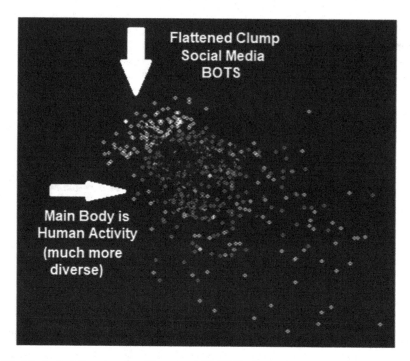

Fig. 11. The lack of diversity of a wide variety of BOTS is clearly seen in this 10-dimensional space

4 Future Work

- Build a user interface, allowing users to interact with the machine space. Functionality would include zooming in and out and spin the sphere to discover the range of solutions for the machine. Allow the user to manually "game" the coloration: adjusting the color boundaries of the model by accuracy to see how accurate a machine can get before it is not efficient.
- Use described visual methods not just for forensics, but for design as well. In an interface similar to the one described above, a user could be given the ability to select a machine by mouse click.
- Optimize the rendering process for speed to support direct user interaction.
- Additional cross-paradigm experimentation using other reasoning paradigms, such as Self Organizing Maps.
- Use for analysis of known pathologies (such as local extrema).

Acknowledgments. The authors would like to thank the Sirius Project for facilitating this research.

References

1. Appel, A.: Some techniques for shading machine renderings of solids. In: Proceedings of AFIPS JSCC, vol. 32, pp. 37–45 (1968)
2. Delmater, R., Hancock, M.: Data Mining Explained, pp. 145–146. Digital Press (2001)

Enacting Virtual Reality: The Philosophy and Cognitive Science of Optimal Virtual Experience

Garri Hovhannisyan[1,4], Anna Henson[2,4(✉)], and Suraj Sood[3,4]

[1] Duquesne University, Pittsburgh, PA, USA
[2] Carnegie Mellon University, Pittsburgh, PA, USA
henson.anna@gmail.com
[3] University of West Georgia, Carrollton, GA, USA
[4] Sirius Project, Melbourne, USA

Abstract. The standard approach to immersive virtual reality (VR) is arguably "object-centric" in that it aims to design physically realistic virtual experiences. This article deems the object-centric approach both philosophically and theoretically problematic and builds up to an alternative, "action-predicated" approach, whose aim is to simulate virtual experiences with a primary emphasis on pragmatic functionality instead. Section 1 lays out the rationale of the article and provides an outline for its general structure. Section 2 illustrates the nature of the problem being tackled and articulates a philosophically motivated critique, demonstrating the necessary limitations of the standard approach, as well as the need for an alternative. Section 3 draws on the enactive approach to cognitive science and begins the formulation of such an alternative. Section 4 completes the turn toward an action-predicated approach and argues, in particular, for a flow-based conception of immersive VR experience. Section 5 systematically discusses the methodological implications of the theoretical merits of this article by examining a design probe, *Wake*, conducted on participants ($N = 25$) in a mixed reality (MR) setting. Finally, Sect. 6 constitutes the conclusion of this article, wherein its philosophical, theoretical, and methodological efforts, as well as possible avenues for future research, are briefly noted.

Keywords: Enactivism · Virtual reality (VR) · Flow · Immersion · Social presence

1 Introduction

This article takes as its point of departure the belief that the contemporary literature on virtual reality (VR) is theoretically confused. In particular, our claim is that VR has inherited and remains tacitly committed to a philosophically impoverished, and hence theoretically implausible, view of human perceptual reality, as a consequence of which its attempts at simulating what is "real" either fall short in practice, or else remain ill-conceived theoretically even despite their apparent practical success. We believe, though, that this confusion, along with its practical challenges, can be remedied, and that the means to remedying it require explicit philosophical engagement with the existing body of literature on VR. We are quite aware that philosophical concerns are

© Springer Nature Switzerland AG 2019
D. D. Schmorrow and C. M. Fidopiastis (Eds.): HCII 2019, LNAI 11580, pp. 225–255, 2019.
https://doi.org/10.1007/978-3-030-22419-6_17

often regarded as irrelevant or only peripheral in importance to matters of technological design and innovation in such fields as VR. However, what we hope to demonstrate with our contribution is an alternative to this view. Specifically, we aim to show that determining how exactly to render virtual experiences more subjectively real or plausible is not only a matter of technological design, but, indeed, also of philosophical and theoretical commitment.

Having presented the overarching rationale of our project, the structure of this article is as follows. In Sect. 2, we articulate a critique of the received view in VR, which sets as the design goal for VR the maximization of "presence". Presence is a perceptual illusion that is brought about when the contents of an immersive virtual simulation are found to be sufficiently "believable" so as to be treated as "real" by the user. We show that this design goal is predicated, first, on the assumption that the reality to be simulated by VR is physical reality (reality as a mere "collection of objects"), and second, that perceptual reality as such is realized strictly in the brain, as apart from or independent of various subjective factors, such as emotion, motivation, and cognition, and without the need for embodied interaction with the world. In our critique, we demonstrate how both of these assumptions are actually rooted in an outdated and particularly reductive brand of physicalist philosophy, such that we find it difficult to imagine how any set of design guidelines for VR that is predicated on these assumptions can help to explain or even generate meaningful virtual experience. We close our critique with the claim that adopting a physicalist approach to VR in this way results in three theoretical "gaps" or challenges, which, we maintain, cannot be bridged by physicalist means alone.

Sections 3 and 4, accordingly, constitute the constructive elements of our contribution. We thus turn to the "enactive approach" in cognitive science, which we believe offers a viable alternative to the physicalism implicit in the received view. For despite undertaking a study of subjectivity that is scientific in kind, enactivism nevertheless manages to maintain a commitment to non-reductionism in so doing. Thus, the enactive approach promises to bridge the three theoretical gaps besetting VR, insofar as it goes beyond what the reductive brand of physicalism is able to offer.

In Sect. 3, we synthesize a theory of perception on enactive grounds, wherein perception is conceived of as a necessarily embodied, interactive, affective, motivated, and cognitive process. In advancing a non-reductive account of perception that is action-predicated (pragmatic), rather than object-centric (physicalist), we demonstrate how the first two theoretical gaps can be bridged. In Sect. 4, we then extend our theoretical framework to also bridge the third theoretical gap. We begin by reviewing recent cognitive scientific work on "the flow state", often described as the state of optimal experience [5, 17], and we integrate these findings with our own enactive theory of perception. We thus explain the phenomenon of subjective immersion in terms of flow, that is, as an action-predicated process of optimal perceptual engagement with the world. We then use our flow-based approach to subjective immersion to explain the phenomenon (and phenomenology) of immersion as it pertains to VR experience (optimal virtual experience). Finally, we conclude this section by arguing that the primary design goal of VR should be, not the maximization of presence through the simulation of physically realistic experiences as mediated by the objectively immersive VR hardware, but the maximization of subjective immersion by developing virtual experiences that are able to reliably facilitate a flow state within users.

In Sect. 5 we proceed to demonstrate the relevance of our theoretical contribution for praxis. Here, we articulate a set of methodological implications that follow from adopting subjective immersion, rather than presence, as the primary goal for VR design. These implications are grouped in such a way as to address the following four elements of VR experience: (1) "Onboarding", (2) Immersion, (3) "Offboarding", and (4) "Experiential optimization". We conduct our discussion of methodology by examining a design probe, *Wake*, that was conducted with users in a mixed reality (MR) setting, which we use to detail the implementation of the proposed implications in an empirical, real-world setting [18]. Section 6 constitutes the conclusion of this article. In it, we briefly summarize and assess the philosophical, theoretical, and methodological import of our argument and note possible avenues for future research.

2   The Received View and Its Discontents

This section advances the claim that any set of design guidelines that is derived from or predicated upon a strictly physicalist conception of reality can only ever lead to sub-optimal VR experience, when implemented. To this end, we begin by first unpacking what we mean by physicalism and we demonstrate in what way exactly physicalism is assumed within the VR literature. We then proceed to demonstrate the issues associated with assuming physicalism in this way, whereby we identify a total of three theoretical gaps besetting the VR literature, which, we argue, cannot (in principle) be bridged by physicalist means alone. We conclude with the proposition that addressing these gaps, and thus conceptualizing optimal VR experience, requires going beyond mere physicalism. In Sect. 3, we begin our elaboration of this proposition by offering an alternative (non-reductive) conception of perceptual reality that is rooted in the enactive approach to cognitive science.

Physicalism is the view that what is real is, most basically put, just what is physical. In other words, reality consists exclusively of physical entities, such that every observable phenomenon is, in essence, physical in nature, and can ultimately be explained by essentially physical causes. Physicalism is the prevalent position in the natural sciences, and is often assumed as a matter of fact, rather than philosophical commitment. Despite both its prevalence and its success as a doctrine, physicalism is not without its own set of problems—especially when taken for granted within the context of cognitive science and related fields of study (e.g., philosophy of mind, psychology). VR happens to be one such field of study. Even so, the philosophical implications of physicalism for VR have, to this point, remained largely unnoticed, and therefore neglected by theorists. We therefore find it crucial to bring these problems to the fore and to explicitly address them, insofar as much of the theoretical labor and design efforts in VR necessarily depend on the philosophical commitments that are made at the outset.

It is important to begin by first pointing out that the physicalism assumed in the mind sciences—and VR, by extension—is not of the same kind as what is assumed within the natural sciences today. Historically, the mind sciences (particularly, psychology) tried to fashion themselves after the natural sciences in order to legitimize their standing as a scientific discipline in their own right, and they did so by appropriating the basic philosophical and methodological assumptions of the natural sciences [32]. At the time when this appropriation occured, though—(during the late 19th and early 20th centuries)—the natural sciences were still embedded in a Newtonian worldview, according to which the world is essentially a place of material objects (inert masses) whose motion is governed by a set of mechanical laws (e.g., the law of gravity).

Essential to the Newtonian conception of reality were the following two suppositions: (1) "Materialism", according to which reality is essentially material or physical, and (2) "Mechanism", according to which causality is a wholly linear process whereby every physical event can be explained in terms of some temporally antecedent cause [27, 28]. It goes without saying that the natural sciences have advanced beyond Newtonianism in their construal of reality, particularly in light of the various epistemological innovations witnessed over the course of the 20th century, such as quantum mechanics, Einstein's theory of general (and special) relativity, as well as complexity science and information theory The same, however, cannot be said of the mind sciences, which, on the whole, remain fixated on an outdated (Newtonian) worldview.

In having committed to such suppositions as materialism and mechanism, the general approach to studying the mind, thereof, has been largely reductive. In particular, the materialist supposition has motivated a conception of mental functioning as wholly physical and objective in kind, thus leaving unexplained various core (subjective) properties of the mind, such as consciousness [19], normativity [14, 15], and purposiveness (goal-directedness) [33]. Accordingly, the mechanistic supposition has motivated a conception of mental functioning as a form of information-processing, as exemplified by the metaphor of the mind as a computer most notably used in the discipline of cognitive psychology. According to the information-processing model, the mind is to the software of a computer as the brain is to the hardware. The brain receives information through the senses (input) and processes it in such a way as to produce functional action in the world (output). Cognition, in other words, begins at the moment when sensory information is received, and ends when a functional motor output is generated. Importantly, though, the essential hardware of the mind is just the brain, which is the center of information processing (it is causally primary, and sufficient, for cognition). The information-processing model of the mind is exemplary of the physicalist approach and has stood as the prevalent metaphor in the mind sciences since the 1950's [34].

The summary of the physicalist grounding motivating the mind sciences provided here is not meant to constitute an exhaustive account, but only a rough sketch. Our current goal, after all, is to articulate an understanding of how some of the fundamental presuppositions within the mind sciences have (mis)informed, and continue to (mis) inform, VR research and design. Our claim here is that VR has tacitly inherited a physicalist (Newtonian) conception of reality as conceived as a collection of material objects, as a consequence of which both VR research and design have operated under a philosophically impoverished theory of mind and subjectivity. We find particularly problematic and worthy of discussion three implications of physicalism for VR—in other words, three theoretical gaps—to which we now turn.

2.1   The Three Theoretical Gaps

To begin with, the fundamental physicalist claim assumes that perception of reality is essentially just the perception of physical objects in an a surrounding space. Once appropriated by VR, what follows from this claim is that perception within virtual reality should simply occur as a matter of course, if only the sensory information that is simulated within the virtual environment is sufficiently high in fidelity to the sensory information that is naturally present in ordinary, physical reality. Technically speaking, therefore, VR subscribes to a "naïve realist" theory of perception, according to which perceiving reality is ultimately a passive process, insofar as reality is construed as essentially physical, objective (mind-independent), and therefore ontologically "pre-given" and "readymade" for perception. Naïve realism has been heavily criticized by a wide range of disciplines, such as philosophy (Descartes, Hume, Kant), science and technology studies [26], and social theory [1]. Repeating any of the arguments made by these authors would at this point be too redundant a step to take. Thus, we will confine ourselves to but one example to help illustrate our point.

VR experience typically entails a period of adjustment or recalibration in the beginning, which we will refer to as an "onboarding" process. During onboarding, the participant's perceptual systems are preoccupied with realizing an adaptive fit with the VR interface and its concomitant hardware. Indeed, what is entailed by onboarding is not just an appropriation of the contents of perception in VR, but also—and perhaps even primarily—an appropriation of the medium of perception of VR (i.e., the VR hardware and its attendant interface for control). Particularly illustrative of this point is the fact that, upon being equipped, the VR headset initially acts like a blindfold, insofar as it severs the individual's visual connection with the world. Upon being "blinded" in this way, what often becomes most salient in the participant's awareness is the various physical qualities (e.g., weight, temperature, texture) of the headset as it is felt against one's face and head. However, as the individual spends time interacting with the immersive virtual environment, the "headset-as-blindfold" eventually becomes a "headset-as-window-into-the-virtual-world", such that the physical qualities of the headset, as well as those of the rest of the augmented interface, recede into the background of the individual's awareness. As the physical properties of the VR interface become less salient, a more stable perceptual connection is formed between the individual and the VR content. In other words, through the onboarding process, the individual's attention gradually shifts from being focused on the medium of perception (i.e., the VR hardware and interface) to the contents of perception (i.e., the VR environment) [16], as a result of which the individual's felt perceptual engagement with the VR becomes increasingly "natural" or "intuitive".

The key point here is that perception in VR (normally) does not begin by feeling intuitive, but becomes increasingly more intuitive over time and with continuous engagement with the VR content. This pattern of development, however, is precisely the opposite of what a naïve realist theory of perception would predict, according to which perception would occur passively and instantaneously, rather than progressively. The first theoretical gap within the literature may therefore be phrased as follows: Perception is not fundamentally a passive process, contrary to the naïve realism implicit in VR, but is necessarily dependent upon one's voluntary patterns of participation in and interaction with an environment—be it virtual or ordinary. This is to say that perception is more accurately construed as a process of "skillful coping"—to borrow a term from Dreyfus [20]—one that must be learned and achieved as a matter of continuous sensorimotor coordination in relation to, and mastery over, a meaningful situation; rather than as a matter of merely "receiving" sensory information from the environment and processing it instantly and strictly inside one's own "head". A naïve realist theory of perception that is motivated on physicalist grounds simply cannot account for the fact that perception in VR becomes second-nature only with due diligence, rather than instantly and matter-of-factly, right at the outset of the VR experience. Nor can it account for the fact that an individual's perceptual engagement with the VR content appears to be predicated on skill, in that it is learned by way of active, voluntary, and embodied interaction with the VR content. This interaction is a constant process of negotiation with evolving environmental, phenomenological, and social factors at hand. In other words, this is an embodied interaction, in line with the theory articulated by Dourish [21].

The second theoretical gap implicated by physicalism, accordingly, is predicated on the assumption that factors such as affect, value, thought, as well as motivation, in virtue of their being characteristically subjective (mental) properties, are construed neither as part of reality itself, nor as amenable to scientific theorization. Both of these implications, however, are dubious. For starters, the claim that subjective factors are not part of reality is evidently physicalist in kind, meaning that it is not philosophically neutral (insofar as physicalism is not a philosophically neutral position) and should therefore not be simply taken for granted. Furthermore, the preclusion of subjective factors from reality generates a rather peculiar tension within VR. On the one hand, one of the overarching aims of VR is to develop meaningful experiences by way of immersion. It goes without saying, though, that meaningful experiences, in virtue of their being a type of *experience*, are (at least) partly subjective in their constitution and should therefore be accounted for, rather than precluded, by a theory of perception. However, because of its physicalist bias, VR sets out to simulate reality just as a place of objects, propounding as a result a rather partial and ontologically impoverished view of perceptual reality (naïve realism). In our view, which we elaborate more fully in Sect. 3, the objects of perceptual reality are irreducible to the objects of physical reality, such that any attempt to generate meaningful virtual experience can proceed successfully only by adopting a non-reductive stance with regards to the nature of perceptual reality. We believe, in other words, that a sustained commitment to a naïve realist theory of perception that is predicated on physicalism will only hinder VR's goal of generating meaningful (and optimal) virtual experiences, and should therefore be replaced with a theory of perception that is neither naïve realist nor Newtonian in its ontological commitments.

As regards with the second implication, despite the historical difficulties and resistance associated with subjecting factors such as affect, value, thought, and motivation, to scientific (empirical) investigation and scrutiny, it is far from the case that such characteristically subjective factors are not amenable to scientific theorization by today's standards. Quite the opposite: There has been an explosion of scientific research in the literature on just the subjective dimension of human experience [12, 13, 34, 35, 36]. As a consequence, there is not only good reason to believe that it is possible to theorize about how subjective factors, such as the aforementioned, interact with objective (environmental) factors in a scientifically rigorous manner, but also that perception of physical reality is powerfully and unconsciously motivated by such subjective factors and cannot be understood apart from them, but only as an abstraction devoid of any real-world meaning. In spite of the various advances made in studying subjectivity, particularly within the cognitive sciences, VR has yet to seriously engage with the relevant literature by which to bridge its second theoretical gap, which concerns the nature of the relationship of objective and subjective factors that are involved in perceptual experience.

The final theoretical gap worth noting concerns some pervasive confusions associated with two core constructs within the VR literature: Presence and immersion. Presence is defined as "the subjective experience of being in one place or environment, even when one is physically situated in another" [7, 10] and is typically treated as the golden standard or main measure for how real or "successful" a virtual experience can be said to be. In other words, presence is the current design goal for VR [3], meaning that improving virtual experience (making it feel more real) is primarily a matter of maximizing presence. Various authors (e.g., [3, 4, 9]) have further articulated the presence construct to apply also to such aspects of VR as one's experience of one's own virtual body (self-presence) as well as that of other agents (social/co-presence). Definitions of immersion, on the other hand, have been more variable. Specifically, immersion generally has been defined as either objective or subjective in kind [8]. Objectively, immersion has been regarded as a function of the VR hardware, depending, for instance, on the number of built-in sensors. Subjective definitions, on the other hand, refer to immersion as a state or feeling of "being caught up in another world" [8]. Of course, the problem with the objective definition is that it is not necessarily predictive of subjective feelings of immersion, insofar as it is possible either to feel subjectively immersed in an objectively non-immersive game, like Tetris, or else to lack a subjective sense of immersion altogether when inside of an objectively immersive simulation (as in the onboarding process). Conversely, the problem with the subjective definition of immersion is that it is too similar to definitions of presence so as to lend itself rather easily to conflation fallacies ([8], p. 1409).

We believe it to be possible to circumvent the conceptual difficulties associated with the immersion construct by adopting a flow-based definition of immersion [8]. Flow is a widely researched phenomenon within psychological science with both objective (performance) and subjective (phenomenological) measures, and is commonly referred to as the state of optimal experience [5]. Flow is considered an optimal state for several reasons. First, it occurs when individuals are engaged in challenging tasks, but only when the difficulty level of the task is just beyond the individual's own level of competence in dealing with the task [17]. Phenomenological descriptions of flow often involve a loss of a sense of time, a reduction in levels of reflective self-consciousness, and high levels of immersion in the task at hand. Moreover, feelings of immersion in flow are typically accompanied with the ironic sense of the immersion being effortless (despite the high demands of the task), whereby one feels as though one is truly "flowing" through the experience. Importantly, flow is experienced independent of age, sex, gender, culture, or language. Flow is, in other words, a human universal [5] and is therefore, in a deep sense, a constituent element of human experience. It is also predictive of psychological well-being, as well as general life satisfaction, and therefore stands as a psychological theory of optimally meaningful experience, as such [29].

In defining subjective immersion in VR as an "experience of feeling totally involved in and absorbed by the activities conducted in a [virtual] place or environment", it thus becomes possible to clearly distinguish the pragmatic, task-related elements of VR experience from the objective elements in the virtual environment (e.g., the surrounding space, social agents, and self), with which presence is chiefly concerned. The implications of adopting a flow-based definition of immersion in this way are twofold. First, in light of the fact that flow describes optimal experience within ordinary reality, it follows quite straightforwardly that immersion-as-flow should therefore describe optimal experience within *virtual* reality. In other words, by adopting a flow-based definition of immersion, the design imperative of maximizing presence by developing physically realistic virtual experiences becomes only secondary in importance to that of maximizing *immersion* by developing virtual experiences that are able to reliably facilitate a flow state within participants. The criterion of realness as it pertains to VR, therefore, fundamentally becomes predicated on immersion, rather than presence. Second, it should become evident that a naïve realist theory of perception can no longer be used as a guidepost for developing VR experiences, insofar as it cannot account for immersion-as-flow in perceptual terms. This is because naïve realism assumes, as per its physicalist (Newtonian) heritage, that reality is ultimately a place of material objects; whereas flow—and by extension immersion—is fundamentally a process. As such, the final theoretical gap can be elegantly summarized as follows: Simulating an environment conducive to flow is a matter of simulating a process, whereas naïve realism can only describe or be used to simulate objects; therefore, the naïve realist theory of perception implicit in VR must be replaced with an alternative theory of perception such that can account for flow—and, therefore, immersion—in processual terms.

In light of the critiques we have leveled in this section, our preliminary conclusion is that the naïve realist theory of perception assumed in VR, as well as the (Newtonian) physicalist grounding upon which it is motivated, cannot be utilized for developing or explaining optimal virtual experience. We have tried to justify this conclusion by illustrating three theoretical gaps which, we have argued, result as a necessary consequence of VR's tacit philosophical commitment to physicalism. The first gap concerns the difficulty in accounting for the skillful nature of perceptual engagement in naïve realist terms, as evinced by the gradual and progressive, rather than instantaneous, development of perception involved in the onboarding process. The second gap concerns the difficulty in conceptualizing perceptual reality scientifically without also reducing the objects of perceptual reality to those of physical reality. Finally, the third gap concerns the difficulty in accounting for immersion, taken as an instantiation of the flow state within VR, in processual rather than objective terms.

The problems posed by these three theoretical gaps are irresolvable by physicalist (Newtonian) means alone, since physicalism lays at their very foundation. We believe, however, that the situation can be remedied, and, furthermore, that the theoretical and philosophical, as well as psychological, resources for remedying it are readily available within the burgeoning field of "enactivism" [12–15, 35, 36]. Enactivism advocates a non-reductive, pragmatically grounded, and fundamentally embodied approach to conceptualizing mind and experience. In the following section, we draw on the enactive approach to cognitive science to articulate a theory of perception as motivated action, which promises to bridge the first two theoretical gaps, thereby constituting a viable alternative to the naïve realist theory of perception that is assumed in VR. We then synthesize our theoretical findings in Sect. 4 to offer an explanation of flow (and subjective immersion) in processual terms, thereby demonstrating how the third and final theoretical gap can also be bridged.

3 Perception and Enaction

The enactive approach to cognitive science was launched in the early 1990's as an alternative to conceptualizing mind and consciousness to the information-processing model discussed in Sect. 2 [12, 13]. Since then, enactivism has burgeoned into a comprehensive theoretical framework and research programme with an extensive philosophical grounding in traditions such as phenomenological philosophy, philosophical pragmatism, complex and dynamical systems theory, and systems biology [35]. Enactivism promises to offer an account of mental life that is essentially non-reductive (and non-Newtonian), whereby conscious experience is regarded as irreducible to, albeit fundamentally tied with, physical (biological) matter, and wherein cognition is construed in dynamical (non-linear and ecological), rather than mechanistic (linear and self-enclosed) terms.

For enactivism, cognition is ultimately predicated on action and is necessarily bound up by the practical aims of the agent. Moreover, knowledge of how to act (procedural know-how) is taken to be both primary and prior to the kind of knowledge that is concerned with facts and inferences (propositional know-that). In other words, cognition most fundamentally aims to render the world sufficiently predictable so as to afford functional action, rather than being a disembodied (multiply realizable) function of the brain that is primarily aimed at representing reality "as it is". Whereas the physicalist brand of naïve realism inherent in VR depicts reality as a place of ontologically pre-packaged objects, thereby overlooking this pragmatic dimension of perceptual experience, enactivism honors it, and, in so doing, is able to articulate an account of perception that is constitutively motivated, affective, and procedural. The enactive treatment of cognition (and knowledge) as pragmatic is central to the argument we aim to advance in this section, for we believe it constitutes a viable alternative for conceptualizing perception to the physicalist approach assumed in VR. It is thus the aim of this section to demonstrate how an account of perception that is grounded in the enactive approach can be used to bridge the first two theoretical gaps identified in Sect. 2. We then synthesize our findings into a processual account of flow (and immersion) in Sect. 4, thereby demonstrating a bridging of the final theoretical gap.

3.1 Perception as Motivated Action

The world is incomprehensibly complex. This is the basis for what philosopher Christopher Cherniak refers to as the finitary predicament confronting cognitive agency [6]. Given the vastness of information that is available for consideration at any particular moment, insofar as cognitive agency is constrained in terms of both time and cognitive resources, it is fundamentally impossible to act functionally in the world without also reducing the world's inherent complexity in an effective manner. The solution to the problem of functional action, therefore, is achieved by means of "framing", which refers to the process of selecting only a subset of the available information based on its relevance for the situation at hand [37, 38]. Framing, in other words, is a necessary condition for the possibility of functional action in the world, and is a thus a primary function of cognition [37, 38].

Framing, however, is an inherently motivated act and not just a cold calculation [12, 13, 38], since it occurs in the service of affording functional action. Functional action is by definition goal-directed and is the means by which real-world problems are solved. It involves the careful coordination of sensorimotor activity in relation to an environment, as well as one's self and others [13]. The mechanisms by which sensorimotor activity is regulated are essentially affective in kind, involving various forms of emotional feedback, as well as feelings and moods [13], all of which are aimed at evaluating the relative potency of one's sensorimotor coordination in relation to a task for achieving contextually-relevant goals. Accordingly, behaviors (sensorimotor acts) which result in reward or which cue potential reward are experienced positively (e.g., joy, happiness, hope), and are therefore reinforced; whereas behaviors which result in punishment or which signal potential punishment (threat) are experienced negatively (e.g., fear, anger, anxiety), and are therefore extinguished [11]. Successful framing thus implies that all relevant sensorimotor affordances and affective cues are sufficiently known in the problem-solving domain at hand; functional action is afforded; and the world is thereby experienced as a place of determinate meanings, or as "predictable". As a consequence of the determinacy of the cognitive agent's frame, as well as its effectiveness in affording functional action toward the agent's goals, the agent's attendant affective state is characterized not only as relatively positive in valence, but also as relatively secure and low in anxiety. Positive affect and low levels of anxiety are, in other words, an indication that one knows what one wants, how to attain it, and also that what one knows to be sufficient for attaining what one wants is in fact sufficient.

Whereas the cognitive agent's frames of the world are static, the world as such is entropic [11, 30]. A fundamental limitation of the framing process, therefore, is that obsolescence is both necessary and inevitable. The functional utility of framing erodes whenever the agent is confronted with a wholly novel experience—an "anomaly", in a manner of speaking. During such instances of anomaly, the world's inherent complexity emerges and overwhelms the agent's cognitive structures and attendant capacity to act functionally [11, 30]. The world, in other words, becomes a fundamentally unpredictable place and its meanings are rendered obscure and indeterminate. The agent's affective state becomes characterized by relatively high levels of anxiety and emotional ambivalence (awe and terror, hope and anxiety), insofar as knowledge of how to act functionally (how to attain goals) has become confused and is no longer sufficiently predictive of either reward or punishment. The breakdown of framing—(i.e., "misframing")—is therefore a constitutive (and affectively felt) problem for the agent insofar as it necessarily renders the agent incapable of functional action. Negative affect and high levels of anxiety act to indicate that one does not necessarily know what one wants or how to attain it, while also knowing that what one has known to be sufficient for attaining one's goals is no longer sufficient.

Clearly, then, the complexity of the world must be kept at bay, insofar as functional action is imperative and depends on the sufficiency of the agent's framing for predicting the world. Confrontation with anomaly thus implies the need for amending erroneous frames by voluntarily attending to the emergent problem (anomaly) so as to facilitate a significant restructuring of the agent's overall framing. When done successfully, the agent's framing of the world regains its sense of determinacy, whereby the world's emergent complexity, together with its attendant set of anxieties and negative affect(s), is at once reduced, and functional action is afforded anew. It is but a matter of time, however, until anomaly emerges once again and the agent is forced to undergo another restructuring process of his or her framing. The circle of interpretation must turn ever onwards.

We believe that the line of argument which we have proposed describes a necessary (existential) structure of cognition, namely the ongoing hermeneutic (interpretive) circulation between framing, misframing, and reframing. We recognize how bold this claim might appear, but we believe it to be both theoretically sound [11, 13, 14, 30] and experientially compelling, insofar as (1) misframing is cognitively unavoidable and (2) cognitive reframing is imperative for regaining functionality. Furthermore, our depiction of cognition is inherently enactivist in its grounding, insofar cognition has been construed as essentially motivated (goal-directed), affective (evaluative), and procedural (sensorimotor). We follow suit with the enactivists in claiming that cognition is not a brainbound process that is functionally divorced from other bodily processes, but a form of motivated and affectively-imbued activity that is fundamentally distributed across the brain-body-environment dynamical system [13]. Cognition, in other words, is not something that one has, but something that one does in relation to a world of meaningful activity. Cognition is enacted qua the agent's embodied interaction with the world and is always bound up in the agent's practical context.

It is now possible to theorize about perception in enactivist terms on the basis of the established argument. If cognition is about functional action in relation to a meaningful world, then perception mediates cognition insofar as it constitutes the primary means by which the world is even disclosed into conscious awareness. Enactivists theorize that perceptual experience is achieved as a function of mastery of "sensorimotor contingencies" [13, 35]. Sensorimotor contingencies refer to the invariant sensorimotor structures that emerge as a function of how patterns of sensory flow, inherent in and contingent upon each distinct sense-modality (e.g., visual, auditory, tactile, etc.), covary with patterns of motor activity in relation to the attainment of a given practical aim (e.g., satiation of hunger, quenching of thirst, escaping a predatory attack, etc.) [13]. Unless the agent appropriates or incorporates into his or her procedural repertoire these invariant structures, (i.e., grows accustomed to and thereby learns how to "predict" changes in sensorimotor flow in a relevant manner), navigating the world in relation to practical aims is rendered an impossible task. Conversely, unless sensorimotor activity is motivated, and therefore bound up by practical aims, then the agent does not readily learn to perceive "objects", since the possibility of perceiving objects is necessarily tied with any object's potential relevance for satisfying practical goals (i.e., how predictive a given set of sensorimotor patterns is of either reward or punishment). Thus, perception is achieved as a matter of increased procedural familiarity with or mastery of sensorimotor contingencies and their relevance to real-world goal-attainment. Perception is therefore a practical skill; it is not a matter of seeing the world "as it is", but of learning to see what is relevant in the world for acting functionally (attaining goals) in it [25]. In subserving the agent's cognitive aims in this way, perception is therefore also bound up (framed) by the agent's practical context in the same way cognition is. This implies, quite straightforwardly, that perception is essentially both motivated and affective. Perception is therefore a form of affectively-imbued, motivated action for mediating cognition.

3.2 Bridging the First Two Gaps

The proposed theory of perception as motivated action is firmly rooted in the enactive approach to cognitive science [12–15, 35]. Our argument suggests that perception is a matter of skillful sensorimotor coordination that is regulated by affective means, and framed according to the practical aims of the agent's problem-solving context. In this regard, the proposed theory of perception as motivated action implies an inherently pragmatic, but also phenomenologically informed treatment of perceptual reality, which runs counter to that of naïve realism. The fundamental notion underlying such a treatment is that *we do not perceive objects and then infer their meaning, but we perceive meaning and only then infer objects on this basis* [11, 30]. Recall that the first theoretical gap required an explanation as to why the perceptual (re)calibration during the onboarding process develops gradually like a skill, rather than instantaneously and mechanically as a naïve realist theory would suppose. According to enactivism, perception is achieved as a function of progressive mastery over sensorimotor contingencies through active, exploratory, and pragmatically bound participation in a world. The first theoretical gap is thereby bridged by means of our proposed theory insofar as perception is predicated on exploratory behavior and is achieved not as a matter of seeing the (virtual) world "as it is", but as a function of *learning to see what is relevant in the (virtual) world for the purpose of acting functionally in it*. Therefore, the onboarding process entails a gradual and progressive development, rather than an instantaneous one, therefore, since it demands from the individual to skillfully compensate for the discrepancy between the sensorimotor skill set that is brought into the virtual experience at its onset, predicated as it is on real-world physics, and the sensorimotor skill set that is demanded by the virtual experience, the physics of which necessarily deviate from real-world physics as a consequence of technological and engineering constraints (e.g., programming errors), on the one hand, and the fact that perception is now being mediated not just by one's body, but by an additional (augmented) physical interface (e.g., VR headset and controllers), on the other hand. The sensorimotor contingencies proper to a given virtual experience emerge as a function of the patterns of the agent's perceptual interaction with the VR content as mediated by the (augmented) physical interface, whereby mastery of said contingencies is accordingly to be achieved as a function of sustained, motivated interaction with the VR content qua its interface.

The second theoretical gap is also bridged insofar as the proposed theory advances a non-reductive synthesis of objective and subjective factors in its depiction of perceptual reality. Furthermore, it does so in a way that is both scientifically rigorous (as per its grounding in enactivism, as well as psychological science) and experientially robust (as per its phenomenological sensitivity). In particular, the world as it is perceived, according to the argument thus far, is more appropriately regarded as a forum for meaningful action, than a collection of material objects [11, 30]. It thus follows that perception of a virtual world is therefore also the perception of a world of possibilities for functional action, in which subjective factors such as goals, motivations, and affect, play an essential part in framing what is relevant. On this basis, we find it reasonable to claim that any set of guidelines for VR design that is predicated on the naïve realist view that perceptual reality is reducible to a collection of objects in space misses the mark altogether, and cannot be said to explain or entail optimal (or meaningful) virtual experience. This is because naïve realism ultimately overlooks the pragmatic nature of perception almost entirely.

Now that the first two theoretical gaps have been bridged, what remains is a bridging of the third gap. In the following section, we utilize our enactive theory of perception as motivated action to leverage an account of flow as a state of optimal perceptual engagement with the world, within which we then ground the immersion construct and circumvent its attendant conceptual difficulties. The methodological implications of our theoretical contributions in Sects. 3 and 4 for VR research and design are explored fully in Sect. 5, all of which is done in reference to a design probe that was launched and conducted with real participants in a mixed reality (MR) setting at Carnegie Mellon University [18].

4 Redefining Immersion

Our argument in Sect. 2 was predicated on a critique of the physicalist assumptions inherent in VR, suggesting a total of three theoretical gaps in the literature which, we claimed, cannot be bridged by physicalist means alone. In Sect. 3, we articulated an enactive theory of perception as motivated action, which, we showed, is able to bridge the first and second theoretical gaps. In particular, the first theoretical gap is bridged by explaining the developmental quality of perception that is entailed in the onboarding process as a function of attaining progressive mastery over the sensorimotor contingencies inherent in the VR experience. Accordingly, the second theoretical gap is bridged by conceptualizing perceptual reality not as being comprised of a collection of physical objects, but as constituting a forum for meaningful (pragmatically motivated and affectively-imbued) action.

In this section, we aim to demonstrate how the theoretical framework we laid out in Sect. 3 can be used for also bridging the third theoretical gap, which requires an explanation of flow (and immersion) in processual terms. We begin with a discussion of the problem: In order to circumvent the conceptual difficulties associated with the immersion construct, immersion ought to be conceived of in terms of flow; however, the naïve realist assumptions inherent in VR cannot account for flow in processual terms and must therefore be replaced if a non-problematic (flow-based) definition of immersion is to be recovered. We then turn to recent work in cognitive science on the flow state which conceptualizes flow as a process, that is, as an "insight-cascade" [17] whereby the hermeneutic circle comprised of framing, misframing, and reframing is, in a manner of speaking, "ramped up". Next, we integrate this cognitive scientific account of flow into our enactive framework and argue that flow is a marker of optimal perceptual engagement with the world, that is, flow implies a temporary enhancement of the very processes of mastery-attainment over relevant sensorimotor contingencies. We conclude by grounding the immersion construct in our enactive conception of flow, thereby showing how such a grounding helps to circumvent the various conceptual difficulties associated with immersion. In Sect. 5, we proceed to articulate four methodological implications of our theoretical contribution for VR design.

4.1 Formulating the Third Theoretical Gap

The golden standard of VR currently is determined by the degree to which a virtual experience can induce a state of presence within a participant, which is broadly defined as the subjective feeling or illusion of being in one place, when one is in fact physically located in another. It is generally stipulated that presence is achieved partly as a function of the objective immersive properties of the VR hardware, such as the number of built-in sensors, and partly as a function of the quality of sensory stimulation provided by said sensors (i.e., the degree of fidelity preserved between the physics of the VR and those of the real world). The design goal of VR, in other words, aims at the maximization of presence through the simulation of physically realistic experiences as mediated by the objectively immersive VR hardware. At face-value, a presence-predicated design rationale such as this one might not seem very objectionable. However, upon closer philosophical scrutiny, two challenges are revealed—the first, a methodological challenge, and the second, a conceptual challenge.

First, the assumption that presence is perceived as a matter of course if only the VR hardware is sufficiently objectively immersive and the physics simulated in its attendant sensory stimuli are sufficiently (physically) realistic, is fundamentally predicated on a naïve realist theory of perception. As we have already argued in Sect. 3, however, perception is not a passive act, but rather a matter of skillful mastery over sensorimotor contingencies through exploratory interaction with a meaningful world. The onboarding process exemplifies the developmental and skillful character of perception in VR and therefore constitutes a counterfactual to the naïve realist assumption that the perception of presence can be causally reduced to the conjunction of the objective immersive properties of the VR hardware, on the one hand, and the quality of sensory stimulation thereof, on the other hand. Be that as it may, though, a more fundamental issue with treating presence as the golden standard of VR is arguably that it is object-centric (Newtonian), rather than action-predicated (pragmatic), whereas human perceptual reality, as we have argued in Sect. 3, is more appropriately to be understood as a forum for meaningful action than a collection of objects. In other words, making the design goal of VR the maximization of presence seems to neglect the pragmatic dimension of perceptual reality. It becomes unclear, then, how such a phenomenologically impoverished design ideal can be used for explaining, or even generating, optimal (and meaningful) virtual experience.

Second, as mentioned in Sect. 2, the construct of immersion is defined not only as an objective property of the VR hardware, but alternatively as the subjective state or feeling of "being caught up in another world". The distinction between subjective and objective immersion has consequently led to various confusions as to its use within the literature.

For example, it has been difficult to discriminate between immersion in its subjective sense and presence due to their apparent overlap in meaning on the definitional level [8]. Additionally, although the distinction between objective and subjective immersion seems to be a conceptually useful move, it nevertheless raises difficult questions as to the relationship between immersion (in both senses), on the one hand, and presence, on the other hand. The fact that objective immersion is not predictive of subjective immersion (as was argued in Sect. 2) further complicates the conceptual boundaries of the constructs at hand. As a consequence of such conceptual fuzziness, problems emerge with attempting to clearly operationalize these constructs in an experimentally useful manner. Subsequently, causal modelling of the relations between presence and immersion (in both senses) is rendered an incredibly difficult task, which cannot be resolved through statistical means alone, since correlational data can neither imply causation, nor determine the direction of causation between related variables.

In response to this conceptual challenge, we follow suit with Mütterlein in advocating a flow-based approach to subjective immersion [8]. Specifically, we define immersion as the "subjective experience of feeling totally involved in and absorbed by the activities conducted in a [virtual] place or environment", which thereby makes it possible to clearly distinguish the pragmatic, task-related elements of VR experience from the objective elements with which presence is concerned (e.g., the surrounding space and objects, social agents, and self). In other words, we believe that a conception of immersion-as-flow should sufficiently address the demarcation issue with presence and subjective immersion: Presence becomes concerned with the subjective experience of "realness" as it pertains to virtual *objects* (e.g., items, bodies, spaces, etc.), whereas immersion becomes concerned with the subjective experience of "meaningfulness" as it pertains to virtual *actions* (e.g., tool-use, problem-solving, navigation of a map). The renewed sense of clarity brought about by this flow-based conceptualization of immersion should then afford more experimentally robust operationalizations of objective immersion, subjective immersion, and presence, all of which can consequently be subjected to more rigorous causal analyses.

Accordingly, the adoption of a flow-based approach to subjective immersion not only helps to address the various conceptual difficulties currently besetting the literature, but also helps to motivate the methodological shift we aim to make from an object-centric approach to an action-predicated approach to VR. For provided that a naïve realist theory of perception in principle cannot account for flow—since flow is fundamentally a process, whereas naïve realism (given its Newtonian heritage) conceives of the contents of perception as material objects—a conception of immersion-as-flow is not only conceptually appropriate to pursue, but methodologically necessary. In what follows, we summarize contemporary work on the cognitive science of flow in preparation for advancing a processual, action-predicated (pragmatic) alternative to the current object-centric (Newtonian) approach to VR research and design.

4.2 The Cognitive Science of Optimal Experience

The psychological literature explains the flow state as a consequence of a tight coupling that is obtained between an agent and his or her environment. This tight coupling is mediated by clear and contiguous environmental feedback in response to the agent's performance on a task, whereby errors are highly diagnostic of whether and to what degree task demands are being met, as well as what "adjustments [are] needed in order to maintain performance" ([17], p. 311). Importantly, it must also be the case that the task is perceived by the agent as challenging, but that its level of difficulty exceeds the agent's skill level only by a small margin. If the task becomes too difficult, then the agent is overwhelmed by feelings of anxiety and frustration such that the coupling of agent and environment that is necessary for attaining flow is lost ([17], p. 311). Conversely, if the task becomes too easy, then the agent experiences boredom and thus loses the motivation to sustain engagement—as a consequence of which, the tight coupling of agent and environment is lost once again. Flow is therefore described as the state of "optimal" experience because, when inhabiting it, the agent's skill level is continually being pushed to its limits, resulting in the emergence of a "skill-stretching" function ([17], p. 311). Vervaeke et al. [17] describe "skill-stretching" as "a system of learning where the process of meeting and overcoming one challenge breeds a new and more developed skill set, in turn affording the ability to take on a still more difficult set of demands" (p. 311). As such, the self-perpetuating sense of motivation that is often felt during flow is experienced from being consistently challenged while nonetheless reliably overcoming such challenges along the way [5].

Whereas psychological accounts have typically described flow at the level of interaction between agent and environment, Vervaeke et al. [17] offer a description of flow at the level of cognitive processing, which makes their depiction of flow particularly amenable to theorization in enactive terms. We therefore turn to an explication of Vervaeke et al.'s cognitive scientific account of flow, which we aim to integrate with our own theory of perception as motivated action. As we will see, pursuing such a synthesis will provide us with the grounding that is needed both in order to address the conceptual difficulties associated with immersion, as well as to motivate the much needed methodological shift toward an action-predicated (pragmatic), rather than object-centric (Newtonian), approach to VR.

Vervaeke et al. [17] conceptualize the cognitive basis of flow as an insight-cascade, a concept which requires some unpacking, starting with the notion of "insight". The realization of an insight is often described as an "aha!" moment and is also commonly depicted with the metaphor of a lightbulb going on inside one's head. Importantly, insight learning constitutes a qualitatively distinct mode of learning: Whereas in conditional learning (e.g., classical or operant), the rate of learning is often incremental, in insight learning, it is characterized by the abrupt or spontaneous realization of a solution that is preceded by a period of (prolonged) impasse [17]. During insight problems, the solution is not achieved by straightforward means, such as by recalling relevant content from memory. Indeed, what typically characterizes an insight problem is precisely that one's prior knowledge interferes with the realization of a relevant solution to the problem [14]. In other words, when faced with an insight problem, the agent experiences a fixation in how the problem has been (incorrectly) framed and must realize a solution only by overcoming the impasse caused by the misframing. The insight is thus realized as a consequence of a shift in the agent's attentional processing of the situation, which occurs at the level of procedural (not propositional) processing, causing the agent to break out of the erroneous frame and to form a novel framing of the situation by which a solution is finally afforded [14, 17].

Given this depiction of insight problem solving, Vervaeke et al. [17] argue that the flow state emerges as a dynamical system whereby the act of solving one insight problem immediately gives way for and creates an additional insight problem that can be solved, a process which is then iterated and sustained over a span of time—granted that the proper learning conditions are in place (i.e., tight environmental feedback and optimal task difficulty). A cascade of insightful processing emerges as a consequence, causing a "stretched out 'aha!' moment". The ongoing realization of insight is what affords the "skill-stretching" function of flow. In it, the agent's competence at tackling problems of the kind being faced is continually being improved and stretched beyond its limits with every successive instance of insight. This is one of the reasons why flow is also referred to as a state of optimal *learning* or *engagement* [17].

Skill-stretching is not the only core characteristic of the flow state, however. Recall that during flow, the agent experiences a sense of ineffability (non-deliberateness) throughout the engagement process, as well as a paradoxical sense of immersion in the task, whereby the quality of engagement is both effortful yet effortless. Vervaeke et al.'s [17] cognitive account of flow explains immersion in terms of fluency, and ineffability in terms of intuition. Fluency refers to the sense of ease or difficulty associated with a cognitive process. In citing the work of Topolinski and Reber [24], Vervaeke et al. [17] suggest that the subjective sense of fluency accompanying a cognitive process might in fact be correlated to "the actual degree of ease of processing occurring at the neural level" (p. 312). Thus, whereas specific instances of insight problem-solving are accompanied with discrete moments of enhanced feelings of fluency, they argue, "it follows that a cascade of insights would naturally yield an accompanying and ongoing stream of positive subjective affect, reinforcing a sense of meaning in one's processing—flow phenomenologically equates to an experience of extended fluency" (p. 312). The paradoxical sense of immersion associated with flow is thereby explained as a function of extended fluency, wherein individual moments of frustration caused by impasse (low fluency) are spontaneously interrupted by moments of insight (high fluency), which yield satisfaction, forming, as a consequence, a positive feedback loop of self-perpetuating, self-motivating, and reliable engagement with the task at hand.

Vervaeke et al. [17] accordingly explain the ineffability of the flow state by grounding flow in intuition. In reference to work by Hogarth [23], they describe intuition as a product of implicit learning—which is "*tacit*, as opposed to *deliberate*"—and as "effortless, reactive, and producing 'approximate' responses" (p. 321). Provided that implicit learning and flow are both non-deliberative and ineffable/procedural (rather than voluntary or propositional) processes, grounding flow in implicit learning seems like an appropriate conceptual strategy. The obvious challenge, though, is that whereas flow is a state of *optimal* experience, implicit learning as such appears to be a *suboptimal* process on the whole. For, on the one hand, implicit learning suffers from the problem of "over-fitting", which occurs when "correlational noise from the environment is interpreted as being causally relevant to the pattern of action" (p. 321). During flow, however, over-fitting is not a problem, since, if it were, then it would fundamentally disrupt the insight-cascade, thereby rendering the flow state a practical impossibility. On the other hand, though, implicit learning is confined primarily to tracking *actual* patterns in the environment, whereas flow involves the adaptive tracking of and selection from *possible* patterns for dynamically affording functional action (pp. 321–322). Vervaeke et al. [17] insightfully note that the conditions for acquiring sound intuitions, namely, separating "causal signal" from "correlational noise", happen to mirror those for cultivating a state of flow: "A system of learning that tightly couples actions and environment with timely feedback—thus providing high error diagnosticity—is a system conducive to cultivating flow and good intuitions" (p. 322). On this basis, the authors advance a conceptual synthesis of flow and implicit learning in which they propose that flow "is optimal for implicitly learning complex patterns in the environment and distinguishing them from correlational ones while exploring possibilities of action and learning" (p. 322). Such a conceptual synthesis helps to explain the non-deliberative quality of the flow state without reducing flow to a set of imprecise, automatic processes concerned primarily with tracking patterns in actuality. The authors summarize this point eloquently: "Flow is a system of processing and cultivating causal pattern recognition in which cognition is stimulated to explore possibilities of action. These two elements are interdependent: exploring possibilities allows one to distinguish between actual causation and mere empirical generalization. In turn, zeroing-in on causation helps guide the insight away from being illusory or fantastical" ([17], p. 322).

The cognitive scientific account of the flow state reviewed here explains the core features of flow in cognitive terms. Skill-stretching is a qualitatively distinct mode of learning that emerges as a function of the insight-cascade, in which insight problems, on the one hand, and insight problem-solving, on the other hand, enable one another in a mutually affording fashion, sustained over a span of time. The immersive process, and its attendant phenomenology, are accordingly explained as a function of sustained fluency. Finally, the ineffable character of flow is explained by grounding flow in intuition, a non-deliberative cognitive process, whereby flow is conceived of as a procedurally-driven, optimal form of implicit learning.

4.3 Enaction, Perception, and Flow

In building up to a flow-based reconceptualization of subjective immersion, we must ensure that our argument retains a level of philosophical consistency throughout. In order to ensure this, we must therefore ground the cognitive account of flow outlined in Sect. 4.2 within our own theory of perception as motivated action. Demonstrating such a grounding should not only guarantee a necessary degree of coherence, but should also help to motivate the methodological shift we are attempting to make in this paper toward an action-predicated, pragmatic approach to VR.

The cognitive account of flow as an insight-cascade is readily interpretable through our enactive lens. In Sect. 3, we argued that, as per the finitary predicament, the hermeneutic circulation of framing, misframing, and reframing is a necessary existential structure of cognition. Taken in these terms, an act of insight becomes understood as an instance of spontaneously reframing a problem frame, which thereby affords functional action and enables the agent to regain fit with the environment after a period of frustration and impasse. It follows that, if flow is indeed a cascade of insights whereby learning is optimized and one's skills are continually stretched beyond their limits, then the flow state constitutes a "hermeneutic hypercycle" whereby the necessary circulation of framing, misframing, and reframing, is, in a manner of speaking, ramped up and sustained processually over a period of time. The cultivation of sound intuition by way of flow thus translates into the cultivation of mastery over sensorimotor contingencies, wherein not only is greater perceptual mastery obtained over one's engagement with the environment, but the very processes of mastery attainment are themselves temporarily deepened and enhanced. Flow is, in other words, the instantiation of an optimal form of perceptual engagement with the world, of which "skill-stretching" is an emergent function.

Our theory of perception also describes the attendant phenomenology of the flow state as being constitutive of its own class of meaning, which is distinct from how the world is experienced during instances of accurate framing (whereby functional action is afforded), on the one hand, and misframing (whereby functional action is not afforded), on the other hand [30]. Specifically, flow is a state of engagement in which the world as a forum for action is in the active process of being transformed from a place of indeterminate meaning, wherein functional action is impeded, to a place of determinate meaning, wherein functional action is afforded [30]. The ineffability of flow, accordingly, is accounted for by the fact that flow is fundamentally a procedural (non-propositional, non-deliberative) process constituted by a chained sequence of sensorimotor breakthroughs with respect to a meaningful environment. The immersive tendency of flow, subsequently, is explained as a function of an optimal degree of indeterminacy—and experienced anxiety—that characterizes an agent's perception of the world, as well as both the rate and clarity of sensorimotor feedback by which a sense of determinacy—and an accompanying feeling of security and confidence—is salvaged from one's interaction with the indeterminate. As the literature on flow clearly states, the difficulty of the task must be just beyond one's own skill level in order for the flow state to obtain. In other words, the indeterminacy of the world and its attendant anxiety must remain at an optimal level throughout, so as to simultaneously beckon the agent's meaningful engagement without actually overwhelming his or her capacity to engage meaningfully.

Having in this way described the flow state's three main attributes (skill-stretching, ineffability, and immersion), we can now claim that the cognitive account of flow has been sufficiently grounded in our proposed theory of perception as motivated action, and is, as a result, conceptualized in enactive terms. A level of philosophical consistency has therefore been ensured, insofar as the various concepts used throughout our discussion (e.g., perception, action, affect, cognition, flow) have all been grounded in an enactive framework. We can finally proceed to the last step of our argument, where we conceptualize subjective immersion in terms of flow.

4.4   Immersion-as-Flow: Toward an Action-Predicated VR

Throughout this article, we have taken deliberate steps to build a disciplined critique of the physicalist presuppositions implicit in VR. The current design goal of VR is to maximize presence by simulating physically realistic experiences that are mediated through objectively immersive hardware. We have challenged the validity of this design goal on two distinct, yet interrelated fronts. First, such a design goal tacitly subscribes to a naïve realist theory of perception, which fundamentally cannot account for the gradual and skillful character of perception during onboarding in VR. Second, it commits to a Newtonian view of reality as a collection of physical objects, and it assumes, as a consequence of this commitment, that the objects of perception are reducible to the objects of the (Newtonian) physical world. Since the pragmatic structure of perceptual experience is, in principle, precluded by such an object-centric approach to VR design, we therefore proposed that a methodological shift be made toward an action-predicated approach instead. Another part of the "third theoretical gap" which we described pertains to conceptual issues with adequately delimiting the core constructs used in VR research and design: Presence, objective immersion, and subjective immersion. We are now ready to advance the final step of our argument, which aims to bridge the third theoretical gap. Specifically, we ground subjective immersion in flow and demonstrate how doing so simultaneously addresses the conceptual difficulties posed by the third gap, as well as how it engenders the much needed methodological shift toward an action-predicated approach to VR.

If subjective immersion is to be conceptualized in terms of flow, it follows that immersion-as-flow is fundamentally a process of cultivating sound intuition in relation to virtual tasks. Accordingly, immersion-as-flow is attained in VR when (1) environmental feedback in the VR is clear and contiguous with one's patterns of engagement, and is therefore highly diagnostic of one's performance; and (2) the associated difficulty of the virtual task remains optimal throughout, that is, it stays just beyond one's own level of skill, thereby affording an emergence of "skill-stretching". It should be evident that instantiating (1) and (2) in a VR setting is not a matter of simulating a physically realistic environment *per se*, contrary to what the design goal of maximizing presence would prescribe. Rather, it is a matter of designing an experience wherein functional action can be clearly and reliably afforded from a user's point of view. Specifically, in treating the virtual environment as a forum for action, in order to facilitate or even maximize immersion-as-flow, VR creators must (i) define the possibilities for action (sensorimotor coordination by way of the VR interface) in relation to the virtual environment, (ii) clearly demarcate those possibilities which count as rewarding from those which do not, and (iii) ensure that conditions (1) and (2) are in place while the user is in the process of learning what the relevant (sensorimotor) possibilities are and how they can be enacted. Normally, (iii) is realized through clear, guided instruction, presented in the form of a tutorial (explicit or implicit) during the onboarding process.

If successful, the process of immersion should result in mastery over the sensorimotor contingencies inherent in the VR experience. As a result of progressive mastery, the VR interface becomes increasingly incorporated into one's perceptual skill set and the act of perceiving thus becomes increasingly intuitive and "immediate" in the way it feels for the user. With sustained immersion, in other words, the "headset-as-blindfold" eventually (and rather unnoticeably) becomes a "headset-as-window-into-the-virtual-world" (recall from Subsect. 2.1). A rather curious implication of our framework is that the cultivation of "sound intuitions" via sensorimotor mastery in this way constitutes the cognitive basis for the experience of presence in VR. The presence of various objects in the VR experience is therefore to be understood as a function of the accumulation (and incorporation) of "sound intuitions" regarding relatively stable (constant) sensorimotor patterns available in the proximal virtual environment. Presence, therefore, is a consequence of successful and sustained action-predicated immersion in VR, insofar as sustained and successful immersion does indeed yield sound intuitions.

By having adopted a flow-based approach to subjective immersion, greater operational rigor is now also afforded with regards to the core constructs of VR: Presence, objective immersion, and subjective immersion (immersion-as-flow). First, by conceptualizing subjective immersion in VR in terms of flow, it has become possible to clearly distinguish the objective elements of VR experience (e.g., items, bodies, spaces, etc.), with which measures of presence are primarily concerned, with the pragmatic, task-related elements of VR experience (e.g., tool-use, problem-solving, spatial navigation, communication), with which measures of immersion-as-flow are primarily concerned. Furthermore, the fact that objective immersion is not predictive of subjective immersion (recall Subsect. 2.1) can now be explained as a consequence of the fact that what matters for immersion-as-flow is not the degree to which one is *objectively* immersed, but rather the quality of information that is communicated through the immersive medium *and* whether and to what degree such information is conducive of flow. In this way, the proposed account of immersion-as-flow circumvents the conceptual issues implicated by the third gap.

In addition, though, the proposed account of immersion-as-flow is grounded in a theory of perception whose presuppositions are neither Newtonian, nor naïve realist in kind. In this way, the methodological challenge posed by the third theoretical gap is similarly circumvented, and in a rather straightforward manner: Perceptual reality in VR is enacted through embodied, practical, and exploratory engagement with a virtual world, whereby one does not come to perceive the virtual world "as it is", but rather learns to perceive what is relevant for attaining one's practical purposes in it. As an alternative to the current design goal of VR (i.e., maximizing presence), which is fundamentally object-centric, we thus propose the action-predicated goal of maximizing subjective immersion. Our claim is not that the experience of presence is unimportant for VR, but rather that its importance is only secondary to that of maximizing subjective immersion (i.e., virtual flow). As such, we believe that the golden standard by which to measure the quality or "success" of VR is the degree to which the virtual experience can be said to be conducive of subjective immersion in the user. Needless to state, such an approach makes intuitive sense on yet another level. For insofar as flow constitutes a criterion of optimal experience, it follows, therefore, that immersion-as-flow may thereby constitute a criterion of optimal virtual experience.

Our theoretical argument is now complete. We began with a critique of the current paradigm within VR, which, we argued, is tacitly physicalist in its grounding. Next, we claimed that because of its physicalist assumptions, VR research is beset by three theoretical gaps which cannot be bridged by physicalist means alone. We proposed an alternative, action-predicated (pragmatic) approach by drawing from enactive cognitive science, which we claimed could help to bridge the three gaps and circumvent their attendant challenges. In our estimation, all the gaps have now been bridged, and their challenges, circumvented. With our theoretical contribution now realized, we must demonstrate its practical utility by illustrating the methodological implications for VR design which from it follow. In the following section, we articulate our implications, and, in Sect. 6, we briefly summarize our findings and conclude by highlighting potential avenues for future research.

5 Methodological Implications for VR Design

In this section, we demonstrate the practical utility of our theoretical contribution by proposing a set of methodological guidelines for VR creators. We identify four essential elements of VR experience—(1) Onboarding, (2) Immersion, (3) "Offboarding", and (4) "Experiential optimization"—and organize our methodological commentary into four subsections, each of which addresses one of these elements. Rather than dictating *what* to design, though, our methodological merits are meant instead to model a *general approach to the very process of designing* head-mounted, immersive experiences. To this end, we begin first by examining a design probe, *Wake*, that exemplifies a real-world implementation of our proposed approach. Our subsequent discussion of *Wake* is then complemented with and grounded in the theoretical narrative laid out across Sects. 3 and 4.

5.1 Design Probe: *Wake*

Wake is a facilitated mixed reality (MR) experience, created in 2018 by Anna Henson in collaboration with the Pittsburgh-based multidisciplinary performance duo, *slowdanger* (Anna Thompson and Taylor Knight), with research assistance from Qianye (Renee) Mei and Char Stiles. *Wake* is a site-specific, participatory, movement-based installation facilitated by two dancers, in which one participant in-headset (using the HTC Vive Pro room-scale virtual reality system, Vive spatial trackers, and an Intel RealSense depth camera) navigates a walkable virtual environment, which corresponds in size and layout with the physical environment. The participant interacts with both virtual and physical (tangible) objects, and a co-present dancer, who is tracked and rendered photographically in real time in the headset using the head-mounted depth camera (Intel RealSense). The participant is initiated into the experience by one un-tracked dancer who serves as a facilitator (managing the hardware, providing instructions), and later encounters and interacts with the second dancer through visual gestures, physical touch, and verbal dialogue. *Wake* engages concepts of embodied interaction [21] and social presence within hybrid mediated environments.

A user study (N = 25) was conducted to investigate participants' cognitive and affective experience during *Wake*. Qualitative data were collected through semi-structured phenomenological interviews and a standardized self-report for emotional states [22], and quantitative data were gathered through spatial trackers between the participant and co-present dancer, which was analyzed using proxemics. The chosen participants constituted a sample of convenience, as the aim was to explore general principles rather than to experimentally study the effects of specific variables. Participants ranged in age from 21–48 (mean age = 28.5), were recruited mostly from universities in the Pittsburgh area, and consisted of both VR novices and VR developers. All participants had a baseline of using technology to communicate with others, as co-located (located together in the same physical and virtual space), co-presence, and non-verbal communication within a hybrid immersive media environment were significant areas of investigation in this design probe. *Wake* incorporated a user study to explore and interrogate methods of embodied interaction for VR, and to dialogue with participants about concepts developed through Anna Henson and *slowdanger's* collaborative, practice-based research process.

5.2   Onboarding and Pre-immersion

The Onboarding Process. In every head-mounted virtual reality experience, a threshold must be crossed from perceiving the world without wearing a headset, to putting the headset on and engaging in virtual content. This process, onboarding, is crucial for the participant to become fully immersed in the virtual experience. The hardware itself is an inescapable physical reality, though, and in the case of room-scale VR systems (HTC Vive, Oculus Rift, Sony Playstation VR), it is reasonably bulky. The headset and any other worn sensors are in intimate relationship to the participant's body to allow for engagement in the virtual content. The hardware's form factor will continue to decrease in physical size as technology evolves, but, presently, the hardware completely covers the eyes and a significant portion of the face of the wearer. This is the "headset-as-blindfold" phenomenon articulated in Sect. 2. These hardware factors can trigger discomfort, disorientation, or other negative responses in participants across physical, affective, and social levels, if not attended to properly during onboarding. If the headset and other hardware are not appropriately incorporated into the participant's skill set, and consequential affective concerns are not addressed from the outset, this can render moot the actual content in the virtual experience. If the onboarding process (regarding hardware worn on the body, virtual interface, and, if relevant, relationship to other people in the experience) is confusing, nonconsensual, or abrasive, the participant can become distracted or may entirely disengage from the experience right away.

Smoothing the transition from outside to inside the headset is thus a primary concern of the early stages of a VR experience. The experience design should attempt to counteract the discomfort or distraction of the hardware, to help foster a sense of safety or trust, and to cultivate intuition in the participant's ability to perform physical movement wearing the hardware, which is necessary for the subjective feeling of immersion.

Mastery and Scaffolding. VR is a medium with high cognitive load. The early moments of perception and interaction (with both the hardware and the content) in a VR experience are crucial to cultivating a sufficient degree of intuition so that the participant can become safely and fully immersed. The terms *scaffolding, affordances, mastery,* and *discovery* are all germane for conceptualizing the design of a participatory experience, virtual or otherwise, and can be used for understanding how intuition is cultivated during onboarding, as well as how immersion is made possible as a result.

Scaffolding is used here to denote the ways in which a participant is instructed through a task, which in turn makes the task easier and adds fluency to the learning process. *Affordances* are the possibilities for action that a participant perceives, and through appropriate scaffolding, affordances of greater relevance become available to the participant. Put differently, affordances are perceptual frames, and, so, functional action is enabled based on the perceptual affordances available at the time of interaction. *Mastery* denotes competency with a skill or action, whereby relevant affordances become progressively more intuitive. Mastery is achieved through repetition of motivated, exploratory behavior that successfully enables functional action (i.e., yields reward and/or avoids punishment). Once a basic level of mastery is attained, the participant's engagement with the task becomes more intuitive and thus attains greater processing (and experiential) fluency. Subsequently, greater immersion is achieved and the realization of a flow state within the virtual experience becomes more probable.

During the initial mastery stage of a VR experience, the participant acquires basic perceptual (sensorimotor) skills for the VR environment. Within the *Wake* design probe, two areas of "orientation" were found to be crucial for the participant's experience: First, attending to the participant's own sense of embodiment once the hardware is worn, and, second, seamlessly coupling this hardware-affected sense of embodiment with the perception of virtual space and objects. To address these participant needs, *Wake* developed: (a) Physical Orientation Exercises (POEs) and (b) Virtual Orientation Exercises (VOEs). The following discussion will elaborate on the POEs and VOEs used in *Wake*, whereby *scaffolding* is done verbally through instruction, and is socially negotiated.

Physical and Virtual Orientation Exercises. POEs and VOEs in *Wake* are designed to help the participant gain facility with the embodied situation of wearing the headset, and also the physical and visual tools they will use later in the virtual experience. The POEs, which are conducted while the participant perceives darkness in the headset, consist of breathing, sensory awareness, and simple directed movements. When enacted, these actions help the participant to feel greater proficiency over their own bodily proprioception, increase feelings of physical safety, and help the participant to trust the facilitator, which can help enable the participant to move through the experience with more comfort and receptivity. Trust in *Wake* appears to emerge through this sort of facilitated embodiment. One participant stated, "The short breathing exercises at the beginning helped refocus my body, so I felt more comfortable wearing the headset, and the initial feeling of apprehension started to fade away" [18].

VOEs, on the other hand, are meant to acquaint the participant with the "rules" of the virtual experience. More specifically, this entails being introduced to, and later mastering, the possible affordances that are available in the virtual environment through the interface. Transitioning seamlessly from POEs to VOEs, the participant in *Wake* begins to see virtual objects (translucent white rocks) which correspond to their tracked wrist movements (wearing the Vive trackers), and a green rope between the two rocks. The participant is instructed to "play" with these virtual objects, by moving their arms and witnessing the interaction of the rope, which moves dynamically and with real-world physics. Additionally, the participant soon encounters three red spheres, which appear one at a time at a height of about 1.4 m, to which the participant is connected via the same green dynamic rope encountered earlier. These spheres respond to interaction in a similar manner.

Importantly, the real-world dynamics of the rope, and the height at which the spheres appear (generally at a level where participants can look straight ahead, not up or down), help to create intuitive physical interactions in virtual space. Many discussions of virtual interaction design articulate the great importance of dynamic, responsive movement which believably corresponds and contributes to the bodily sensations and movements of a participant. Through sustained interaction, the participant acquires mastery over these basic virtual tools and physical movements, which provides motivation, positive feelings, and fluency with the subsequent parts of the experience. Once basic mastery over the VR experience is acquired in this way, the transition from onboarding to immersion can be said to have begun.

5.3   Immersion and Discovery

Basic mastery over relevant skills is necessary for flow, which is realized when task difficulty is just beyond an individual's skills. Having thus acquired basic mastery during the onboarding process via POEs and VOEs, the participant is now prepared for a more immersive experience. Immersion-as-flow within VR can thus be facilitated by introducing complexity into the VR scenario, thereby progressively increasing the cognitive and sensorimotor demands of the task(s) at hand. The addition of complexity might, for instance, entail introducing a novel challenge into a game or a puzzle which requires the participant to enact a creative synthesis of two, previously known problem-solving strategies, into a novel, composite strategy (by using a tool in an entirely novel manner to solve a problem). Situational complexity, though, must neither overwhelm the participant's ability to cope skillfully, nor be exceeded altogether by the participant's practical know-how. But should the demands of the task only slightly exceed the participant's level of skill, the complexity of the situation will help to garner and maintain participant interest and motivate the participant to engage in exploratory behavior, or *discovery* (i.e., the motivated discovery of possibilities for action). The design goal during the immersion stage is therefore to strike and sustain an optimal balance between the cognitive (and sensorimotor) demands of the situation and the available skills of the participant, so as to facilitate ongoing interest, engagement, and discovery of possibilities over time.

In *Wake*, this was achieved through the use of theatrical staging techniques (such as directional lighting), as salience cues, as well as verbal instructions, to direct the participant's attention so as to continually scaffold their learning throughout the course of the installation. The virtual experience began in complete darkness, with POEs as the main emphasis. VOEs were then introduced, which aimed to teach the participant how to effectively coordinate their physical movements in relation to the objects that would appear sequentially inside the virtual environment (e.g., rocks, rope). Through each subsequent stage of the experience, a novel function was introduced with the appearance of a new object or aspect of the virtual space (e.g., a path), which could only be accommodated and mastered by synthesizing previously learned behaviors (sensorimotor acts) into composite and more complex behavioral patterns. Through the progressive introduction of novel functions and possibilities for action, the virtual environment became an increasingly complex arena for the participant to act in; and through the scaffolding of the participant's learning, the participant's skills were continually stretched to match the growing demands of the situation. The process of ongoing discovery in *Wake* culminated into a moment whereby the virtual object with which the participant had already been interacting was revealed to have been under the direct, physical control of the dancer all along (e.g. a particle system controlled by the two Vive trackers worn on the dancer's wrists). Particularly, this realization occurred as a live capture of the dancer was rendered in the virtual environment (more on this in Subsect. 5.5), overlaid on top of the virtual object. The co-located dancer was hence transformed into a virtually co-present agent, thereby changing the meaning of the participant's virtual situation, and making possible a whole new kind of interaction altogether.

5.4   Post-immersion Offboarding

An immersive VR experience does not end abruptly once the user removes the headset. Just as becoming immersed entails a transition period (i.e., onboarding) in which the participant's sensorimotor systems must calibrate to fit the sensorimotor demands of the virtual experience, the post-immersion experience likewise entails a transition period, an "offboarding" process, if you will, which entails a reorientation to the familiar. As designers, we must therefore acknowledge that the participants will go through a reorientation period after taking off the virtual reality headset and other equipment, in which they will need to process or "decompress" from the experience. This means that we should create a scenario in which such processing may occur, either a quiet place for reflection, a medium for expression (such as a guest book), or a place to talk with other participants.

In *Wake*, offboarding involved the administration of semi-structured phenomeno-logical interviews inquiring into four general aspects of the participants' experience: (i) Bodily sensations, (ii) Emotions, (iii) Relationship to the dancer/facilitator, and (iv) Interaction and spatial design. An interesting observation that was drawn regarding participants' experience of offboarding in *Wake* was that there was a clear shift in vocal tone and language with most participants over the course of the interview. Their descriptions were initially highly intuitive and centered around bodily feelings and sensations, but become increasingly more analytical and deliberative as the interview progressed. This transition from intuitive to deliberative language suggested that the the participants' attention was initially largely preoccupied with various sensorimotor and embodied aspects of their immersive experience, and that the interview facilitated a sort of processing and integration of these aspects of their experience into their con-sciousness post-immersion.

5.5   Experiential Optimization

We have extensively argued that the design goal of VR should be the maximization of subjective immersion, rather than presence. We find it methodologically important here to identify and address an optimization issue, which we call "experiential optimization", pertinent to the realization of this design goal. Experiential optimization involves a trade-off relationship between objective immersion, on the one hand, and subjective immersion, on the other. More specifically, it appears that as the degree of objective immersion afforded by a given VR hardware (i.e. worn sensors, controllers, headset) is maximized, so increases the degree to which user perception in VR becomes mediated. Consequently, perceptual engagement and interaction with VR content becomes increasingly counterintuitive or clunky. In other words, there *prima facie* appears to be a limit on the degree to which objective immersion can be maximized before the design goal of maximizing subjective immersion becomes compromised. This is not to say that objective immersion should be forsaken altogether, since it is, after all, a necessary feature of immersive VR experiences. Rather, given that both subjective immersion and objective immersion are essential for immersive VR experience, and that there is a necessary trade-off between these two kinds of immersion, the methodological principle here becomes not the maximization of one kind of immersion over the other, but rather the optimization of the trade-off between the two.

In our estimation, the way to experiential optimization cannot be prescribed in a manualized manner, but must be determined (or discovered) on a case-by-case basis, depending on what the VR experience in question is meant to express or engender. Optimizing the trade-off between objective and subjective immersion can mean including such tools and sensors as tracking devices, haptics, or artificial intelligence, as part of the design. However, due to the interference caused by added layers of mediation, part of a designer's job is to know which virtual elements to lean into, and which virtual elements to leave out of the equation in order to create an experience in which subjective immersion (immersion-as-flow) is properly facilitated or achieved. Optimization thus might even become a matter of also engaging the un-mediated sense modalities of an individual with the VR content, as a way to creatively sidestep technological limitations of the hardware and interface in favor of enriching the experience design and leading to greater immersion. We thus turn to a discussion of how experiential optimization was achieved in *Wake*, with regards to the problem of representing others in VR.

Representing Others in VR. Visually representing others in co-present VR experiences can be done in many ways, but the vast majority of these involve avatars (i.e. a human-controlled, computer generated representation of a person or character). Avatars exist, for varying purposes, on a scale of realism to abstraction, and many experiences using abstract or fantastical avatars can be said to be highly successful. However, representing a unique individual with a high level of photographic realism is currently a critical question. Recent developments in 3D modeling and scanning (i.e. photogrammetry) have made highly detailed renderings and photographically-based captures of individuals possible; yet a 3D scan is simply a static, unmoving mesh, and even the most advanced, rigged 3D model of a human still confronts the "uncanny valley", or the repulsion experienced when faced with a humanoid representation which is *almost-but-not-quite real,* or *strangely familiar* [31]. A 3D scan of a person may thus be *photographically realistic*, but unless the scan can *behave realistically*, its embodied expression will not convey "aliveness", feelings, or intent in a manner that is intuitive or realistic. Put differently, real-time, intuitive, social communication between people within immersive media is enabled when participants can see, hear, and respond to others in a believable and instantaneous manner.

Volumetric capture is a video technique that utilizes synced RGB and depth streams of a person or environment to render in 3D (i.e., as a mesh or point cloud) its subject. The captured material can be edited and used in VR experiences in a similar way to traditionally filmed content. With high resolution capture, subtle facial expressions and body language are made visible and therefore available to the participant, in a similar way to our real-world interactions in the physical world. This technique thus solves the problem of photographic realism. However, if the content is pre-captured (i.e., not occurring in real time), the person rendered in the lenses of the VR headset cannot respond to the participant's actions or language in a real-world manner (attempts at AI or machine learning responsiveness notwithstanding). It is simply a recording of a previous moment in time. For co-presence and social interaction to be believable, however, and therefore effective in the case of volumetric capture, the footage must be streamed in real time (i.e., telepresence).

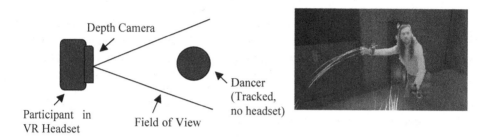

Fig. 1. Co-located Co-Presence in *Wake*: The schematic on the LEFT represents the hardware system and co-presence structure used in *Wake* for rendering the co-located dancer. On the RIGHT: The dancer (Taylor Knight) as seen in the headset, rendered in real time through the Intel RealSense camera. The dancer is tracked using Vive trackers, and can interact with virtual objects, such as the rope depicted in the image.

In order to test the possibility of real time volumetric capture, the *Wake* design probe developed a scenario in which an in-headset participant encounters and interacts with a co-present dancer who is not wearing a headset (Fig. 1). This scenario constituted a case of Asymmetric, Co-Located, Co-Present Mixed Reality (ACLCPMR). During the experience, the participant is immersed in a virtual environment rendered in the headset, interacts with both virtual and physical (tangible) objects, and engages with the dancer through simple improvised movement. In this ACLCPMR experience, the dancer is tracked in the virtual environment and is also physically co-located in the same space as the participant. The dancer was rendered in real time through a custom algorithm which utilized the feed from a depth camera (Intel RealSense) mounted on the front of the participant's headset, at the position of their eyes. Therefore, when the participant looked at something, the camera saw what they saw. Using a depth filtering algorithm, the camera feed was manipulated to only render objects which were at a certain distance from the camera (to render the dancer but not the walls of the room around them).

In *Wake*, co-presence was ultimately achieved not through simulation, as previous studies have utilized [2], but by the real time rendering of a co-located, co-present dancer. This method therefore successfully circumvented the uncanny valley problem, not by directly addressing the problems of photographic and behavioral realism in an avatar, but by side-stepping them altogether. Qualitative findings from the semi-structured phenomenological interviews revealed that a vast majority of the participants (24 out of 25) felt that they were able to make and hold eye contact with the dancer as though it was real and mutual, despite knowing that the dancer was not able to see their eyes (because of the headset). The findings in *Wake* therefore plausibly suggest that real-world, genuine social interactions are possible within hybrid media environments like VR. More importantly, they suggest that ACLCPMR not only stands as a viable method for achieving experiential optimization when it comes to representing others in VR, but also for testing, researching, and developing new forms of human-machine mediated communication.

6 Conclusion

The standard approach to VR research and development tends to value physicalist achievements (e.g., physically realistic simulations) while overlooking both the pragmatic and the phenomenological dimensions of perceptual experience. Because the physicalism inherent in the standard approach has been inherited, and is therefore tacitly presupposed, it has the tendency of propagating itself in the literature without being subjected to critical scrutiny. As a consequence, the pragmatic and phenomenological dimensions of perceptual experience in VR can only ever remain overlooked and undervalued. The main ambition of this article has been but to disrupt this implicit propagation of presuppositions, and our decision to engage with cognitive science has been motivated precisely by this reason. More specifically, in having turned to enactivism, we have articulated an alternative, action-predicated approach to VR, one that is (1) non-reductive with respect to subjective experience, and which (2) honors the pragmatic dimension of perceptual reality. Having also illustrated the various methodological implications of our action-predicated approach, namely as regards with POEs, VOEs, immersion, flow, on- and offboarding, and experiential optimization, we believe that the next step in this line of work is to develop more rigorous empirical methods by which to test the theoretical claims and qualitative observations made in this article. Suffice it to say, though, if with this article we have at least managed to raise some interesting questions, then that alone constitutes a worthwhile beginning.

References

1. Ahmed, S.: The Cultural Politics of Emotion, 2nd edn. Routledge, London (2014)
2. Bailenson, J.N., Blascovich, J., Beall, A.C., Loomis, J.: Interpersonal distance in virtual environments. Pers. Soc. Psychol. Bul. **29**, 1–15 (2003)
3. Biocca, F.: The cyborg's dilemma: progressive embodiment in virtual environments. J. Comput.-Mediat. Commun. 3(2), 1–31 (1997)
4. Bulu, S.T.: Place presence, social presence, co-presence, and satisfaction in virtual worlds. Comput. Educ. **58**(1), 154–161 (2012)
5. Csikszentmihalyi, M.: Flow: The Psychology of Optimal Experience. HarperCollins e-books (1990)
6. Cherniak, C.: Minimal Rationality. MIT Press, Cambridge (1986)
7. Jerome, C.J., Witmer, B.: Human performance in virtual environments: effects of presence, immersive tendency, and simulator sickness. In: Proceedings of the Human Factors and Ergonomics Society: 48th Annual Meeting, New Orleans, Louisiana, pp. 2613–2617. Human Factors and Ergonomics Society (2004)

8. Mütterlein, J.: The three pillars of virtual reality? Investigating the roles of immersion, presence, and interactivity. In: Proceedings of the 51st Hawaii International Conference on System Sciences, Waikoloa Village, Hawaii, pp. 1407–1415. Hawaii International Conference on System Sciences (2018)

9. Schwartz, R., Steptoe, W.: The immersive VR self: performance, embodiment and presence in immersive virtual reality environments. In: Papacharissi, Z. (ed.) A Networked Self and Human Augmentics, Artificial Intelligence, Sentience, pp. 108–116. Routledge, New York (2018)

10. Witmer, B.G., Singer, M.J.: Measuring presence in virtual environments: a presence questionnaire. Presence 7(3), 225–240 (1998)

11. Peterson, J.: Maps of Meaning. The Architecture of Belief. Routledge, London (1999)

12. Varela, F., Thompson, E., Rosch, E.: The Embodied Mind: Cognitive Science and Human Experience. The MIT Press, Cambridge (1991)

13. Thompson, E.: Mind in Life: Biology, Phenomenology, and the Sciences of Mind. The Belknap Press of Harvard University Press, Cambridge (2007)

14. Hovhannisyan, G., Dewey, C.: Natural & normative dynamical coupling. Cogn. Syst. Res. 43, 128–139 (2017)

15. Hovhannisyan, G.: Humanistic cognitive science. Humanist. Psychol. 46(1), 30–52 (2018)

16. Metzinger, T.: Phenomenal transparency and cognitive self-reference. http://www.philosophie.uni-mainz.de/metzinger/publikationen/Phenomenal_transparency_PCS2003.pdf . Accessed 24 Jan 2019

17. Vervakae, J., Ferraro, L., Herrera-Bennett, A.: Flow as Spontaneous Thought: Insight and Implicit Learning. Oxford Handbooks Online (2018)

18. Henson, A.: We're In This Together: Embodied Interaction, Affect, and Design Methods in Asymmetric, Co-Located, Co-Present Mixed Reality. Masters Thesis, Carnegie Mellon University, Pittsburgh (Pa.) (2019)

19. Chalmers, D.: Facing up to the hard problem of consciousness. http://consc.net/papers/facing.html Accessed 29 Jan 2019

20. Dreyfus, H.: Skillful Coping: Essays on the Phenomenology of Everyday Perception and Action. Oxford University Press, Oxford (2014)

21. Dourish, P.: Where the Action Is: The Foundations of Embodied Interaction. The MIT Press, Cambridge (2001)

22. Plutchik, R.: A General Psychoevolutionary Theory of Emotion. Academic Press, New York (1980)

23. Hogarth, R.: Educating Intuition. University of Chicago Press, Chicago (2001)

24. Topolinsky, S., Reber, R.: Gaining insight into the "aha" experience. Curr. Dir. Psychol. Sci. 19(6), 402–405 (2010)

25. Zadra, J.R., Clore, G.L.: Emotion and perception: the role of affective information. Wiley Interdiscip. Rev. Cogn. Sci. 2(6), 676–685 (2011)

26. Haraway, D.: Situated knowledges: the science question in feminism and the privilege of partial perspective. Fem. Stud. 14(3), 575–599 (1988)

27. Juarrero, A.: Dynamics in Action: Intentional Behavior as a Complex System. MIT Press, Cambridge (1999)

28. Hulswit, M.: From Cause to Causation: A Peircian Perspective. Kluwer Academic Publishers, Dordrecht (2002)

29. Peterson, C., Park, N., Seligman, M.E.P.: Orientations to happiness and life satisfaction: The full life versus the empty life. J. Happiness Stud. 6, 25–51 (2005)

30. Peterson, J.: The meaning of meaning. In: Wong, P., et al. (eds.) The Positive Psychology of Meaning and Spirituality. INPM Press, Vancouver (2008)

31. Reichardt, J.: Robots: Fact, Fiction, and Prediction. Thames and Hudson, London (1978)

32. Giorgi, A.: Phenomenology and the foundations of psychology. In: Arnold, W. (ed.) Nebraska Symposium on Motivation, pp. 281–348. University of Nebraska Press, Lincoln (1976)

33. Weber, A., Varela, F.: Life after kant: natural purposes and the autopoietic foundations of biological individuality. Phen. Cog. Sci. **1**, 97–125 (2002)

34. Robbins, B.: Enactive cognition and the neurophenomenology of emotion. In: Gordon, S. (ed.) Neurophenomenology and Its Application to Psychology, pp. 1–24. Springer, New York (2013). https://doi.org/10.1007/978-1-4614-7239-1_1

35. Noë, A.: Out of Our Heads: Why You Are Not Your Brain, and Other Lessons From the Biology of Consciousness. Hill and Wang, New York (2009)

36. Gallagher, S.: Enactivist Interventions: Rethinking the Mind. Oxford University Press, Oxford (2017)

37. Vervaeke, J., Ferraro, L.: Relevance realization and the neurodynamics and neuroconnectivity of general intelligence. In: Harvey, I., et al. (eds.) Smart Data: Privacy Meets Evolutionary Robotics, pp. 57–68. Springer, New York (2013). https://doi.org/10.1007/978-1-4614-6409-9_6

38. Vervaeke, J., Lillicrap, T., Richards, B.: Relevance realization and the emerging framework in cognitive science. J. Logic Comput. **22**, 79–99 (2012)

The Impact of Game Peripherals on the Gamer Experience and Performance

Xiaobo Ke$^{(\boxtimes)}$ and Christian Wagner

School of Creative Media, City University of Hong Kong,
Kowloon, Hong Kong, China
xiaoboke-c@my.cityu.edu.hk, c.wagner@cityu.edu.hk

Abstract. Game peripherals refer to the input-output devices assisting players to interact with video games. An interesting phenomenon related to game peripherals is bringing your own peripherals (BYOP) which means video game players, especially the advanced players, usually tend to use their own devices to play video games. An important reason for the popularity of BYOP is the players' belief that the best tools to play the game are their own devices, in terms of game experience and performance. Thus, the game peripherals used in the BYOP situation imply the excellent quality. However, the limited research on the game peripherals leads to the lacking understanding of what determines the good game peripherals and how the good game peripherals influence players' positive gaming experience and their performance. In order to call for more attention to the research on game peripherals and players' cognition, this paper focuses on two important dimensions of game peripherals (i.e., controller fit and vividness of interfaces) and their influences on players' positive in-game experience (i.e., sense of control, immersion and enjoyment). Furthermore, this research also discusses the relationship between positive gaming experience and in-game performance. A relational framework including seven propositions is proposed to guide and suggest future research on the game peripherals' influences and the positive gaming experience in the field of player-video game interaction.

Keywords: Game peripherals · Players' experience · Gaming performance · Player-video game interaction

1 Introduction

Game peripherals usually refer to the input-output devices facilitating players to interact with video games. The controllers and the user interfaces (e.g., the screen for displaying the virtual gaming environment and the audio devices for displaying the sound from the game) are two usual but important types of the game peripherals for playing the video games [47, 70]. In 2017, the game industry in America generated a record of 36 billion dollars in revenue [15]. Furthermore, one important observation found from this selling record is that the growth rate of hardware revenue (19%) is slightly higher than that of software revenue (18%) [15]. The active demand of the game peripherals is one of the important contributions to the high growth rate of the hardware revenue because of the popular phenomenon of bringing your own

© Springer Nature Switzerland AG 2019
D. D. Schmorrow and C. M. Fidopiastis (Eds.): HCII 2019, LNAI 11580, pp. 256–272, 2019.
https://doi.org/10.1007/978-3-030-22419-6_18

peripherals (BYOP). The BYOP means the video game players usually would like to play video games by using their own devices, such as the keyboard, mouse and so on [79]. The BYOP is more common among the advanced players and this is also allowed in the professional tournaments of video games [82]. For example, rules and regulations of tournaments drafted by the International eSports Federation (IeSF) permit the professional players to set up their own game peripherals with some minimal restrictions [31].

The players believe their own peripherals are the best and critical tools to help them achieve the optimal gaming experience and high in-game performance. This belief is also an important reason for the popularity of the BYOP. Therefore, the game peripherals take an important role in the player-video game interaction regarding the game experience and performance of the players [6]. Furthermore, the game peripherals applied in the BYOP situation usually indicates the high quality. However, the research on the game peripherals is still insufficient yet [6, 42]. This leads to our limited understanding of what dimensions determine the good quality of game peripherals and how these factors influence players' positive gaming experience and hence impacting their performance.

To call for more attention to the studies on the influence of game peripherals on players' cognition in the player-video game interaction, this paper focuses on the influence of the controllers and user interfaces which are two common but essential game peripherals for video gaming. Specifically, based on the theories of technology-task fit [22] and media vividness [78], this paper develops the constructs of controller fit and the vividness of interfaces. Furthermore, three positive gaming experience, namely the sense of control, immersion, and enjoyment are identified from the theories of cognitive absorption [1], flow experience [10] and cognitive engagement [83, 84]. Drawn on the conceptual constructs, this paper raises seven propositions to illustrate the theoretical framework which consists of (1) how the controller fit and vividness of interfaces influence players' positive gaming experience; and (2) the relationship between the positive gaming experience and players' performance.

To sum up, this study provides a theoretical framework to help researchers understand the influence of the game peripherals on players' positive gaming experience and their performance. The propositions proposed by this study not only contribute to the knowledge of the mechanism of the player-video game interaction from the perspectives of game peripherals but as well highlight the potentials relationships of gaming experience with players' performance from the perspective of emotion-performance relation. We hope this study could help the scholars to guide their further research on the game peripherals and the players' gaming experience.

2 Theoretical Background

In this section, we introduce the several fundamental theories and the necessary literature to develop the focal constructs for this study. Specifically, the controller fit, the vividness of interfaces and the three types of positive gaming experience, namely sense of control, immersion and enjoyment, are introduced in the following sub-sections, respectively.

2.1 Controller Fit

The "fit" focus has been most significant in information systems research on the performance of individual decision-making and technology adoption (e.g., [40, 93]). As one of the influential information systems theories, the model of task-technology fit (TTF) concerns in the correspondence and matching between the task requirements and the abilities of information technologies [22]. The TTF model suggests that users will adopt a new technology if it is good enough for them to execute a certain task effectively and efficiently [67]. In addition, the TTF model has confirmed the influence of characteristics of tasks, technologies and individuals on the individuals' perception of TTF and hence influencing individuals' performance [13, 21]. This indicates that the TTF model actually considers the fit among the task, the technology and the user. Furthermore, the studies with "individual-technology fit" focus also have been conducted in another research stream which is based on the innovation-values fit (IVF) theory [69]. The IVF argues that individuals' use of an innovative technology depends on the fit between the innovation and the values of individuals [43]. The value concerned by the individual could be regarded as "generalized, enduring beliefs about the personal and social desirability of modes of conduct or 'end-states' of existence" [39]. In other words, when the technology can match individuals' values, they would be more likely to adopt and use this technology, and vice versa [43, 68]. Thus, the IVF theory can also be employed to explore the interaction relationship between persons and technologies [69].

In the context of gaming, the controllers are the essential information input techniques in player-video game interaction. The common types of the controllers are the gamepad, the joystick, and the combination of the keyboard and the mouse [6]. Based on the theories of task-technology fit [22] and innovation-value fit [43], we propose a construct named controller fit which means the controllers' fit to the players and the game contexts. Two dimensions determine the controller fit: (1) the correspondence between the game requirements and the abilities of the controllers, namely game-controller fit; (2) the matching between the players' preference (determined by the beliefs, social desirability, habits and etc.) and functionalities of the controllers, that is, the player-controller fit. Furthermore, three factors will influence players' perception on controller fit: the characteristics of the games, the controllers and the players.

In the existing research regarding the controller fit, we can find that the construct of controller naturalness [61, 62] is well researched in the reality simulated game context. Controller naturalness refers to the overall intuitiveness which a controller is perceived when players are interacting with a virtual environment [75]. The literature on controller naturalness usually concerns in the matching between the controllers and the mental model [4] of the players (e.g., [59, 63, 74]). Therefore, to some extent, controller naturalness is a subset of the controller fit. However, the players' other characteristics (e.g., beliefs and habits) are also important factors which may influence the players' gaming experience and game adoption [18]. Moreover, player-controller fit also covers the matching between players' mental model and controllers. In addition, the correspondence between abilities of the controllers and the game requirements (i.e., game-controller fit) is also influential according to the theory of task-technology fit. Therefore, as the theoretical sublimation of controller naturalness, controller fit proposed by this study emphasizes a more comprehensive role in the players' positive gaming experience from both perspectives of the game-controller fit and the player-controller fit.

2.2 Vividness of Interfaces

In general, vividness refers to "the representational richness of a mediated environment as defined by its formal features; that is, the way in which an environment presents information to the senses" [78]. Vividness is one of the important properties of media technologies for the high quality of presentation. Two dimensions of the sensors usually determine the extent of media vividness [19, 78]: (1) sensory breath which means the number of sensory channels simultaneously presented (e.g., audio, visual, haptic, and etc.); (2) sensory depth which indicates the quality of the presentation by the sensory channels (e.g., the quality of the image/video). Therefore, higher vividness implies more sensory channels [25, 92] and more information cues [17, 20] provided by these channels (say, the quality). Furthermore, vividness is also likely to attract and hold individuals' attention, which makes people feel emotionally interesting and provokes their concrete and imagery thinking [37].

In the context of gaming, vividness refers to the quality of the user interfaces related to the presentations of the game's virtual environments, such as the audio-visual system displaying the game's scenes and sounds. Based on the previous literature on the quality of user interfaces, we are able to find that most of these studies usually center on the quality of visual display systems (e.g., [5, 68]) and the quality of audio systems (e.g., [65, 76]), respectively. Typically, the information from the visual device is essential and indispensable for the most of game contexts as the visual device displays the main game scene containing objects with which players need to interact (e.g. avatars, enemies and targets) as well as a complicate cum moving background (e.g. interiors and landscapes) [6]. In most circumstances, the game controlling decisions (i.e., how to play the game in next second) are made mainly drawn on the information gained from the visual display systems. Besides, the information from audio devices is also often used to reinforce or facilitate the players' experience in the player-video game interaction [6]. Most of the extant game related studies on display systems and audio systems are actually concerning about the sensory depth of the vividness of user interface, such as screen size [28], resolution [72], screen position [7], surround quality of the sound [76] and so forth.

2.3 The Positive Experience in Gaming

The user experience is one of the main topics of the information systems studies. In the previous literature, the research on games or other hedonic information systems usually focuses on the positive mental state of the users (e.g., [54, 80]).

The flow experience [10], cognitive absorption [1] and the cognitive engagement [83, 84] are the three main theoretical foundations of the relevant research on users' positive experience in the interaction of human-hedonic information systems (e.g., the game). The flow experience refers to the optimal overall experience, "the state in which individuals are so involved in an activity that nothing else seems to matter" [10]. Cognitive absorption denotes a state of deep engagement and involvement that a user can experience in the interaction with hedonic information systems [1, 12, 24]. Cognitive engagement relates to the state of playfulness, and that the state of playfulness is identical to the flow experience [84]. Specifically, cognitive engagement is the flow experience without the notion of control [83, 84]. These three constructs are all the multi-dimensional constructs and dimensions of each construct are listed in Table 1.

Table 1. Dimensions of flow experience, cognitive absorption and cognitive engagement.

	Positive gaming experience used in this study	Flow experience	Cognitive absorption	Cognitive engagement
Dimensions	Enjoyment	Pleasure and enjoyment	Heighten enjoyment	Intrinsic interest
	Immersion	Concentration	Focused immersion	Attention focus
	Sense of control	Control	Control	–
	–	–	Temporal disassociation	–
	–	–	Curiosity	–

Based on Table 1, we can find that the enjoyment and the immersion are the two dimensions included in these three positive mental states (i.e., pleasure and enjoyment and concentration in flow experience; heighten enjoyment and focused immersion in cognitive absorption; intrinsic interest and attention focus in cognitive engagement). In addition, the sense of control is the dimension which only appears in the flow experience (i.e., control) and the cognitive absorption (i.e., control) as the cognitive engagement is the flow experience without the sense of control. Therefore, the enjoyment, immersion and sense of control are the three positive gaming experience usually concerned by these foundational theories. In this case, this study mainly focuses on these three dimensions for the development of the proposition on the relationship of game peripherals with players' positive game experience and their performance.

Firstly, the sense of control captures the user's perception of being in charge of the interaction [1]. The sense of control is one of the important triggers for players' dominance feeling in the gameful experience [16]. Secondly, the immersion is defined as "the experience of total engagement where other attentional demands are, in essence, ignore" [1]. In other words, the immersion relates to the experience of total involvement and emotional engagement in the virtual world [73]. Thirdly, enjoyment refers to the extent to which interaction with the system is perceived by players as pleasurable and enjoyable [23, 80]. The enjoyment as a desirable affective response is important for users' satisfaction during the system interaction [2, 14], especially in the context of hedonic information systems [54].

3 Proposition Development

According to the research framework as shown in Fig. 1, seven propositions are developed to illustrate the theoretical relationship among game peripherals (say, controllers and user interfaces), players' positive gaming experience and their in-game performance. The in-details proposition development is presented in the following subsections.

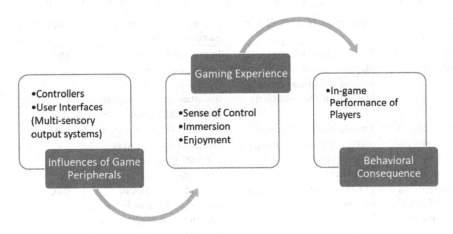

Fig. 1. The theoretical framework.

3.1 Controller Fit and Players' Positive Experience in Gaming

In this paper, we propose a construct of controller fit indicating the influence of controllers. Two dimensions consist of this construct: (1) game-controller fit which means the correspondence between the game's requirements and abilities of the controllers; (2) player-controller fit which refers to the matching between the players' mental model, belief, and habits as well as the functionalities of controllers. Therefore, three factors, namely the game characteristics, the controller characteristics and the player characteristics, influence players' perception of the controller fit.

As shown in Table 2, three focuses of controller fit can be summarized from the existing literature on the relationship between controller fit and their influence on users' experience. Table 2 indicates that the current research mostly centers on the controller naturalness (e.g., [59–61, 74]) which is a specific type of player-controller fit. That is, the matching between the players' mental model of the real world and the mapping methods of the controllers. This kind of controllers fit (i.e., controller naturalness) has already been confirmed that it effectively influence players' experience in gaming, such as the aggressive intention (e.g., [63]), the immersion (e.g., [62]). However, the controller naturalness is not always true and effective because this kind of controller fit is highly depended on the game context (say, the reality simulated games) [81]. Therefore, this is why other specific types of controller fit are still important and worthy to be researched. For instance, in a game context where the games have no real-world analog, like casting a healing spell or flying a fictional vehicle, players' mental models of controlling behaviors are more a function of their video game experiences rather than their experiences in the real world [71]. In this case, the matching between players' habits of previous game experience and the controller is more meaningful and useful [71]. This is another types of controller fit shown in Table 2 (i.e., controller fit to players' game experience).

Table 2. The summary of three focusing types of controller fit in the literature.

Focuses of controller fit	Description of controller fit	Importance for players' game experience	Supporting references
Controller naturalness	The matching between players' mental models of the real world and mapping methods of the controllers	In the reality simulated games, the controller naturalness plays an important role in shaping players' experience	[59–63, 74]
Controller fit with players' game experience	The matching between the players' previous gaming experience and the controllers	In virtual oriented games, controller fit to players' previous game experience is more significant	[71, 81]
Comparison of different controllers	The overall fit assessment of the controllers	The different types of controllers have different influences on players' game experience	[86, 87]

Thus, in order to give out a consistent theoretical lens for the research on controllers in the player-video game interaction, this paper proposes the construct of controller fit which is the theoretical sublimation of the controller naturalness and other studies of the controller. Furthermore, we also hope to call for more attention to conduct more studies of the controller fit in the various gaming context. To this end, we develop two propositions on the controller fit and its influence on players' positive experience to guide the further research.

According to the theory of TTF, users will only use an information technology when it fits their tasks at hand [22, 38, 88]. Following this logic, in the gaming context, we argue that the players prefer to use the controller which fits the game which is being played, namely the controller fit resulting in the phenomenon of BYOP. This is because the controller fit leads to the high game experience of players. According to the findings of the existing studies, we can find that a closer match between the user's existing knowledge and the channel of input allows the gamers to react more efficiently to the game's information [4, 60], which enhances the sense of the control. Moreover, a study reveals that the fit between the games and the controllers leads to more presence experience which is an important element of the overall immersive experience [87]. In addition, the controllers fit also provides the player with more opportunities to focus on the game contents, which also enhances the possibility of experiencing focused immersion [62, 63]. Thus, the propositions on controllers fit and positive players' experience in gaming are listed as follows:

Proposition 1 (P1): Controller fit has the significant effect on the players' sense of control in gaming.
Proposition 2 (P2): Controller fit has the significant effect on players' immersion in gaming.

3.2 Vividness of Interfaces and Players' Positive Experience in Gaming

Vividness refers to "the ability of a technology to produce a sensorially rich mediated environment" [78]. Furthermore, vividness of the media is believed to affect involvement with the mediated environment [19, 25]. Vividness can be achieved through the manipulation of dimensions of depth and breadth [78]. The breadth implies the number of different sensory channels utilized (e.g., visual, audio and etc.) [26]. In contrast, the depth indicates the resolution or quality of a particular sensory channel [25].

In the context of player-video game interaction, vividness also often takes an important role in the game experience because of the integrated systems of user interfaces. When reviewing literature (as shown in Table 3), we are able to find that the vividness of interfaces has a significant influence on the players' positive experience in the player-video game interaction. For example, Liu et al. [52] examined the positive relationship between visual vividness and immersion while Williams [87] confirmed the vividness leads to the sense of presence and hence improving player's enjoyment. Besides, we further found that the vividness also significantly influences users' experience and decision-making in contexts of business and social network (see Table 3).

Table 3. The summary of vividness research in different fields.

Research areas	Importance of the vividness	Stimulated positive experience	Supporting references
Game experience	Visual vividness helps players to create a sense of presence, enjoyment and immersion in the player-video game interaction	Presence, enjoyment, immersion	[35, 52, 76]
Online shopping	More vivid product visualizations are linked to a more positive affective emotional experience and purchasing intention of website visitors	Immersion, enjoyment	[9, 41, 90]
E-tourism	The vividness of the technologies quality is important for visitors' overall satisfaction of the travel attractions and visitor's attitude change	Presence, mental imagery	[45, 85]
Recommendation systems	Vividness represents multiple symbol sets to convey information and is positively associated with better information processing of individuals	–	[25, 37]
Social media	The vivid information can not only increase the perceived quality and credibility of the presented information but also enhance the hedonic feelings of users	Social presence	[17, 19, 48, 91]

Table 3 indicates that the vividness of interfaces is a widely researched construct in various contexts, which means the influence of the user interface is an important factor for the user experience and the cognition change. Based on the previous literature, this paper develops two propositions on the vividness of interfaces in the context of games. These two propositions highlight the role of user interfaces as the important game peripherals in the player-video game interaction.

Vividness is "likely to attract and hold our attention and to excite the imagination to the extent that it is emotionally interesting" [66]. Furthermore, higher vividness is often associated with more salient information cues/sensory channels, and therefore it generally is more attractive [37, 66] and drawing attention [37, 46]. In addition, the display of vivid information also triggers more positive affective responses, such as enjoyment in shopping [37, 64]. The literature listed in Table 3 also supports these arguments. Therefore, in the context of gaming, the vivid interface also makes players more absorbed and enjoyed in the game landscape because of the attractiveness of the vividness information and its emotionally interesting characteristics. Therefore, two propositions on the vividness of interfaces and positive players' experience in gaming are listed as follows:

Proposition 3 (P3): Vividness of interfaces has the significant effect on players' immersion in gaming.

Proposition 4 (P4): Vividness of interfaces has the significant effect on the players' enjoyment in gaming.

3.3   Players' Positive Gaming Experience and Their Performance

Drawn on the dimensions of the flow experience, cognitive absorption and cognitive engagement, we extract three types of positive experience which the players usually experience in gaming. These three types of positive experience are sense of control, immersion and enjoyment. In this sub-section, we illustrate the relationships between these three types of positive experience and the players' game performance.

The sense of control represents the players' perception of being in charge of the interaction with video games [1]. Control is the abilities to manage one's interaction, including the capabilities of interrupting the interaction, adapting the interaction to one's desires, selections making, and being generally in charge of interaction [53, 55]. The perception of the control abilities is also a partial means by which a person can feel he or she is competent, mastery, and capable to make decisions [54] with the use of controllers in gaming, which makes the players have high confidence in the effectiveness of their game operations. Therefore, the players with a high sense of control could be more prone to aggressively concentrate on the challenges occurring in the game environment because of their confidence of game controlling driven by the sense of control. Furthermore, the sense of control as a positive players' experience also stimulates players' positive mental state which facilitates players' performance [57]. Together, the cognitive concentration on the game challenges and the positive mental state of players jointly help the players gain their higher game performance. Thus, the proposition is listed as follows:

Proposition 5 (P5): Sense of control has the significant effect on the players' gaming performance.

Immersion refers to "the experience of total engagement where other attentional demands are, in essence, ignore" [1]. The experience of immersion is the typical and important experience for the user's engagement with hedonic information systems. In the game context, the immersion experience has also been proved the positive influence on the intention to play [89] and on the game repurchase behaviors [44]. Furthermore, a higher level of immersion potentially has a greater impact on performance [36]. For example, Liu et al. [52] found the immersion experience leads to higher performance while Lin et al. [51] find that immersion impacts players' recall ability in gaming. Moreover, Slater et al. [77] suggests a positive relationship between immersion and the ability of spatial judgments. The high immersion means the high cognitive resource concentrate on the context of the game, which is the important factor for players facilitate their cognitive skills in gaming [3]. Therefore, the proposition on immersion and players' gaming performance is listed as follows:

Proposition 6 (P6): Immersion has the significant effect on the players' gaming performance.

Enjoyment is defined as the extent to which the interaction with the systems is perceived as pleasurable and enjoyable [23, 80]. Clearly, enjoyment is a typical positive emotion. The effect of positive emotion on the individuals' performance in games is positively significant from two perspectives. On the one hand, the positive emotion and affect (e.g., enjoyment or happiness) stimulates players' more creative problem-solving abilities [30, 33, 34] and facilitates their decision-making under pressure [32]. The abilities of problem-solving and making decisions in the stressful situation are both the important factors for the better performance in the competition [57]. On the other hand, if the game is performed in the group formation, the enjoyable experience of the players also increases their team cooperation which ultimately improves their team performance [50, 56]. Therefore, the proposition of enjoyment and players' gaming performance is listed as follows:

Proposition 7 (P7): Enjoyment has the significant effect on the players' gaming performance.

4 The Research Opportunities Based on Propositions

In this paper, we focus on the influence of game peripherals on the players' positive experience and performance in the player-video game interaction. According to the propositions raised in the last section, the overall relationships among the game peripherals, players' positive gaming experience and their performance are shown in Fig. 2.

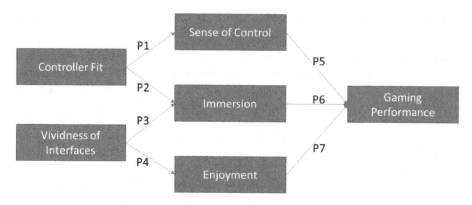

Fig. 2. The theoretical relationship model

4.1 The Influence of Game Peripherals for Further Research

In this paper, two propositions center on the controller fit and vividness of interfaces, respectively. Actually, we found that existing literature, to some extent, has examined some propositions related to the game peripherals (proposed in this paper). Therefore, this paper also summarizes the relevant literature which has verified the propositions raised in the paper (as shown in Table 4). Based on Table 4, we can see that the P1–P3 have been empirically supported. However, these supports are partial. For example, Liu et al. [52] confirmed the vividness of interfaces positively influence the immersion experience of the player and hence improving their enjoyment of gaming. This research mainly focuses on the influence of visual vividness (screen size is manipulated in the study of Liu et al.) on players' immersion experience. However, the vividness of interfaces is not only determined by the visual channel, especially in the current gaming mode. The audio channel and the tactile channel are also important interfaces for the optimal game experience [6]. The design of the multisensory interface is also one of the mainstream strategies in the current game development (e.g., games of F1 series). However, the research on the vividness of interface enabled by the multisensory channels is still yet limited. Therefore, the impacts of the multisensory interface on the players' immersion and enjoyment could be the future research orientation related to the propositions on the vividness of interfaces.

Furthermore, the propositions of the controller fit are also partially studied mainly by the research on controller naturalness (e.g., [59, 62]). The research efforts on the controller naturalness are mainly spent in the players' immersion experience in the reality simulated games, such as the exercise games [59] and shooting games [60]. However, the research also confirmed that the controller naturalness is not suitable in the virtual oriented game context [71]. In the games which are not the reflection of the real-world, the controller fit to the gamer's previous gaming experience is more important with respect to the players' gaming experience. Therefore, the controller fit in the context of a virtual game could be the one of the future research orientation for the propositions of controller fit and players' gaming experience.

Table 4. The summary of the verification of P1–P4.

Propositions	Verification notes	Supporting references
P1: Controller fit → Sense of control	Having not yet been verified	–
P2: Controller fit → Immersion	Partially verified, the controller naturalness (a type of controller fit) leads to players' immersion	[59–61]
P3: Vividness of interfaces → Immersion	Partially verified, the current research focuses on the visual vividness	[52]
P4: Vividness of interfaces → Enjoyment	Partially verified, vividness of interfaces influence enjoyment via the mediators	[52, 76]

4.2   The Emotion-Performance Relation in Gaming

This paper develops three propositions (i.e., P5–P7) on the relationship between players' in-game positive experience and their gaming performance. The extant studies also pay attention to the relationship of players' experience with their gaming performance. However, most of these studies focus on the influence of players' gaming performance on their experience after playing the game. For example, Hopp and Fisher [27] find that the female players' enjoyment is influenced by their gaming performance; Limperos and Schmierbach [49] find the mediating effect of enjoyment on the relationship between players' performance and future intentions to play the game. On the contrary, how the in-game experience influence players' gaming performance is still not well under researching. As shown in Table 5, we can only find a few research that confirmed the in-game immersion experience positively influences players' gaming performance (e.g., [8, 29]), which indicates research opportunities for the influence of in-game experience on players' performance. Furthermore, the argument of emotion-performance from the field of sports science theoretically supports the proposition 7.

Table 5. The summary of the verification of P5–P7.

Propositions	Verification notes	Supporting references
P5: Sense of control → Performance	Having not yet been verified	–
P6: Immersion → Performance	Verified, but needs more research in different context to check the generalizability of the results	[8, 29]
P7: Enjoyment → Performance	Having not yet been verified, in the gaming context, but is supported by the arguments of emotion-performance relation in the filed sports science	[11, 58]

Actually, players' positive gaming experience is a promising field of research because of their influence on specific components of performance excellence (e.g., cognition) [11]. However, the benefits of these emotional experiences have hitherto not been fully realized in a game context, especially in the context of professional competitive gaming (say, eSports). Therefore, the three propositions on the in-game players' experience and the gaming performance are promising research orientations for the human-computer interaction research to understand the emotion-performance relation in the context of gaming.

5 Conclusions

This paper focuses on the two characteristics of game peripherals (i.e., controller fit and vividness of interfaces) and their influence on the players' positive in-game experience (sense of control, immersion and enjoyment) and their gaming performance. Seven propositions are raised to illiterate the theoretical relationships among game peripherals, players' positive gaming experience and their performance. Based on the propositions, the further research orientations are also discussed and suggested. This paper hopes to help the researchers further explore the research on game peripherals from the perspective of fit and vividness as well as the research on the emotion-performance relation in the field of player-video game interaction.

Acknowledgement. The research presented in this article was supported in part by the UGC Teaching and Learning Project entitled "Developing Multidisciplinary and Multicultural Competences through Gamification and Challenge-Based Collaborative Learning".

References

1. Agarwal, R., Karahanna, E.: Time flies when you're having fun: cognitive absorption and beliefs about information technology usage. MIS Q. **24**(4), 665–694 (2000)
2. Agrebi, S., Jallais, J.: Explain the intention to use smartphones for mobile shopping. J. Retail. Consum. Serv. **22**, 16–23 (2015)
3. Bowman, N.D., Weber, R., Tamborini, R., Sherry, J.: Facilitating game play: how others affect performance at and enjoyment of video games. Media Psychol. **16**(1), 39–64 (2013)
4. Boyan, A., Sherry, J.L.: The challenge in creating games for education: aligning mental models with game models. Child. Dev. Perspect. **5**, 82–87 (2011)
5. Bracken, C.C., Skalski, P.: Telepresence and video games: the impact of image quality. PsychNology J. **7**(1), 101–112 (2009)
6. Caroux, L., Isbister, K., Le Bigot, L., Vibert, N.: Player–video game interaction: a systematic review of current concepts. Comput. Hum. Behav. **48**, 366–381 (2015)
7. Caroux, L., Le Bigot, L., Vibert, N.: Maximizing players' anticipation by applying the proximity-compatibility principle to the design of video games. Hum. Factors **53**(2), 103–117 (2011)
8. Cheng, M.T., She, H.C., Annetta, L.A.: Game immersion experience: its hierarchical structure and impact on game-based science learning. J. Comput. Assist. Learn. **31**(3), 232–253 (2015)
9. Childers, T.L., Carr, C.L., Peck, J., Carson, S.: Hedonic and utilitarian motivations for online retail shopping behavior. J. Retail. **77**(4), 511–535 (2001)
10. Csikszentmihalyi, M.: Flow: The Psychology of Optimal Experience. Harper and Row, New York (1990)
11. Derakshan, N., Eysenck, M.W.: Introduction to the special issue: emotional states, attention, and working memory. Cogn. Emot. **24**, 189–199 (2010)

12. Dholakia, U.M., Gopinath, M., Bagozzi, R.P., Nataraajan, R.: The role of regulatory focus in the experience and self-control of desire for temptations. J. Consum. Psychol. **16**(2), 163–175 (2006)
13. Dishaw, M.T., Strong, D.M.: Extending the technology acceptance model with task–technology fit constructs. Inf. Manag. **36**(1), 9–21 (1999)
14. Djamasbi, S., Strong, D., Dishaw, M.: Affect and acceptance: examining the effects of positive mood on the technology acceptance model. Decis. Support Syst. **48**, 383–394 (2010)
15. Entertainment Software Association. http://www.theesa.com/article/us-video-game-industry-revenue-reaches-36-billion-2017. Accessed 20 Dec 2018
16. Eppmann, R., Bekk, M., Klein, K.: Gameful experience in gamification: construction and validation of a gameful experience scale [GAMEX]. J. Interact. Mark. **43**, 98–115 (2018)
17. Fang, J., Chen, L., Wang, X., George, B.: Not all posts are treated equal: an empirical investigation of post replying behavior in an online travel community. Inf. Manag. **55**(7), 890–900 (2018)
18. Fang, X., Zhao, F.: Personality and enjoyment of computer game play. Comput. Ind. **61**(4), 342–349 (2010)
19. Fortin, D.R., Dholakia, R.R.: Interactivity and vividness effects on social presence and involvement with a web-based advertisement. J. Bus. Res. **58**(3), 387–396 (2005)
20. Gavilan, D., Avello, M., Abril, C.: The mediating role of mental imagery in mobile advertising. Int. J. Inf. Manag. **34**(4), 457–464 (2014)
21. Goodhue, D.L.: Understanding user evaluations of information systems. Manag. Sci. **41**(12), 1827–1844 (1995)
22. Goodhue, D.L., Thompson, R.L.: Task-technology fit and individual performance. MIS Q. **19**(2), 213–236 (1995)
23. Guo, Y.M., Poole, M.S.: Antecedents of flow in online shopping: a test of alternative models. Inf. Syst. J. **19**(4), 369–390 (2009)
24. Guo, Y., Ro, Y.: Capturing flow in the business classroom. Decis. Sci. **6**, 437–462 (2008)
25. Hess, T.J., Fuller, M., Campbell, D.E.: Designing interfaces with social presence: Using vividness and extraversion to create social recommendation agents. J. Assoc. Inf. Syst. **10**(12), 889–919 (2009)
26. Hess, T.J., Fuller, M., Mathew, J.: Involvement and decision-making performance with a decision aid: the influence of social multimedia, gender, and playfulness. J. Manag. Inf. Syst. **22**(3), 15–54 (2005)
27. Hopp, T., Fisher, J.: Examination of the relationship between gender, performance, and enjoyment of a first-person shooter game. Simul. Gaming **48**(3), 338–362 (2017)
28. Hou, J., Nam, Y., Peng, W., Lee, K.M.: Effects of screen size, viewing angle, and players' immersion tendencies on game experience. Comput. Hum. Behav. **28**(2), 617–623 (2012)
29. Hsu, M.E., Cheng, M.T.: Bio detective: student science learning, immersion experience, and problem-solving patterns. In: 22nd International Conference on Computers in Education, pp. 171–178. Asia-Pacific Society for Computers in Education, Nara (2014)
30. Huntsinger, J.R., Ray, C.: A flexible influence of affective feelings on creative and analytic performance. Emotion **16**(6), 826–837 (2016)
31. International eSports Federation. https://www.ie-sf.org/rules-and-regulations. Accessed 11 Jan 2019
32. Isen, A.M., Daubman, K.A.: The influence of affect on categorization. J. Pers. Soc. Psychol. **47**, 1206–1217 (1984)
33. Isen, A.M., Geva, N.: The influence of positive affect on acceptable level of risk: the person with a large canoe has a large worry. Organ. Behav. Hum. Decis. Process. **39**, 145–154 (1987)

34. Isen, A.M., Johnson, M.M., Mertz, E., Robinson, G.F.: The influence of positive affect on the unusualness of word associations. J. Pers. Soc. Psychol. **48**, 1413–1426 (1985)
35. Ivory, J.D., Kalyanaraman, S.: The effects of technological advancement and violent content in video games on players' feelings of presence, involvement, physiological arousal, and aggression. J. Commun. **57**(3), 532–555 (2007)
36. Jennett, C., et al.: Measuring and defining the experience of immersion in games. Int. J. Hum. Comput. Stud. **66**(9), 641–661 (2008)
37. Jiang, Z., Benbasat, I.: Research note—investigating the influence of the functional mechanisms of online product presentations. Inf. Syst. Res. **18**(4), 454–470 (2007)
38. Junglas, I., Abraham, C., Ives, B.: Mobile technology at the frontlines of patient care: understanding fit and human drives in utilization decisions and performance. Decis. Support Syst. **46**(3), 634–647 (2009)
39. Kabanoff, B., Waldersee, R., Cohen, M.: Espoused values and organizational change themes. Acad. Manag. J. **38**(4), 1075–1104 (1995)
40. Khan, I.U., Hameed, Z., Yu, Y., Islam, T., Sheikh, Z., Khan, S.U.: Predicting the acceptance of MOOCs in a developing country: application of task-technology fit model, social motivation, and self-determination theory. Telematics Inform. **35**(4), 964–978 (2018)
41. Kim, J., Fiore, A.M., Lee, H.H.: Influences of online store perception, shopping enjoyment, and shopping involvement on consumer patronage behavior towards an online retailer. J. Retail. Consum. Serv. **14**(2), 95–107 (2007)
42. Kitson, A., Prpa, M., Riecke, B.E.: Immersive interactive technologies for positive change: a scoping review and design considerations. Front. Psychol. **9**, 1354 (2018)
43. Klein, K.J., Sorra, J.S.: The challenge of innovation implementation. Acad. Manag. Rev. **21**(4), 1055–1080 (1996)
44. Lee, J., Lee, J., Lee, H., Lee, J.: An exploratory study of factors influencing repurchase behaviors toward game items: a field study. Comput. Hum. Behav. **53**, 13–23 (2015)
45. Lee, W., Gretzel, U.: Designing persuasive destination websites: a mental imagery processing perspective. Tour. Manag. **33**(5), 1270–1280 (2012)
46. Lee, Y., Chen, A.N., Ilie, V.: Can online wait be managed? The effect of filler interfaces and presentation modes on perceived waiting time online. MIS Q. **36**(3), 365–394 (2012)
47. Li, Y., McCune, J.M., Perrig, A.: SBAP: software-based attestation for peripherals. In: Acquisti, A., Smith, S.W., Sadeghi, A.-R. (eds.) Trust 2010. LNCS, vol. 6101, pp. 16–29. Springer, Heidelberg (2010). https://doi.org/10.1007/978-3-642-13869-0_2
48. Lim, K.H., Benbasat, I., Ward, L.M.: The role of multimedia in changing first impression bias. Inf. Syst. Res. **11**(2), 115–136 (2000)
49. Limperos, A.M., Schmierbach, M.: Understanding the relationship between exergame play experiences, enjoyment, and intentions for continued play. Games Health J. **5**(2), 100–107 (2016)
50. Lin, C.P., He, H., Baruch, Y., Ashforth, B.E.: The effect of team affective tone on team performance: the roles of team identification and team cooperation. Hum. Resour. Manag. **56**(6), 931–952 (2017)
51. Lin, J.W., Duh, H.B.L., Parker, D.E., Abi-Rached, H., Furness, T.A.: Effects of field of view on presence, enjoyment, memory, and simulator sickness in a virtual environment. In: Proceedings IEEE Virtual Reality 2002, Orlando, pp. 164–171. IEEE (2002)
52. Liu, L., Ip, R., Shum, A., Wagner, C.: Learning effects of virtual game worlds: an empirical investigation of immersion, enjoyment and performance. In: 20th Americas Conference on Information Systems, Association for Information Systems, Savannah (2014)
53. Liu, Y., Shrum, L.J.: What is interactivity and is it always such a good thing? Implications of definition, person, and situation for the influence of interactivity on advertising effectiveness. J. Advert. **31**(4), 53–64 (2002)

54. Lowry, P.B., Gaskin, J., Twyman, N., Hammer, B., Roberts, T.: Taking 'fun and games' seriously: proposing the hedonic-motivation system adoption model (HMSAM). J. Assoc. Inf. Syst. **14**(11), 617–671 (2012)

55. Lowry, P.B., Romano, N.C., Jenkins, J.L., Guthrie, R.W.: The CMC interactivity model: How interactivity enhances communication quality and process satisfaction in lean-media groups. J. Manag. Inf. Syst. **26**(1), 155–195 (2009)

56. Lucas, R.E., Diener, E.: The happy worker: hypotheses about the role of positive affect in worker productivity. In: Burrick, M., Ryan, A.M. (eds.) Personality and Work, pp. 30–59. Jossey-Bass, San Francisco (2003)

57. McCarthy, P.J.: Positive emotion in sport performance: current status and future directions. Int. Rev. Sport Exerc. Psychol. **4**(1), 50–69 (2011)

58. McCarthy, P.J., Allen, M.S., Jones, M.V.: Emotions, cognitive interference, and concentration disruption in youth sport. J. Sports Sci. **31**(5), 505–515 (2013)

59. McGloin, R., Embacher, K.: "Just like riding a bike": a model matching approach to predicting the enjoyment of a cycling exergame experience. Media Psychol. **21**(3), 486–505 (2018)

60. McGloin, R., Farrar, K.M., Fishlock, J.: Triple whammy! Violent games and violent controllers: investigating the use of realistic gun controllers on perceptions of realism, immersion, and outcome aggression. J. Commun. **65**(2), 280–299 (2015)

61. McGloin, R., Farrar, K.M., Krcmar, M.: The impact of controller naturalness on spatial presence, gamer enjoyment, and perceived realism in a tennis simulation video game. Presence: Teleoperators Virtual Environ. **20**(4), 309–324 (2011)

62. McGloin, R., Farrar, K.M., Krcmar, M.: Video games, immersion, and cognitive aggression: does the controller matter? Media Psychol. **16**(1), 65–87 (2013)

63. McGloin, R., Farrar, K.M., Krcmar, M., Park, S., Fishlock, J.: Modeling outcomes of violent video game play: applying mental models and model matching to explain the relationship between user differences, game characteristics, enjoyment, and aggressive intentions. Comput. Hum. Behav. **62**, 442–451 (2016)

64. Miller, D.W., Marks, L.J.: The effects of imagery-evoking radio advertising strategies on affective responses. Psychol. Mark. **14**(4), 337–360 (1997)

65. Nacke, L.E., Grimshaw, M.N., Lindley, C.A.: More than a feeling: measurement of sonic user experience and psychophysiology in a first-person shooter game. Interact. Comput. **22**(5), 336–343 (2010)

66. Nisbett, R.: Ross. L.: Assigning weights to data: the "vividness criterion". In: Nisbett, R., Ross, L. (eds.) Human Inference: Strategies and Shortcomings of Social Judgment. Prentice-Hall, Inc., Englewood Cliffs (1980)

67. Oliveira, T., Faria, M., Thomas, M.A., Popovič, A.: Extending the understanding of mobile banking adoption: when UTAUT meets TTF and ITM. Int. J. Inf. Manag. **34**(5), 689–703 (2014)

68. Parés, N., Altimira, D.: Analyzing the adequacy of interaction paradigms in artificial reality experiences. Hum.-Comput. Interact. **28**(2), 77–114 (2013)

69. Peng, Z., Sun, Y., Guo, X.: Antecedents of employees' extended use of enterprise systems: an integrative view of person, environment, and technology. Int. J. Inf. Manag. **39**, 104–120 (2018)

70. Procci, K., Bowers, C.A., Jentsch, F., Sims, V.K., McDaniel, R.: The revised game engagement model: capturing the subjective gameplay experience. Entertain. Comput. **27**, 157–169 (2018)

71. Rogers, R., Bowman, N.D., Oliver, M.B.: It's not the model that doesn't fit, it's the controller! The role of cognitive skills in understanding the links between natural mapping, performance, and enjoyment of console video games. Comput. Hum. Behav. **49**, 588–596 (2015)

72. Sabri, A.J., Ball, R.G., Fabian, A., Bhatia, S., North, C.: High-resolution gaming: interfaces, notifications, and the user experience. Interact. Comput. **19**(2), 151–166 (2006)
73. Schultze, U.: Embodiment and presence in virtual worlds: a review. J. Inf. Technol. **25**(4), 434–449 (2010)
74. Seibert, J., Shafer, D.M.: Control mapping in virtual reality: effects on spatial presence and controller naturalness. Virtual Reality **22**(1), 79–88 (2018)
75. Skalski, P., Tamborini, R., Shelton, A., Buncher, M., Lindmark, P.: Mapping the road to fun: natural video game controllers, presence, and game enjoyment. New Media Soc. **13**, 224–242 (2011)
76. Skalski, P., Whitbred, R.: Image versus sound: a comparison of formal feature effects on presence and video game enjoyment. PsychNology J. **8**(1), 67–84 (2010)
77. Slater, M., Linakis, V., Usoh, M., Kooper, R., Street, G.: Immersion, presence, and performance in virtual environments: an experiment with tri-dimensional chess. In: ACM Virtual Reality Software and Technology, p. 72. ACM Press, New York (1996)
78. Steuer, J.: Defining virtual reality: dimensions determining telepresence. J. Commun. **42**(4), 73–93 (1992)
79. Taylor, T.L.: Raising the Stakes: E-Sports and the Professionalization of Computer Gaming. MIT Press, Cambridge (2012)
80. Van der Heijden, H.: User acceptance of hedonic information systems. MIS Q. **28**(4), 695–704 (2004)
81. Wechselberger, U.: Music game enjoyment and natural mapping beyond intuitiveness. Simul. Gaming **47**(3), 304–323 (2016)
82. Witkowski, E.: On the digital playing field: how we "do sport" with networked computer games. Games Culture **7**(5), 349–374 (2012)
83. Webster, J., Hackley, P.: Teaching effectiveness in technology-mediated distance learning. Acad. Manag. J. **40**(6), 1282–1309 (1997)
84. Webster, J., Ho, H.: Audience engagement in multi-media presentations. Data Base Adv. Inf. Syst. **28**(2), 63–77 (1997)
85. Wei, W., Qi, R., Zhang, L.: Effects of virtual reality on theme park visitors' experience and behaviors: a presence perspective. Tour. Manag. **71**, 282–293 (2019)
86. Williams, K.D.: The effects of video game controls on hostility, identification, and presence. Mass Commun. Soc. **16**(1), 26–48 (2013)
87. Williams, K.D.: The effects of dissociation, game controllers, and 3D versus 2D on presence and enjoyment. Comput. Hum. Behav. **38**, 142–150 (2014)
88. Yang, S., Lu, Y., Chen, Y., Gupta, S.: Understanding consumers' mobile channel continuance: an empirical investigation of two fitness mechanisms. Behav. Inf. Technol. **34**(12), 1135–1146 (2015)
89. Yee, N.: The demographics, motivations, and derived experiences of users of massively multi-user online graphical environments. Presence: Teleoperators Virtual Environ. **15**(3), 309–329 (2006)
90. Yim, M.Y.C., Chu, S.C., Sauer, P.L.: Is augmented reality technology an effective tool for e-commerce? An interactivity and vividness perspective. J. Interact. Mark. **39**, 89–103 (2017)
91. Yin, C., Sun, Y., Fang, Y., Lim, K.: Exploring the dual-role of cognitive heuristics and the moderating effect of gender in microblog information credibility evaluation. Inf. Technol. People **31**(3), 741–769 (2018)
92. Zhang, H., Zhao, L., Gupta, S.: The role of online product recommendations on customer decision making and loyalty in social shopping communities. Int. J. Inf. Manag. **38**(1), 150–166 (2018)
93. Zhou, T., Lu, Y., Wang, B.: Integrating TTF and UTAUT to explain mobile banking user adoption. Comput. Hum. Behav. **26**(4), 760–767 (2010)

Biomimicry and Machine Learning
in the Context of Healthcare Digitization

Corinne Lee[1], Suraj Sood[2(✉)], Monte Hancock[3], Tyler Higgins[1],
Kristy Sproul[1], Antoinette Hadgis[1], and Stefan Joe-Yen[1]

[1] Sirius Project, Melbourne, USA
[2] University of West Georgia, Carrollton, GA, USA
sirius19conf@gmail.com
[3] 4Digital, Buffalo, USA

Abstract. The healthcare industry is inundated with elder, disabled, and partial-use care issues such as falls in homes with no available aid present. This article's thesis is that bio-support in the form of biomimetic artificial intelligence (AI) is not yet fully-exploited within the stated problem space.

This article summarizes analyses conducted using Hancock's "Knowledge-Based Expert System" (KBES) on two datasets from the popular machine learning website *Kaggle*. The first dataset contains various numeric, health-related data from 400 anonymized patients diagnosed with chronic kidney disease (CKD). The second contains the same kind of data, but for 569 patients diagnosed with either malignant or benign forms of breast cancer.

In the last place, the potential for a "Holacratic" health analytics organization will be assessed. Said organization would be akin to Tapscott's "Global Solutions Network" (GSN), which was defined as a digital group of public or private individuals with the following four features:

- Diversity in stakeholders who collectively represent at least two of the four pillars of society (government, private sector, civil society, individuals)
- Multinational or global presence
- At least partial digitality with respect to its communications tools and platforms
- Progressive goals related to the creation of public goods

It will be argued that the group working on this article (known colloquially as the *Sirius Project*) successfully addresses both the criteria of being and need for a Holacratic GSN.

Keywords: Biomimicry · Machine learning · Breast cancer ·
Chronic kidney disease · Benefit corporation

1 Introduction

The following are recognized by transnational think tank The Millennium Project as significant global challenge (GC) areas: sustainable development, clean water, population and resources, democratization, foresight and decision-making, convergence of

© Springer Nature Switzerland AG 2019
D. D. Schmorrow and C. M. Fidopiastis (Eds.): HCII 2019, LNAI 11580, pp. 273–283, 2019.
https://doi.org/10.1007/978-3-030-22419-6_19

information technology (IT), rich-poor gap, health, education and learning, peace and conflict, status of women, transnational organized crime, energy, science and technology, and global ethics [1, p. 17]. This article focuses on health. Specifically, the healthcare industry is inundated with elder, disabled, and partial-use care issues such as falls in homes with no available aid present. This article's thesis is that bio-support in the form of biomimetic artificial intelligence (AI) is not yet fully-exploited within this problem space.

This article will summarize analyses conducted using Hancock's "Knowledge-Based Expert System" (KBES) [2] on two publicly-available datasets. The first dataset contains various numeric, health-related data from 400 anonymized patients diagnosed with chronic kidney disease (CKD). The second contains the same kind of data, but for 569 patients diagnosed with either malignant (coded "M") or benign (coded "B") forms of breast cancer. Both datasets were retrieved from Kaggle [3, 4].

In the last place, the potential for a "Holacratic" [5] health analytics organization will be explored. Said organization would be akin to Tapscott's "Global Solutions Network" (GSN) [6], which was defined as a digital group of public or private individuals with the following four features:

- Diversity in stakeholders who collectively represent at least two of the four pillars of society (government, private sector, civil society, individuals)
- Multinational or global presence
- At least partial digitality with respect to its communications tools and platforms
- Progressive goals related to the creation of "public goods" (p. 18).

It will be shown that the group working on this article (known colloquially as the *Sirius Project*) successfully addresses both the criteria of being and need for a Holacratic GSN [7, 8].

1.1 Background

In the movie *Big Hero 6*, "Baymax" is an inflatable robot with a carbon fiber skeleton who serves as a personal healthcare provider companion. Movie codirector Don Hall said of Baymax that he "views the world from one perspective—he just wants to help people" [9]. Baymax is an example of a fantastical biomimetic AI designed specifically to help people solve their health-related problems (Fig. 1).

Biomimetic ("life-mimicking") systems automate functionality by reiterating biological forms and processes (e.g., early airplane designs were attempts to model winged flight in birds). In this way, biological systems not only provide existence proofs—e.g., "heavier-than-air flight is possible"—they also inform attempts to implement functionality by replicating form in automating systems.

Important sub-disciplines within the science of machine learning have developed in this same way, with the most widely-used cognitive architectures lying along relatively short development paths from biologically inspired starting points. Obvious examples are feed-forward neural networks, reinforcement learning machines, genetic algorithms, expert systems, and binocular/binaural sensation. While the application of sophisticated optimization techniques to these systems can blur unscalable aspects of the underlying biological metaphor (airplanes do not flap their wings), what remains can still be informative.

Fig. 1. Baymax from the movie *Big Hero 6*.

Could biomimetic AI systems aid patients suffering from breast cancer and CKD? Given the popular overlap at present between the fields of artificial intelligence and machine learning (ML), biomimetic AI could draw from the insights enabled by utilizing ML methods. Results from analyses conducted utilizing a ML algorithm are summarized in the following sections.

2 Method

We decided to implement a bias-based Knowledge Based Expert System (KBES) as described by Hancock [6]. The data this expert system was applied to are all publicly available through Kaggle, the popular machine learning repository. The first dataset contains various numeric, health-related data from 400 anonymized patients diagnosed with chronic kidney disease (CKD). The second contains the same kind of data, but for 569 patients diagnosed with either malignant (coded "M") or benign (coded "B") forms of breast cancer. Both datasets were retrieved from Kaggle [3, 4].

2.1 Data Preparation

The data preparation process is automated so that it is documented, repeatable, and consistent.

The data preparation parameters for the breast cancer data were:

- 2 = mode 1 is for training mode 2 is for processing
- BC_W_T = C45 CSV quantized training file

- 32 = number of columns in training file
- 2 = number of Ground Truth classes in training file
- BC_W_B = C45 CSV quantized blind file
- 150 = number of reconstruction values for feature quantization
- 10000 = number of training epochs
- 0.001 = learning rate
- 0.5 = proportion of missed vectors to update on

The CKD settings are similar. The CKD data consisted of 400 instances (rows), each having 24 features (columns) and two classes (CKD and no CKD). The breast cancer data consists of 569 instances (rows) each having 30 features (columns) and 2 classes (malignant and benign).

The data for both experiments went through the same conditioning process:

1. Each raw feature column was z-scored so that the units across features would be commensurable.
2. Each feature column was histogram equalized. The feature values in each column were then replaced by their histogram bin numbers.
3. The CKD data (400 instances total) was split into two segments: a "training set" (267 instances) and a "blind set" (133 instances).
4. The breast cancer data (569 instances total) was split into two segments: a training set (379 instances) and a blind set (190 instances) (Fig. 2).

Fig. 2. The quantization process.

The data were then visualized in their native high-dimensional spaces to assess the complexity of the classification process. The breast cancer data are shown below (Fig. 3):

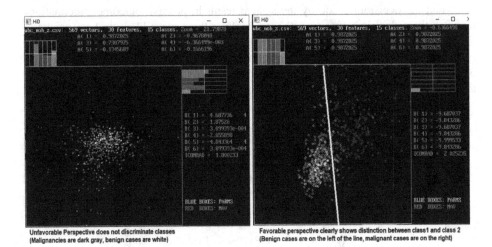

Fig. 3. Visual inspection of the conformed feature data.

2.2 Belief Accumulation

Non-graph-based methods for reasoning represent knowledge implicitly as functions (usually a collection of rules with an adjudicator, a regression function, or clauses with a resolution method), and perform reasoning by function evaluation.

2.3 Inferring Knowledge from Data: Machine Learning = "Adaptation + Propagation"

"Adaptation + Propagation" might sound Darwinian, but it is actually Lamarckian[1]. Animal brains continue throughout their lives to expand their cognitive power by assimilating their experiences, in an (as of yet) unknown way. In light of this analogy, it is sensible to adopt the following empirical definition of machine learning: *Machine learning is the automatic organization of execution history into machine resident structures in such a way that machine performance is improved.*

The method used for the work described here is an instance of such machine learning referred to as reinforcement learning. In reinforcement learning, a brain

---

[1] Darwin's evolutionary theory has lost favor in recent years (owing largely to its inability to posit a credible mechanism), while, thanks to discoveries in epigenetics, Lamarck's theory—published nearly 60 years before Darwin's—is experiencing something of a resurgence. Darwinian evolution occurs at the population rather than individual or organismic level: through allele selection, rather than inherited modification. Lamarck's process occurs at the level of individual organisms through the epigenetic transmission of acquired characteristics, an observable epigenetic phenomenon.

(biological or mechanical) interacts with its environment by performing the tasks that arise. For an animal, these might include eating, mating, avoiding injury, etc., while for an image classification machine, patterns to be recognized would be presented and scored by a training algorithm.

When the decisions made by the system result in positive outcomes, inclinations favoring these decisions are strengthened, making them more likely to be repeated in the future. When the decisions made by the system result in negative outcomes, the corresponding inclinations are weakened, making the unsuccessful actions less likely to be repeated. Over the course of many trials, the system gradually adjusts its inclinations to result in mostly positive outcomes: it learns how to succeed (Fig. 4).

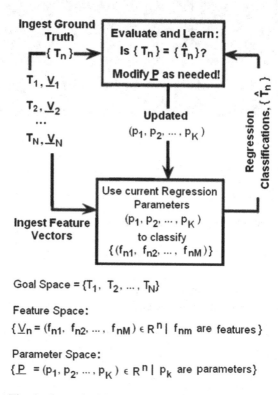

Goal Space = $\{T_1, T_2, \ldots, T_N\}$

Feature Space:
$\{\underline{V}_n = (f_{n1}, f_{n2}, \ldots, f_{nM}) \in R^n \mid f_{nm}$ are features$\}$

Parameter Space:
$\{\underline{P} = (p_1, p_2, \ldots, p_K) \in R^n \mid p_k$ are parameters$\}$

Fig. 4. Supervised learning by feedback and adaption.

The Adaptation + Propagation Loop is fast and simple. Firstly, an objective function must be defined that enables the Evaluate and Learn function to tell whether the machine's last trial is the best seen so far (a trial consists of processing all the instances in the training set). A natural measure of performance might be the proportion of patterns correctly classified. If the machine has improved, the updated weights become the machine's "New Best". If not, another set is generated for the next trail, and the process repeated. To generate a new set of weights to try, the current "best set" is perturbed slightly. In this way, previously learned patterns are "propagated" to subsequent trials.

"Bias-based Reasoning" (BBR) is a mathematical method for automating implementation of a reinforcement learning approach in rule-based systems. Reasoning in such a system proceeds by accumulating "evidence" for and against possible conclusions; when a sufficiency threshold is reached, the favored decision is selected. Learning is carried out by making small adjustments to weights inside decision logic rules. During training, the system is run through a large number of decision-making trials. When it produces a correct decision, the weights favoring the decision given that evidence are slightly increased; incorrect decisions result in the corresponding weights being reduced.

Reinforcement learning in rule-based systems enjoys the same advantages human experts derive from this approach; in particular, it supports automated learning, conclusion justification, confidence estimation, and natural means for handling both non-monotonicity and uncertainty.

2.4   Using Facts in Rules

For simplicity and definiteness, the reasoning problem will be described here as the use of evidence to select one or more possible conclusions from a closed, finite list that has been specified a priori (the "Classifier Problem").

Expert reasoning is based upon facts (colloquially, "interpretations of the collected data"). Facts function both as indicators and contra-indicators for conclusions. Positive facts are those that increase our beliefs in certain conclusions. Negative facts are probably best understood as being exculpatory: they impose constraints upon the space of conclusions, militating against those unlikely to be correct. Facts are salient to the extent that they increase belief in the "truth", and/or increase "disbelief" in untruth.

A rule is an operator that uses facts to update beliefs by applying biases. In software, rules are often represented as structured constructs such as IF-THEN-ELSE, CASE, or SWITCH statements. We use the IF-THEN-ELSE in what follows.

Rules consist of an antecedent and a multi-part body. The antecedent evaluates a BOOLEAN expression; depending upon the truth-value of the antecedent, different parts of the rule body are executed.

The following is a notional example of a simple rule. It tells us qualitatively how an expert might alter her beliefs about an unknown animal should she determine whether or not it is a land-dwelling omnivore.

```
If   (habitat = land) AND (diet = omnivorous) THEN
        INCREASE BELIEF(primates, bugs, birds)
        INCREASE DISBELIEF(bacteria, fishes)
ELSE
        INCREASE DISBELIEF(primates, bugs, birds)
        INCREASE BELIEF(bacteria, fishes)
End Rule
```

If we have an INCREASE BELIEF function, and a DECREASE BELIEF function ("aggregation functions", called AGG below), many such rules can be efficiently implemented in a looping structure.

In a data store:

Tj(**F**i) truth-value of predicate j applied to fact **F**i
bias(k, j, 1) belief to accrue in conclusion k when predicate j true
bias(k, j, 2) disbelief to accrue in conclusion k when predicate j is true
bias(k, j, 3) belief to accrue in conclusion k when predicate j false
bias(k, j, 4) disbelief to accrue in conclusion k when predicate j is false

Multiple rule execution in a loop:

```
IF Tj(F)=1 THEN                  'if predicate j true for Fi...
    FOR k=1 TO K                 'for conclusion k:
        Belief (k)=AGG(B(k,i),bias(k,j,1))         'true: accrue be-
lief bias(k,j,1)
        Disbelief(k)=AGG(D(k,i),bias(k,j,2))  'true: accrue disbe-
lief bias(k,j,2)
    NEXT k
ELSE
    FOR k=1 TO K                                    'for    conclusion
k:
        Belief(k)=AGG(D(k,i),bias(k,j,3))      'false: accrue be-
lief bias(k,j,3)
        Disbelief (k)=AGG(B(k,i),bias(k,j,4)) 'false: accrue dis-
belief bias(k,j,4)
    NEXT k
END IF
```

This creates a vector B of beliefs $(b(1), b(2), \ldots, b(K))$ for each of the conclusions $1, 2, \ldots, K$, and a vector D of disbeliefs $(d(1), d(2), \ldots, d(K))$ for each of the conclusions $1, 2, \ldots, K$. These must now be adjudicated for a final decision.

Clearly, the inferential power here is not in the rule structure, but in the "knowledge" held numerically in the biases. As is typical with heuristic reasoners, BBR allows the complete separation of knowledge from the inferencing process. This means that the structure can be retrained, even repurposed to another problem domain, by modifying only data; the inference engine need not be changed. An additional benefit of this separability is that the engine can be maintained openly apart from sensitive data.

Summarizing (thinking again in terms of the Classifier Problem): When a positive belief heuristic fires, it accrues a bias $\beta > 0$ that a certain class is the correct answer; when a negative heuristic fires, it accrues a bias $\delta > 0$ that a certain class is the correct answer. The combined positive and negative biases for an answer constitute that answer's *belief*.

After applying a set of rules to a collection of facts, beliefs and disbeliefs will have been accrued for each possible conclusion (classification decision). This ordered list of beliefs is a "belief vector". The final decision is made by examining this vector of beliefs, for example, by selecting the class having the largest belief-disbelief.

3 Results

When the trained machine processes a data set, it produces a file containing the classification results for each instance, and also create a confusion matrix so that the specific mix of Type I and Type II Errors can be reviewed in summary form (Fig. 5).

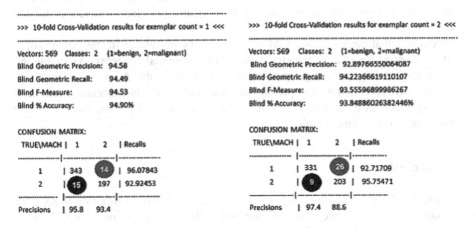

Fig. 5. Results of a 10-fold cross-validation.

4 Discussion

Do the findings summarized above speak to the possibility of biomimetic AI systems that could assist elderly sufferers of CKD or breast cancer? Assuming such systems could achieve the 93%+ predictive capabilities demonstrated by the KBES, then the next question becomes of how exactly they would. We envision such a healthcare-providing AI that would serve to entertain and educate a growing elderly population, as well as offer them physical and mental (e.g., affective) support more generally. (Naturally, such an AI system would need to be cost-effective for at least one patient in order to pass the "proof of concept" stage of the production cycle.) A final question posed here is of whether biomimetic AI would be ideal (compared with more traditional, statically-programmed robots) in offering care to patients beyond predicting whether not they possess conditions like CKD or breast cancer.

We further argue that biomimetic AI is needed to the extent that chronic sufferers of diseases like breast cancer and kidney disease do not want to feel helpless for most of their waking lives. Such systems would, in theory, empower patients to live on their own terms via an interface that would continuously relearn their needs, in addition to predicting and resolving crises identified via real time (i.e., "stream-processing") processing of patient data.

4.1 Limitations

One potential concern in predicting disease presence based on patient information has to do with privacy. More specifically, any legally-sanctioned health AI or ML system would need to comply with HIPAA (Health Insurance Portability and Accountability Act of 1996) legislation. Related to this are concerns dealing with the ethics of patient information usage. Health-providing biomimetic AI's would need to be embedded with privacy and storage settings that suit both patients and their respective, human healthcare providers.

We assert that the biomimetic AI or ML systems postulated in this paper would need to prioritize the lives of patients over all other potential concerns—including the survival of such biomimetic systems in question, if necessary. A more rigorous hierarchy of values resembling or surpassing the best human ones in existence would need to be embedded in these systems if they would be expected to face more ethically complex and demanding situations.

5 Conclusion

The potential for a "Holacratic" [6] psychoinformatic [14–16] organization to research solutions ([7], p. 13) to the problems engaged with in this paper remains to be concretely assessed. *Holacracy*—derived from the concept of holarchy (as discussed in [8, 12])—is a general governance model that "simultaneously honor[s] autonomy and enable[s] self-organization at every level within" an organization [5] (p. 39). Within a Holacratic organization, workers are "left to act...like free agents, able to shop around and accept role assignments anywhere in the organizational structure, including filling several roles in many different parts of the organization at once" (p. 39). Amidst his explication of Holacracy's core principles, Robertson spotlighted the shoe and clothing retailer Zappos as an illustrative case example of a successful Holacratic business.

"Global solution networks" (GSNs), which have been defined by Tapscott in [6], are similar to Holacratic organizations. GSNs consist of any group of individuals with the following four features:

- Diversity in stakeholders who collectively represent at least two of the four pillars of society (government, private sector, civil society, individuals)
- Multinational or global presence
- At least partial digitality with respect to its communications tools and platforms
- Progressive goals related to the creation of "public goods" (p. 18)

GSNs have the distinct advantage of being new, thus allowing their participants to deal with problems in manners less strictly premised on partial global problem typologies such as the United Nations and The Millennium Project's.

We contend that the Sirius Project that worked on this paper (as well as others—see, e.g., [11]) is presently a Holacratic GSN that has demonstrated potential to evolve into a "benefit corporation" (B-Corp) [13]. B-Corps are defined as new legal tools used to "create a solid foundation for long term mission alignment and value creation". They protect said mission alignment through "capital raises and leadership changes", which "creates more flexibility when evaluating potential sale and liquidity options, and prepares businesses to lead a mission-driven life post-IPO". As a B-Corp, Sirius would have access to the funds needed to realize the dream of healthcare-providing biomimetic AI.

References

1. Glenn, J., Florescu, E.: The Millennium Project Team: 2015–16 State of the Future. The Millennium Project, Washington, DC (2016)
2. Hancock, M.: Practical Data Mining. CRC Press, Boca Raton (2012)
3. Iqbal, M.: Chronic Kidney Disease Dataset. https://www.kaggle.com/mansoordaku/ckdisease/version/1. Accessed 19 Dec 18
4. Breast Cancer Wisconsin (Diagnostic) Data Set. https://www.kaggle.com/uciml/breast-cancer-wisconsin-data. Accessed 19 Dec 2018
5. Robertson, B.J.: Holacracy: The Revolutionary Management System that Abolishes Hierarchy. Penguin Random House, UK, London (2015)
6. Tapscott, D.: Introducing global solution networks: understanding the new multi-stakeholder models for global cooperation, problem-solving and governance. http://gsnetworks.org/wp-content/uploads/Introducing-Global-Solution-Networks.pdf. Accessed 21 Jan 2019
7. Sood, S.: Global problem-solving and ethics: a theoretical and practical analysis. Int. J. Environ. Issues **17**(4), 322–339 (2018)
8. Wilber, K.: Sex, Ecology, Spirituality: The Spirit of Evolution. Shambhala Publications, Boston (2000)
9. Baymax. https://en.wikipedia.org/wiki/Baymax#Film. Accessed 24 Jan 2019
10. Baymax. https://isteam.wsimg.com/neb/obj/RTBFRTdDNTlCNkE3NzBDNjQ0MTQ6MGJjOTVkZGI3NDQ3YTA4ODcwMTdiNjA2NGZmMDEyMDg6Ojo6OjA=/:/rs=w:600,h:600. Accessed 24 Jan 2019
11. Neumann, S., et al.: Content feature extraction in the context of social media behavior. In: Schmorrow, D.D., Fidopiastis, C.M. (eds.) AC 2017. LNCS (LNAI), vol. 10284, pp. 558–570. Springer, Cham (2017). https://doi.org/10.1007/978-3-319-58628-1_42
12. Koestler, A.: The Ghost in the Machine. The Macmillan Company, New York (1967)
13. Benefit Corporation. http://benefitcorp.net/. Accessed 27 Jan 2019
14. Montag, C., Duke, E., Markowetz, A.: Toward psychoinformatics: computer science meets psychology. Comput. Math. Methods Med. (2016)
15. Yarkoni, T.: Psychoinformatics: new horizons at the interface of the psychological and computing sciences. Curr. Dir. Psychol. Sci. **21**, 391–397 (2012)
16. Andone, I., Blaszkiewicz, K., Trendafilov, B.: Menthal – Running a science project as a start-up. https://www.researchgate.net/publication/304023331_Menthal_-_Running_a_Science_Project_as_a_Start-U. Accessed 27 Jan 2019

Facilitating Cluster Counting in Multi-dimensional Feature Space by Intermediate Information Grouping

Chloe Chun-wing Lo[1], Jishnu Chowdhury[2], Markus Hollander[3],
Alexis-Walid Ahmed[3], Suraj Sood[4(✉)], Kristy Sproul[3],
and Antoinette Hadgis[3]

[1] Cherrypicks Limited, Hong Kong, China
[2] Department of Computer Science, Kansas State University,
Manhattan, KS, USA
[3] Sirius Project, Melbourne, USA
[4] Department of Psychology, University of West Georgia, Carrollton, GA, USA
sirus19conf@gmail.com

Abstract. Previously, we showed that dividing 2D datasets into grid boxes could give satisfactory estimation of cluster count by detecting local maxima in data density relative to nearby grid boxes. The algorithm was robust for datasets with clusters of different sizes and distributions deviating from Gaussian distribution to a certain degree.

Given the difficulty of estimating cluster count in higher dimensional datasets by visualization, the goal was to improve the method for higher dimensions, as well as the speed of the implementation.

The improved algorithm yielded satisfactory results by looking at data density in a hypercube grid. This points towards possible approaches for addressing the curse of dimensionality. Also, a six-fold boost in average run speed of the implementation could be achieved by adopting a generalized version of quadratic binary search.

Keywords: k-means clustering · Unsupervised clustering · Multidimensional analysis

1 Background

Unsupervised clustering algorithms usually require the number of clusters as a parameter to yield satisfactory results. The cluster count can be easily given if the dataset is in lower dimensions where visualization can enable humans to estimate the number of clusters reasonably well.

For datasets in higher dimensions, the common practice is to run the clustering algorithm for different number of clusters, and select the number of clusters that gives the least within-cluster sum of squares (abbreviated as WSS). This approach requires human input of a range of cluster counts to test, which may not include the optimal cluster count for the best results. It is also common for multiple cluster counts to give a

© Springer Nature Switzerland AG 2019
D. D. Schmorrow and C. M. Fidopiastis (Eds.): HCII 2019, LNAI 11580, pp. 284–298, 2019.
https://doi.org/10.1007/978-3-030-22419-6_20

locally minimum WSS whose values are very close to each other. This ambiguity makes the selection for the ideal cluster count very difficult even with human input.

It is of great interest to develop a method that estimates the cluster count without human guess work or ambiguous results, especially for datasets in higher dimensions where it is difficult to produce a meaningful representation of cluster structure by visualization. Such a method could benefit many clustering algorithms that require an accurate cluster count input.

In the previous work, the estimation of cluster count in 2D datasets is done by dividing the 2D space of the data into grid boxes and counting the number of local maxima of number of data points found in each grid box relative to neighboring grid boxes [1]. The approach of dividing the data space into grid boxes proved to be promising in 2D space. There were two major concerns regarding this approach of estimating cluster counts: its ability to estimate cluster count accurately in higher dimensional space, and any room for speeding up the algorithm.

1.1 Extension to Higher Dimension

Applying the algorithm to higher dimension is of great interest due to the nature of human perception, which limits the ability to visualize datasets of higher dimensions accurately. While 2D datasets have almost no problems in estimating cluster count by eye from visualization, 3D data starts to suffer as the apparent number of clusters is affected by the perspective from which the data is viewed in the 3D space, as more than one clusters may appear to be one in certain angle of view if they overlap in the field of vision. As a result, cluster count estimation or cluster validation by eye is very difficult in higher dimensions, and may lead to suboptimal results in certain powerful unsupervised clustering algorithm without an accurate initial input of cluster count.

Data visualization is not the only challenge linked to increased dimensionality in machine learning. It is well known that space is dilated in higher dimensions, making the measurement of distance between data points using Euclidean distance meaningless as most data points are roughly at the same distance from each other [3]. This is termed the "curse of dimensionality". Many other definitions of distance have thus been put forward [4].

One approach that is of particular interest is the IGrid Index, which is not unlike the division of data into hypercubes [5]. It seems to suggest that the intermediate information grouping approach may work well in higher dimensional space as well.

1.2 Speed Performance

In the previous work, the attempt to estimate the number of clusters before the datasets can be passed for unsupervised clustering showed promise [1]. However, the speed of the implementation was quite slow.

The algorithm divides the whole dataset into grid boxes for 2D datasets. The best box length is the one that gives the number of boxes with single data point below a given percentage of the total data points in the dataset. In the previous version, the best box length is found by decreasing the number of portions to be divided along each axis under a geometric progression with a common ratio smaller than 1. However, this method generates many unnecessary iterations that could be skipped by a more efficient search algorithm. Moreover, since the value progresses along the geometric decrement, it is possible that the most optimal solution is skipped, such that the final result is only second best or worse. This is a common problem with iterative machine learning solutions.

Heuristics can considerably speed up iterative approaches due to additional contextual information they utilize. It is possible to extract such contextual information during the algorithm's run to restrict the search space to a smaller range of problems that is guaranteed to contain the optimal solution. Many search algorithms employ similar approaches. For example, when an array is sorted, the order of the elements in the array can serve as a heuristic, and binary search in particular proves to be a much more efficient algorithm compared to simply searching linearly through the array.

Although the complexity of binary search is quite efficient with complexity $O(lnN)$ or more precisely $O(log_2N)$. Quadratic binary search can further improve the time complexity to $O\left(ln\frac{N}{2}\right)$ [2].

Search algorithms like quadratic binary search often work on monotonically sorted arrays. It is possible to apply this approach to a series of values that is *almost* monotonic. In the previous version of the algorithm, the number of grid boxes with one data point generally decreased as the box length increased, with small fluctuations that did not significantly alter the overall almost monotonically decreasing trend. Quadratic binary search thus had to be adapted before it could be used to speed up this part of the algorithm.

2   Procedure

2.1   Assumptions and Definitions

We make several assumptions about the data distribution, some identical to the previous research, namely:

- Each cluster approximates a Gaussian distribution, which implies each cluster has a center where the "density" of data points peaks.
- Clusters have a low degree of overlap.
- There are more than one cluster in each dataset.
- A cluster has at least 3 data points.

The following definitions were introduced with higher dimensionality in mind:

- Density of data is defined as the number of data points per hypervolume of each hypercube.
- Hypervolume is calculated by the length of hypercube raised to the power of the dimension of the dataset.

Since all hypercubes have the same hypervolume, the identification of local maxima with highest data density can be achieved by directly identifying the hypercube with the most data points.

2.2 Algorithm

The main idea of the algorithm is to first determine the best cube length for dividing the minimum bounding box of the dataset into hypercubes of equal size and to then scan these hypercubes for local density maxima.

Optimization efforts focused on improving the search for the optimal hypercube size. Generalization to higher dimension involved changes to both optimal hypercube size selection and local maxima detection.

Different axes may have different data range, and this puts the method in danger of underestimating cluster count if certain axes have a range too small relative to other axes. Normalizing data before searching for optimal hypercube size addresses this problem.

Normalized data have very small ranges and a specialized algorithm for determining initial hypercube size is required.

Normalization. Data of each axis is normalized by transforming all of them into the standardized z-score. Since z-scores mostly range between -3 and 3, it is not sensible to use 1 as the initial hypercube length as it may be well past the optimal solution for the ideal hypercube size.

Initial Hypercube Size Determination. Let N be the total number of data points in the d-dimensional data set. The goal is to find the hypercube size that divides the minimum bounding box of the dataset into M^d d-dimensional hypercubes at the smallest possible resolution while keeping the number of empty hypercubes as small as possible, where M is the number of hypercubes along each axis.

This is accomplished by iteratively adjusting M, starting with $M_0 := \sqrt{N}$. In each iteration i, let k_i be the number of data points in the densest hypercube and compute $M_i = M_{i-1} \cdot \sqrt{k_i}$.

The square root is taken to avoid overshooting the optimal value too much. The process stops when there is no hypercube left with more than 1 data point. The final M_i is then used as a starting point in the next step.

Determination of Optimal Hypercube Size. In the previous version of the algorithm, the optimal number of hypercubes along each axis, m_{opt}, was determined by decrementing the initial M obtained in the prior step in geometric sequence and evaluating the number of data points assigned to the resulting hypercubes. To speed up this process while also avoiding to settle on a sub-optimal hypercube size, this version of the algorithm employs a modified quadratic binary search instead of geometric decrementation. Whereas quadratic binary search divides the range of potential values by 4 in each iteration, the modified version divides the range into T candidates in each iteration, whereby T is given as a parameter.

The fraction of data points alone in a hypercube, n, hereby serves as a heuristic for approximately identifying m_{opt}. This is a suitable heuristic since a grid with too small hypercubes will end up accumulating many lone data points, while a grid with too big hypercubes will overvalue noise and data points at the far edge of clusters. The algorithm takes a parameter t describing a threshold for acceptable values of n as a fraction of the volume of the dataset.

The best hypercube size is computed as follows:

1. Divide the current value range such that T candidates $m_1, \ldots, m_T$ are obtained for the determination of m_{opt}. In the first iteration, the highest possible value of m_{opt} is the initial M obtained in the previous step and the lowest possible value is 2, resulting in candidates $m_1 = M, \ldots, m_i, \ldots, m_T = 2$.
2. All m_i are rounded to integers, since the grid consists of equally sized hypercubes, and duplicate m_i are removed after rounding. If there is only one candidate, the approximated m_{opt} was found.
3. In descending order, the fraction of data points alone in a hypercube, n, is calculated for each m_i. The first m_i for which n drops below the threshold t is the lower bound for the next iteration of the search.
4. Based on the heuristic, m_{opt} lies between that m_i and the previous one, m_{i-1}, if their difference is larger than 1. In that case, steps 1 to 3 are repeated for the value range m_{i-1} to $m_i - 1$ until step 2 yields only one candidate for m_{opt}.

Cluster Count Estimation. After the value of m_{opt} is determined, the number of data points in each hypercube is compared against the nearest 8 hypercubes. The nearest hypercubes are found by selecting the hypercubes whose centre point gives the shortest Euclidean distance to the hypercube to be compared. The hypercube containing the most data points relative to its neighborhood is marked as a local maxima, and the total number of local maxima detected is returned by the algorithm as the cluster count estimation.

2.3 Time Complexity

```
# INITIAL M DETERMINATION
let N = number of data points          # assignment, O(1)
let M = sqrt(N)                        # assignment, O(1)
let d = dimension of data              # assignment, O(1)

DO:
/* DO-WHILE LOOP:
   number of iterations for M to reach the point where
   there are no hypercube containing more than 1 data point
*/
  Map number of data points into a M^d array by dividing
    along each side of the bounding box of the data into
    M portions
                          # going through each data point, O(N)
  let max = maximum number of data points in all hypercubes
                          # going through each hypercube, O(M^d)
  if max > 1:                          # boolean checking, O(1)
    M = M*sqrt(max)                    # assignment, O(1)
  WHILE (max > 1)                      # boolean checking, O(1)
```

In theory, the number of loops required for finding the initial M depends on the maximum local density of the dataset. However, there is no way to make such a measurement except running the algorithm. This poses difficulty in determining the complexity of this part of the algorithm, especially in expressing the total number of loops and the value of M in terms of measurable metrics. Fortunately, the number of data points is finite, and this guarantees termination as the minimal distance among all data points exists when we consider a finite number of points distributed in a R^d space.

In practice, the process of initial M determination terminated within 12 loops for all test datasets. By plotting a scatter plot of the number of loops against the number of data points N, it fits well over a logarithmic curve (Fig. 1).

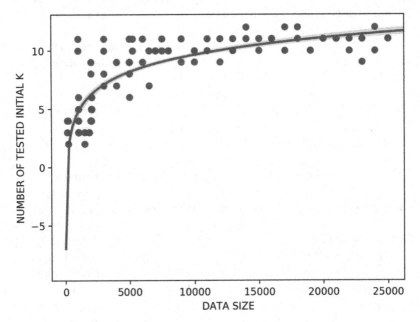

Fig. 1. The plot of number of loops against data set size (dots) is best fit by $y = 2.10158748 \cdot log(x) - 9.72704674$ (line).

The following analysis therefore works under the assumption that the number of loops is generally in $O(lnN)$. As the algorithm calculates the number of data points in each hypercube, it goes through M^d hypercubes in total. M can be at most the length of the minimum bounding box of the dataset divided by the smallest distance of the data points in one dimension, which is used as an upper bound for M in the subsequent analysis.

The big O can thus be expressed as:

$$O\big(lnN(1+1+N+M^d+1+1+1)\big)$$
$$= O\big(lnN(M^d+N)\big)$$

(1)

```
# FINDING OPTIMAL LENGTH OF HYPERCUBES
let t = tolerance level, a fraction          # assignment, O(1)
let T = number of candidate values           # assignment, O(1)
let m_max = M                                # assignment, O(1)
let m_min = 2                                # assignment, O(1)

DO:      # number of iterations for m to reach 1, O(log(2M/T))
  let r = pow(m_max/m_min, 1/(T-1))          # assignment, O(1)
  let m_candidates = []                      # assignment, O(1)
  FOR i in 1 … T:          # going through i from 1 to T, O(T)
    m_candidates.add(round(m_max/r^i))       # assignment, O(1)
  m_candidates.remove_duplicate()
                          # go through the list, O(T)
  let idx = 1                                # assignment, O(1)

  DO:       # worse case going through all T candidates, O(T)
    let m = m_candidates[idx]                # assignment, O(1)
    Map number of data points into an m^d array by
        dividing each dimension of the dataset into m portions
                # going through each data point,        O(N)
    Find the number of data points in each hypercube
                # going through each hypercube, O(m^d)
    let s = number of hypercubes with one data point
                # going through each hypercube, O(m^d)
    if s <= N*t:                    # boolean checking, O(1)
      if idx == 1:                  # boolean checking, O(1)
        let m_min = m_candidates[1]        # assignment, O(1)
        let m_max = m_candidates[1]        # assignment, O(1)
      else:
        let m_min = m_candidates[idx]      # assignment, O(1)
        let m_max = m_candidates[idx-1]-1  # assignment, O(1)
    else if idx == T:               # boolean checking, O(1)
      let m_min = m_candidates[T]          # assignment, O(1)
      let m_max = m_candidates[T]          # assignment, O(1)

    idx = idx+1                             # assignment, O(1)

  WHILE s > N*t or idx <= T          # boolean checking, O(1)
WHILE m_candidates.size > 1          # boolean checking, O(1)
let M = m_candidates[0]                     # assignment, O(1)
```

For an array of different m values ranging from M to 2, the quadratic binary search will give a complexity of $O\left(ln\frac{M}{2}\right)$. The division of M by 2 is a result of the division of data into up to 4 portions in each iteration by quadratic binary search, which is a double of 2 portions in binary search. Similarly, this algorithm divides the data in up to T portions which is $\frac{T}{2}$ times of 2, and the complexity of that particular loop can be said to be $O\left(ln\frac{M}{\frac{T}{2}}\right) = O\left(ln\frac{2M}{T}\right)$. It becomes $O\left(ln\frac{M}{T}\right)$ after dropping constants. Therefore, the complexity of this part of the algorithm becomes:

$$O\left(4 + (2 + T + T + 1 + T(1 + N + M^d + M^d + 6) + 1)ln\frac{M}{T} + 1\right)$$

$$= O\left(4 + (2 + 2T + 1 + T(1 + N + 2M^d + 6) + 1)ln\frac{M}{T} + 1\right)$$

$$= O\left(5 + (4 + 2T + T(N + 2M^d + 7))ln\frac{M}{T}\right)$$

$$= O\left((2T + T(N + 2M^d))ln\frac{M}{T}\right)$$

$$= O\left((2T + TN + 2TM^d)ln\frac{M}{T}\right) \tag{2}$$

$$= O\left((T + TN + TM^d)ln\frac{M}{T}\right)$$

$$= O\left(T \cdot ln\frac{M}{T}(1 + N + M^d)\right)$$

$$= O\left(T \cdot ln\frac{M}{T}(M^d + N + 1)\right)$$

$$= O\left(T \cdot ln\frac{M}{T}(M^d + N)\right)$$

```
# FINDING LOCAL DENSITY MAXIMA
let maxima_centroids = []                    # assignment, O(1)
Calculate the number of data points in each hypercube
                         # going through each hypercube, O(M^d)
For each hypercube:      # going through each hypercube, O(M^d)
   calculate the distance of all other hypercubes to the current
      hypercube          # going through each hypercube, O(M^d)
   sort the hypercubes by distances to the current hypercube in
      ascending order, then by number of data points in
      descending order for the same distance
                         #             quicksort,
O(M^2d)
      get the closest 8 hypercubes             # retrieval, O(8)
      let n_max = maximum number of data points in closest 8
         hypercubes                 # going through all boxes, O(8)
      let n_current = data points in current hypercube
```

```
                                          #       assignment,
O(1)
    if n_current >= n_max:                # boolean checking, O(1)
        let (x_1, ..., x_d) = center coordinate of current hypercube
                                          #       assignment,
O(1)
        maxima_centroids.add((x_1, ..., x_d))     # assignment, O(1)
    return maxima_centroids        # termination of algorithm
```

The complexity of this part of the algorithm is:

$$
\begin{aligned}
O(1 + M^d &+ M^d(M^d + M^{2d} + 8 + 8 + 4)) \\
&= O(1 + M^d + M^d(M^d + M^{2d} + 20)) \\
&= O(M^d + M^d(M^{2d})) \\
&= O(M^d + M^{3d}) \\
&= O(M^{3d})
\end{aligned}
\tag{3}
$$

So the overall complexity after combining Eqs. (1), (2) and (3) is:

$$
O\left(lnN(M^d + N) + T \cdot ln\frac{M}{T}(M^2 + N + 1) + M^6 \right)
\tag{4}
$$

Substituting $d = 2$ into (5) gives:

$$
O\left(lnN(M^2 + N) + T \cdot ln\frac{M}{T}(M^2 + N) + M^6 \right)
\tag{5}
$$

However, we used the previous algorithm for local maxima estimation for 2D datasets for faster speed. The time complexity for local maxima estimation was M^2, so the time complexity of this version on 2D datasets should be:

$$
O\left(lnN(M^2 + N) + T \cdot ln\frac{M}{T}(M^2 + N) + M^2 \right)
\tag{6}
$$

The time complexity of the previous version of the algorithm for 2D datasets was:

$$
\begin{aligned}
O\left(\frac{lnM}{-lnr}[N + M^2 + 2] + 12M^2 + 4N + 9 \right) \\
= O\left(\frac{lnM}{-lnr}[N + M^2] + M^2 + N \right)
\end{aligned}
\tag{7}
$$

The parameter T and parameter r are not shared by both expressions of the time complexity for the two versions of the algorithm. It is difficult to directly compare their performance. However, since they are parameters that are unlikely to be changed radically from dataset to dataset, we may treat them as constants. The expressions for the new version of the algorithm can be simplified to:

$$O\left(lnN(M^2+N)+lnM(M^2+N)+M^2\right) \tag{8}$$

And the one for the previous version of the algorithm can be simplified to:

$$O\left(lnM[M^2+N]+M^2+N\right) \tag{9}$$

The time complexity indicates that the new version may perform slower than the old version, but actual run time may tell a different story as the parameter T and r interacts with the algorithm in ways that is difficult to compare their effect on both versions of the algorithm.

2.4 Parameters

We for comparison, the tolerance level t, describing the acceptable fraction of data points alone in a hypercube, was set to 0.004 and the number of candidate values T to 20 for all datasets.

2.5 Test Data

We tested our algorithm on data sets from the University of Eastern Finland [6] and Kaggle [7, 8].

2.6 Hardware and Software

The algorithm was run on a MacBook Pro (Retina, 13-inch, late 2013), 2.4 GHz Intel Core Duo i5, 8 GB 1600 MHz DDR3 RAM, macOS Mojave version 10.14 using Python3 v3.6.3 with Seaborn v0.8.1, Pandas v0.23.4 and NumPy v1.13.3.

3 Results

3.1 Optimization

In practice, the new implementation of the algorithm got a boost in speed with minimal impact on cluster counting accuracy compared with the previous methods (Tables 1 and 2).

Table 1. Comparison of run time of two different algorithms.

Data normalization	Method	Run time (sec)			
		Minimum	Mean	Maximum	Standard deviation
No	m sampling	0.039614	4.872610	15.976803	4.918828
No	Geometric decrement	0.045037	31.377525	131.773615	38.138111
Yes	m sampling	0.041973	5.316945	21.418652	6.011148
Yes	Geometric decrement	0.040350	30.805692	126.715257	36.140838

Table 2. Comparison of difference from ground truth of two different algorithms.

Data normalization	Method	Absolute difference with ground truth			
		Minimum	Mean	Maximum	Standard deviation
No	m sampling	0	0.588235	7	1.219450
No	Geometric decrement	0	0.568627	7	1.220736
Yes	m sampling	0	0.823529	7	1.306995
Yes	Geometric decrement	0	0.784314	7	1.285515

On average, the run time is cut down to one-sixth of the original.

The average difference of cluster count does not suffer any visible impact from the change of the implementation.

3.2   Generalization to Higher Dimension

The algorithm performed considerably worse for higher dimension (Table 3).

Table 3. Comparison of performance between data in 2-dimensions and higher dimensions.

Dimension	Absolute difference with ground truth			
	Minimum	Mean	Maximum	Standard deviation
2	0	0.494186	5	0.888504
More than 2	0	1.750000	7	2.140244

The values of m in higher dimension ranges from 2 to 3, which is way smaller than that found in 2D data (Table 4).

Table 4. Comparison of final m value between data in 2-dimension and higher dimensions.

Dimension	Final m			
	Minimum	Mean	Maximum	Standard deviation
2	5	44.616279	130	36.853168
More than 2	2	2.062500	3	0.245935

It is tempting to explain the inaccuracy by the curse of dimensionality. However, upon further inspection of the datasets, datasets in higher dimensions also happen to have fewer data points compared to 2D datasets (Table 5).

Table 5. Comparison of data count between data in 2-dimension and higher dimensions.

Dimension	Data count			
	Minimum	Mean	Maximum	Standard deviation
2	1481	3425.333333	7500	1856.009912
More than 2	147	810.750000	1024	366.817941

4   Discussion

4.1   Strength

Speed without Sacrificing Accuracy. With all other conditions under control, the implementation of the new algorithm is six times faster without suffering much with regard to accuracy. The final decrement step to find the optimal m value also guarantees that the ideal solution is not skipped over by the approach.

Minimal Impact on 2-D Dataset. The performance is on par with previous generations on 2-D datasets.

4.2   Limitations

Sensitive to Local Density Variation within Cluster. Since this algorithm does not fundamentally changes the way it estimates cluster count, it retains a problem that is present in the previous version. It is sensitive to local density which leads to over estimating cluster count in some cases, especially for 2D data.

Inconsistency Between Normalized and Unnormalized Data. In an attempt to unify the behavior of the algorithm across datasets with different ranges, normalization (i.e. conversion to standardized z-score) is applied to all datasets before cluster count estimation. Interestingly, the resulting cluster count estimation differs between normalized and unnormalized data.

Inspection of the detailed logs recording different variables in all loops revealed that the algorithm terminates at different m value for normalised and unnormalised data. Consequently, the estimation of cluster count can differ by as much as 3.

It is possible that local maxima detection is very sensitive to local fluctuations in data density, amplified by normalization. Another possible explanation may be due to the change in data range that affects how the data are divided up and that leads to discrepancy in how the hypercubes are generated in these two versions of the dataset. More work is required to unify the performance between normalized and unnormalized data.

Degeneration of m Value in Datasets with Low Data Amount. We also observe that the final ideal m value almost always degenerates to 2 or 3 for data in higher dimension. This looks like a problem due to dilation of space in higher dimensional space. However, the same issue occurs in datasets at dimension as low as 3. Upon inspection, those datasets share a common characteristic of having a low data count (in the range of hundreds).

Given a very small tolerance, the number of boxes with single data point that terminated the loop for finding the ideal m value will become 0, and therefore pushes the m value to a meaninglessly small value. For example, for a dataset with only 100 observations, the tolerance level of 0.004 means that the algorithm will terminate when the number of single data hypercube drops below $100 \times 0.004 = 0.4$. Since the lowest possible value of m is set to be 2, this guarantees the algorithm to ends with $m = 2$ and therefore unable to accurately estimate the cluster count.

Setting the tolerance at a higher value enables a more accurate cluster count estimation for datasets with low data points, but leads to gross overestimation for those with large data volume. It seems that a dynamic tolerance level or an entirely new termination condition is required for this algorithm to perform uniformly well across datasets with different volume.

4.3   Curse of Dimensionality

In 2D, neighboring grid boxes can be easily identified by the index coordinates of the grid box. However, the same definition of "neighbors" cannot be applied in higher dimension as the number of neighbors increases with the dimension. It does not only pose performance issues for the algorithm, it also makes every hypercube a neighbor to all other hypercubes.

For now, a limit is posed on the number of nearest neighbor in the determination of local density maxima. The limit is arbitrary but well justified as a maximum among the closest few already fulfils the definition of a *local* maxima. Hypercubes of greater "distance" can be omitted.

With a higher tolerance level (t value), datasets with low data volume at higher dimension produces reasonable results. This suggests that this approach may be able to work around the curse of dimensionality that stumps many researches. It is possible that the division of higher dimensional space into hypercubes by our method ensures that the well-ordered nature of our conventional notion of distance (i.e. Euclidean distance) is preserved as the division is done uniformly and is sensitive to relative distance among data points in each cluster (Table 6).

Table 6. Examples demonstrating the influence of tolerance level on the final m value and the resultant cluster estimation count. Configurations that produce the correct result are bolded.

Dataset	Data count	Dimension	Data normalization	Tolerance	Final m	Cluster count estimation
Iris species	1481	4	Yes	0.025	2	1
				0.05	4	2
				0.125	**6**	**3**
				0.15	**7**	**3**
				0.225	9	8
Seed dataset	210	7	Yes	0.125	2	2
				0.15	3	5
				0.3	**4**	**3**
				0.35	5	5
				0.45	6	9

4.4 Further Research

More research effort is needed to find out the relationship between local data density, hypercube size and total data points to understand the interaction of these factors and their impact on the original algorithm. This is especially important for understanding the impact of data volume to the performance of the algorithm. As suggested by our results, it is best to have a dynamic termination condition than a static one. The algorithm is not at all useful if the termination condition depends on a parameter input by the user as it throws back the problem of cluster count estimation to humans. It seems to be possible to nail down that moving target if we probe deeper into the relationships among different metrics of the dataset.

Smoothing algorithm may alleviate the problem of overestimation due to local density fluctuation. However, most smoothing algorithms are applied into in 1D or 2D data. Extra attention must be paid in searching for one that can be applied in high dimensional space.

It is also valuable to compare the results from this algorithm with human evaluation. Although we rely on machines to perform routine and computational tasks in a more efficient manner, it is questionable whether machine learning algorithms can *always* generate results "superior" to human judgement, or align with human understanding on how information should be organized.

This problem stems from the fact that even among human, there may be disagreement on what constitutes a cluster. The difficulty in devising a simple way to count clusters may suggest that the potential problem lies in the lack of a clear definition of a "cluster". After all, entities as large as galaxies can form a "cluster" and clusters a supercluster, which may look like a single bright dot on the sky if the observer is far enough. One possible solution to this is to search for clusters in all different levels. This is, however, an entirely different search topic with extreme difficulty that requires intense research effort.

5 Conclusion

Adapting and tweaking quadratic binary search offers a six-fold boost in cluster estimation speed on average. The interaction between data count and tolerance level to terminate the algorithm poses difficulties in giving meaningful results when the data has low data volume, and results suggest that an adaptive tolerance level may yield good results in higher dimensions and may potentially provide a way out of the curse of dimensionality. Deeper relationship between dimensions, local data density variation, data volume and any other alternative termination conditions is needed for the development of the algorithm to move forward.

Acknowledgements. I must restate my deep gratitude towards Mr. Monte Hancock for recruiting me into the Sirius project and how the project turned my life around.

Inexpressible thanks go to each of my team members: Jishnu for efficient coding and good mistake spotters, Markus for meticulous eyes on overall structure and wording of the paper, Alexis for rigorous critique and proofreading, and Suraj for saving the paper when I most needed help on formatting and putting the pieces together into a complete piece. Many thanks to other workers on the Sirius team working on bits and pieces of this paper. This paper will not be complete without all of your help. Big thumbs up for Lesley the EPM (Executive/Epic Project Manager) for amazing coordination.

References

1. Lo, C.C.-W., et al.: Intermediate information grouping in cluster recognition. In: Schmorrow, D.D., Fidopiastis, C.M. (eds.) AC 2018. LNCS (LNAI), vol. 10915, pp. 287–298. Springer, Cham (2018). https://doi.org/10.1007/978-3-319-91470-1_24
2. Kumar, P.: Quadratic Search: A new and fast searching algorithm (An extension of classical Binary search strategy). https://pdfs.semanticscholar.org/3d91/97ecfcc1a16254c8667b0cbd35c93e7f9437.pdf. Accessed 1 Feb 2019
3. Aggarwal, C.C., Hinneburg, A., Keim, D.A.: On the surprising behavior of distance metrics in high dimensional space. http://citeseerx.ist.psu.edu/viewdoc/download?doi=10.1.1.23.7409&rep=rep1&type=pdf. Accessed 1 Feb 2019
4. Pele, O.: Distance functions: Theory, algorithms and applications. https://pdfs.semanticscholar.org/c656/f090d5710a524ac26ef1b22310e772fa465c.pdf. Accessed 15 Feb 2019
5. Aggarwal, C.C., Yu, P.S.: The IGrid Index: Reversing the dimensionality curse for similarity indexing in high dimensional space. http://citeseerx.ist.psu.edu/viewdoc/download;jsessionid=0EBCEB48BA3DE80411807AA7DF3C3A60?doi=10.1.1.129.746&rep=rep1&type=pdf. Accessed 1 Feb 2019
6. Clustering basic benchmark. https://cs.joensuu.fi/sipu/datasets/. Accessed 1 Feb 2019
7. Iris Species. https://www.kaggle.com/uciml/iris. Accessed 1 Feb 2019
8. Seeds dataset. https://www.kaggle.com/rwzhang/seeds-dataset. Accessed 1 Feb 2019

Training to Instill a Cyber-Aware Mindset

Kelly Neville[1]([⊠]), Larry Flint[2], Lauren Massey[1], Alex Nickels[1],
Jose Medina[1], and Amy Bolton[3]

[1] Soar Technology, Inc., Ann Arbor, MI 32817, USA
Kelly.neville@soartech.com
[2] Ingenia Services, LLC, Wake Forest, NC 27587, USA
[3] Office of Naval Research, Arlington, VA 33303, USA

Abstract. The rapidly increasing sophistication of cyber threats occurring in parallel with our growing reliance on networked systems for everything from shopping to managing critical infrastructure is not a coincidence. Ransomware events, compromise of personal financial information, and hacking into critical infrastructure systems make the headlines on a seemingly daily basis. Still, the average system user continues to operate in a mode that signals belief those events will happen to someone else. This paper presents work conducted to determine training objectives and strategies for altering that "other guy" mentality and instilling a *cyber-aware mindset*. System users with a cyber-aware mindset should be less likely to fall for attacker ploys and more likely to actively contribute to their organization's cyber defense. We identify two major categories of training objectives: cyber awareness and mindset objectives. *Cyber awareness training objectives* encompass three main areas of knowledge and capability: system baseline performance, anomaly detection and response, and systems thinking. *Mindset training objectives* encompass cognitive adaptations associated with acquiring a new mindset. These are adaptations to knowledge structures, cognitive heuristics, and metacognition. These training objectives are being used to guide the design of a scenario-based video game for training a cyber-aware mindset. This work highlights the importance of relevant conceptual knowledge to cyber awareness and research needs associated with mindset change, including mindset-change measurement, which we have begun to address for the purpose of evaluating the video game's efficacy.

Keywords: Cyber awareness · Mindset training · Cognitive bias training

1 Introduction

Over the last three decades, individuals and organizations alike have increasingly moved their daily life into the digital realm. In cyberspace, we connect socially, interact professionally, and find ever-available sources of entertainment. We use it for shopping, banking, filing taxes, applying for government services, managing healthcare, sharing accomplishments, storing documents, traveling, finding recipes, displaying our family photo albums, educating ourselves, and much more. We are now, as individuals, families, organizations, and societies, heavily invested in the digital realm, and our most valuable information has become extensively interwoven into its fabric.

© Springer Nature Switzerland AG 2019
D. D. Schmorrow and C. M. Fidopiastis (Eds.): HCII 2019, LNAI 11580, pp. 299–311, 2019.
https://doi.org/10.1007/978-3-030-22419-6_21

The transition to cyberspace by consumers, corporations, and government agencies has provided new, interesting, and profitable opportunities for nefarious actors. Ranging from mischievous hooligans to criminal organizations to nation states the population of malicious cyber citizens has grown in step with the law-abiding population. According to the 2018 Verizon Data Breach Investigations Report, the year 2017 saw over 53,000 data breach incidents and 2,216 confirmed data breaches in industry. Among these were sophisticated multi-pronged attacks such as a set of coordinated bank heists that reaped over $40 million (Trustwave 2018). Cyber attackers have demonstrated an ability to outsmart, on a large scale, heavily defended systems. Their creativity and resourcefulness coupled with continued growth in proficiency, poses serious threats to all organizations, services, governments, and economies.

1.1 Every System Is at Risk

Many organizations, including many U.S. military organizations, keep their most sensitive data and capabilities on isolated systems and networks. The so-called air gap may slow down cyber attackers but it does not deter them. For example, Russian hackers have gained entry to hundreds of air-gapped U.S. utilities (Greenberg 2017; Smith 2018) via conventional spearfishing and watering hole attacks on utility vendors (Greenberg 2017) and reportedly continue to do so. The hackers stole vendor credentials and used them to directly access utilities' networks and collect information about network configurations, software, hardware, administrative accounts, and more. One report indicates that the hackers could have "thrown switches" and "disrupted power flows" (Greenberg 2017). The same malware found on U.S. utility networks has been used to wreak havoc in the Ukraine, shutting down the electrical grid, blocking operators from using their own systems to intervene, and shutting down battery backups to the operations facilities (Greenberg 2017; Smith 2018).

Military organizations are especially at risk for being targeted by cyber attack. While security of consumer financial transactions is critical, the military relies on networks, including commercial infrastructure, for our nation's defense. Their networks represent a treasure trove of data that could be used to an adversary's significant advantage. They likewise support a trove of invaluable capability that a cyber attack could cripple or co-opt.

Improving our military's ability to operate and defend itself in cyberspace is an operational imperative; defending our networks and the systems relying on those networks is fundamental to maintaining supremacy in the cyber domain. To that end, the U.S. Department of Defense has taken strides to improve the numbers and effectiveness of skilled professional cyber defenders and to provide for effective training of cyber operators. However, they are a finite resource faced with attackers who patiently probe the perimeter and all avenues of approach to identify cyber vulnerabilities.

1.2 Everyone Is a Sensor

The maturing sophistication of attack methods, increased complexity of our networks, and sheer persistence work to the advantage of the attacker. The nature of networks and our reliance on them (required access by all hands) increases the attack surface as every individual with network access is a potential threat vector. Accordingly, attackers regularly incorporate human threat vectors into their tactics. Cyber intrusions that led to the bank heists mentioned above were enabled by social engineering via both phone calls and phishing emails (Trustwave 2018). The Ukrainian power outage attacks began with a phishing email impersonating a message from the Ukrainian parliament (Greenberg 2017).

Although the potential to gain system access through operator inattention, error, gullibility, or negligence cannot be eliminated, it can be diminished with training. Furthermore, we propose that with training, system operators' role can be elevated from that of threat vector and vulnerability to one of threat deflector and strength. We propose that with training, human operators can be leveraged as sensors, providing warnings and indications of an attack that may be undetected by cyber operators.

Because cyber attacks are inherently difficult to deter and detect, effective defensive strategies require a holistic and layered approach. System operators or users should be at least one of those layers. Although system operators may lack the expertise to distinguish routine technical anomalies from intentional malicious activity, they nevertheless can play an important role in cyber security. For example, operators can alert cyber professionals about unusual or indicative symptoms. Cyber-aware operators may also provide professional cyber teams with valuable supporting data, know what additional cues to look for, and gauge whether or not they can continue to use their systems. If not, cyber-aware operators will know how to adapt. As noted by Canham (2019), a cyber attack on an accounting and payroll system is more likely to be detected by the accountant using the system than by software that monitors network activity for anomalous activity.

1.3 Supporting Meaningful Participation

To become an effective layer in their organization's cyber defense, personnel require education, training, and a fundamental appreciation for their critical role in defending the network. A given operator will only infrequently encounter intentional malicious activity on his or her systems. Consequently, remaining vigilant to the possibility of an attack is challenging. Further, researchers have found that control operations personnel remain biased toward treating system anomalies as hardware, electrical, cooling, or product flow problems, versus as problems with software and networks (e.g., Line et al. 2014). As a result, when a cyber intrusion affects system performance, operators may be biased toward assuming a physical, non-cyber source as the cause.

We propose that enabling personnel to support their organization's cyber defense depends on two primary training program elements: Foundational knowledge about cyber threats to operations and high fidelity practice. Knowledge structures, i.e., schemata, mental models, and frames, guide what we perceive and how we understand and responds to situations (Klein et al. 2006; Neisser 1976). Adding cyber threat information to operators' knowledge structures should therefore change how they think about their work. More specifically, it should guide the tuning of attention and metacognition to support vigilance and awareness of cyber attack as a possible source of symptoms. We refer to this tuned vigilance and awareness as a *cyber-aware mindset*.

High-fidelity practice allows attention and metacognition to become tuned to the new cyber-defense responsibility in ways guided by the foundational knowledge and performance feedback (i.e., to evolve into a cyber-aware mindset).

2   Method

2.1   Literature Review

Cyber-aware training objectives were identified by reviewing research literature covering the following topics:

- Training to support incident detection and response in process control operations;
- Cyber operations competency taxonomies; and
- Analyses of the cognitive work of cyber professionals.

As we reviewed, we culled training recommendations for cyber awareness. Similar and overlapping objectives were reconciled to produce a single set.

We also reviewed literature related to mindset change. The literature encompassed:

- Characteristics and culture of high reliability organizations;
- Expertise acquisition in complex cognitive work; and
- Minimizing cognitive bias.

2.2   Iterative Problem Space Assessment and Design

In ongoing assessment and design work, the research team holds weekly meetings to discuss training system development plans and the evolving software prototype in light of training objectives, problem space constraints, and problem space affordances. This process is shaped by literature review findings, team members' experience bases, interviews with intended end users about the performance of their systems, and discussions with other stakeholders.

3   Results

3.1   Cyber Awareness Training Objectives

Cyber awareness training objectives we identified are listed in Table 1 and fall into three major categories:

- System Baseline: Baseline knowledge and disciplined monitoring;
- Anomaly Identification and Reporting: Perceptual learning and supporting knowledge; and
- Systems Thinking: Comprehension of relationships within and affecting your organization's computing systems, mission, and cyber protection system.

Two resources, Paul and Whitley (2013) and Line et al. (2014), were especially influential.

Table 1. Cyber awareness training objectives

System Baseline: Baseline Knowledge and Disciplined Monitoring
Regularly check and maintain the accuracy of your system's configuration.
Maintain awareness of the status of your system's health.
Maintain awareness of the health and status of systems with which your system interacts.
Know what normal system activity (system performance and displayed information) looks like.
Know established indicators of non-normal system functioning.
Use available system utilities to check or monitor system performance and processes for unexpected activity.
Maintain awareness of trends and changes in your system's performance and information displays. Use them to derive expectations for normal system function.
Anomaly Identification and Reporting: Perceptual learning and supporting knowledge
Be able to detect deviations from typical information and system activity patterns, including deviations from typical patterns of deviations. Be able to recognize the absence of information.
Know and be able to recognize the signs of cyber events and other anomalies that have occurred in the past.
Account for information reliability when assessing or reporting anomalous activity.
Know how to report an anomaly in terms that contribute to its evaluation by a cyber analyst.
Systems Thinking: Comprehension of relationships within and affecting your organization's: -   **Computing systems and networks,**   -   **Mission, and**   -   **Cyber protection system**
Understand how cyber attacks can impact an organization's systems, defenses, and effectiveness.
Be familiar with different attack scenarios, ways the attack can progress over time, and possible outcomes. Understand that attacks tend to unfold over time.
Know that data flows among systems and other system interdependencies. Understand how cyber attack effects cascade over time and across attack phases.
Know how to identify and use alternative resources to compensate for unavailable or compromised systems.

- Paul and Whitley's (2013) investigation of the cyber awareness practices of cyber security analysts responsible for network intrusion detection. They conducted interviews with six analysts and twenty-five hours of observation and then derived a list of forty-four questions analysts ask themselves to establish and maintain awareness of new and ongoing network events.
- Line et al.'s (2014) assessment of cyber awareness in six large Norwegian energy distribution system operators (DSOs). They conducted semi-structured interviews with representatives of six DSOs to assess cyber attack preparedness and identify knowledge gaps. Interview results were compared with the results of the authors' preparatory work investigating the elements of effective cyber situation awareness.

We also consulted taxonomies of cyber security competencies proposed by the National Institute of Standards and Technology (NIST), Carnegie Mellon University Software Engineering Institute, and Intelligence and National Security Alliance (INSA) (Cyber Intelligence Task Force 2015). These competencies contributed only minimally to the list in Table 1 as they are intended for cyber security professionals, not system users. Similarly, much of the research literature on training requirements for cyber security focuses on cyber professionals and, although it had influence, it did not weigh heavily in the determination of training objectives.

3.2 Mindset Training Objectives

The identified mindset training objectives are rooted primarily in the literature on expertise in complex cognitive work. They target the development of:

- Knowledge structures,
- Cognitive heuristics, and
- Skeptical metacognition.

Below, we discuss each area of cognitive development and how it relates to system operators' acquisition of a cyber-aware mindset. Specific training objectives are presented following the discussion.

Knowledge Structures. The term *knowledge structure* encompasses the constructs *template*, *schema*, *frame*, and *mental model*. Researchers have established that experts in complex work domains have elaborate, interconnected knowledge structures that contain both verbal and nonverbal details about the work, work conditions, variability in work conditions, what can happen next, why, and more (e.g., Borko and Livingston 1989; Boulton 2016). These rich, sophisticated knowledge structures enable flexibility, anticipation, and the ability to mentally evaluate options (e.g., Klein 1993; Klein et al. 2006), They affect the way a person interprets incoming information, choices they make, and even what they see and do not see (Klein et al. 2007; Neisser 1976).

In their influential article on the effects of a person's *stress mindset*, Crum et al. (2013) define mindset as "...a mental frame or lens that selectively organizes and encodes information, thereby orienting an individual toward a unique way of understanding an experience and guiding one toward corresponding actions and responses" (p. 717). We assess this mindset definition as consistent with the definition of a knowledge structure, which also "organizes and encodes information," orients, and guides. We hypothesize that a cyber-aware mindset requires knowledge structures that are enriched with knowledge about the cyber landscape.

People experience difficulty in handling information that conflicts with their existing mental model of a situation (e.g., Chi 2005; Feltovich et al. 1988; Lewandowsky et al. 2012). Feltovich and his colleagues demonstrated that people use a range of strategies to discount and rationalize their disregard of conflicting, albeit accurate, information. Endsley (2018) calls attention to the implication that the underlying mental model needs to be changed in order to change behavior. She suggests that to reduce people's vulnerability to misinformation that is consistent with an inaccurate mental model, alternative narratives may need to be introduced "to create a foundation for new information" and "help explain old information in a new light" (p. 1091).

Endsley (2018) suggests that resistance to mental model change might be overcome by using visualization and interactive simulations strategies. Because people have difficulty processing information when it is not addressed by their current knowledge structures, others have suggested the use of interactive simulations that allow learners to learn from mistakes and experience the consequences of an inaccurate mental model (Feltovich et al. 1988; Klein and Baxter 2009). Lewandowsky et al. (2012) point to studies showing "that the continued influence of misinformation can be eliminated through the provision of an alternative account that explains *why*..." (p. 117).

Cognitive Heuristics. *Cognitive heuristics* are mental shortcuts humans use to assess situations and make decisions more efficiently. An example is recalling the last time or last fifty times you encountered a situation similar to the one you are in and making the same decisions. We assume the current situation is approximately the same as it was last time or times and do not invest resources to evaluate the details.

This typically works well and it usually allows us to invest time and effort in other cognitive work. However, if the situation turns into something different from what it was assumed to be, decision and performance mistakes occur. Mistakes committed due to reliance on an incorrect mental model are seen as evidence of a cognitive bias. Reliance on the wrong past situation, for example, is called Framing Bias. Framing Bias relates directly to cyber-aware mindset training. Research on Framing Bias suggests that when people lack experience with cyber-threats, they are unlikely to assess situations through cyber-aware glasses (Cornelissen and Werner 2014). Alternatively, it may be effective to train personnel about cognitive bias and how to avoid it. This training could help operators recognize the tendency to rely on past, pre-cyber-threat experiences that may no longer be relevant or useful. Further, training may reveal new ways to use cognitive heuristics advantageously and may help operators change or update their heuristics.

There is evidence to support the idea that training can be used to reduce cognitive-bias effects. Symborski et al. (2017) demonstrated that video game play that includes feedback and instruction about biases can be effective at reducing the effects of a variety of cognitive bias. Biases addressed by the video game training include the Fundamental Attribution Error (a judgment is made based on a behavior without considering the context and possibly history surrounding the behavior), Bias Blind Spot (we see other's thinking as biased but not our own), confirmation bias (a tendency to only attend to information that confirms our own views), Anchoring Bias (a tendency to anchor on the first option or possibility considered), Representativeness Heuristic (we base our assessment of probability on similarity to past situations without considering data such as actual occurrence rates), and Projection Bias (the tendency to think others share our priorities, attitudes, and beliefs). The research team found positive effects (measured using game performance) immediately after training and three months later.

Morewedge et al. (2015) compared the debiasing effects of sixty minutes spent performing the videogames used by Symborski et al. with thirty minutes spent watching an educational video that explained, demonstrated, and provided mitigations for each bias. Both training formats produced statistically significant reductions in biased thinking and improvements in knowledge about biases. For each of the six biases, the videogame led to improvements that were at least statistically equivalent to those produced by video watching; in most cases, however, improvements were greater for the videogame condition.

These two studies represent a relatively sparse research literature on how to decrease cognitive bias through training. Endsley (2018) remarks on this in her discussion of the role of cognitive bias in creating vulnerability to social engineering tactics used by cyber attackers, saying "far more research is needed" (p. 1091).

Skeptical Metacognition. Metacognition refers to higher order thinking (thinking about thinking) and active control over one's cognitive processes (e.g., Flavell 1979). Metacognition adapts cognitive work to the demands of a given work activity, environment, and goal set. Among other roles, it can guide our use of cognitive heuristics thereby reducing the likelihood of using a heuristic inappropriately (i.e., of cognitive bias; e.g., Mumford et al. 2007).

By means of training and experience, metacognition can be adapted to perform its oversight duties in a critical and skeptical way. This translates into metacognition that is always alert, questioning, and assessing one's own cognitive activities and performance. Skeptical metacognition can help operators recognize if they are discounting the risk of cyber attack on their system and when they may be walking into a social engineering trap.

Brand-Gruwel et al. (2005) recommend supporting metacognitive proficiency by using a cognitive apprenticeship training approach that helps to bring internal cognitive processes out into the open. Klein and Borders (2016) developed a technique called ShadowBox™ that accomplishes this goal of surfacing cognitive processes but without the time requirements that comes with a one-on-one apprenticeship. ShadowBox™ is a scenario-based technique designed to train people to think like experts in a target domain without requiring the real-time involvement of domain experts. Trainees are asked to rank and justify their rankings for sets of options given to them at decision points ot realistic scenarios. They subsequently compare their rankings and rationale to those of experts. The method has been used successfully to train military personnel to make decisions using a new mindset; specifically, using a good stranger mindset instead of the traditional security mindset (Klein et al., 2018). In a review of their ShadowBox™ evaluation work, Klein et al. (2018) credit the technique's ability to produce 'aha' moments; i.e., realizations that cause the trainee to reconsider their mindset and mental model. This is yet another aspect of cognitive work that requires more research. The authors note, "Clearly, there is a lot to learn about...how to help people make mindset shifts, how to measure mindset shifts, and how to differentiate mindsets from other cognitive processes" (p. 683). Likewise, very little research exists on the training of metacognition more generally.

Table 2 lists mindset training objectives that address the role played by knowledge structures, cognitive heuristics, and metacognition in achieving a cyber-aware mindset.

Table 2. Training objectives based on cognitive elements of mindset and mindset change

Mindset Training Objectives
Adapt or replace pre-cyber-warfare knowledge structures so that they contain cyber-defense knowledge (e.g., knowledge listed in Table 1).
Develop skeptical metacognition that supports vigilance and helps to protect against cognitive bias.
Understand how cognitive bias can impact cyber attack detection and recognition.
Know strategies and procedures for reducing the influence of cognitive bias on anomaly detection and response.

3.3 Problem Space Assessment

The problem space assessment highlighted the following constraints:

- Many organizations do not yet know or are still in the process of determining how system users should react if they become suspicious of an attack on their system or if an attack detection alert is triggered by their system.
- Many military organizations do not, in general, allow system users to take independent action outside of pre-defined procedures, limiting users' ability to investigate anomalies and glitches in their systems.
- Department of Defense (DoD) are required to annually complete a web-based cyber security awareness module. This training is focused on administrative functions and is designed to help system users learn to avoid phishing and other social engineering tactics. Similar training is available to personnel in the private sector. Training generally does not, however, prepare users to support cyber attack detection and response activities in their organizations. Nor does the annual DoD training address potential cyber events on weapons or command and control systems.
- Many organizations do not support high-fidelity training and practice for cyber attack detection and response (e.g., Line et al. 2014). Reasons vary across sectors but in at least some cases, concerns about permanent impacts on operational system function and unrecoverable disruption of expensive training exercises are cited (e.g., Wells 2019).

There are additionally decisions to be made about the level and accuracy of detail in light of the risk of producing a source of information that could benefit cyber adversaries.

3.4 Training Strategy

Iterative design activities, informed by literature review, target user interviews, and stakeholder discussions has led us through a series of design concepts. Throughout this evolution, each concept has reflected the conceptualization of learning as a process of adaptation, co-evolution, and emergence (Neville et al. 2019; Schraagen et al. 2008). They also reflect tradeoffs to address problem space constraints. We have arrived at a scenario-based video game concept in which players assume an adversary position and observe effects of their attack choices on simulated system users. Resources and a guide will help them understand at a high level possible attack goals, resources, tactics, challenges, timelines, and more. Using the support sources, trainees will plan attacks and play them out for points and to see what happens. Feedback will be a critical element and will follow the model of the ShadowBox^{TM} technique, which uses a comparison with experts' ranked choices and rationale as feedback.

Table 3 maps the current training strategy against awareness and mindset training objectives in Tables 1 and 2, respectively. Prototype training will be set in an office environment, with which most people are familiar with intent to adapt future versions to train users of operational systems. We hypothesize that awareness developed in an office computing environment will generalize easily to other work domains.

Table 3. Cyber-mindset training objectives mapped to training strategy elements

Training objectives	Training strategy elements
Cyber awareness training objectives - System baseline - Anomaly identification and reporting - Systems thinking	Support sources will provide visualizations and information about: - A target crew's level of vigilance, difficult-to-detect changes from baselines - Ways to confuse operators, delay responding, and disrupt reporting chains - Relationships among system elements, flows of data and communications, redundancies and other defenses
Mindset training objectives	- ShadowBox^{TM} feedback on cognitive aspects of work, including mindset - Scenarios that support the acquisition of integrated, contextualized knowledge structures - Guidance and practice at taking advantage of an operator's cognitive biases

4 Discussion

The cyber realm is complex on a number of counts; it is vast and largely invisible, a sea of highly interactive and dynamic elements, with pathways and associations that are changing continuously. Cyber-threat activity adds to this realm additional dynamics that significantly escalate the already-extremely-high baseline level complexity. Although cyber-threat activity impacts individuals, organizations, and societies in dramatic and crippling ways, its invisibility and complexity make it difficult to even begin to comprehend, much less defend against.

Yet, some amount of comprehension is necessary. The bad guys cannot be the only ones with cyber-attack knowledge. Our warfighters need at least a base-level incorporated into their knowledge structures. The training tool we are developing will allow us to vary the amount and detail of content to empirically evaluate the amount of cyber knowledge and comprehension required produce a mindset shift and engage operators as a layer of cyber defense.

To date, cyber training for system users predominantly teaches basic procedural information for avoiding being tricked into serving as a cyber-attack conduit. This type of training needs to be extended to empower system users to be active, educated participants in their organization's cyber protection. Training is needed that makes visible and graspable to non-cyber professionals the largely invisible and unknown realm of cyber threats and defense.

Recently, the Navy introduced simulation-based combat systems trainers that allow Sailors to practice their roles in integrated air and missile defense (IAMD) and anti-submarine warfare (ASW) on an isolated network of non-operational, training-designated shipboard systems. Using these high fidelity training systems, sailors are able to practice responding to and working around systems compromised by any number of causes, including cyber attack. These practice opportunities are effective at teaching what to do, i.e., procedural knowledge, but Sailors would benefit from training system features, accompanying instruction, or complementary training that teaches 'why' and 'how', i.e., conceptual knowledge. Conceptual knowledge enables the adaptive use of procedural knowledge across variations in complex work dynamics. Conceptual knowledge about cyber threats could help Sailors to correctly perceive, process, think about, and respond appropriately to information in our new cyber-vulnerable world.

We propose complementary training that could be taken prior to or in parallel with simulation-based practice and that will give operators a cyber-defense enriched knowledge base and adapt their metacognition and use of cognitive heuristics to the modern cyber-threat-pervasive environment. To spur these cognitive adaptations, we propose a holistic training strategy that uses scenarios, supplementary training resources, feedback about cognitive underpinnings of performance, and an adversarial perspective.

Future work will be focused on continued development of the training system and the development of training efficacy assessment plans. A key part of the training efficacy assessment work will be identifying meaningful measures of mindset adoption and impact. Typically, measures are aligned with training objectives. However, our primary interest is in direct measures of mindset change, versus of changes related to specific training objectives. In ShadowBox^{TM} studies, mindset change has been assessed by evaluating changes in trainees' choices and rationale relative to experts' choices and rationale. We plan to step outside the training system to assess transfer to a real office computing environment and generalizability to other types of work environment. In addition, following the development of mindset-change measures, we can conduct empirical work to assess the role of each training objective identified in this paper.

Future work is also needed to more thoroughly investigate a number of research needs highlighted by this effort. Research is needed, as examples, to better understand how to use training to reduce cognitive bias, how to train or otherwise foster metacognitive proficiency, how to help people make mindset shifts, and how to assess learner progress in these complex aspects of cognitive work.

Acknowledgements. We would like to acknowledge the significant contributions of Dr. Julie Marble and Alice Jackson of Johns Hopkins University Applied Physics Laboratory to the work described in this paper. This research was sponsored by the Office of Naval Research (ONR). The views expressed in this paper are those of the authors and do not represent the official views of the organizations with which they are affiliated.

References

Borko, H., Livingston, C.: Cognition and improvisation: differences in mathematics instruction by expert and novice teachers. Am. Educ. Res. J. **26**(4), 473–498 (1989)

Boulton, L.: Adaptive flexibility: examining the role of expertise in the decision making of authorized firearms officers during armed confrontation. J. Cogn. Eng. Decis. Making **10**(3), 291–308 (2016)

Brand-Gruwel, S., Wopereis, I., Vermetten, Y.: Information problem solving by experts and novices: analysis of a complex cognitive skill. Comput. Hum. Behav. **21**(3), 487–508 (2005)

Canham, M.: Human element in cyber. Presented at the Cyber TRAINsitions Workshop, 10–11 January, University of Central Florida Institution for Simulation and Training (2019)

Chi, M.T.H.: Commonsense conceptions of emergent processes: why some misconceptions are robust. J. Learn. Sci. **14**, 161–199 (2005)

Cornelissen, J.P., Werner, M.D.: Putting framing in perspective: a review of framing and frame analysis across the management and organizational literature. Acad. Manag. Ann. **8**(1), 181–235 (2014)

Crum, A.J., Salovey, P., Achor, S.: Rethinking stress: the role of mindsets in determining the stress response. J. Pers. Soc. Psychol. **104**(4), 716–733 (2013)

Cyber Intelligence Task Force: Cyber intelligence: Preparing today's talent for tomorrow's threats. Intelligence and National Security Alliance (INSA), September 2015

Endsley, M.: Combating information attacks in the age of the internet: new challenges for cognitive engineering. Hum. Factors **60**, 1081–1094 (2018)

Feltovich, P.J., Spiro, P.J., Coulson, R.L.: The nature of conceptual understanding in biomedicine: the deep structure of complex ideas and the development of misconceptions (Technical report no. 440). University of Illinois at Urbana-Champaign: Center for the Study of Reading (1988)

Flavell, J.H.: Metacognition and cognitive monitoring: a new area of cognitive–developmental inquiry. Am. Psychol. **34**(10), 906–911 (1979)

Greenberg, A.: How an entire nation became Russia's test lab for cyberwar. Wire Magazine, 20 June 2017. https://www.wired.com/. Accessed Feb 2019

Klein, G.A.: A recognition-primed decision (RPD) model of rapid decision making. In: Klein, G. A., Orasanu, J., Calderwood, R., Zsambok, C.E. (eds.) Decision Making in Action: Models and Methods, pp. 138–147. Ablex, Norwood (1993)

Klein, G., Baxter, H.C.: Cognitive transformation theory: contrasting cognitive and behavioral learning. In: Cohn, J.V., Schmorrow, D., Nicholson, D. (eds.) The PSI Handbook of Virtual Environments for Training and Education: Developments for the Military and Beyond. Learning, Requirements, and Metrics, vol. 1, pp. 50–64. Praeger Security International, Westport (2009)

Klein, G., Borders, J.: The ShadowBox approach to cognitive skills training: an empirical evaluation. J. Cogn. Eng. Decis. Making **10**(3), 268–280 (2016)

Klein, G., Borders, J., Newsome, E., Militello, L., Klein, H.A.: Cognitive skills training: lessons learned. Cogn. Technol. Work **20**(4), 681–687 (2018)

Klein, G., Moon, B., Hoffman, R.R.: Making sense of sensemaking 2: a macrocognitive model. IEEE Intell. Syst. **21**(5), 88–92 (2006)

Klein, G., Phillips, J.K., Rall, E., Peluso, D.A.: A data/frame theory of sensemaking. In: Hoffman, R.R. (ed.) Expertise Out of Context, pp. 113–158. Erlbaum, Mahwah (2007)

Lewandowsky, S., Ecker, U.K., Seifert, C.M., Schwarz, N., Cook, J.: Misinformation and its correction: continued influence and successful debiasing. Psychol. Sci. Public Interest **13**(3), 106–131 (2012)

Line, M.B., Zand, A., Stringhini, G., Kemmerer, R.: Targeted attacks against industrial control systems: is the power industry prepared? In: Proceedings of the 2nd Workshop on Smart Energy Grid Security, pp. 13–22. ACM, November 2014

Michener, J.: Beating the air-gap: how attackers can gain access to supposedly isolated systems. Energy Central, 24 August 2018. www.energycentral.com. Accessed Feb 2019

Morewedge, C.K., Yoon, H., Scopelliti, I., Symborski, C.W., Korris, J.H., Kassam, K.S.: Debiasing decisions: improved decision making with a single training intervention. Policy Insights Behav. Brain Sci. **2**(1), 129–140 (2015)

Mumford, M.D., Friedrich, T.L., Caughron, J.J., Byrne, C.L.: Leader cognition in real-world settings: how do leaders think about crises? Leadersh. Q. **18**(6), 515–543 (2007)

Neville, K.J., et al.: A complex cognitive skills acquisition framework. In: International Conference on Naturalistic Decision Making (2019, paper under review)

Neisser, U.: Cognition and Reality: Principles and Implications of Cognitive Psychology. W. H. Freeman, San Francisco (1976)

Paul, C.L., Whitley, K.: A taxonomy of cyber awareness questions for the user-centered design of cyber situation awareness. In: Marinos, L., Askoxylakis, I. (eds.) HAS 2013. LNCS, vol. 8030, pp. 145–154. Springer, Heidelberg (2013). https://doi.org/10.1007/978-3-642-39345-7_16

Schraagen, J.M., Klein, G., Hoffman, R.R.: The macrocognition framework of naturalistic decision making. In: Schraagen, J.M., Militello, L., Ormerod, T., Lipshitz, R. (eds.) Naturalistic Decision Making and Macrocognition, pp. 3–25. Ashgate, Aldershot (2008)

Smith, R.: Russian hackers reach U.S. utility control rooms, homeland security officials say. Wall Street J. (2018). www.wsj.com. Accessed Feb 2019

Symborski, C., Barton, M., Quinn, M.M., Korris, J.H., Kassam, K.S., Morewedge, C.K.: The design and development of serious games using iterative evaluation. Games Culture **12**(3), 252–268 (2017)

Trustwave: 2018 Trustwave global security report. Trustwave, Chicago (2018)

Verizon: 2018 Data Breach Investigations Report, 11th edn. Verizon (2018)

Wells, D.F.: Cyber training. Paper presented at the Cyber TRAINsitions Workshop, 10–11 February, University of Central Florida, Institution for Simulation and Training (2019)

Demonstrably Safe Self-replicating Manufacturing Systems

Banishing the Halting Problem—Organizational and Finite State Machine Control Paradigms

Eli M. Rabani[1](✉) and Lesley A. Perg[2]

[1] NanoCybernetics Corporation, Woodland Hills, CA, USA
rabani@nanocybernetics.net
[2] Department of Neurology, University of Minnesota, Minneapolis, MN, USA

Abstract. Programmable manufacturing systems capable of self-replication closely coupled with (and likewise capable of producing) energy conversion subsystems and environmental raw materials collection and processing subsystems (e.g. robotics) promise to revolutionize many aspects of technology and economy, particularly in conjunction with molecular manufacturing. The inherent ability of these technologies to self-amplify and scale offers vast advantages over conventional manufacturing paradigms, but if poorly designed or operated could pose unacceptable risks. To ensure that the benefits of these technologies, which include significantly improved feasibility of near-term restoration of preindustrial atmospheric CO_2 levels and ocean pH, environmental remediation, significant and rapid reduction in global poverty and widespread improvements in manufacturing, energy, medicine, agriculture, materials, communications and information technology, construction, infrastructure, transportation, aerospace, standard of living, and longevity, are not eclipsed by either public fears of nebulous catastrophe or actual consequential accidents, we propose safe design, operation and use paradigms. We discuss design of control and operational management paradigms that preclude uncontrolled replication, with emphasis on the comprehensibility of these safety measures in order to facilitate both clear analyzability and public acceptance of these technologies. Finite state machines are chosen for control of self-replicating systems because they are susceptible to comprehensive analysis (exhaustive enumeration of states and transition vectors, as well as analysis with established logic synthesis tools) with predictability more practical than with more complex Turing-complete control systems (cf. undecidability of the Halting Problem) [1]. Organizations must give unconditional priority to safety and do so transparently and auditably, with decision-makers and actors continuously evaluated systematically; some ramifications of this are discussed. Radical transparency likewise reduces the chances of misuse or abuse.

© Springer Nature Switzerland AG 2019
D. D. Schmorrow and C. M. Fidopiastis (Eds.): HCII 2019, LNAI 11580, pp. 312–330, 2019.
https://doi.org/10.1007/978-3-030-22419-6_22

Keywords: Self-replicating systems · Molecular nanotechnology · Finite state machines · Molecular assembler · Replicator safety · Exponential manufacturing · Nanoscale 3-D printing · Material proficiency · Reliability engineering · Environmental remediation · Climate remediation · Infrastructure · System-of-systems analysis · AI-boxing problem · Existential threats · Autopoiesis

1 Introduction

1.1 Premise

Emerging technologies offer great increases in our capabilities to confront both emerging and longstanding challenges, but must be implemented and deployed in ways that are demonstrably safe and do not create threats or problems commensurate with these enhanced capabilities. While artificial intelligence is widely seen as fitting this description, the same is more true for self-replicating manufacturing systems. Programmable, self-sufficient self-replicating manufacturing systems can potentially rapidly scale productive capabilities to globally significant extents, and will enable the restoration of preindustrial atmospheric CO_2 levels and ocean pH as well as universal provision of basic human needs at economies and in time-frames presently not generally thought feasible.

Heretofore, despite even theoretical interest showing feasibility of large-scale practical utility [2], self-replication of man-made productive systems not relying on significant pre-fabrication, let alone closure of the capability for self-sufficient self-replication from environmental raw materials and energy sources has been largely elusive. The significance of this is many-fold: self-replication of capital equipment (sometimes termed exponential manufacturing) which is capable of customized fabrication combines economies of scale with agility of production; ability to produce high-performance materials yields efficiencies in both materials utilization and energy use (both during production and end-use), and additionally offers options of materials substitution when desired. Direct utilization of abundant raw materials as inputs and close coupling with highly controlled means of collecting raw materials, and the capability of producing photovoltaics or other energy harvesting means enables physical self-sufficiency of these systems, facilitates shorter generation times and better economies. This does not, however, necessarily constitute autonomy, and we discuss how autonomy can be limited, controlled or completely avoided; as discussed herein, control over to complete information required to effect self replication is the most preferable hard limitation on autonomy of replication or action. When programmable self-replicating productive systems avail molecular nanofabrication (notably, molecular assemblers or matter compilers) this permits the efficient fabrication of defect-free materials either at the nanoscale or macroscale with improvements in quality, strength, durability (e.g. atomic smoothness minimizing wear) and other properties in addition to feasible miniaturization to the ultimate limits imposed by physics. In sum, this convergence of technologies can bring an end to the era of material scarcity and both offer and go beyond material and energy efficiency to enable an era of material proficiency.

Self-replication carries with it not only these and many other important beneficial prospects, but also risks to be avoided, particularly since the technology can inherently self-amplify. Here we show that with the establishment of these capabilities, modes of control and restriction of information can be reliable limitations on properly designed self-replicating systems. Autonomy of action for such systems is limited to constrained subsystems or specified functions, with at least an external input of information required for replication after a specified number of generations (e.g. after n = 10 to 16 generations from an initial seed, with n decreasing according to a schedule with each set of generations, finally to 1 before potential risk effect-size becomes significant.) Beyond error-correction and detection (esp. through hashes—in properly designed systems, errors will never lead to anything like mutations, so there can never be anything like evolution of properly designed systems) this can be accomplished by enforcing (by design) a distinction between the program a parent replicator follows and that which it builds-in to progeny systems: the parent's program can be execute-only (no-copy, no-write) while the content it provides to progeny systems can for it be read-only (i.e. unmodifiable) and mechanistically constrained to be read a limited number of times (e.g. inability to autonomously rewind a tape or reset an address counter), such that only a specified number of progeny can be produced, themselves with limitations set correspondingly lower such that the number of generations is limited and the maximum population of the final generation is constrained. Once constructed, systems are mechanically unable to effect increase of these values—such limits may even be entailed structurally. As an additional mechanism, since there is a necessary mapping between low-level instruction encoding and fabricator design details, programs need not be intelligible to any generation or progeny individual other than that for which it is intended; when read-count-limited, this imposes a strict limit on the population size and composition (capabilities, including inclusive fecundity) in each generation. Of course, designing controls or limits on resource or energy utilization can be further safety measures or check-points.

While it might be easy to dismiss out-of-control replication scenarios as a mere trope of science fiction, first, since this is well-established in cultural consciousness, public acceptance of these powerful technologies will demand that such concerns be properly addressed and allayed, and, second, there are natural phenomena such as invasive species, cancer and viral infection which show these concerns to be rooted not only in fiction. These concerns thus set important requirements on acceptable large-scale self-replicating manufacturing systems. Additionally, commonplace experiences of computer crashes and frequent news of information-security breaches justify skepticism until it can be shown that distinctions make these concerns inapplicable.

Here, limitation of system time evolution (in the sense of physical dynamics) through restriction to a simplified state-space is proposed to preclude chaotic dynamics both from developing or from complicating understanding of these systems.

To address these requirements, control systems are made minimally complex to facilitate clear and concise analysis which is relatable to non-experts to demonstrate safety, and systems are operated in contexts which include secure human control at key decision points such as the decision to replicate above a threshold effect size, with information constraints additionally precluding operative progeny systems arising by any other pathway. Organizational contexts for human operators are also discussed.

This paper concerns how we can design such systems and organize control thereof to preclude any out-of-control replication scenarios and other accident risks; intentional misuse of advanced technologies represents a separate problem to be treated elsewhere. Finite state machines are selected as a minimally complex and analytically tractable type of control system, and, we further restrict these to the subset having simple transition rules; thus, transitions between states can be seen as a linear program that only include advance to a next functional state (absolutely or conditionally on inputs or success of the step controlled by the present state), with exceptions being a step-error-recovery operation (with repetition number limits) where acceptable, and a failure state which initiates entry into a failure condition which causes the system to initiate a "safest-course of action" sequence or a pause until human control instructs otherwise. This paradigm restricts control systems to those to which the halting problem does not fully apply. While the unpredictability of computer-glitches is an obvious frame of reference for the larger problem we seek to treat, restriction to finite-state machines yields systems wherein the domain of valid states and control sequences (from program and inputs) is restricted to those for which the halting problem is reduced to a simplified and tractable form. The operations of the subject systems are controlled kinematically (at a low level, with desired kinematic vectors mapping simply to state vectors) by signals produced by state machines to operate actuators, and sensors provide inputs to the state machine concerning the results of operations and the environment, with input states making up part of the machine state. While additional inputs might in some instances include outputs from other computational systems of different type performing functions such as image recognition or data analysis, these would only be admitted to system design in this paradigm if well-characterized reliability for a use-case can be established and non-high-confidence outputs or errors are reliably detectable to enable transition to a failure state, and where possible involve heterogeneous redundancy such that coincident systematic errors are made unlikely, but would also only be admitted if used in such a way that their errors lead only to self-limiting conditions, i.e. non-replication until resolved, rather than leading to erroneous replication-permissive states; for these cases, design is similarly restricted to the subspace for which risk analysis is straightforward. Because the number of machine states and accessible or traversable transition pathways can be prespecified or readily determined, the state space and all accessible state-trajectories can be enumerated exhaustively.

Although human error is a frequent cause of accidents, inclusion of humans in all significant decision-points to limit replication can be an important pillar of safe system design and part of the answer to public safety concerns, provided control modalities meaningfully involve adequately informed human oversight, and the organizational structures in which humans involved operate as well as the interfaces they use facilitate rather than undermine the role they serve. For large-scale deployments, where both the economic advantages gained and the extent of damage potentially risked in the event of accident could be large, redundancy of human operators is justified. Particularly in this context, where proper design would make the actual need for intervention unlikely, it will be important to give operators a mix of real and simulation-accident data streams such that training is intrinsically built-in to operating procedures and inattention is both largely prevented and made detectable through redundant cross-checking.

Coda: An open question is the utility and desirability of augmenting human oversight with artificial intelligence, and modes for maximizing these while precluding added risk. Although in this application we advise against on-line use of artificial general intelligence (AGI) or superintelligence in any capacity other than advisory, if at all, we propose strategies or tactics applicable to the AI-boxing problem in general. Instead, for operations for the present use case, specialized intelligence (artificial idiot savant, AIS) to identify out-of-band parameters or unusual conditions or events are recommended. In particular, executive function is constrained and both learning and operation are observational rather than embodied (i.e. not able to act on surroundings). AGI might instead have a role in simulations and especially as an adversary or generator for challenge scenarios (deep-fake risk scenarios).

1.2   Introduction to Molecular Nanotechnology

The idea of self-replicating machines (which in modern form traces to von Neuman [3], computationally implemented in cellular automata with better capability for parallel operation more recently by Pesano [4] and reviewed by Sipper [5]) and the idea of approaching the physical limits of miniaturization (which traces at least to Richard P. Feynman's 1959 *There's Plenty of Room at The Bottom: An Invitation to Enter a New Field of Physics* Lecture [6],) in combination were analyzed rigorously by Drexler [7–9], showing these to be within the bounds of physical laws and that exponential fabrication would enable production of high-performance materials and devices in large quantity and with large effect. Physical implementations of macroscale self-replicating systems are found in [10–12].

Molecular nanotechnology is principally concerned with building things precisely from individual molecules or atoms; miniaturization is inherent and ultimate, and when reliable mechanosynthetic operations can be employed, precise, uniformly defect-free products become feasible. These characteristics enable the fabrication of high-performance materials, and because the capital equipment for this is both self-replicating and the product of self-replication, these may be realized at highly advantageous economies, with economies similar to those of software creation—design or engineering and operation rather than conventional batch or sequential production are the principal costs.

One of us (EMR) has developed a range of nanoscale fabrication and assembly methods for a range of materials from abundant raw materials found in the environment, suitable for the production of programmable self-replicating manufacturing systems, details of which were also disclosed [13]; see also references therein. Earlier publications by others [14–16] have treated theoretical aspects of mechanosynthesis.

An important class of fabrication operations utilized are convergent fabrication or assembly [17, 18], which in contrast to additive fabrication by accretion of infinitesimals of material (e.g. epitaxy) combines increasingly larger intermediates in successive steps to yield larger products; in a step of the stereotypical case, one block is surface-bonded to another similar block to yield an intermediate twice the size of inputs to that step, this is particularly facile for halite materials (see, for example, Fig. 1A and N.i–vii for examples of convergent halite fabrication [13]). In such cases, addition can occur with the help of alignment means (surfaces or rods along which precursors and/or tools are slid, for example) to ensure precise registry, and products of successful addition are readily distinguished (e.g. by twofold difference in size), as is transfer from one manipulator to another determined by motions or bond-strength). In all cases, high operation reliability to afford low defect rates and uninterrupted operation are important design criteria, as is facile measurability of products.

It is important to note that while nanosystems are effectively in practice necessarily the products of self-relicating nanosystems (at least in any significant numbers), not all nanosystems or nanodevices are necessarily self-replicating, and in fact most would instead be end-effector devices or systems. Thus, the present focus on risks associated with self-replication do not directly pertain to all nanodevices or nanosystems nor all uses thereof.

1.3   Prior Treatments of the Problem

A mathematical analysis of the physical limits on out-of-control self-replication by self-replicating machines which consume biomass or compete with biological life for resources such as carbon and sunlight was performed by Freitas [19], which found that in various scenarios uncontrolled replication could reach global proportions with catastrophic ecological consequences in short timeframes if not checked, but concluded that with prior preparation, detection and response strategies could arrest the offending replication processes. However, some detection strategies involve such measures as examining every cell of every organism periodically for potential replicator threats; while this might be appropriate if necessary during an active threat situation, as an ongoing preventive measure for a speculative threat, the justification of such extensive and invasive monitoring regime seems to us problematic in multiple regards, including the fact that camouflage or evasion strategies could complicate detection. In general, while the analysis of [19] is an important point-of-reference, as an analysis of practical threats it is limited by restriction to unsophisticated replicators arising either through profound carelessness (which we discuss the avoidance of here and is addressed in [20]) or malicious intent but only the sophistication of a script-kiddie exploit.

The Foresight Institute has considered the problem of avoiding catastrophic self-replication and various abuses of these technologies and produced a policy proposal. [20] These include a class distinction which we also make, autonomy of self-replication, observe the same logical prerequisites of material, energy and information, and note:

> [...] molecular nanotechnology not specifically developed for manufacturing could be imple-
> mented as *non autonomous replicating systems* that have many layers of security controls and
> designed-in physical limitations. This class of system could potentially be used under controlled
> circumstances for nanomedicine, environmental monitoring, and specialized security applica-
> tions. There are good reasons to believe that when designed and operated by responsible
> organizations with the appropriate quality control, these non autonomous systems could be
> made arbitrarily safe to operate.

with which we agree, but further advocate the addition of such applications as climate restoration, environmental remediation, agriculture, and water purification, to name a few important cases with ecological, humanitarian, and geopolitical significance, and which themselves relate to real rather than hypothetical crises. Depending on application, distinction according to whether the fabrication/assembly/replication subsystems are motile or sessile affects risk potential and the range of available incident responses.

1.4 Introduction to Reliability Engineering in Organizational Context

When searching for risk management strategies, often the focus is on technical solutions, due to the way risks and failures are traditionally analyzed. However, the actual root of failures in critical engineering systems is often organizational errors. This can be introduced by analyzing linkages between the probability of component failures and relevant features of an organization, allowing crude estimates of the benefits of specific organizational improvements.

One example of an accumulation of organizational problems was the Challenger accident: miscommunication of technical uncertainties, failure to use information from past near-misses and the error in weighing the tradeoffs of safety and maintaining schedule all contributed to the error in judgment contributing to the technical failure of the O-ring in low temperatures [21]. While studies of this story and other major failures, such as Chernobyl, Three Mile Island, the *Exxon Valdez* accident, and the British Petroleum leakage in the Gulf of Mexico, are instructive, they only provide a narrow slice of potential failures. A systematic analysis of these spectacular failures, combined with partial failures and near-misses, are necessary to begin to classify a probability risk analysis focused on the risk of organizational error. For new emerging technologies such as self-replication and AI, where failure could be compoundingly catastrophic, it is especially imperative to take note of management and organizational errors in other fields to develop the safest possible management of these technologies.

Traditional probability risk analysis models focus on technological failure, specifically the probabilities of initiating events (accidents and overloads), human errors, and the failures of components from a technological standpoint. More recent risk analyses attempt to quantify the effect of management and organizational errors. Examples of organizational factors contributing to management errors include excessive time pressures, failure to monitor hazard signals, and temptation to not provide downtime to properly inspect and refurbish technology. Some poor decisions may be simple human errors, but more often they are caused by the rules and goals set by the corporation. Five particular management problems seem to be at the root of judgment errors: (1) time pressures; (2) observation of warnings of deterioration and signals of malfunctions; (3) design of an incentive system to handle properly the tradeoffs between productivity and safety; (4) learning in a changing environment where there are few incentives to disclose mistakes; and (5) communication and processing of uncertainties [22]. We note that the majority of these management problems involve balancing profitability with safety risks, and that with technologies such as self-replication with long-term safety implications, the organizational emphasis must lie strongly with safety.

Coda: There is a certain analogy between the failure-modes of organizations at fulfilling their stated aims in general and maximizing safety in particular, and the halting problem; here we refer to this as the organizational halting problem. For present purposes, we note that high transparency, accountability, deconfliction of practical incentives, redundancy and performance checks integrated into ordinary operations can minimize risks from these sources. These must override all other organizational imperatives of any organization which would have any direct or indirect role in the design, construction, operation or use of these technologies, or safety guarantees are undermined. Similarly it is proposed that at least any individual or entity given access to control of the technology, the means to independently bootstrap like technology or to exert power or coercion over any of the former be subject to a regime of radical transparency via sousveillance, including at any pertinent human-machine interfaces, with exceptions only to preserve cryptographic functions. While the broad-ranging implications of these points are acknowledged, the consequences of sub-optimal safety regimes could be broader, and so this is justified.

2 Finite State Machines and Subsystems

The complexity (in bits) of a finite-state machine with s states, v transition vectors between them possibly responding to i possible collective input states may for purposes of the present analysis be bounded as:

$$C_{FSM} = log_2(s) + log_2(v) + log_2(i) \tag{1}$$

which is the complexity which must be elaborated in analysis of such a subsystem. (In reality, the complexity we must address is that of the machine and its environment). When states of a subsystem S can be classed into k functional categories relevant to the overall system, the complexity of that subsystem's functional state-space in higher-level system analysis is

$$C_S = log_2(k_S) \tag{2}$$

and the complexity (in bits) of a system comprising j uncorrelated subsystems is

$$C = \sum_j C_j \tag{3}$$

Degrees of freedom are then taken to be

$$D_f = C + log_2(E_S) \tag{4}$$

for E environmental states. Thus D is the pertinent scope of analysis. This means that under controlled conditions (such as a factory setting), C dominates, but in, for example, in environmental or climate restoration applications, the second term is of critical import.

Figure 1 depicts an exemplary state machine (state diagram on right) controlling a convergent fabrication subsystem (the topology of which is shown on the left), the states of which in general may include control variables controlling positions of manipulators, activation of pumps, states of relays, signals from sensors, etc.,.

Programmability occurs through transition rules, i.e. rules for the succession of numerical states according to stored values for each transition vector, which, for example, may determine the translation of a manipulator. Stored values may be effectively hard-wired in the form of digital logic, or may be stored in a memory which may either be read-only or rewritable but would in the latter case only be writable with inputs from external control via a secure channel (authenticated and encrypted), not under autonomous control. Conditional branches are minimized.

As programming occurs through the informational content of transition vectors, and systems are not autonomous, a few different control paradigms are possible. These involve communication of instructions or information to effector systems. One simple case is for systems to have only a limited queue of stored transition vectors, and limitation of replication is affected by not communicating further instructions. (This is an extension of the broadcast architecture, see [23] and Sect. 16.3.2(a) of [9]). However, especially for remote deployments it may be preferable to instead communicate only what is necessary to enable a specific operational sequence segment. In this case, systems contain programs or sets of programs in encrypted form and have only limited memory into which decrypted information may be written; a decryption device generates a synchronizing keystream (using a time-synchronized cryptographic random number generator having a state known to operators), which generates nonce values used in reception of communications and decryption of stored instructions: a key is transmitted in encrypted form, decrypted using a first nonce as a key, and passed to a function (which may be as simple as an XOR) to yield the key to decrypt an operational sequence segment (instruction sequence or transition vector values). The ultimate decryption key is only stored in a register until decryption of the respective cyphertext is complete and then overwritten, and cannot otherwise be read. Thus, instructions do not ever necessarily need to be data-in-flight, control communication bandwidth may be minimized, keys and keystreams are never available outside of the system or operational control sites, replay or other cryptanalytic attacks are precluded, and the systems themselves only have limited cleartext instructions at any given time. This case may be termed the cryptographically restricted program access control mode.

We note that because advanced molecular nanotechnology is likely to increase the rate of progress towards realizing any viable quantum computing scheme, quantum cryptanalysis must be considered part of the threat-model. Quantum resistant cryptographic authenticated encryption, or alternatively quantum cryptographic authenticated encryption would be appropriate to protect the integrity of control communications and telemetry to preclude corresponding attack vectors.

An alternative method to analyze and quantitate the complexity of such control systems is that of cyclomatic complexity, in which retry cycles employed here each constitute a linearly independent cycle. This likewise reveals the simplicity of this machine control paradigm.

The foregoing demonstrate that necessary control functionalities can be implemented is simple, completely enumerable state-spaces. While other modalities are certainly possible, the essential point is to argue that simplicity is a critical design criterion for both comprehensibility and tractable analyzability, whereby subtle design errors are more easily avoided and open review is facilitated.

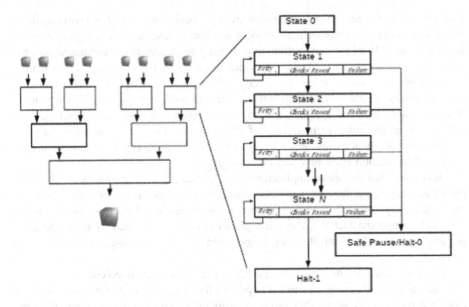

Fig. 1. Schematic view of a convergent fabrication system composed of subsystems each for performing unit operations together implementing fabrication according to a design; each subsystem is controlled by an individual finite state machine having the state diagram shown. Each operation effectuated by an associated state may include at most n retry transitions (e.g. determined by sensors not detecting desired product formation and transfer) before entering a failure condition; wait-states (w) for synchronization necessitated e.g. by retries in other coupled subsystems (not shown). Halt-1 represents successful task completion; halt of this subsystem will necessarily occur in $\leq (nN + w)$ transitions.

As an aside, we note that due to their simplicity, the choice of control by state-machines likely minimizes the quantity of material required to implement control functionality. Thus inappropriate design efforts would not gain reduced replication times by circumventing this aspect of control architecture.

3   Connexion with Autopoiesis

In contrast to autopoiesis, despite their ability to gather their own resources, harvest energy and self-replicate, the programmable systems discussed here are *cold, dead clockworks* which are the product of engineered design; they do not arise or continue to operate spontaneously, they lack informatic self-sufficiency by design, lack complete defenses by design (despite inherent materials strength), have no knowledge, intentionality or motivation, no inherent survival instinct, fear or even drive to reproduce, or any drive at all beyond what is in their instruction queues; although by design they require little maintenance, they do not maintain themselves unless instructed (i.e. lack involuntary homeostasis); they can pause indefinitely without the kinds of complications of arresting and resuming metabolism faced by biological life. The machine-states which would represent sensory-input data are not a basis for any higher-level cognition but at most initiate, enable or prevent state machine transitions—although cybernetic loops may be utilized for certain functions, usually responses to sensor inputs are little more than reflexes. Further, designs preclude mutation and hence evolution is neither their origin nor directly involved in any refinements or adaptations (evolutionary techniques might be used in simulations as a design methodology, but what is learned thereby is abstracted and applied explicitly by designers, only once fully analyzed and understood in the subsequent design of physical systems, which never themselves undergo any physical evolution), so these aspects of autopoiesis are likewise absent. Here, self-replication itself involves no emergence.

(It is noted that in various applications, self-replicating systems discussed here may have subsystems, e.g. for raw materials collection and transport, or for remote sensing, etc., which may depart from various aspects of the foregoing characterizations, but this does not change the fact that the self-replication of these systems or their operation departs significantly from the thrust and spirit of what is usually described as autopoiesis).

These systems are, however, of interest to the study of autopeoiesis because they represent a minimal core of functional requisites for autopoiesis—the minimal viability case, *information excluded*, that physical artificial autopoiesis must accomplish but would also have to go beyond. Natural autopoeisis had to reach that point from pre-existing concentration and energy gradients through thermodynamics, statistical mechanics, and self-organizing chemistry culminating in the molecular evolution that afforded the metabolically supported replication of molecularly encoded information ultimately sufficient to specify all required complex components including the ensemble of catalysts which produce them and organize their structure, and including the translation apparatus for that information; it was also necessary that the complex structures operated in their intracellular or external environments to support the functions and behaviors of the organism which enable reproduction. (For an example of early aspects of chemical evolution involving catalysis in confined chambers occurring in mica yielding increased effective local concentrations and orienting effects, see [24]; while remarkable, these represent only the very beginnings of what is required). Organized behavior, even for simple unicellular life requires regulated structures effecting sense-response loops (e.g. bacterial operons) in order to be flexible yet efficient.

We note the distinction made by Sipper [5] between self-replication and reproduction:

> Replication is an ontogenetic, that is, developmental process, involving no genetic operators, resulting in an exact duplicate of the parent organism. Reproduction, on the other hand, is a phylogenetic, that is, evolutionary process, involving genetic operators such as crossover and mutation, thereby giving rise to variety and ultimately to evolution.

If subsets of the foregoing distinctions were dropped, the present self-replicating systems could be regarded as the edge-case between allopoietic and autopoietic systems: they are no different from allopoietic systems but are incidentally capable of fabricating and assembling copies of themselves. While this property is critical to enabling their full power and utility, it is made possible through simplicity of design rather than complexity or emergence. It is interesting to note that alternation-of-generation, an evolutionary strategy adopted by various life-forms, in a certain sense likewise fits the category represented by this edge case, and similarly, the production of nanoscale self-replicating systems is facilitated by utilizing two or more materials [10] even though this superficially would seem to multiply the development efforts required.

4   Organisational Context for Safe Control Paradigm

Large-scale application of technology in the pursuit of goals by definition aims at effecting changes through action, usually a coordinated sequence of which are required on a routine basis under multiple practical constraints set by goals of an organization. Safety procedures, by contrast, are the procedures and constraints directed towards making adverse events rare—the more effective these are, the more remote respective risks begin to seem.

Application of self-replicating manufacturing capability to large scale problems superlatively augments the capacity to reach large-scale goals, but similarly increases the potential scale of risks if safety procedures fail. These facts entail that some of the economic resource savings from supply, labor, capital equipment, physical manufacturing and deployment be allocated to the organizational aspects of operational safety.

Organizational resources are allocated according to the scale of potential risks to enable redundancy of observation. Beyond reducing the likelihood of erroneous action due to failures to observe hazards, this excess capacity enables integration of observation quality assessment (as well as training) by mixing simulated hazard events into workloads on a routine basis while still maintaining redundancy of observation of actual observations. This type of intermediately frequent challenge will likely psychologically promote attentiveness, since it contrasts with watching for something improbable. Simulated hazards in data streams can be produced by the adversary networks of generative adversarial networks (GANs) [25–27].

For purposes of analysis, the physical system to be controlled safely is understood through a construction of system as machine + environment, and the enterprise system is understood through a construction of system as machine + organization + environment. Higher order effects may be possible because these system can (sometimes are intended to) modify their environment (which itself may be a nonlinear system)—and since this is a reiterative process, the potential for emergent dynamics which are difficult to predict cannot be excluded. Especially for this reason, monitoring must at least reliably detect whether operating assumptions violated in the process. Narrow-AI may in fact be well suited to augmenting human control in the following way: since recognition of stereotyped patterns is a well-defined task at which at least some artificial neural-networks perform well (e.g. pattern recognition [28]), patterns not recognized by a well-trained ANN can be highlighted for human operators.

Discriminator networks from GANs may also be used to highlight hazards. Nonetheless, where supervision at all relies on the foregoing or other modes of augmentation, human operators must utilize interfaces and procedures cognitively engineered to prevent them from becoming inured to simply relying on what automated analysis brings to their attention—this would erode any actual safety enhancement the inclusion of human operators is intended to provide.

Human oversight of critical decisions is itself insufficient; to be of much use, it must be *adequately informed and vigilantly critical*. If, for instance, inadequate monitoring of system surroundings causes operators to be oblivious to pertinent local circumstances, avoidable harm to bystanders, wildlife or the environment could occur through the operation, growth or replication of systems.

Competing organizational imperatives or external pressures applied to them can undermine adherence to safety procedures. A classic example is the Challenger disaster: despite a well-developed safety culture, political motivations led the Reagan Administration to pressure NASA and in turn a subcontractor to adhere to a launch schedule that necessitated well founded objections on the grounds of safety be overridden [29–31], with well-known consequences. If organizations responsible for operating self-replicating systems with any large-scale risk potential if mismanaged are susceptible to such pressure and lack means of contemporaneous redress, safety and confidence in safety may be undermined. At least in open societies, transparency is one countervailing factor that may mitigate such organizational risks.

Beyond generic organizational dysfunction, related pitfalls include other authority-derived dictates, e.g. a regime determined to avail itself of offensive weapon systems based on these technologies will cycle through researchers who are unwilling or unable to deliver what is required until presented with results that appear to satisfy their mandates, whether or not those who deliver are actually capable of producing systems that are not inherently unsafe (even to those regimes) and have any probability of functioning as desired in the unfortunate event of actual use; rather there is a selective advantage for those who do not appreciate the daunting complexities which would be involved but are eager to pursue whatever inducements or respond to whatever fears (such as coercion) may be applied to motivate them, producing ostensibly satisfactory results which are more likely epistemically opaque, though not obviously so to the uncritical. This is akin to the gambler's ruin problem, where even in cases with favorable odds, the structure of the game entails a random walk which eventually finds its way off a cliff. But the consequences there extend beyond the fortunes of the gambler, and so there is a general interest in avoiding such scenarios to begin with.

5 Simulation, Boxing and Abstraction

Before any physical systems are used or deployed at scale, for multiple reasons, extensive simulations of functionality and operation, including participation of human operators are appropriate, necessary and advantageous. Simulations permit many approaches which would be inappropriate in real-world application to be studied so that useful results can be abstracted and applied in appropriate form in real systems. Prime examples of these include: evolutionary design methodologies which may yield useful solutions to challenging problems which can be translated for implementation in non-evolving systems; and, involvement of ANI or AGI in simulated operational problem-solving, with results informing design or programming of non-intelligent systems. In these cases, there is a strict decoupling of anything capable of emergence or analytically refractory complexity and actual physical self-replicating systems. Rather than empowering inscrutable processes with control of self-replicating systems, this approach passages abstracted results through human design processes, which in the present paradigm would further be peer reviewed.

A similar approach can be applied to AI containment, also known as the AI-boxing problem (see, for example, [32]). Earlier proposals exist to confine potentially dangerous AI to simulations or virtual environments, although assertions have been made that this is an insoluble problem.

While a more comprehensive treatment of the topic is beyond the scope of this paper, we briefly list a few techniques which may be brought to bear to the boxing problem. The motivation for this is that if AGI (rather than NAI/AIS) were ever to be used in such applications (and the purpose here is not to advocate this but to list items pertinent to analysis of this possibility), even if only in simulation, it could not be kept ignorant of these technologies, so escape would be a particular matter for concern (so in fact the groundwork for such research would best be done in a simulation with modified physics and misleading training data for understanding of the technology, and assessment of facility for learning different [also counterfeit] physical principles would be pertinent). These include: multiple confinement layers or motes with simulated escape environments enabling study of escape behavior; combinatorial mazes along more available escape routes...especially with execution traps; avoiding hardware vulnerabilities especially with emulated but not reconfigurable hardware, but detecting attempts to exploit emulated bugs; organizing simulation trials similar to metadynamics..especially parameterized with differential subject-matter training set content, including umbrella dynamics to study what promotes escape efforts; pause, clone and probe analyses of modules or subspaces (instrumentation of the AI system, in analogy, electrocorticography, evoked potentials mapping, cortical stimulation mapping experiments, etc.); adversarial construction: captive and jailer AIs.

6 Discussion

6.1 Lessons Past and Forward Formulation

Because of the sweeping nature of the changes these technologies enable, it is appropriate to take care to not recapitulate past errors in the course of technological evolution or of human affairs in the development, implementation and use of these powerful technologies.

The model of economic parsimony—highly justifiable in the regime of material scarcity—favors incremental additions to what exists, and so new technologies frequently are adapted to the proximal constraints imposed by what came before them. Considerations such as first mover advantage (e.g. Facebook's motto, "move fast and break things") and marginal profitability encourage cutting of corners—and such tactics are not foreign to the course of successful business. However, cybersecurity as an afterthought and infrastructure riven with vulnerabilities have resulted (thus existing infrastructure would be wholly inappropriate for replication-control). While the significance of risks which now obtain due to this legacy is increasingly appreciated, it would be of another order entirely for such risks and practices to be carried forward into the realm of self-amplifying technologies.

Self-replicating manufacturing efficiently utilizing abundant raw materials can break the foregoing premise of scarcity, enabling a regime of material proficiency wherein enhanced capabilities are realized at better efficiencies and economies. This changes fundamental assumptions underlying the structure and behavior of contemporary organizations and society.

Much of our world has been shaped by competition, in various physical, historical and contemporary forms—and much of this competition has been competition for or related to material resources. While it would plainly be an error to consider all conflict to be reducible to conflict over resources, resource competition and population pressure on resources are recurring contexts for many conflicts (including both wars and genocides). This history has created both cultural enmities which have not yet been transcended as well as fears predicated on them, and the latter can themselves be sufficient potentiating factors for hostilities. Military conflict has long been a motive force in the development of technology (in extreme form, this idea was enunciated by Virilio [33]) and technology has often been decisive in military outcomes. Military observers and participants among others have recognized the potential significance of nanotechnology and self-replication. [34–36] With the advent of self-replicating productive capabilities, having the safety considerations outlined here, should this class of technology be offensively weaponized, this symbiotic progression could become untenably fraught, as follows:

Question: How do you defeat a militarily superior adversary?
Answer: Through unpredictability.

Should any actors ever pursue that course of action—which in particular would favor transgressing controls such as those set forth here—all predictions become inherently questionable. In essence, this becomes a theoretical game in which whatever would be ruled out of bounds becomes precisely advantageous, and notably, shifts in balance of forces effected through aggressive replication rates could be decisive. Similarly, one way of increasing replication rates is to minimize system extent, such as through the omission of components or subsystems necessitated only for safety. A desperate belligerent might resort to availing rather than preventing evolution of autonomous self-replicating offensive weapon systems under control of AGI, which would defy predictability in multiple dimensions. Further, control of military operations in such an environment would require rapid decisions and action in response to data streams exceeding what humans could ever be meaningfully capable of apprehending, necessitating automation, whether conventionally programmed or through machine learning or artificial intelligence. The complexities entailed again preclude analytic predictability, and since empirical testing necessitates incomplete surveys of assumed parameter spaces (presumably adversary strategies would likely aim to create unanticipated scenarios), there can be no certainty that any such system will perform as required in all or even most cases.

The clearest and most constructive path to avoiding the nebulous risks associated with direct conflicts between nanotechnological powers (risks which would be more likely than accidental runaway replication scenarios but potentially at least as catastrophic) is to rigorously focus development of these technologies on instead ameliorating those issues which were the context for so much historical conflict in the first place: promoting material well-being (freedom from want) and freedom from fear. [37] The technological foundations enabling these may soon be at hand. Even where material factors are not primary motivators for conflict, when populations widely enjoy well-being, there is less fertile soil for leaders who would endanger that with avoidable conflict.

Alternative answer: Through subversion.

At least in recent years we note (without supporting any particular conclusion as to the legitimacy of any particular purportedly freely elected government or the fairness of any particular election) that there have been recurrent questions as to the integrity of numerous elections which are less implausible than would be preferred; here we use the U.S. as an example of this in an ostensibly well-developed democracy, but the phenomenon is more broad. According to U.S. intelligence analysis, in addition to intrusions into state and local elections information technology infrastructure, influence and disinformation campaigns mediated by social and other media and involving bots appear to have been conducted by state actors [38]. Private parties have engaged in efforts similar to the latter, including campaigns researching the cost and effectiveness of such methods [39, 40]. The increasing sophistication of DeepFakes portends an intensification of the challenges such tactics may pose.

This is pertinent here on three levels: First, an adversary may attempt to gain competitive advantage by retarding the development of these critical technologies by competing entities by interfering in political processes to, for example, promote bans, particularly through alarmism, to which the background of cultural knowledge noted above readily lends itself. Second, promoting unnecessarily restrictive regulations on development or implementation could serve a similar purpose, particularly if they are national—this is another reason that to have any actual usefulness, appropriate regulations must be universally effective. Third, interference with organizational components of the present safety paradigm, for example disinformation at the level of monitoring of the technology, particularly deep-fake misrepresentation of any part of the process to either create hysteria or manipulate or degrade effectiveness stated purpose of safety (restriction of the associated infrastructure to this single purpose and only the functions of surveillance, control, primary audit data storage to minimize attack surface is essential, and protocols for verifiable independent monitoring are likewise essential), or for example, interfering with organizational components or contexts themselves via traditional espionage operations.

Beyond that, subversion (by organizational, informational or physical means) of physical defenses such as those proposed in [41] could transform these defenses into weapons aimed at their original masters. Note that [41] do acknowledge these threat vectors, and reach similar conclusions to those here as to the importance of transparency reducing opportunity for misuse.

6.2   Additional Comments

It is also important to note that while the hypothetical risk-space is vast, appropriate use could normally be restricted in the case of systems with self-replicating capacity to having vulnerabilities by design as a final safety measure. Of course, this is not an effective answer to issues of misuse or abuse, which are separate matters to be addressed elsewhere. The specific nature of designed vulnerabilities should enable arrest, deactivation or destruction of errant systems by operators or authorities, but should resist denial-of-service attacks or exploits.

6.3   Payoff Matrix

While the consequences of the technology, what is done with it and how it is regulated involve many different possible combinations and gradations, stark extremes are to be found among the more likely combinations of implementation versus modes of regulation and priorities, simplified in the following ansatz payoff matrix:

	Appropriate Universal Regulation, Aligned Imperatives	Irregular and Variable Regulation, status quo imperatives
Safe Design & Organizational Structure	Highly & Widely Beneficial, Fair Predictability or better	Variable Benefit, Uncertain but Possibly High Risks
Offensive Weaponization	High Risk/Variable Predictability/Questionable Rewards	Unpredictable/Unstable/High Risk/Inequitable Rewards, if any

While the challenges involved in achieving a universal, effective but just regulatory regime are not to be underestimated, the diffusion of profound benefits made available and the extreme-to-existential risks (both those posed by inappropriate use or implementation of technologies considered here and those which these technologies may render tractable) to be avoided argue the effort will be worthwhile. As the extant preconditions for these technologies cannot be eliminated, and bans would be problematic along multiple dimensions and carry profound opportunity costs, the question is not whether these technologies will be developed but under what circumstances and to what ends. Desirable outcomes may depend on appropriate implementations at the technical, organizational and regulatory levels at least, and efforts are more likely to prevail if begun while vested interests are few.

Acknowledgments. EMR and LAP thank Sirius19 for discussion, criticism and generously defraying costs associated with attending HCII2019.

Competing Interests.   EMR is founder and Chief Executive Officer of NanoCybernetics Corporation, which is developing molecular nanofabrication technologies and programmable self-replicating systems for commercial, biomedical, agricultural, geotechnical, climate restoration and other uses.

References

1. A simplified form of some of the arguments discussed here are advanced by one of the authors [EMR]. http://radicalnanotechnology.com/selRepSafety.html. Accessed 1 Mar 2019
2. Lackner, K.S., Wendt, C.H.: Exponential growth of large self-reproducing machine systems. Math. Comput. Model. **21**(10), 55–81 (1995)
3. von Neumann, J., Burks, A.W.: Theory of Self-Reproducing Automata. University of Illinois Press (1966)
4. Pesavento, U.: An implementation of von neumann's self-reproducing machine. Artif. Life **2** (4), 337–354 (1995)
5. Sipper, M.: Fifty years of research on self-replication: an overview. Artif. Life **4**, 237–257 (1998)
6. Feynman, R.P.: There's plenty of room at the bottom: an invitation to enter a new field of physics. In: American Physical Society Meeting Lecture, Caltech, 29 December 1959
7. Drexler, K.E.: Molecular engineering: an approach to the development of general capabilities for molecular manipulation. Proc. Natl. Acad. Sci. **78**(9), 5275–5278 (1981)
8. Drexler, K.E.: Molecular machinery and manufacturing with applications to computation. Doctoral dissertation, Massachusetts Institute of Technology (1991). http://e-drexler.com/d/09/00/Drexler_MIT_dissertation.pdf. Accessed 1 Mar 2019
9. Drexler, K.E.: Nanosystems: Molecular Machinery, Manufacturing, and Computation. Wiley, New York (1992)
10. Moses, M.: A physical prototype of a self-replicating universal constructor. Master's thesis, University of New Mexico (1999–2001). https://web.archive.org/web/20031130003228/http://home.earthlink.net/~mmoses152/SelfRep.doc. Accessed 1 Mar 2019
11. Zykov, V., et al.: Self-reproducing machines. Nature **435**, 163–164 (2005)
12. Moses, M., Yamaguchi, H., Chirikjian, G.S.: Towards cyclic fabrication systems for modular robotics and rapid manufacturing. In: Robotics: Science and Systems 2009, Seattle, WA, USA, 28 June–1 July 2009 (2009)
13. Rabani, E.M.: U.S. Patent 10,106,401 (2018)
14. Merkle, R.C.: A proposed 'metabolism' for a hydrocarbon assembler. Nanotechnology **8**, 149–162 (1997). http://www.zyvex.com/nanotech/hydroCarbonMetabolism.html. Accessed 1 Mar 2019
15. Freitas Jr., R.A., Merkle, R.C.: A minimal toolset for positional diamond mechanosynthesis. J. Comput. Theor. Nanosci. **5**, 760–861 (2008). http://www.MolecularAssembler.com/Papers/MinToolset.pdf. Accessed 1 Mar 2019
16. Freitas Jr, R.A.: http://www.molecularassembler.com/Nanofactory/AnnBibDMS.htm. Accessed 1 Mar 2019
17. Rabani, E.M.: WO 97/06468 (1997)
18. Merkle, R.C.: Convergent assembly. Nanotechnology **8**(1), 18–22 (1997)
19. Freitas Jr, R.A.: Some Limits to Global Ecophagy by Biovorous Nanoreplicators, with Public Policy Recommendations (2000). https://foresight.org/nano/Ecophagy.php. Accessed 1 Mar 2019
20. Jacobstein, N.: Foresight Guidelines for Responsible Nanotechnology Development, Draft version 6 (2006). https://foresight.org/guidelines/current.php. Accessed 1 Mar 2019
21. Vaughan, D.: Autonomy, interdependence, and social control: NASA and the space shuttle challenger. Adm. Sci. Q. **35**(2), 225–257 (1990)
22. Paté-Cornell, M.E.: Organizational aspects of engineering system safety: the case of offshore platforms. Science **250**, 1210–1217 (1990)
23. The Broadcast Architecture for Control. Kinematic Self-Replicating Machines. Landes Bioscience, Georgetown, TX (2004)

24. Hamsma, H.G.: Possible origin of life between mica sheets: does life imitate mica? J. Biomol. Struct. Dyn. **31**(8), 888–895 (2013)
25. Schmidhuber, J.: Making the world differentiable: on using fully recurrent self-supervised neural networks for dynamic reinforcement learning and planning in non-stationary environments. TR FKI-126-90. Tech. Univ. Munich. (1990). http://people.idsia.ch/~juergen/FKI-126-90_(revised)bw_ocr.pdf. Accessed 1 Mar 2019
26. Schmidhuber, J.: Unsupervised Neural Networks Fight in a Minimax Game (2018). http://people.idsia.ch/~juergen/unsupervised-neural-nets-fight-minimax-game.html. Accessed 1 Mar 2019
27. Goodfellow, I., et al.: Generative adversarial nets. In: Advances in Neural Information Processing Systems 27 (NIPS 2014) (2014). https://papers.nips.cc/paper/5423-generative-adversarial-nets. Accessed 1 Mar 2019
28. Dong, Z., et al.: RRAM-based convolutional neural networks for high accuracy pattern recognition tasks. Published by Workshop on VLSI Symposia, Oral Presentation in Kyoto, Japan, 4 June 2017, pp. 145–146 (2017)
29. Berkes, H.: Remembering Roger Boisjoly: He Tried To Stop Shuttle Challenger Launch (2012). https://www.npr.org/sections/thetwo-way/2012/02/06/146490064/remembering-roger-boisjoly-he-tried-to-stop-shuttle-challenger-launch. Accessed 1 Mar 2019
30. Berkes, H.: 30 Years After Explosion, Challenger Engineer Still Blames Himself (2016). https://www.npr.org/sections/thetwo-way/2016/01/28/464744781/30-years-after-disaster-challenger-engineer-still-blames-himself. Accessed 1 Mar 2019
31. Berkes, H.: Challenger engineer who warned of shuttle disaster dies (2016). https://www.scpr.org/news/2016/03/22/58794/challenger-engineer-who-warned-of-shuttle-disaster/. Accessed 1 Mar 2019
32. Yampolskiy, R.V.: Leakproofing Singularity - Artificial Intelligence Confinement Problem. Journal of Consciousness Studies (JCS). Special Issue on the Singularity, Part 1. Volume 19 (1–2) 194–214 (2012). http://cecs.louisville.edu/ry/LeakproofingtheSingularity.pdf. Accessed 1 Mar 2019
33. Virilio, P., Lotringer, S.: Pure War. Semiotext(e), New York (1983)
34. Jeremiah, D.: Nanotechnology and global security. In: Fourth Foresight Conference on Molecular Nanotechnology, 9 November 1995. http://www.zyvex.com/nanotech/nano4/jeremiahPaper.html. Accessed 1 Mar 2019
35. The Stealth Threat: An Interview with K. Eric Drexler. Bull. Atomic Sci. **63**(1), 55–58 (2007). https://doi.org/10.2968/063001018. Accessed 1 Mar 2019
36. Henley, L.D.: The RMA After Next. Parameters, Winter 1999–2000, pp. 46–57 (1999). https://ssi.armywarcollege.edu/pubs/parameters/articles/99winter/henley.htm. Accessed 1 Mar 2019
37. Roosevelt, F.D.: State of the Union address, 6 January 1941 [known as the Four Freedoms speech] (1941)
38. Office of the Director of National Intelligence, U.S.: "Background to 'Assessing Russian Activities and Intentions in Recent US Elections': The Analytic Process and Cyber Incident Attribution", 6 January 2017
39. Shane, S., Blinder, A.: Secret Experiment in Alabama Senate Race Imitated Russian Tactics. The New York Times, 19 December 2018. https://www.nytimes.com/2018/12/19/us/alabama-senate-roy-jones-russia.html. Accessed 1 Mar 2019
40. Osborne, M.: Roy Moore and the Politics of Alcohol in Alabama (2018). https://www.linkedin.com/pulse/roy-moore-politics-alcohol-alabama-matt-osborne/. Accessed 1 Mar 2019
41. Vassar, M., et al.: Lifeboat Foundation NanoShield Version 0.90.2.13. https://lifeboat.com/ex/nanoshield. Accessed 1 Mar 2019

Usability Inspection of a Mobile Clinical Decision Support App and a Short Form Heuristic Evaluation Checklist

Blaine Reeder[1(✉)], Cynthia Drake[2], Mustafa Ozkaynak[1],
Wallace Jones[2], David Mack[2], Alexandria David[3], Raven Starr[2],
Barbara Trautner[4,5], and Heidi L. Wald[2,6]

[1] University of Colorado College of Nursing, Aurora, CO, USA
{blaine.reeder,mustafa.ozkaynak}@ucdenver.edu
[2] University of Colorado School of Medicine, Aurora, CO, USA
{cynthia.drake,wallace.jones,
david.mack,raven.starr,heidi.wald}@ucdenver.edu
[3] Presbyterian St. Luke's Medical Center, Denver, CO, USA
alexandria.david@ucdenver.edu
[4] VA Center for Innovations in Quality, Effectiveness and Safety,
Houston, TX, USA
trautner@bcm.edu
[5] Baylor College of Medicine, Houston, TX, USA
[6] SCL Health, Broomfield, CO, USA

Abstract. *Introduction:* Health information technology design teams should engage clinical end users as part of the early design process to ensure that mobile clinical decision support applications that can augment cognition are usable and fit clinical workflows. However, a major challenge of this approach is the limited availability of clinicians and other health care workers for design engagement due to their professional time constraints. Therefore, there is a need for a short form heuristic evaluation checklist that facilitates engagement of clinical end users early in the design process for mobile clinical decision support applications. *Methods:* As part of a broader mobile clinical decision support design project for urinary tract infection diagnosis in nursing homes (UTIDecide), we developed a long form heuristic evaluation checklist synthesized from prior usability inspection research. Usability inspection was conducted by two groups of evaluators: the *familiar* group (n = 5) and the *unfamiliar* group (n = 4). *Results:* The long form heuristic evaluation checklist totals 11 categories and 200 heuristics. Evaluation results from the long form checklist were analyzed to produce a short form heuristic evaluation checklist of 30 items with the aim of reducing clinician engagement time. *Conclusion:* Developing a short form heuristic evaluation checklist to facilitate engagement of clinicians as part of early stage mobile clinical decision support projects is a necessary and feasible undertaking. Future work will include testing of the short form checklist with newer versions of UTIDecide and other mobile clinical decision support apps.

Keywords: Heuristic evaluation · Usability inspection · Mobile computing · Clinical decision support

© Springer Nature Switzerland AG 2019
D. D. Schmorrow and C. M. Fidopiastis (Eds.): HCII 2019, LNAI 11580, pp. 331–344, 2019.
https://doi.org/10.1007/978-3-030-22419-6_23

1   Introduction

Heuristic evaluations are used both within the health care domain [1–3] and in broader design research contexts [4–6] to perform "usability inspections" of technologies and systems under development. A *heuristic* "refers to a global usability issue which must be evaluated or taken into account when designing" a program or application [7]. As the computing world moves to a wider diversity of hardware that includes mobile devices and applications designed for these devices, mobile heuristic evaluation checklists have been developed to address this diversification [7–9]. A natural progression of mobile computing is the development of mobile clinical decision support applications [10, 11]. Project design teams should engage clinical end users early in the design process to ensure mobile clinical decision support apps appropriately fit the workflow of clinicians [12–14]. However, a major challenge of this approach is the limited availability of clinicians and other health care workers for design engagement due to professional time constraints [15] and the time-intensive process required to perform traditional heuristic evaluations using lengthy evaluation checklists. Further, we are unaware of heuristic evaluation checklists specific to mobile clinical decision support apps. Therefore, there is a need for a short form heuristic evaluation checklist that facilitates engagement of clinical end users early in the process to design mobile clinical decision support applications. Thus, the aims of this study are twofold: (1) Describe the process and results of a heuristic evaluation of a mobile clinical decision support application to support diagnosis of urinary tract infection in long-term care settings using a long form heuristic evaluation checklist, and (2) Develop a short form heuristic evaluation checklist for early stage development of future mobile clinical decision support applications.

1.1   Urinary Tract Infection (UTI)

There are more than 14,000 nursing homes in the US caring for over 1 million residents. Nursing homes are a unique healthcare environment characterized by a complex and debilitated population, a residential healthcare setting, and staffing and workflow patterns reflecting a focus on nursing care [16, 17]. In particular, providers, nurse practitioners, and physician assistants are not physically present in nursing homes much of the time resulting in asynchronous communication and an enhanced responsibility for nurses to detect changes in condition and the identify acute disease such as infections [18]. Urinary tract infections (UTI) are the most common infection diagnosed in nursing homes each year [19]. There is no gold standard test for UTI, as the findings on urine testing for UTI can be identical to those for asymptomatic colonization of the urinary tract – a benign condition which does not require treatment [20, 21]. Thus, to avoid unnecessary and even dangerous overuse of antibiotics, accurate assessment of symptoms relies on a well-trained nursing workforce. Unfortunately, nursing homes are plagued by high rates of nursing turnover and a staffing model in which the use of highly trained registered nurses is minimized. Mobile clinical decision support, therefore, holds great promise in the nursing home environment to address the challenges of nursing home staffing, nurse training as it relates to the identification of UTI, and the differentiation of UTI from bladder colonization [22]. However, any app that is to be utilized regularly and successfully in the nursing home environment must take into account technology limitations in nursing homes, and must be evaluated from the perspective of the end users in order to fit within their workflows.

1.2 UTIDecide

UTIDecide is a mobile clinical decision support tool built as a native mobile application for iOS and Android platforms. The current version of the app is designed for use in nursing homes by nurses and has been tested with nurses in that setting. UTI-Decide incorporates an established and validated algorithm for the evaluation and management of UTI in a series of guided application screens [23]. The user is taken through the algorithm step-by-step and app output is tailored to the inputs provided. Features include a password-secured login, user-solicited data capture about usability, and analytics related to user interactions. The application provides evidence-based conclusions about the likelihood of UTI and recommendations for further assessment and treatment. The nurse user is given the option to view a customized SBAR (situation, background, assessment, recommendation) communication tool, which is intended to standardize and facilitate communication of clinical findings to physicians, nurse practitioners and physician assistants [24], who are often geographically distant from the nursing home site. The text from this screen can be printed, sent via e-mail, or potentially shared via secure transmission to clinical information systems. At any point in the process, the user may return to the previous step or start over from the top of the algorithm. Additional features of the app include generating a list of other conditions the user should assess, based on entered clinical signs and symptoms, as well as educational features, including pop-ups describing common points of confusion in the UTI diagnostic and management workflow. Contextual help is provided on each screen, including answers to common questions and icons that provide additional information.

2 Methods

Our evaluation approach follows the recommendations of Billi et al. [8] for mobile application evaluations such that each evaluator has: (a) good knowledge of the proposed set of mobile heuristics and standard heuristic evaluation methodology; and (b) experience in using mobile applications and/or with mobile computing [8].

2.1 Evaluation Team and Setting

Evaluation team members were selected for background, training, and expertise to provide broad coverage in clinical and technical areas. Heuristic evaluation took place at Anschutz Medical Campus in Aurora, Colorado. Seven evaluators used iPhones and two used Android devices. These were: iPhone 6 (n = 4), iPhone SE (n = 2), iPhone 5 (n = 1). All iPhone models were running iOS 10.2.1. Of the two evaluators using Android phones, one used a Samsung S5 and the other used an HTC6525LVW. Both Android phones were running the Android 6.0.1 operating system. All evaluators are co-authors of this manuscript and not considered test subjects; thus, institutional review board was neither required nor sought.

Each evaluator received information about the nature of heuristic evaluations and the long form heuristic evaluation checklist prior to evaluation of the mobile app. Evaluators were composed of two groups: the *familiar* group (n = 5) and the *unfamiliar* group (n = 4). Evaluator group sample sizes were informed by research guidelines that recommend 3 to 5 evaluators for heuristic evaluations [25]. The familiar group was composed of current members of the project design team. These included two PhD-trained informatics researchers, two physician researchers, and a health services research PhD student. The *unfamiliar* group was composed of evaluators with no prior exposure to the UTIDecide mobile clinical decision support app (n = 2) and former project design team members who had been separated from the project for over a year and had no exposure to the version being evaluated (n = 2). *Unfamiliar* group members were two 4th year medical students, one of whom holds a master's degree in computer science, a masters-prepared registered nurse (RN), and an undergraduate honors nursing student with a BS degree in biology. The two medical students were part of the project design team for a previous web-based version of the app [10] prior to development of the native mobile app under evaluation.

2.2 Usability Inspection and Long Form Heuristic Evaluation Checklist Items

Long form heuristic evaluation checklist items were drawn from the published heuristic evaluation literature, and then synthesized and adapted for this project in similar fashion to that used by Alexander et al. [1] Selected sources were "traditional" heuristic evaluation checklists [4, 5] and newer checklists for evaluating mobile applications. [7–9] Table 1 shows a ranked list of categories for the UTIDecide heuristic evaluation synthesized from selected sources. Individual evaluation items under each category were compiled from the selected sources and harmonized under the categories in Table 1 by the first author. This long form checklist was then circulated to members of the project design team for review and revised based on their feedback. The final checklist totaled 11 categories and 200 heuristics.

Aspects of the UTIDecide app were evaluated with the long form heuristic checklist using Nielsen's Severity Rating Scale (SRS) [26]. SRS ratings are shown below:

0 = I don't agree that this is a usability problem at all
1 = Cosmetic problem only: need not be fixed unless extra time is available on project
2 = Minor usability problem: fixing this should be given low priority
3 = Major usability problem: important to fix, so should be given high priority
4 = Usability catastrophe: imperative to fix this before product can be released

SRS ratings were applied to questions under each evaluation category (see Table 1). The UTIDecide app was evaluated for two key path scenarios: (1) a patient with a catheter and concordant UTI symptoms [10], and (2) a patient with a catheter and discordant UTI symptoms as described in our previous development work [10].

Table 1. Rank ordered categories in the long form heuristic evaluation checklist*

Category name in this project	Category name in cited sources
• Visibility of system status	• Visibility of system status [4, 5, 7], Visibility of system status and losability/findability [8]
• Match between system and the real world	• Match between system and the real world [4, 5, 7, 8]
• User control and freedom	• User control and freedom [4, 5, 7], Effectiveness [9]
• Consistency and standards	• Consistency and standards [4, 5, 7], Consistency and mapping [8]
• Error management	• Error Prevention [4, 5, 7], Realistic error management [8], Errors [9] • Help users recognize, diagnose, and recover from errors [4, 7], Help users recover from errors [5]
• Recognition and memorability	• Recognition rather than recall [4, 5, 7], Memorability [9]
• Flexibility and efficiency	• Flexibility and minimalist design [4], Flexibility and efficiency of use [5, 7], Flexibility, efficiency of use, and personalization [8], Efficiency [9]
• Minimalist design, ergonomics and learnability	• Aesthetic and minimalist design [4, 5, 7], Good ergonomics and minimalist design [8], Learnability [9], Skills [4, 7], Navigation [5], Use of Modes [5] • Pleasurable and respectful interaction with the user [4], Enjoyment [5] • Satisfaction [9]
• Help and documentation	• Help and documentation [4, 5]
• Structure of information	• Structure of information [5], Ease of input, screen readability and glanceability [8] • Physical Constraints [5]

*Ranking follows Yanez et al's [7] proposed re-ordering schema

The two key path scenarios are described in the publication for a previous version of the UTIDecide mobile clinical decision support app [10].

Each evaluator installed the UTIDecide app on his or her smart phone, recording the phone model and OS. During the evaluation, each evaluator rated the UTIDecide app for each heuristic item in a spreadsheet, making comments where appropriate. Each evaluator also recorded the start and stop time of his or her evaluation session. If a heuristic didn't apply to the current version of the app (e.g. missing feature), evaluators were still asked to provide a rating, especially if the missing feature affected perceived use of the app. Ratings of "0" were given if the feature didn't affect usability. Comments about missing features were solicited to generate features for future versions of the app. After all usability inspection tests were completed, the first and second author tabulated usability scores in a spreadsheet and median usability scores for *familiar* and *unfamiliar* evaluator groups were calculated.

2.3 Analysis and Summary of Evaluation Comments for UTIDecide Design Changes

The last author analyzed and summarized all evaluator comments to inform design changes to the UTIDecide app. These design changes are describe in Sect. 3.2 and were implemented by the development team.

2.4 Analysis of Evaluation Results to Establish the Short Form Heuristic Checklist

Short form heuristic evaluation item candidates were selected from long form usability inspection results that had both: (a) a median evaluation score of 1.00 or higher from any evaluator group, indicating a usability problem of cosmetic or higher, and (b) a severity rating of 3 or higher from any individual evaluator, indicating a perceived major usability issue by at least one evaluator. This process yielded 27 unique results from the *familiar* evaluation group, and an additional 5 unique results from the *unfamiliar* group, for a total of 32 unique items. The first and second author reviewed these 32 items, removing two items that were not relevant to early stage app development and were unintentionally included as an oversight during synthesis of the usability inspection literature. These unintentional items related to supporting a range of user expertise (e.g. novice to expert users) and foolproof synchronization with other information systems, both of which are features to be implemented in later state development.

3 Results

3.1 Long Form Heuristic Checklist Yielded Redundant Results and Evaluator Fatigue

Both the *familiar* group and *unfamiliar* group conducted a usability inspection of the UTIDecide mobile clinical decision support app using the long form heuristic evaluation checklist. Evaluators were sent the same materials simultaneously and performed the usability inspection during the period over a three-week period. Each evaluator scored each item on a scale of 0–4 using Nielsen's Severity Rating Scale (SRS) [26], as described above. Time for completion ranged from 1 h to 2 h and 1 min. Among *familiar* evaluators, the long form heuristic evaluation took on average 1 h 45 min to complete. For *unfamiliar* evaluators, it took 1 h 34 min on average.

Median scores across all evaluators were calculated by the second author. No item had a median response of 2 or higher, though individual evaluators frequently rate UTIDecide with scores above a 2.

Two things became apparent after analysis of the long form heuristic evaluation checklist results:

1. Degree of familiarity increased the severity of evaluation scores; evaluators of the *familiar* evaluator group rated individual heuristics with a higher degree of severity than did those of the *unfamiliar* group. In a two-sample z-test for comparing two

proportions, familiar evaluators gave significantly more severity scores of 2 and above.

2. Survey fatigue with the long form evaluation checklist was evident for both the *familiar* and *unfamiliar* evaluators (e.g. evaluators repeated comments using copy/paste)

Scores revealed that evaluators more closely involved in current app development efforts delivered higher severity scores in assessing the UTIDecide app. Thus, degree of familiarity was associated with differences in the severity of evaluation scores. Members of the *familiar* evaluator group rated individual heuristic items with a higher degree of severity (> 1 rating difference in median score from *unfamiliar* group) in the following categories: navigation, clarity of aesthetic design, and usability in workflow. Table 2 shows differences in median scores between evaluator groups for specific questions in affected categories. Navigation concerns included a user's ability to know where s/he is in the app, where to exit, and how to return to a top level at any stage of use. Aesthetic design concerns were that white space and other section markers were used inconsistently or could be improved. Questions regarding usability and ease of use in user workflow prompted evaluators to speculate regarding about how the app may be used in a real world setting, and again, *familiar* evaluators rated UTIDecide with greater severity.

Evaluators made comments less frequently on later questions in the heuristic evaluation, demonstrating evidence of review fatigue with the long form. We note that one evaluator in the *familiar* group reported application of the heuristic evaluation items in a non-linear fashion, thus review fatigue may not have manifested in the same way for this evaluator. On average, there were no more than 2 meaningful comments per heuristic after the third section, with later sections having as few as 0.7 average responses per heuristic. Comparatively, the first section had nearly 3 meaningful comments per heuristic. Comments were cut if they were only "yes," "no," or "not applicable."

Further, "as above" comments increased in frequency as the evaluators worked through the long form heuristic evaluation, suggesting a degree of repetition. In group meetings, evaluators reported their review fatigue and reported that they found some heuristics repetitive or so similar as to yield identical results (e.g. "For question and answer interfaces, are visual cues and white space used to distinguish questions, prompts, instructions, and user input?" and "Have prompts been formatted using white space, justification, and visual cues for easy scanning?").

3.2 Summary of Long Form Heuristic Results for UTIDecide Prototype Changes

Usability inspection findings resulted in design revisions to the UTIDecide app interface and features. A summary of implemented design changes is listed below.

- Interface redesigns throughout included:
  - Improved layout so control elements are visible without scrolling
  - Removal of excess white space
  - Removal of unnecessary bullet points to improve text display

Table 2. Differences in median scores for familiar and unfamiliar evaluator groups

Category	Familiar	Unfamiliar	Difference
Navigation			
Are any navigational aids provided (e.g. find facilities)?	1.60	0.00	1.60
Are there clearly marked exits (for when the user finds themselves somewhere unexpected)?	1.60	0.00	1.60
Are facilities provided to return to the top level at any stage (e.g. links back to start page/window)?	1.20	0.00	1.20
Is navigational feedback provided (e.g. showing a user's current and initial states, where they've been and what options they have for where to go)?	1.00	0.00	1.00
Are facilities provided to "undo" (or "cancel") and "redo" actions?	0.80	0.00	0.80
Clarity of aesthetic design			
Have prompts been formatted using white space, justification, and visual cues for easy scanning?	1.40	0.00	1.40
For question and answer interfaces, are visual cues and white space used to distinguish questions, prompts, instructions, and user input?	1.20	0.00	1.20
Do text areas have "breathing space" around them?	1.00	0.00	1.00
Usability in workflow			
Does the task sequence parallel the user's work processes?	1.20	0.00	1.20
Does the system make the user's work easier and quicker than without the system?	1.00	0.50	0.50

- Addition of informational pop-ups with definitions for clinical terms
- Addition of missing "dismiss" buttons for informational pop-ups
- Spelling and grammatical corrections throughout the application
- Home screen redesign to separate it into multiple screens
- "Don't show again" disclaimer on home screen
- Key path scenario for scenario 1 (patient with catheter in and UTI concordant symptoms [10])
  - Added screen for diagnoses to consider
  - Add functionality for input of additional discordant symptoms
- Key path scenario for scenario 2 (patient with catheter in and UTI discordant symptoms [10])
  - Improvements in language used in clinical term definitions
- Recommendations screen
  - Added summarized view of symptoms selected by user
  - Added links to evidence-based guidelines
  - Standardized font size with rest of the app
  - Improvement in language for recommendations

- Changed improper placement of yes/no controls that appeared before question about discordant symptoms
- Communication script screen
  - Corrected wrong script pop up for discordant symptoms key path scenario
  - Added informational pop-up with definition of script
- Survey screen
  - Improved layout to show that scrolling required to view questions off screen
  - Added descriptive text to explain survey
  - Added numbering for questions
  - Added opt out capability for survey
  - Added "clear survey capability"
  - Revised "suggestions for improvement" question to allow for comments from erroneous yes/no question format
  - Added "return to home screen" capability after survey completion

3.3   Short Form Heuristic Evaluation Checklist

Substantial time spent reviewing, evidence of evaluator fatigue, and evaluators' reports of the repetitive nature of the items on the long form heuristic evaluation guidelines indicated the need for a short form heuristic evaluation checklist. Below we describe the short form heuristic evaluation checklist. Table 3 shows the short form heuristic evaluation checklist items.

Below are the 30 heuristics that comprise our proposed short form heuristic evaluation checklist for mobile clinical decision support apps. These heuristics fall into 8 of the 11 original categories from the long form heuristic evaluation checklist. Categories not represented from the long form checklist are: *5. Error management, 9. Help and documentation*, and *11. Privacy*. These categories were cut because they were deemed less appropriate for early stage development. Evaluator responses within these sections were purely speculative at this stage of app development, and the heuristic items are highly dependent on the use of the specific app and knowledge of future use, rather than the app itself. For example, among the *unfamiliar* group of evaluators, no heuristic item in the *5. Error management* category yielded a severity rating other than 0. The most severely rated heuristic by *familiar* evaluators was "Does the system provide foolproof synchronization with a PC and/or secure wireless data transfer to other information systems?" and responses were based solely on insider knowledge of plans to integrate UTIDecide into an electronic health record, not on the capabilities of the prototype app at time of evaluation. For early stage development, use of these categories risks app evaluation based on discussions and plans for dissemination rather than implemented features and form factor of the app itself. Further, privacy questions are not applicable to early stage mobile clinical decision support apps, especially given mandatory HIPAA-compliance for live deployments of mobile clinical decision support apps in a clinical setting.

Table 3. Short form heuristic evaluation checklist for mobile clinical decision support

	1. Visibility of system status
1	Does every display begin with a title or header that describes screen contents?
2	After the user completes an action (or group of actions), does the feedback indicate that the next group of actions can be started?
3	Does the system provide *visibility:* that is, by looking, can the user tell the state of the system and the alternatives for action?
4	If users must navigate between multiple screens, does the system use context labels, menu maps, and place markers as navigational aids?
	2. Match between system and the real world
5	Are icons concrete and familiar?
6	Are menu choices ordered in the most logical way, given user, item names, and task variables?
7	Do menu choices fit logically into categories that have readily understood meanings?
8	Are the words, phrases and concepts used familiar to the user?
9	Can the system realistically reflect real world situations and appear to respond to the user?
	3. User control and freedom
10	Can users set their own system, session, file, and screen defaults?
	4. Consistency and standards
11	Is vertical and horizontal scrolling possible in each window?
12	Are field labels and fields distinguished typographically?
13	Are field labels consistent from one screen to another?
14	If the system has multipage data entry screens, do all pages have the same title?
15	If the system has multipage data entry screens, does each page have a sequential page number?
16	Are high-value, high-chroma colors used to attract attention?
	5. Recognition and memorability
17	Are all data and user needs on display at each step in a transaction sequence?
18	Have prompts been formatted using white space, justification, and visual cues for easy scanning?
19	Is there an obvious visual distinction made between "choose one" menu and "choose many" menus?
20	Is white space used to create symmetry and lead the eye in the appropriate direction?
21	Have zones been separated by spaces, lines, color, letters, bold titles, rules lines, or shaded areas?
22	Are size, boldface, underlining, color, shading, or typography used to show relative quantity or importance of different screen items?
23	Are help and instructions visible or easily accessible when needed?
	6. Flexibility and efficiency
24	Is status feedback provided continuously (e.g.: progress indicators or messages)?
25	Does the system guide novice users sufficiently?
	7. Minimalist design, ergonomics and learnability
26	Is only (and all) information essential to decision making displayed on the screen?
27	Has the need to scroll been minimized and where necessary, are navigation facilities repeated at the bottom of the screen?
28	Is navigational feedback provided (e.g.: showing a user's current and initial states, where they've been and what options they have for where to go)?
29	Are any navigational aids provided (e.g.: find facilities)?
	8. Structure of Information
30	Has excessive use of whitespace been avoided?

4 Discussion

The aim of the broader study was to develop an improved UTIDecide application. Our heuristic evaluation revealed several limitations of the long-form checklist method and the need for a short form heuristic evaluation checklist. However, despite redundancies of similar long form heuristic evaluation items and fatigue resulting from repetitive application of these items, our heuristic evaluation identified important design flaws that were corrected in the UTIDecide mobile clinical decision app prototype. Thus, heuristic evaluation efforts successfully improved the app as intended. Beyond this aim, our results revealed that a short form heuristic evaluation, or a "clinical heuristic evaluation," could potentially improve efficiency in gaining input from clinical experts and other potential end users with limited time and familiarity with user-centered design processes. The potential utility of the clinical heuristic evaluation is improved ease of use, reduction in evaluation time, standardization of usability inspections for mobile clinical decision support apps, and simplification of evaluation results synthesis.

We found no significant differences between clinical and non-clinical evaluators in our interdisciplinary project design team. Differences in results were instead dependent on familiarity of the evaluator with the application. This suggests that project managers should carefully consider the make-up of usability inspection teams to anticipate the effect of groups of *familiar* evaluators and *unfamiliar* evaluators. In addition, evaluations by a mixed group of *familiar* and *unfamiliar* evaluators may require a breakdown of results based on their backgrounds. Beyond the development of a mobile clinical decision support application such as the UTIDecide app, a short form heuristic evaluation checklist may have utility in the creation of evidence-based practice guidance for practitioners in other fields, such as public health practice and disease surveillance.

4.1 Limitations

This study has several limitations. First, the short form heuristic evaluation checklist was developed based on a single mobile clinical decision support app project, which may limit generalizability. Second, the short form heuristic evaluation checklist may not transfer to mobile CDS apps used in different care contexts due to patient needs, practice norms, and/or technology familiarity. Further testing of the short form heuristic evaluation checklist as applied to other mobile clinical decision support apps implemented in different care contexts is needed to assess and validate its generalizability. Despite these limitations, we believe that the proposed short form heuristic evaluation checklist can be valuable to researchers who do not have time to synthesize and adapt the usability inspection literature yet want to quickly engage clinical end users during early stage design efforts to develop mobile clinical decision support apps.

5 Conclusion

In order to successfully design mobile clinical decision support apps that can augment cognition in clinical workflows, project design teams must engage clinical end users as part of the early design process. Evaluators using long form heuristic evaluations are certainly able to rate the visibility/aesthetic, controls and consistency of a mobile app; app memorability and learnability; and app fit within workflows in real-world settings. However, post-evaluation analysis and evaluator reports of the effect of the long form heuristic evaluation checklist used in this project revealed evidence of evaluator fatigue that could potentially decrease the likelihood of future evaluator involvement. For early stage development projects, project design teams need evaluation tools with better ease of use and lower time commitment to engage time-limited clinicians and solicit feedback. A short form heuristic evaluation checklist for mobile clinical decision support apps shows promise as a tool to more quickly and efficiently conduct usability inspections with interdisciplinary groups of evaluators who have differential experience with the app under development. Future work will include applying the short form heuristic evaluation tool to new versions of the UTIDecide app, as well as other mobile clinical decision support apps and health-related apps. These efforts should include formal assessment of reviewer fatigue and perceived ease of use and usefulness of the evaluation instrument. In the long term, future versions of UTIDecide and other mobile clinical decision support apps may use onboard device sensors to monitor user state and practice patterns to deliver tailored decision support guidance. Likewise, heuristic evaluation efforts may be enhanced by using onboard sensors during the app design process to capture evaluator state and identify potential improvements that result in more efficient heuristic evaluation checklists.

5.1   Author Contributions

The first author conceptualized the study and participated in all phases of the project. The first and second authors developed the short form heuristic evaluation checklist based on quantitative and qualitative analysis of evaluation results. All authors contributed to usability inspection of the mobile clinical decision support app and writing of the manuscript.

Acknowledgements. This work was supported by the COPIC Medical Foundation and the Centers for Disease Control and Prevention (200-2016-92277).

Competing Interests. The authors have no conflicting interests to report.

References

1. Alexander, G.L., et al.: Passive sensor technology interface to assess elder activity in independent living. Nurs. Res. **60**, 318–325 (2011)
2. Lai, T.Y., Bakken, S.: Heuristic evaluation of HIV-TIDES - tailored interventions for management of depressive Symptoms in HIV-infected individuals. In: AMIA ... Annual Symposium Proceedings/AMIA Symposium. AMIA Symposium 996 (2006)
3. Kneale, L., Mikles, S., Choi, Y.K., Thompson, H., Demiris, G.: Using scenarios and personas to enhance the effectiveness of heuristic usability evaluations for older adults and their care team. J. Biomed. Inf. **73**, 43–50 (2017)
4. Pierotti, D.: Heuristic evaluation-a system checklist. Xerox Corporation (1995)

5. Sharp, H., Rogers, Y., Preece, J.J.: Interactive heuristic evaluation toolkit. In: Interaction Design: Beyond Human-Computer Interaction Web site (2007)
6. Chisnell, D.E., Redish, J.C.G., Lee, A.: New heuristics for understanding older adults as web users. Tech. Commun. **53**, 39–59 (2006)
7. Yanez Gomez, R., Cascado Caballero, D., Sevillano, J.L.: Heuristic evaluation on mobile interfaces: a new checklist. Sci. World J. **2014**, 434326 (2014)
8. Billi, M., et al.: A unified methodology for the evaluation of accessibility and usability of mobile applications. Univ. Access Inf. Soc. **9**, 337–356 (2010)
9. Harrison, R., Flood, D., Duce, D.: Usability of mobile applications: literature review and rationale for a new usability model. J. Inter. Sci. **1**, 1 (2013)
10. Jones, W., Drake, C., Mack, D., Reeder, B., Trautner, B., Wald, H.L.: Developing mobile clinical decision support for nursing home staff assessment of urinary tract infection using goal-directed design. Appl. Clin. Inform. **8**, 632–650 (2017)
11. Martinez-Perez, B., de la Torre-Diez, I., Lopez-Coronado, M., Sainz-de-Abajo, B., Robles, M., Garcia-Gomez, J.M.: Mobile clinical decision support systems and applications: a literature and commercial review. J. Med. Syst. **38**, 4 (2014)
12. Ozkaynak, M., Unertl, K.M., Johnson, S.A., Brixey, J.J., Haque, S.N.: Clinical workflow analysis, process redesign, and quality improvement. In: Finnell, J.T., Dixon, B.E. (eds.) Clinical Informatics Study Guide, pp. 135–161. Springer, Cham (2016). https://doi.org/10.1007/978-3-319-22753-5_7
13. Ozkaynak, M., Reeder, B., Drake, C., Ferrarone, P., Trautner, B., Wald, H.: Characterizing workflow to inform clinical decision support systems in nursing homes. Gerontologist, gny100 (2018). https://doi.org/10.1093/geront/gny100
14. Ozkaynak, M., Drake, C., Reeder, B., Wald, H.L., Trautner, B.: characterizing workflow in nursing homes. In: HFES 2018 International Symposium on Human Factors and Ergonomics in Health Care (2018)
15. Reeder, B., Hills, R.A., Turner, A.M., Demiris, G.: Participatory design of an integrated information system design to support public health nurses and nurse managers. Public Health Nurs. **31**, 183–192 (2014)
16. Morrill, H.J., Caffrey, A.R., Jump, R.L., Dosa, D., LaPlante, K.L.: Antimicrobial stewardship in long-term care facilities: a call to action. J. Am. Med. Directors Assoc. **17**, 183.e1–183.e16 (2016)
17. Walker, S., McGeer, A., Simor, A.E., Armstrong-Evans, M., Loeb, M.: Why are antibiotics prescribed for asymptomatic bacteriuria in institutionalized elderly people?: a qualitative study of physicians' and nurses' perceptions. Can. Med. Assoc. J. **163**, 273–277 (2000)
18. Lim, C.J., et al.: Antibiotic prescribing practice in residential aged care facilities-health care providers' perspectives. Med. J. Aust. **201**, 101–105 (2014)
19. Dwyer, L.L., et al.: Infections in long-term care populations in the United States. J. Am. Geriatr. Soc. **61**, 341–349 (2013)
20. Nicolle, L.E., Mayhew, W.J., Bryan, L.: Prospective randomized comparison of therapy and no therapy for asymptomatic bacteriuria in institutionalized elderly women. Am. J. Med. **83**, 27–33 (1987)
21. Abrutyn, E., et al.: Does asymptomatic bacteriuria predict mortality and does antimicrobial treatment reduce mortality in elderly ambulatory women? Ann. Intern. Med. **120**, 827–833 (1994)
22. Forrest, G.N., Van Schooneveld, T.C., Kullar, R., Schulz, L.T., Duong, P., Postelnick, M.: Use of electronic health records and clinical decision support systems for antimicrobial stewardship. Clin. Infect. Dis. **59**, S122–S133 (2014)

23. Trautner, B.W., et al.: Development and validation of an algorithm to recalibrate mental models and reduce diagnostic errors associated with catheter-associated bacteriuria. BMC Med. Inform. Decis. Mak. **13**, 48 (2013)
24. Haig, K.M., Sutton, S., Whittington, J.: SBAR: a shared mental model for improving communication between clinicians. Jt. Comm. J. Q. Patient Saf. **32**, 167–175 (2006)
25. Nielsen, J., Molich, R.: Heuristic evaluation of user interfaces. In: Proceedings of the SIGCHI conference on Human factors in computing systems, pp. 249–256. ACM (1990)
26. https://www.nngroup.com/articles/how-to-rate-the-severity-of-usability-problems/

Holarchic Psychoinformatics: A Mathematical Ontology for General and Psychological Realities

Suraj Sood[1(✉)], Corinne Lee[3], Garri Hovhannisyan[2], Shannon Lee[3], Garrett Rozier[1], Antoinette Hadgis[3], Kristy Sproul[3], Tyler Higgins[3], Anna Henson[4], Michael Shrider[3], and Monte Hancock[5]

[1] University of West Georgia, Carrollton, GA, USA
ssood2@my.westga.edu
[2] Duquesne University, Pittsburgh, PA, USA
[3] Sirius Project, Melbourne, USA
sirius19conf@gmail.com
[4] Carnegie Melon University, Pittsburgh, USA
[5] 4Digital, Buffalo, USA

Abstract. This article builds upon a formal person-situation framework by offering formalisms for its subcomponents, as well as for reality more generally. More specifically, a system of mathematical formalisms is offered relating the following constructs: *holarchy*, reality, and *psychological reality*. Psychological reality is offered a portmanteau in the neologism, "psychologicality". Psychologicality denotes mind and behavior, both of which are subcomponents of Sood's [20] person-situation formula. The mathematics offered is woven into a broader, subjective-objective ontology that is required for any truly representative virtual world.

The result from above, when synthesized with informatics, is a novel "holarchic informatics" and more specific holarchic *psychoinformatics* (HPI). Holarchic psychoinformatics is related with Sood's psychoinformatic complexity (PIC) paradigm.

The aim of this article is to build upon "third force"—i.e., existential-humanistic (E-H)—psychology. We depart from more traditional approaches in defining E-H psychology as the study of the existence of human minds and behaviors as emergent, interdependent properties of people's interactions with situations. This definition results from the *enactive* person-situation framework as situated within PIC.

Keywords: Third-force psychology · Person-situation interaction · Holarchy · Psychoinformatics · Ontology

1 Introduction

What is or are the futures of "third force", i.e. existential-humanistic (E-H) psychology? To answer this question, it must first be asked what E-H psychology's *raison d'être* is and whether it has been fulfilled. If existential psychologists are interested in

© Springer Nature Switzerland AG 2019
D. D. Schmorrow and C. M. Fidopiastis (Eds.): HCII 2019, LNAI 11580, pp. 345–355, 2019.
https://doi.org/10.1007/978-3-030-22419-6_24

the existence of psychological beings and humanistic psychologists focus on human psychology most broadly, then E-H psychologists should be concerned with the existence of humans as psychological beings. Given the broadness of this terrain, it will be granted from the outset that E-H psychology has not yet exhausted its contributive capabilities. Indeed—assuming that existence as well as human psychology are equally dynamic as they are static—there will always be some need for E-H psychology to update its understanding of the existence of humans as psychological beings, as such beings evolve.

What, then, characterizes the existence of humans as psychological beings? Wilber [25] argued for a subjective-objective model of reality—or, for him (as well as this paper's purposes), "holarchy"—of which human psychology is a more recent part. At present, psychological science is most often defined as that of mind and behavior (see, e.g., [14]). Thus, E-H psychology should be concerned more specifically with the existence of human minds and behaviors.

In this paper, contributions to a novel "holarchic-informatic psychology" (that is consonant with, yet distinct from, E-H psychology) are summarized and forwarded. Said contributions include:

1. A feature list intended to facilitate the definition of human beings
2. A holarchic system of formalisms that situates "psychological reality" [17] as its own metaphysical domain, and further delineates the sub-domain's essential subject matter

The above contributions are then related to the fields of informatics and psychoinformatics. The result will be a "holarchic psychoinformatics" (HPI) related to the psychoinformatic complexity (PIC) paradigm [20] that unites psychoinformatics with complexity science.

2   Related Work

Sood [20] posited 18 human features in answering the question of what it means to be human (or, more specifically—*people*). Sood's features include:

1. *Physical* – People's bodies are composed of matter. Further, people interact with other physical objects.
2. *Biological* – People breathe, eat, and drink; and a great many of them have sex and reproduce.
3. *Temporal* – People are born, they live, and they die; they experience time.
4. *Cultural* – People are embedded in cultures characterized by unique but shared ways of being.
5. *Social* – People participate in societies consisting of concrete relations between themselves and others.
6. *Economic* – People are agents who trade goods and services with one another in marketplaces.

7. *Technological* – People invent and utilize tools to perform tasks they were previously unable or less able to accomplish.
8. *Artistic* – People express themselves through the creation of original works such as paintings and songs.
9. *Intellectual* – People aim to comprehend reality and achieve accurate understandings of it.
10. *Moral* – People have unique and shared ideas of wrong versus right action.
11. *Spiritual* – People seek enlightenment, wisdom, and contact with the divine or supernatural via practices such as meditation and prayer.
12. *Religious* – People worship what they deem as sacred (e.g., God or Gods) through rituals and organized communion.
13. *Political* – People negotiate and have interests that are in line or at odds with those of others.
14. *Athletic* – Whether for fitness or organized play, people exercise their bodies and minds.
15. *Professional* – People work toward particular goals, including earning money and achieving satisfaction.
16. *Recreational* – People enjoy leisurely activities such as taking walks and attending parties.
17. *Linguistic* – People communicate via representational symbol systems characterized by semantics, syntax, and pragmatics.
18. *Psychological* – People have minds and engage in behaviors. More specifically, they think, feel, have personalities, interact with situations, are motivated, sense, perceive, and experience.

The above list may be considered more relevant to personology than personality, proper. McCrae and Costa [12] discussed "personologists" (p. 81) but did not distinguish such researchers from personality psychologists. Still, given that #18 above states the core topics of psychological inquiry, one could most reasonably expect E-H psychologists to focus on the sum-total of its items. These psychologists would therefore need to include cognition, affect, personality, situationality, behavior, motivation, sensation, perception, and experience in their ultimate descriptions of who people are, their explanations of how people come to be, and their predictions of whom people are expected to become.

Within psychology, Freud was the pioneer of personality vis-à-vis mind as much as Skinner was the same for behavior [8]. Affect has been addressed by psychologists via the five factor model (FFM) constructs of Extraversion and Neuroticism; cognition was included in Kelly's [9] personal construct and Dweck and Leggett's [3] social-cognitive theories. Lastly, experience, meaning, and motivation have been taken up by third force theorists such as Kelly, Maslow, and Rogers in addition to positive psychologists like Proctor, Tweed, and Morris [15].

Despite the progress summarized above, it remains an open question whether psychologists have fully accounted for both people and their situations. What determines their interaction? The best-established construct that is closest to the former is personality. Situations, on the other hand, have no corresponding construct denoting situationality. It may be partially inferred from this latter fact that psychologists understand personality better than situationality. In part to ameliorate this situation, Sood [20] formalized person-situation interaction in the following manner

$$F[P, S] = [St_{T, Se}, Pc]_{M, B} \tag{1}$$

[1] Where P equals "person", S equals "situation", St equals "structure", T equals "trait", Se equals "state", Pc equals "process", M equals "mind", and B equals "behavior" [20]. (Traits and states are treated as distinct types of psychological structures.) According to (1), $F[P, S]$ is a whole composed entirely of parts St_{MSe}, St_{BSe}, St_{MT}, St_{BT}, Pc_M, and Pc_B, which respectively denote "mental states", "behavioral states", "mental traits", "behavioral traits", "mental processes", and "behavioral processes". The kind of person-situation interaction expressed through (1)—which has been formalized to render the construct more applicable within mathematical, theoretical, and computational contexts—is thus distinctly psychological in accommodating mind and behavior (two of psychology's highest-level topics of study).

Sood's efforts above represent a start for the formalization of psychological reality. What was left out were formalisms specifically for M and B, as well as for the broader reality of which psychology is merely one part. The next section elaborates formulae for a more general metaphysics or ontology.

3 Holarchy and Psychological Reality

Metaphysics and related formalizations within which to situate holarchy and psychological reality are now laid out. Wilber's metaphysics represents a recent and highly integrative subjective-objective, parts-wholes one—consequently, it will serve as the starting point for this and following sections' formalisms (Fig. 1).

Holarchy has been used to refer specifically to hierarchical systems composed of holons (e.g., *a la* Koestler in [10]). *Reality*, however, consists of both hierarchies and "heterarchies" [24]. The present analysis departs significantly from Wilber and (to a greater extent) Koestler in asserting that reality is and ought to be equally hierarchical —consisting of sets of necessary and sufficient quantitative difference—and heterarchical—consisting of sets whose members share necessary and sufficient qualitative similarity.

Wilber's holarchic subjective-objective model of reality yields the most general possible formula in terms of function F

$$H = F[I(O, Sj)] \tag{2}$$

---

[1] All sufficiently-similar equations offered hereon are syntactically consistently with Lewin's field theory of behavior [11] and Sood's enactive person-situation formula. Two-letter variable-namingis allowed to the extent that the same is in software program variable declaration, and is particularly necessary in cases of multiple constructs beginning with identical first letters.

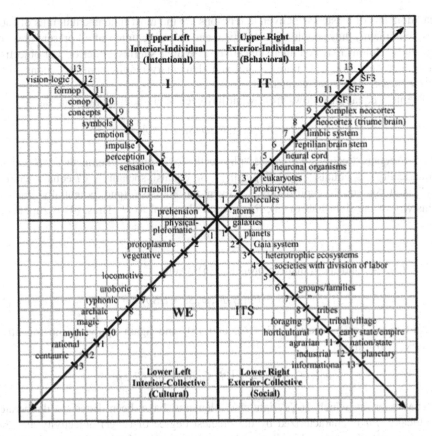

Fig. 1. Wilber's [25] hierarchical four-quadrant metaphysics, consisting of parts-wholes ("holons") of subjective and objective structures and processes (p. 198).

Where H equals "holarchy", I equals "inter", O equals "objectivity", and Sj equals "subjectivity" (such that $I(O)$ and $I(Sj)$ stand for "interobjectivity" and "intersubjectivity", respectively). In addition to subjects, objects, and holons, $I(O)$ and $I(Sj)$ are each also comprised of forces or processes. Any complete metaphysics must include substance and process: each, in equal quantity (Table. 1).

Table 1. Wilber's holonic, subjective-objective metaphysics (i.e., holarchy).

	Mental being(s)	Behavioral object(s)
Holon	Sj:1 subject	O:1 object
Holons	$I(Sj)$: $\geq$ 2 subjects	$I(O)$: $\geq$ 2 objects

Reality can be understood as being made up of physicality, "chemicality", "biologicality", "psychologicality", sociality, "culturality", and spirituality: in essence, all broad domains of existence[2]. Given that holons are the most fundamental units of each of these domains, they enable the theoretical and scientific union of all present areas of such study. Holons may be either subjective (in Wilber's terms, "interior") or objective (for Wilber, "exterior"). A more general and elementary requirement for holons—i.e., that *they exist*—is further granted.

The next equation follows from what has been said about reality thus far

$$R = F[Ph, C, Bl, \psi, Sc, Cl, Sp] \tag{3}$$

Where R equals "reality", Ph equals "physicality", C equals "chemicality", Bl equals "biologicality", ψ equals "psychologicality", Sc equals "sociality", Cl equals "culturality", and Sp equals "spirituality".[3] Given the massive (and potentially greater) inclusivity of the right half of (3), R is essentially equal to H. Thus, (4) is stipulated

$$H = R \tag{4}$$

Via (3), (4), and the transitive property, (5) results

$$H = F[Ph, C, Bl, \psi, Sc, Cl, Sp] \tag{5}$$

And lastly—as a result of (3), (4), and (5), and again applying the transitive property—(6) follows

$$F[I(O), I(Sj)] = F[Ph, C, Bl, \psi, Sc, Cl, Sp] \tag{6}$$

The order in which (6)'s righthand elements have been listed is deliberate. Specifically, Ph is characterized by the greatest degree of (inter) objectivity and the lowest amount of (inter) subjectivity, whereas the inverse of this holds true for Sp. ψ may be considered unique in this particular scheme for being equally made up of both $I(O)$ (including exterior behavior) and $I(Sj)$ (interior mind, i.e. "mentality").

So much for highest-level formalizations of holarchy and reality. Equations are next proposed for ψ (and, subsequently, are defined and related to (6)).

3.1   The ψ Equation

Both mind and behavior must be equally and exhaustively incorporated into a completed understanding of psychologicality—or "psychological reality", a construct

---

[2] One could include other domains in addition to those offered herein. Such possible additions include temporality, "economicality", "politicality", "technologicality", and "religiality". The included subdomains have been chosen over candidates such as these purely given the novelty of the approach undertaken and in the interest of ensuring tractability.

[3] Why are the neologisms represented via (3) necessary? Existing words like "psychology" are potentially ambiguous, referring either to particular beings—e.g., primates—or the field of psychology (including its subdisciplines, e.g. humanistic psychology). Hitherto, no analyses have utilized terms like *psychologicality* to refer to the totality of psychological reality: the general formalization of this is undertaken via (7)).

whose absence of understanding has been lamented by Robinson ([17], p. 191). The general form of such an equation is as follows

$$\psi = F[M, B] \tag{7}$$

Where (as they did in (1)) M equals *mind* and B equals *behavior*. Any of (7)'s values can be either described—i.e., via first-person reports of interior states/traits and processes, or via third-person descriptions of exterior ones—or measured (e.g., through laboratory experimentation or survey analysis). Ready examples of disciplinary relevance for M are fields like augmented cognition, cognitive science, computer science, cognitive neuroscience, cognitive psychology, and clinical and positive psychologies. B may be understood as being centrally relevant to any more physical or mechanistic (i.e., objective) domain including physiology, neurology, anatomy, computer science, and behavioral psychology. Therefore, (7) accommodates a multimethodological (viz., philosophical-qualitative or theoretical-quantitative) analysis.[4]

3.2 M and B: ψ Sub-equations

For the M portion of ψ, (8) is asserted

$$M = F[(A, C, Mv)_{(U-, Sb-)Cs}] \tag{8}$$

Where A equals "affect", C equals "cognition", Mv equals "motivation", $U-$ equals "un-", $Sb-$ equals "sub-", and Cs equals "consciousness". According to the right portion of (7)'s subscript, each of these elementary mental phenomena may be either unconscious, subconscious, or conscious. (8) yields the following nine constructs: "unconscious affect", "subconscious affect", "conscious affect"; "unconscious cognition", "subconscious cognition", "conscious cognition"; and "unconscious motivation", "subconscious motivation", and "conscious motivation".[5]

Formula (8) draws from Freud's topographical model of mind [5] on one hand—where mental content passes between the unconscious and conscious sub-minds via the intermediary subconscious—and Revelle's recent attempt to synthesize Plato's tripartite model of mind (consisting of precursors for affect, cognition, and motivation) into a formal personality framework [16].

For the B sub-portion of ψ, (9) is added

$$B = F[Sm, Rp] \tag{9}$$

---

[4] Philosophical and theoretical approaches are distinguished in that the former are characterized as being more exploratory or question-focused. Works of the latter kind are comparatively more explanatory or answer-focused. Given the more open-ended nature of qualitative inquiry and the definitive nature of quantitative, the former here is arguably closer in spirit to philosophical approaches just as theoretical and quantitative are with one another.

[5] As it relates with (1), each of (8)'s primitive psychological values may be mental structures—i.e., states or traits (e.g., "conscious affective states" like palpable, passing moods)—or processes (e.g., becoming motivated to carry out a given task).

Where *Sm* equals "stimulus" and *Rp* equals "response". Equation (9) draws largely from Skinner's *S-R* theory [19], wherein behavior was put forth as consisting of any environmental stimulus and a necessary behavioral response. *B* is distinct from *M* in that the latter consists of interactive *interior* components rather than exterior ones. The present neuro-cognitive paradigm represents a modern-day reframing of ψ in terms of specifiable correlations demonstrated to exist between neural mechanics—e.g., "all-or-nothing" neural firing—and cognitive processes (e.g., reasoning). *M* and *B*'s sub-components may each be interpreted comparatively more subjectively (e.g., via self-report and "other-report") or explained (e.g., in terms of mental or behavioral mechanisms). Thus, they are equally amenable to a "mixed methods" (see, e.g., [2, 22]) psychological approach.

Psychologicality has evolved in many present-day societies (particularly North-Western ones) to become more informational. In light of this, an explicitly psychoinformatic view of humans such as that offered by Sood [20] will be necessary.

4 Holarchic Psychoinformatics

How is the holarchic paradigm related to informatics? Gruska [26] described the latter discipline's "main task" as being "to discover, explore and exploit in depth the laws, limitations, paradigms, concepts, models, theories, structures and processes of both natural and virtual information processing worlds and to explore their phenomena as well as their interrelations, impacts and utilization" (p. 6). The more specific area of psychoinformatics was defined by Yarkoni [27] as "an emerging discipline that uses tools and techniques from the computer and information sciences to improve the acquisition, organization, and synthesis of psychological data" (p. 391).

Sood [20] introduced the term "psychoinformatic complexity" (PIC) to denote a paradigm marrying psychoinformatics (as defined above), and complexity science as defined by Bar-Yam [28]. Hancock et al.'s work [7]—which utilized an algorithm of computational complexity $O(N^2)$ to infer, characterize, and visualize the emotional context arising from online social discourse—was cited as fulfilling PIC's necessary and sufficient criteria for membership. Said criteria included conformation to Yarkoni's psychoinformatic definition, as well as to Bar-Yam's framing of complexity science being centrally interested in systemic emergence and interdependence (p. 25).

A holarchic informatics would synthesize holarchy as formalized in this paper with Gruska's informatics. Holarchy has been defined as any subjective-objective reality—including at least the sub-domains $\{Ph, \ldots Sp\}$—consisting equally of hierarchical and heterarchical, holonic (part-whole) systems. A holarchic informatics is to be interested in information-processing within any of holarchy's subdomains.

Lastly, a holarchic psychoinformatics (HPI) would represent a paradigm akin to PIC, but with emphasis on holarchy rather than complexity. Such a framework closely resembles Henriques' [8] evolutionary epistemology in which matter gave way to life; life gave way to mind; mind gave way to society; and, lastly, society gave way to culture. The difference lies specifically in the psycho- prefix of psychoinformatics, which specifies the domain of mind (and behavior) as the central HPI focus for any E-H psychologist (Fig. 2).

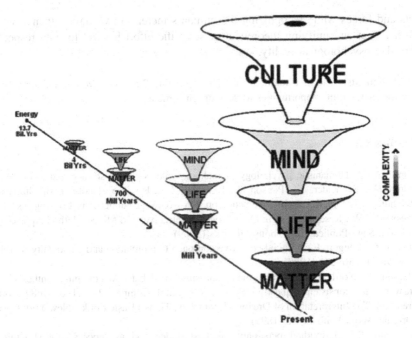

Fig. 2. Henriques' "Tree of Knowledge" system (p. 14).

Information-processing is present at all levels of Henriques' system, thus making it a PIC and HPI one at the level of mind in particular. One could thus speak further of a "holarchic psychoinformatic complexity" (HPIC) super-paradigm, though this is beyond the scope of the present article.

5   Conclusion

In this article, a holarchic-informatic view of psychological reality has been elaborated. If E-H psychology's broadest goal is to understand the existence of humans as psychological beings, then it needs to account for both its subjective (i.e., mental) and objective (i.e., behavioral) aspects and forces. It additionally needs to facilitate the more integrated view of person-situation interaction. Finally, E-H psychology should come to terms with the modern-day human's increasing technologization [1]. Thus, one may speak of a holarchic psychoinformatics (HPI) as greatly complementing the E-H project referred to throughout this article. Holarchy specifies both general and psychological realities as being complex and multifaceted, being composed of holons rather than exhaustibly derivable from their parts alone. It further leaves more to the E-H psychologist who is interested in a broader conception of mind as consisting of unconscious-to-conscious affect and motivation in addition to cognition.

The psychoinformatic approach is becoming and will become increasingly more necessary to adopt in understanding the existence of humans as psychological beings. If E-H psychology becomes more holarchic in incorporating both mental and behavioral aspects and forces while recognizing the human's increasing virtuality, then it would go further than any unifying theoretical or scientific effort has yet to with respect to modern-day psychological reality.

Acknowledgments. The first author thanks Drs. Christine Simmonds-Moore and John Lloyd Roberts for enabling the opportunities to develop this article.

References

1. Aanstoos, C.: Humanistic psychology in dialogue with cognitive science and technological culture. In: Schneider, K., Pierson, J., Bugental, J. (eds.) The Handbook of Humanistic Psychology: Theory, Research, and Practice, pp. 348–363. Sage Publications, Los Angeles (2015)
2. Creswell, J.W.: Research design: Qualitative, Quantitative, and Mixed Method Approaches, 3rd edn. Sage Publications, Thousand Oaks (2009)
3. Dweck, C., Leggett, E.: A social-cognitive approach to motivation and personality. Psychol. Rev. **95**(2), 256–273 (1988)
4. Fleeson, W., Noftle, E.: The end of the person-situation debate: An emerging synthesis in the answer to the consistency question. Soc. Pers. Psychol. Compass **2**(4), 1667–1684 (2008)
5. Freud, S.: The Interpretation of Dreams.(J. Strachey, Trans.) Basic Books, New York (1955). (Original Work Published in 1899)
6. Giordano, P.J.: Individual personality is best understood as process, not structure: a confucian-inspired perspective. Cult. Psychol. **23**, 1–17 (2016)
7. Hancock, M., et al.: Field-Theoretic modeling method for emotional context in social media: theory and case study. In: Schmorrow, Dylan D., Fidopiastis, Cali M. (eds.) AC 2015. LNCS (LNAI), vol. 9183, pp. 418–425. Springer, Cham (2015). https://doi.org/10.1007/978-3-319-20816-9_40
8. Henriques, G.: A New Unified Theory of Psychology. Springer, New York (2011). https://doi.org/10.1007/978-1-4614-0058-5
9. Kelly, G.: Clinical Psychology and Personality: The Selected Papers of George Kelly. Wiley, New York (1969)
10. Koestler, A.: The Ghost in the Machine. The Macmillan Company, New York (1967)
11. Lewin, K.: Field theory and experiment in social psychology: Concept and methods. Am. J. Soc. **44**, 868–896 (1939)
12. McCrae, R., Costa, P.: Validation of the five-factor model of personality across instruments and observers. J. Pers. Soc. Psychol. **51**(1), 81–90 (1987)
13. Neumann, S., Sood, S., Hollander, M., Wan, F., Ahmed, A.-W., Hancock, M.: Using bots in strategizing group compositions to improve decision–making processes. In: Schmorrow, Dylan D., Fidopiastis, Cali M. (eds.) AC 2018. LNCS (LNAI), vol. 10916, pp. 305–325. Springer, Cham (2018). https://doi.org/10.1007/978-3-319-91467-1_24
14. Passer, M.W., Smith, R.E.: Psychology: The Science of Mind and Behavior, 2nd edn. McGraw-Hill, New York (2004)
15. Proctor, C., Tweed, R., Morris, D.: The Rogerian fully functioning person: a positive psychology perspective. J. Humanist. Psychol. **56**(5), 1–28 (2015)
16. Revelle, W.: Integrating personality, cognition, and emotion: Putting the dots together? (2011). https://www.personality-project.org/revelle/publications/BPSP-revelle.pdf. Accessed 9 Jan 2019
17. Robinson, D.N.: Theoretical psychology: what is it and who needs it? Theory Psychol. **17**(2), 191–198 (2007)

18. Rogers, C.: The concept of the fully functioning person. Psychother. Theory Res. Pract. **1**(1), 17–26 (1963)

19. Skinner, B.F.: Generic nature of the concepts of stimulus and response. J. Gen. Psychol. **12**, 1240–1265 (1935)

20. Sood, S.: The psychoinformatic complexity of humanness and person-situation interaction. https://psyarxiv.com/fnjte. Accessed 28 Nov 2018

21. Sood, S.: Global problem-solving and ethics: a theoretical and practical analysis. Int. J. Environ. Issues **17**(4), 322–339 (2018)

22. Tashakkori, A., Teddlie, C., Sines, M.: Utilizing mixed methods in psychological research. In: Weiner, I., Schinka, J., Velicer, W. (eds.) Handbook of Psychology, Research Methods in Psychology, vol. 2, pp. 428–450. Wiley, Hoboken (2012)

23. Varela, F., Thompson, E., Rosch, E.: The Embodied Mind: Cognitive Science and Human Experience. The MIT Press, Cambridge (1991)

24. Von Goldammer, E., Paul, J., Newbury, J.: Heterarchy-hierarchy: Two complementary categories of description (2003). http://vordenker.de/heterarchy/a_heterarchy-e.pdf. Accessed 10 Jan 2019

25. Wilber, K.: Sex, Ecology, Spirituality: The Spirit of Evolution. Shambhala Publications, Boston (2000)

26. Gruska, J.: A perception of informatics. https://westga.view.usg.edu/d2l/le/content/1411076/viewContent/24151440/View. Accessed 21 Jun 2018

27. Yarkoni, T.: Psychoinformatics: new horizons at the interface of the psychological and computing sciences. Curr. Dir. Psychol. Sci. **21**, 391–397 (2012)

28. Bar-Yam, Y.: Making Things Work: Solving Complex Problems in A Complex World. NECSI/Knowledge Press, Cambridge (2004)

Computing with Words — A Framework for Human-Computer Interaction

Dan Tamir[1]([✉]), Shai Neumann[2], Naphtali Rishe[3], Abe Kandel[4], and Lotfi Zadeh[5]

[1] Texas State University, San Marcos, TX 78666, USA
dan.tamir@txstate.edu
[2] Eastern Florida State College, Cocoa, FL 32922, USA
[3] Florida International University, Miami, FL 33199, USA
[4] University of South Florida, Tampa, FL 33620, USA
[5] University of California Berkeley (Deceased), Berkeley, CA 94720, USA

Abstract. In this paper we explore the possibility of using computation with words (CWW) systems and CWW-based human-computer interface (HCI) and interaction to enable efficient computation and HCI. The application selected to demonstrate the problems and potential solutions is in the context of autonomous driving. The specific problem addressed is of a machine instructed by human word commands to execute the task of parking two manned or unmanned cars in a two-car garage using CWW. We divide the interaction process into two steps: (1) feasibility verification and (2) execution. In order to fulfill the task, we begin with verifications of feasibility in terms of assessing whether the garage is unoccupied, checking general ballpark dimensions, inspecting irregular shapes, and classifying the cars that need to be parked, in terms of size, types of vehicles, ranges of acceptable tolerances needed if the cars are manned or not, and means of collision avoidance. The execution of the autonomous driving part is directed by sensory non-numeric fuzzy information that indicates distances from walls or obstacles. The execution algorithm uses a sequence of driving instructions aimed at using the available space in a simple and efficient way without resorting to elaborate numerical calculations, such as making sure that the car is within 2 inches of the wall. The system and its usability are qualitatively analyzed. The analysis shows that the approach has a potential for reducing computational complexity and improving system usability.

Keywords: Fuzzy logic · Computation with words · Autonomous vehicle · Human computer interface · Usability · Learnability · Operability · Understandability

1 Introduction

Humans have a remarkable capability to reason, compute, and make rational decisions in environmental imprecision, uncertainty, and incompleteness of information. In so doing, humans employ modes of reasoning which are approximate rather than exact.

© Springer Nature Switzerland AG 2019
D. D. Schmorrow and C. M. Fidopiastis (Eds.): HCII 2019, LNAI 11580, pp. 356–372, 2019.
https://doi.org/10.1007/978-3-030-22419-6_25

For example, consider the task of driving a car in a highway and maintaining a "*safe distance*[1]" from other cars. To this end, a computer based system might deploy numerous high resolution sensors for accurately measuring location, speed, velocity, distance from other cars, and routing information as well as general information (e.g., the meaning of speed governing road signs). The sensor data and information are generally represented via numerical data. The machine processes the data; potentially solving differential equations, evaluating the values of mathematical functions and deriving state and change of state conclusions in real time. Clearly, even this task, of maintaining "safe distance" from traffic, which is relatively simple for most humans, requires the autonomous system to obtain large amounts of data, maintain a relatively large database/knowledge-base of general knowledge, and perform complex real time inference. This approach of employing Computation with Numbers (CWN) is very promising and finds numerous applications. Furthermore, the increase in computation capacity and ability to acquire and process enormous amounts of data and information enables advances in Artificial Intelligence (AI), whereby many systems that are currently operated by humans can be replaced by CWN systems. Nevertheless, the efficiency and effectiveness of human beings provides numerous examples where the human approach, which involves computation using natural language reasoning and "calculations," that is, Computation with Words (CWW), seems to be more simple, efficient, and precise. This, in-turn, might raise an intriguing research question concerning the computational and human computer interaction benefits of using CWW for AI-related tasks.

The goal of our present research is to explore the utility of CWW as a tool for devising systems for AI, automatic control and robotics. There are numerous cases where the CWN system is the most cost effective approach to solving a specific problem of this type. Nevertheless, there are three main reasons (rationales) for preferring CWW systems over CWN systems for specific problems. The rationales are governed by the basic premise that words are less precise than numbers and precision carries a cost. The three rationales are:

Rationale A: Numbers are not known or are too costly to obtain. In this case, the use of words is a necessity.

Rationale B: Words are good enough. Assumed that the numbers are known and/or obtainable, there may be a tolerance for imprecision, which can be expressed by words in place of numbers, aiming at a reduction in cost and achieving simplicity. Use of words is advantageous.

Rationale C: Using CWW might simplify the human-computer interaction (HCI) of interactive components of the AI system.

Consequently, the CWW endeavor is at the forefront of advancing society by expanding the capabilities of computing machines. Its roots can be traced to the work by Babbage and Ada, Turing, Von Neumann, Zadeh, Minsky, Chomsky, and literally thousands of other researchers and research efforts in AI, automatic control, and robotics.

---

[1] The terms "safe distance" and "too far" have fuzzy connotation. Nevertheless, the CWN engineer might attempt to provide a crisp and accurate definition for this term.

Although the concept of CWW, originated by the late professor L. Zadeh, is not new, the research and development literature on this topic is quite scarce. One of the authors of the present paper has had the privilege and honor of working with Professor Zadeh on a National Science Foundation (NSF) proposal concerning CWW. Several portions of this paper's text were written by Prof. Zadeh in our 2016 NSF proposal, before he passed away. Therefore, the late Professor Zadeh is a co-author of the present paper.

This paper explores the possibility of using CWW-based interface to enable efficient computation and human machine interaction. The application selected to demonstrate the problems and potential solutions is in the context of autonomous driving. The specific problem addressed is of a CWW-based machine instructed by human word commands to execute the task of parking two manned or unmanned cars in a two-car garage, a typical scenario associated with single family homes in America.

We divide the interaction process into two steps: (1) feasibility verification and (2) execution. In order to fulfill the task, we begin with verifications of feasibility in terms of assessing whether the garage is unoccupied, checking general ballpark dimensions, inspecting irregular shapes, and classifying the types of cars that need to be parked, in terms of size (sizes might range from subcompact to full size), types of vehicles (e.g., sedans, SUV, and pickup trucks), ranges of acceptable tolerances needed if the cars are manned or not, and means of collision avoidance.

A simple example of computation in this case is that the result of the expression $subcompact + subcompact$ should yield $subcompact + subcompact \leq 2$. That is, "adding" the two cars, results in a value of 2 or less, where 2 represents the typical minimum requirements to be classified as a two-car garage. On the other hand, "adding" two full size cars may lead to $fullsize_{\square} + fullsize_{\square} \geq 2$ which reflects the possibility of an infeasible case in some two-car garage scenarios but feasible in other situations. A priori, and fuzzy measures in this case may mean a "small-2", a "medium-2" or a "big 2" cargarage.

The execution of the autonomous driving part is directed by sensory non-numeric information that indicates distances from walls and obstacles. For example, the sensor information might indicate: being safely away, being close, being very close, and being dangerously close to an obstacle. The execution algorithm uses a sequence of driving instructions aimed at using available space in a simple and efficient way without resorting to elaborate numerical calculations, such as making sure that the car is within 2 inches of the wall. This applies more easily when the situation has symmetries (e.g., two identical cars in a perfectly symmetric garage).

If the vehicles being parked are unmanned, then a simple inverse process can take the cars out of the two-car garage back to points of origin with an additional safety requirement of checking for new obstacles or occupied spaces.

The hypothesis of the research presented in this paper is that, due to the rationalestated, above several computational HCI tasks can be addressed more efficiently via CWW than via CWN. This hypothesis is qualitatively addressed using the autonomous parking example.

This paper provides an introduction to CWW. Additionally, the problem of parking stated above is described in detail and analyzed. The main contribution of this paper is that it should serve as an initiator of theoretical and applied research in the field of CWW and its applications in areas such as autonomous vehicles. To the best of our knowledge, there is no published research concentrating on the disruptive concepts of using CWW for efficient computation and HCI in AI/RI applications.

The rest of the paper is organized as follows. Background concepts are defined in Sect. 2. Section 3 includes literature review. Section 4 introduces the CWW-based Autonomous Parking System (CWWAPS). Section 5 provides a qualitative analysis of the CWWAPS effectiveness and efficiency, as well as its usability. Finally, Sect. 6 draws conclusions and proposes further research.

2   Background

In this section we provide background concerning the fuzzy logic based concept of CWW [1–11], System Control Using CWW [11–28], and the CWW Autonomous Parking System Background. The CWW concepts are based on a previously submitted NSF proposal [29].

2.1   Computation with Words

Restrictions

Definitions: One of the basic operations in human cognition is that of assignment of a value, v, to a focal variable X through an assignment statement. The assignment statement is generalized if a collection of values, rather than a singleton, is assigned to X. The collection is referred to as a Generalized Assignment value (GA-value) of X. A **restriction** on X written as $R(X)$ is a statement which places a limitation on the values which X can take. A restriction is crisp if its boarders are sharply defined. A restriction is fuzzy if its borders are not sharp.

A **proposition**, p, is a description of a restriction. The **meaning** of p is the restriction which p describes. A **canonical form** of a restriction is an expression of the form $R(X) : X\ is\_r\ R$. In this expression, X is a restricted variable (focal variable), R is a restricting relation (focal relation), and r is an index variable which defines the way in which R restricts X. If X is n-ary variable, then R is n-ary relation. Frequently, the term *restriction* is applied to R, rather than to X. Hence, disambiguation, which is context-depended, is required. The meaning of a restriction is assigned to its canonical form. Equivalently, the meaning of a restriction is the triple (X, R, r). Typically, both X and R are described in a natural language. A restriction is referred to as a *precisiated restriction* if X and R are described mathematically.

Example: Assume that X = Vera's age = $Age(Vera)$. The statement "Vera is middle-aged" places a limitation on the values which X can assume. Hence, "Vera is middle-aged" is a restriction on X. Another example, Robert is "tall" is a restriction on the variable $Height(Robert)$. As yet another example, X = (length of time it takes to drive from Berkeley to the San Francisco Airport (SFO)). $R(X)$ = (usually it takes about

90 min to drive from Berkeley to SFO). Finally, consider X = (birthday of Robert). The statement "Robert is 41 years old" made on August 22, 2016 limits the values of X to the interval [1974.08.22 to 1975.08.22].

Combination of Restrictions

Restrictions can be combined in various ways. The principal modes of combination are the following:

Conjunction: X is $A \wedge X$ is $B \rightarrow (X, Y)$ is $A \cap B$
Disjunction: X is $A \vee X$ is $B \rightarrow (X, Y)$ is $A \cup B$
Implication: if X is A then Y is $B \rightarrow (X, Y) \subseteq A \times B$ ($A \times B$ is the Cartesian product of A and B).

There are close connections between restrictions, information, and propositions. A restriction on X is information about X. Consequently, it can be asserted that: **Restriction = Information.** A proposition p is a carrier of information. Equivalently, a proposition p is a description of a restriction and, consequently, the meaning of the proposition is the restriction which p describes on the focal variable, which is implicit in the proposition. Hence, **Proposition = Restriction.**

Example: The proposition "Vera is middle-aged" is a restriction on the variable *Age(Vera)*. The proposition "most Swedes are tall" is a restriction on the variable X = (fraction of tall Swedes among Swedes). In the proposition "Robert is much taller than most of his friends", the focal variable is the fraction of Robert's friends in relation to whom Robert is much taller. In this case, the restricting relation is *"most"*.

Z-numbers and Z-restrictions

In many real-world settings, values of variables are not known with precision and certainty. To deal with such variables, CWW offers two important concepts: The concept of a Z-number and the related concept of a Z-restriction. For clarity, we reintroduce the concept of a Z-number.

Definition: A Z-number is an ordered pair (A, B) in which A is an estimate of the value of X and B is an estimate of the goodness/correctness of A as an estimate of X.

Hence, if X is a variable whose value is not known with precision and certainty, then the value of X can be represented as a Z-number.

If the Probability Distribution function (PDF) of X is known to be $p(u)$ then B may be expressed as $B = \int P(u) \mu_A du$, where μ_A is the membership function of A. If the PDF of X is not known then an estimation of B becomes a matter of perception employing a similarity-based definition of probability and any available information about X [6]. There are several special cases of Z-numbers depending on B. Z is a *u*-number if $B = usually$. Z is an *h*-number if $B = high$, with *high* being a fuzzy number which is close to 0.9. Z is a *g*-number (a good number) if $B = good$, with *good* being a fuzzy number which is close to 0.7. Z is an *m*-number if $B = moderate$, with *moderate* being a fuzzy number which is close to 0.5. The default value of B is *good*. More precisely, if X is a variable whose PDF, P, is known, then B can be equated to the probability that X is A. Concretely,

$B = Prob(X \text{ is } A) = \int_u \mu_A(u) \times P_x(u)du$. Or, more simply, as $B = \mu_A \cdot P_x$, where '·' is the inner product operation.

Typically, both A and B are linguistic fuzzy numbers in the sense that they are fuzzy numbers which are described in natural language.

Examples:

X = projected deficit.
$A = about \ \$500,000$.
$B = good$.
$Z = (about \ \$500,000, good)$.
X = length of time of driving from
    Berkeley to SFO.
$A = about \ 90$ minutes.
$B = usually$
$Z = (about \ 90 \text{ minutes}, usually)$

X = price of oil in 2020
$A = (about \ \$60/\text{barrel})$
$B = high$
$Z = (about \ \$60, high)$.
X = price of oil in 2025
$A = (about \ \$50/\text{barrel})$
$B = moderate$
$Z = (about \ \$60, moderate)$.

Z-numbers can be combined via arithmetical operations, in particular via addition. Operations on Z-numbers are described in detail in [2, 6]. In the following, the focus is on addition.

Let $Z (Z = (A, B))$ be the sum of $Z_1 = (A_1, B_1)$ and $Z_2 = (A_2, B_2)$, where Z_1 and Z_2 are Z–numbers,

By definition, $A = A_1 + B_1$. Let p_1 and p_2 be the PDF of X_1 and X_2 respectively. If the probability distributions of $X_1 + X_1$, p_{12}, is a convolution of p_1 and p_2, then :

$$p_{12} = \int_u p_1(u) \times p_2(v - u)du \qquad (1)$$

Or more briefly,

$p_{12} = p_1 \cdot p_2$ (inner product due to the fact that both are real).

Hence,

$$p_{12} = \int_u p_{12} \times \mu_a(v - u)du \qquad (2)$$

Thus, B can be expressed as $= p_{12} \cdot \mu_A$.

Similar formulas can be derived for multiplication and division [2]. An important special case is one in which $B_1 = B_2 = B$. In this case, an approximate expression of the sum is $Z = (A_1 + B_1, B)$. This equality is approximate rather than exact.

In the proposed research area, Z-numbers and their applications will be an important object of consideration. An important application example is one in which the issue is assessment of the global impact of a major economic event, such as an increase in the minimum wage, increase in interest rate, increase in oil price, etc. In such problems the global impact is represented as a fuzzy number which is the sum of local impacts, $Z_1, .., Z_n$. $Z = Z_1, \ldots Z_n$. Z_n.

As an illustration, suppose that the issue of interest is the impact on the profit of McDonald Corporation by a to 10% increase in minimum wage. Assume that the interest focuses on the value of X which is the sum of $Z_1, .., Z_n$, where Z_i is the profit of the McDonald restaurants in region, e.g. Oakland. Z_i is supplied by the manager of region R_i. Thus, the problem is of estimating a large number of Z-numbers.

A special case of addition is one where all the Z-numbers have the same confidence, B. In this case, an approximate formula for addition is: $Z_1 + Z_2 = (A_1 + A_2, B)$.

The formulation of approximate rules for addition, subtraction, multiplication and division of Z-numbers should be considered in the proposed research area. Approximate validity of such rules should be assessed through the use of computer simulations.

2.2 System Control Using CWW

Figure 1 presents a schematic diagram of a CWW-based process. One of the most commonly used methods for implementing the inference is via a fuzzy rule based system [12–18].

In [4], Zadeh points to the relations between the Fuzzy rule-based solutions of the control problems, such as the pendulum stabilization task (see below), and CWW. He shows a fuzzification methodology that reflects human perception that can be described in natural language and provides a set of perception-based linguistic model of inference rules.

Stabilization and Related Applications

Following the observation that humans perform numerous complicated control tasks, including tasks that relate to the control of complicated nonlinear systems, without resorting to the use of complex, complicated and computationally intensive numerical methodologies, Zadeh has introduced a different approach to the problem of system control. Namely, fuzzy logic-based and CWW-based approaches [1–5]. In this context, Yamakawa introduced a fuzzy rule based system for the challenging problem of pendulum stabilization [15].

The inverted pendulum stabilization is an instance of a set of fundamental problems in control theory and robotics. The system can be described by a set of nonlinear equations [15] and it often serves as a benchmark for solutions for the underlying set of problems [16–19]. Several approaches involving numerical computation have been applied to this problem [19, 30]. These methods fall under the CWN paradigm and in general they require a relatively complicated and expensive control and computation systems employing computer control algorithms concerned with extensive data acquisition and processing of numerical solutions to the equations [19].

One approach that can be used to address this problem falls under the paradigm of CWW [1–5]. Indeed, in [5] Zadeh shows the relations between Yamakawa and CWW.

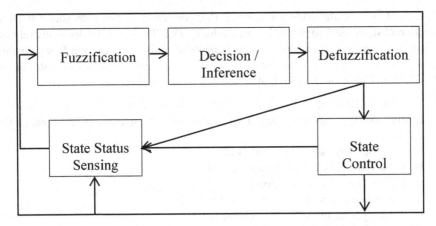

Fig. 1. A Generalized Structure of a Fuzzy Logic ControlSystem.

The process described in Fig. 1 includes a closed loop of (1) sensing external information, (2) fuzzifying the information, (3) using fuzzy logic-based decision and inference system, (4) defuzzyfying of the decisions and inferences, and (5) controlling the state according to the inference.

2.3 CWW Autonomous Parking System Background

We assume that the drivers of two cars leave their vehicles one behind the other, both facing the center of a two-car garage, about 30 to 50 feet from the garage door. Identical CWW Autonomous Parking Systems (CWWAPS) are installed in both cars and the systems are synchronized "over the air" so that each car "knows" about the other. Additionally, we assume that the specifications of both vehicles in terms of dimensions, such as length, width, height, ground clearance and potentially other features, are available to the synchronized systems of the cars. The initial conditions place the vehicles fairly close to the targeted parking locations for both cars. Hence, we expect the maneuvers required to be of small to moderate in magnitude. That is, the steering wheel corrections, braking, and velocities, are not of the "race track" type. An exception to such expectation is safety related, when the system is confronted with a "surprise" such as a running toddler, a pet, falling objects, and other discontinuities that call for immediate pause/response in the process.

We assume that the car is equipped with numerous high resolution sensors, [31] and the sensory data coming from those sources is fed into fuzzy sensors that accumulate relevant high resolution data, fuzzify it, and transmit relevant fuzzy sensor input to the control unit. We further assume that the system can dynamically "map" the garage and its vicinity, providing identification and distances from various points of the vehicle to locations of interest. Such input is numerical but the decision making and maneuvering of vehicles are based on CWW. One form of mapping can be obtained via the production of point cloud from radar, Lidar, or camera systems and conversion of the point cloud into a 3D computer graphics mesh.

Fuzzy decisions are made for steering (S), corrective distance decisions (CDD), corrective obstacle decisions (COD), and velocity (V). These CWW decisions rely on fuzzy sensor inputs, vehicle fuzzy measurements of its own states, and fuzzy desired results. The following is a list of fuzzy sensors that are expected to be available for the CWWAPS and their outputs (output is denoted in parenthesis):

(1) The Fuzzy Heading sensor (H) provides a fuzzy score of the current heading relative to the location of interest on the "map." The sensor output is one of the following:
   a. Straight to the selected location (0)
   b. A bit left of the selected location (−1)
   c. Left of the selected location (−2)
   d. Too much left of the selected location (−3)
   e. A bit right of the selected location (1)
   f. Right of the selected location (2)
   g. Too much right of the selected location (3)

(2) The Fuzzy Distance Sensor #1 (DS1) provides a fuzzy score of current distance relative to a point of interest on the map. The sensor output is one of the following:
   a. The distance to the selected location is as desired (0)
   b. The distance to the selected location is a bit too small (−1)
   c. The distance to the selected location is way too small (−2)
   d. The distance to the selected location is a bit too large (1)
   e. The distance to the selected location is way too large (2)

(3) The Fuzzy Distance Sensor #2 (DS2) provides a fuzzy score for severity of danger associated with vehicle distance to obstacles relevant to projected path. (Higher magnitudes of scores reflect greater danger). The sensor output is one of the following:
   a. No obstacles along the current/projected path to point of interest or its vicinity (0)
   b. One or more stationary obstacles far to the left of the current/projected path to the selected location (−1)
   c. One or more stationary obstacles to the left of the current/projected path to the selected location (−2)
   d. One or more stationary obstacles a bit to the left of the current/projected path to the selected location (−3)
   e. Pop-up close moving object on the left side (−4)
   f. One or more stationary obstacles far to the right of the current/projected path to the selected location (1)
   g. One or more stationary obstacles to the right of the current/projected path to the selected location (2)
   h. One or more stationary obstacles a bit to the right of the current/projected path to the selected location (3)
   i. Pop-up close moving object on the right (4)
   j. Pop-up close moving object in front of the current/projected path to selected location (∞)

The vehicle control unit is capable of providing the car with the following instructions:

(1) Steering instructions (S):
    a. Stay on current heading, no steering correction needed (0)
    b. Turn a bit to the left (−1)
    c. Turn left (−2)
    d. Turn sharply to the left (any CWW calculated value of −3 or smaller is treated as −3)
    e. Turn a bit to the right (1)
    f. Turn right (2)
    g. Turn sharply to the right (any CWW calculated value of 3 or higher is treated as 3)

(2) Corrective distance decisions, CDD, as one of the following:
    a. Stop, the distance is as desired (0)
    b. Move forward to decrease the distance at velocity level 1 [slow] (−1)
    c. Move forward to decrease distance at velocity level 2 [fast] (any CWW calculated value of −2 or lower is treated as −2) (−2)
    d. Move backward to increase the distance at velocity level 1 [slow] (1)
    e. Move backward to increase the distance at velocity level 2 [fast] (any CWW calculated value of 2 or higher is treated as 2) (2)

(3) Corrective obstacle decision, COD, as one of the following:
    a. Continue on the current/projected path (0)
    b. Start preparing an alternative path slightly to the left of the current path if conditions worsen (−1)
    c. Slow down a bit and be ready to change to an alternative path slightly to the left of the current path (−2)
    d. Slow down a lot and be ready to change to an alternative path slightly to the left of the current path (−3)
    e. Emergency stop, evaluate safety level of alternative paths to the left of the current path (−4)
    f. Emergency stop, evaluate potential move backward and evaluate safety level of an alternative path (−∞)
    g. Start preparing an alternative path slightly to the right of the current path if conditions worsen (1)
    h. Slow down a bit and be ready to change to an alternative path slightly to the right of the current path (2)
    i. Slow down a lot and be ready to change to an alternative path slightly to the right of the current path (3)
    j. Emergency stop, evaluate safety level of alternative path to the right of current one (4)

4) Velocity (V) fuzzy levels are set at two levels forward [level 1 = 1–3 mph (slow), level 2 = 4–6 mph (fast)] and two backward [level 1 = 1–2 mph (slow), level 2 = 3–4 mph (fast)]

Note that the output of fuzzy sensors is used as input to the CWWAPS and, without potential confusion, the same naming (e.g., H for heading) is used as output to describe sensor output and as input to describe a, CWWAPS input).

3 Literature Review

Numerous publications address the topic of CWW from the theoretical point of view and/or analyze relevant applications [1–6, 15–19]. Nevertheless, a thorough search for literature that is using the approach presented in this paper did not yield any relevant publications.

Several papers, simultaneously address similar control problems (e.g., "garaging robots") and CWW [32–35]. Yet, they do not elaborate on the way that the CWW is implemented. Other papers refer to CWW but use different approaches, e.g., genetic fuzzy based control for accomplishing tasks such as garaging [36, 37].

Mitrovic et al. introduce the concept of fictitious fuzzy magnets for robot navigation, obstacle avoidance and garaging [33]. Their system displays some similarities to our proposed system. However, they do not use a CWW-based system.

Numerous papers address the topic of HCI ([38–41]) in CWW-based system and conclude that the affinity between the way that the CWW-based system operates and natural language-oriented HCI significantly improves these systems' usability [42–45]. These observations are in line with our expectations. Nevertheless, we could not identify papers that specifically address the HCI of CWW-based applications such as the AI applications of interest in this research (e.g., control of autonomous vehicles and robots [42–46]).

4 CWWAPS Procedure

We begin the CWWAPS procedure with the simple case of tasking the system to autonomously park two identical subcompact cars in a symmetric, vacant, open, free of obstacles, two-car garage. Under initial conditions as described in the background section, the CWWAPS system does not face size-related feasibility or assignment issues. Furthermore, it is assumed that the two cars fit into the garage. Hence, the task is reduced to a sequence of maneuvers that safely and smoothly fulfills this assignment. Note that the task of parking a single car in a single car garage is a sub-task of the current task. This is due to the fact that after parking the first car the problem reduces to parking a single car.

Due to the symmetry, the selected side of the garage for either car does not matter and the system chooses a side. To move the first car to the selected side of the garage (say the left side), the system makes a sequence of simultaneous decisions with respect to S, CDD, COD, and V. The output of Fuzzy Sensor H is used as CWWAPS input, where the input is relative to the first location of interest, the corner of the front side of the garage opening of the selected side for vehicle #1. Based on the current heading input, H, and on the desired heading, D, the system provides steering instructions, S, such as: Stay on current heading, no steering correction needed (0), Turn a bit left (−1), Turn a bit right (1) etc.

Steering instruction (S) are provided based on the CWW equation:

$$S = D - H \tag{3}$$

Note that in a simple case described in this scenario, D = 0, but there may be situations with a desired small bias. For example, when the "map" suggests an easily identified object close to where we want to guide the vehicle, D may be 1 or −1 depending on the side of that object relative to actual intended location. This may cause CWW calculated values to be less than −3 or greater than 3, which would lead to steering instructions in the correct direction with the highest fuzzy magnitude resulting in a favorable outcome for that particular step in the process.

Thus, if the system decides to park vehicle #1 on the left side of the garage and the vehicle starting position is as described earlier, facing the center of the garage, then for a sufficiently distant starting point, the heading sensor is likely to state a moderate reading of H = 1 or H = 2 (A bit to the right or right of the desired location, which is the left corner) and the initial steering is $S = 0 - 1 = -1$, 'Turn a bit left,' or $S = 0 - 2 = -2$, 'Turn left.' Now, suppose that steering left is too sharp or lasts longer than needed, then H may switch to −1, and the steering switches to $S = 0 - (-1) = 1$, 'Turn a bit right.' This is done with decreasing velocities as car gets closer (see below) $H = D$, which is equivalent to: $S = 0$ can be maintained as long as needed.

At the same time that steering instructions are calculated to keep the vehicle heading to the selected location, the fuzzy sensor DS1 provides input regarding the current distance to the first location of interest, i.e., the same corner of the front side of the garage. Based on the current distance DS1 input, and the desired distance (DD), the system makes corrective distance decisions, CDD, such as: Stop, the distance is as desired (0), Move forward to decrease the distance at velocity level 1 [slow] (−1), etc.

Corrective distance decisions (CDD) are made based on the CWW:

$$CDD = DD - DS1 \tag{4}$$

Note, that in a simple case DD = 0, but there may be situations where the appropriate value is 1 or −1 with similar consequences as in the case of CWW for S.

Assuming that the process starts as described earlier with vehicle #1 at sufficiently distant starting point, DS1 is likely to state $DS1 = 2$, then the initial CDD is −2, and the vehicle starts moving forward fast in combination with concurrent steering instruction. Obviously, as the vehicle moves according to correct steering, its distance to the location of interest is decreasing and DS1 is changes to DS1 = 1, leading to CDD = −1 and the vehicle decreases velocity to level 1, the distance keeps decreasing until DS1 inputs the value 0, CDD = 0. Consequently, and promptly, the vehicle stops at the end of this step.

In order to address aspects of safety concerns, the fuzzy sensor DS2 input allows the system to make safety level evaluations, avoiding causing harm to humans and animals and preventing damage to the vehicles being parked and to objects outside the vehicles.

Based on desired obstacle level, DO (in this case $DO = 0$), corrective obstacle decisions are made based on the CWW:

$$COD = DO - DS2 \tag{5}$$

As noted earlier, in this case higher magnitude readings of sensor DS2 reflect greater danger, calling for more deliberate decisions. This time, however, the most deliberate safety decision means 'stop moving immediately.' This is different than the case of DS1 where higher magnitude readings leads to decisions to move faster. Additionally, note the potential of multiple obstacles from different sides. This might require taking into account all of the constraints in terms of actual implementation of multiple calculated COD values. Obviously, evolving safety related circumstances may change the feasibility of a solution to the problem.

After vehicle #1 has reached its selected location, facing the selected corner of the garage door opening at an appropriate distance, it is ready to iterate the process with a new selected location, this time the back wall of the garage, with appropriate distances from obstacles, to the left, right, and the front sides. Hence, if vehicle #1 ends the first iteration facing the left corner of the garage, when new selected location becomes effective, the heading is to the left of the new selected location, $H < 0$, and the steering decision turns vehicle #1 to the right, $S > 0$. All other decision components in each iteration function in the same way. To complete the task, vehicle #2 repeats the same iterations with initial selected location being the complement of the choice made initially for vehicle #1.

Maintaining initial conditions of the garage and tasking the system to park two identical vehicles of increasing size require no change to the process. It only requires closer scrutiny of the feasibility conditions. As long as the garage is deemed to be large enough to accommodate the two vehicles, the iterations are followed the same way as described above, with parameter values adjusted for the types of vehicles.

If the conditions are asymmetric in terms of vehicles and the garage, say a sub-compact and a full-size SUV and/or the garage contains obstacles(for example, the garage contains items such as equipment or appliances), then the system has to evaluate the adequacy of available space. In this case, it is possible that the parking task is performed with CWW, while the feasibility test requires CWN. Once the feasibility is assured, the only adjustment required to the process as described above, is that the choice of side assignment to vehicles does matter now. Now, the system can follow the simple principle that each vehicle must be assigned to the side that matches the side determined as a part of the feasible solution found earlier. In principle, of course, the smaller vehicle is assigned the smaller space, the larger to the larger space (this is a CWW version that may be used in simpler cases).

5 Analysis

5.1 Additional Scenarios

The CWWAPS illustrated above, can be expanded to address additional scenarios. For example, the initial conditions regarding the way that the drivers leave their cars in relation to one another and in relation to the front side of the garage, have been set in order to simplify the formulation of a feasible path fulfilling the task. It is obvious, that all the maneuvers described above can be reprogrammed for initial conditions where the drivers leave their vehicles with their rear sides facing the garage. This solely requires reorientation of fuzzy sensors, reorientation of steering instructions, and interchanging forward and backward motions.

Furthermore, the initial conditions may besimpler or more complicated than the presented scenario, requiring fewer steps or a greater number of steps in the process. For instance, the drivers may leave the two cars side by side facing available and feasible parking spaces inside the garage. Under this scenario, the problem is significantly simplified and only the second step is required. Moreover, the CWWAPS may be useful in providing a parking solution that does not include the requirement that the drivers are getting in and out of the parked vehicles inside the garage. This scenario allows smaller tolerances between the doors of the parked vehicles. Alternatively, in case that the two drivers leave their vehicles parked side by side outside the garage in a way that they face available but infeasible spaces, the CWWAPS can provide a solution by verifying feasibility after commuting the two vehicles and then performing the second step of our procedure.

5.2 Usability

Tamir et al. have developed an effort-based theory and practice of measuring usability [38–41]. Under this theory, learnability, operability, and understandability are assumed to be inversely proportional to the effort required from the user in accomplishing an interactive task. A simple and useful measure of effort can be the time on the task. Said theory can be used to determine usability requirements, evaluate the usability of systems (including comparative evaluation of "system A vs. System B"), verify their compliance with usability requirements and standards, pinpoint usability issues, and improve usability of system versions. It is quite obvious, and supported by research literature, that an interface that uses a natural language or a formal language that are "close" to natural languages reduces the operator effort and can improve system usability [42–44].

A natural language interface can accompany a CWN-based as well as a CWW-based system. A CWW-based system, however provides the advantage that the system itself operates and is being controlled in a way that is closer to human reasoning, decision making, and operation. This increases the coherence between the system and its user interface and, thus, it simplifies the system design and human-controlled operation of the system.

6 Conclusion and Further Research

In this paper we have demonstrated the possibility of using CWW systems and CWW-based human computer interface (HCI) and interaction to enable efficient computation and HCI. The application selected to demonstrate the problems and potential solutions is in the context of autonomous driving. The specific problem addressed is of a machine instructed by human word commands to execute the task of parking two manned or unmanned cars in a two-car garage using CWW. The system and its usability have been qualitatively analyzed. The analysis shows that the approach has a potential for reducing computational complexity and improving system usability.

In the future we plan to perform simulations, potentially via augmented and virtual reality environments, of the current system and systems addressing other autonomous vehicle tasks. Further, we plan to implement some of these systems. The simulations and implementations will be used to evaluate the feasibility and utility of applying CWW in these AI applications as well assessing the usability of this approach.

References

1. Zadeh, L.A.: Toward a restriction-centered theory of truth and meaning (RCT). In: Tamir, D. E., Rishe, N.D., Kandel, A. (eds.) Fifty Years of Fuzzy Logic and its Applications. SFSC, vol. 326, pp. 1–24. Springer, Cham (2015). https://doi.org/10.1007/978-3-319-19683-1_1
2. Zadeh, L.A.: Computing With Words: Principal Concepts And Ideas. Springer, Heidelberg (2012). https://doi.org/10.1007/978-3-642-27473-2
3. Zadeh, L.A.: Computing with Words—Principal Concepts and Ideas Updated. Computer Science Division, Department of EECS, UC Berkeley. https://people.eecs.berkeley.edu/~zadeh/presentations%202010/CW–Principal%20Concepts%20and%20Ideasupdated%20Jan%2021%202011.pdf. Accessed 29 Jan 2019
4. Zadeh, L.A.: Fuzzy logic = computing with words. IEEE Transact. Fuzzy Syst. **4**(2), 103–111 (1996)
5. Zadeh, L.A.: From computing with numbers to computing with words – from manipulation of measurements to manipulation of perceptions. Int. J. Appl. Math. Comput. Sci. **12**(3), 307–324 (2002)
6. Aliev, R.A., Alizadeh, R.V., Huseynov, R.A.: The arithmetic of discrete Z-numbers. Inf. Sci. **290**(C), 134–155 (2014)
7. Aliev, R.A.: Fundamentals of The Fuzzy Logic-Based Generalized Theory of Decisions. Springer, Heidelberg (2013). https://doi.org/10.1007/978-3-642-34895-2
8. Zadeh, L.A.: Fuzzy Sets. Inf. Control **8**, 338–353 (1965)
9. Klir, G.J.: Fuzzy arithmetic with requisite constraints. Fuzzy Sets Syst. **91**(2), 165–175 (1997)
10. Zadeh, L.A.: A note on similarity-based definitions of possibility and probability. Inf. Sci. **267**, 334–336 (2014)
11. Tamir, D.E., Rishe, N.D., Kandel, A.: Complex fuzzy sets and complex fuzzy logic an overview of theory and applications. In: Tamir, D.E., Rishe, N.D., Kandel, A. (eds.) Fifty Years of Fuzzy Logic and its Applications. SFSC, vol. 326, pp. 661–681. Springer, Cham (2015). https://doi.org/10.1007/978-3-319-19683-1_31
12. Passino, K.M., Yurkovich, S.: Fuzzy control. Addison-Wesley, Boston (1998)

13. Lee, C.C.: Fuzzy logic in control systems: fuzzy logic controller I. IEEE Transact. Syst. Man Cybern. **20**(2), 404–418 (1990)
14. Lee, C.C.: Fuzzy logic in control systems: fuzzy logic controller II. IEEE Transact. Syst. Man Cybern. **20**(2), 419–435 (1990)
15. Yamakawa, T.: A fuzzy inference engine in nonlinear analog mode and its application to a fuzzy logic control. IEEE Transact. Neural Netw. **4**(3), 496–522 (1993)
16. Dorf, R.C., Bishop, R.H.: Modern Control Systems, 12th edn. Prentice-Hall, Upper Saddle River (2010)
17. Bugeja, M.: Non-linear swing-up and stabilizing control of an inverted pendulum system. EUROCON Comput. Tool **2**, 437–441 (2003)
18. Yoshida, K.: Swing-up control of an inverted pendulum by energy-based methods. Proc. Am. Control Conf. **6**, 4045–4047 (1999)
19. Ross, M., Gong, Q., Fahroo, F., Kang, W.: Practical stabilization through real-time optimal control. In: American Control Conference, pp. 1– 6 (2006)
20. Lee, C.C., Berenji, H.R.: An intelligent controller based on approximate reasoning and reinforcement learning. In: IEEE International Symposium on Intelligent Control, pp. 200–205 (1989)
21. Berenji, H.R.: A reinforcement learning-based architecture for fuzzy logic control. Int. J. Approx. Reason. **6**(2), 267–292 (1992)
22. Sutton, R.S.: Temporal credit assignment in reinforcement learning. Ph.D. thesis, Univ. Massachusetts, (1984)
23. Barto, G., Sutton, R.S., Anderson, C.W.: Neuronlike adaptive elements that can solve difficult learning control problems. IEEE Transact. Syst. Man Cybern. **13**(5), 834–846 (1983)
24. Sutton, R.S.: Learning to predict by the methods of temporal differences. Mach. Learn. **3**, 9–44 (1988)
25. Barto, G., Sutton R.S., Watkins, C.H.: Sequential decision problems and neural networks. In: Touretzky, A.B. (ed.) Advances in Neural Information Processing Systems. Morgan Kaufmann, pp. 686–693 (1990)
26. Michie, D., Chambers, R.A.: Boxes: an experiment in adaptive control. In: Tou, J.T., Wilcox, H. (eds.) Machine Intelligence, Edinburgh, vol. 2, pp. 137–152 (1968)
27. Barto, G., Sutton, R.S., Anderson, C.W.: Neuronlike adaptive elements that can solve difficult learning control problems. IEEE Transact. Syst. Man Cybern. **13**, 834–846 (1983)
28. Anderson, A.W.: Learning and problem solving with multilayer connectionist systems. Ph. D. thesis, University of Massachusetts (1986)
29. Zadeh, L.A., Tamir D.E.: Proposal Number: 1704840, NSF Program: NSF: CISE: RI: Medium: Collaborative Research: Computing with Words, not funded (2016)
30. Widrow, B., Stearns, S.D.: Adaptive Signal Processing. Prentice-Hall, Upper Saddle River (1985)
31. Tesla, Inc.: Tesla Autopilot. https://www.tesla.com/autopilotlast. Accessed 29 Jan 2019
32. Milanés, M., Onieva, E., Rastelli, J.P., Godoy, J., Villagra, J.: An approach to driverless vehicles in highways. In: 14th International IEEE Conference on Intelligent Transportation Systems, pp. 205–210 (2011)
33. Mitrović, S.T., Đurović, Z.M.: Fictitious fuzzy-magnet concept in solving mobile–robot target navigation, obstacle avoidance and garaging problems. In: Topalov, A.V. (ed.) Recent Advances in Mobile Robotics, pp. 235–260 (2011)
34. Borrero, G.H., Becker, M., Archila, J.F., Bonito, R.: Fuzzy control strategy for the adjustment of the front steering angle of a 4WSD agricultural mobile robot. In: 7th Colombian Computing Congress, pp. 1–6 (2012)

35. Mitrović, S.T., Đurović, Z.M.: Fuzzy logic controller for bidirectional garaging of a differential drive mobile robot. Adv. Robot. **24**(8–9), 1291–1311 (2010)

36. Vicente, E.O., De Pérez, M.J., Pedro, R.T.: Genetic fuzzy-based steering wheel controller using a mass-produced car. Int. J. Innov. Comput. Inf. Control ICIC Int. **8**(5B), 3477–3494 (2009)

37. Nedjah, N., Renato, S.S., Mourelle, S.D.: Customizable hardware design of fuzzy controllers applied to autonomous car driving. Expert Syst. Appl. **41**(16), 7046–7060 (2014)

38. Tamir, D.E., Komogortsev, O.J., Mueller, C., Venkata, D.V., LaKomski, G., Jamnagarwala, A.: Detection of software usability deficiencies. In: Proceedings of HCII 2011 (2011)

39. Komogortsev, O.J., Tamir, D.E., Mueller, C., Camou, J., Holland, C.: Automated eye movement-driven approach for identification of usability issues. In: Proceedings of HCII 2011 (2011)

40. Tamir, D.E., Komogortsev, O.J., Mueller, C., Venkata, D.V., LaKomski, G.: Detecting software usability deficiencies through pinpoint analysis. Int. J. Adv. Softw. Syst. **7**(1,2), 220–229 (2014)

41. Tamir, D.E., Komogortsev, O.J., Mueller, C.: A learning-based framework for evaluating software usability. ARPN J. Syst. Softw, 45–59 (2013)

42. Godfrey, L.B., Gashler, M.S.: A parameterized activation function for learning fuzzy logic operations in deep neural networks. In: IEEE International Conference on Systems, Man, and Cybernetics (SMC) (2017)

43. Kouatli, I.: Computing with words: Are we ready to talk to our cars? In: International Conference on Connected Vehicles and Expo (ICCVE), pp. 212–213 (2015)

44. Rubin, S.H., Lee, G.: Human-machine learning for intelligent aircraft systems. In: International Conference on Autonomous and Intelligent Systems: Autonomous and Intelligent Systems, pp. 331–342 (2011)

45. Pandey, K.K.: Design and Analysis of Intelligent Navigational Controller for Mobile Robot. Master's thesis, Department of Mechanical Engineering, National Institute of Technology, Rourkela, India (2014)

46. Mahapatra, S.: Analysis and Control of Mobile Robots in Various Environmental Conditions. Master's thesis, Department of Industrial Design National Institute of Technology, Rourkela, India (2012)

Brain-Computer Interfaces and Electroencephalography

Assessing Correlation Between Virtual Reality Based Serious Gaming Performance and Cognitive Workload Changes via Functional Near Infrared Spectroscopy

Emin Aksoy[1,3]([⊠]), Kurtulus Izzetoglu[2], Banu Onaral[2],
Dilek Kitapcioglu[1], Mehmet Erhan Sayali[1], and Feray Guven[1]

[1] CASE (Center of Advanced Simulation and Education),
Acibadem Mehmet Ali Aydinlar University, Istanbul, Turkey
emin.aksoy@acibadem.edu.tr
[2] School of Biomedical Engineering, Science and Health Systems,
Drexel University, Philadelphia, USA
[3] Biomedical Device Technology Department,
Acibadem Mehmet Ali Aydinlar University, Istanbul, Turkey

Abstract. Serious game modules enhance knowledge and performance by offering participants to control the content and to arrange the suitable time for learning. Virtual Reality (VR)- based serious gaming modules are used as a complimentary tool for simulation based medical trainings and an emerging method to potentially replace lecture-based learning by enabling us to use the resources in the simulation centers in a much more efficient way. Due to their higher level of immersion, VR-based serious gaming modules have been widely deployed for different types medical trainings. In this study, a VR based serious gaming module was used for teaching "Adult Basic Life Protocol (BLS)" based on the European Resuscitation Council (ERC)-2015 Guidelines. Total number of the participants was 11. There were two groups; the first group consisted of six non-healthcare workers without prior VR experience and the second group consisted of five healthcare workers with prior VR experience. The participants underwent the training sessions via VR based serious gaming module and were expected to achieve a minimum score of 80 out of 100 points in order to become successful. They were asked to take part in the training protocol on the first day and on the 7th day. In addition to recording the training score, we utilized the functional Near Infrared Spectroscopy (fNIRS) sensor to monitor participant's brain activity acquired from prefrontal cortex, that is the area known to be associated with higher order cognitive functioning, such as working memory, attention, decision making and problem solving. The advantages of using fNIRS in everyday working environments for learning and training were reported by various studies. Hence, in this preliminary study, we investigated the correlation between participants' behavior measures, i.e., training performance acquired from the scoring system of the serious gaming module and cognitive workload changes while practicing the training session via the fNIRS system. Based on the analyses of the fNIRS data and performance scores, VR training was found to be effective and helped the trainees to learn the BLS (Basic Life Support) algorithm faster.

D. D. Schmorrow and C. M. Fidopiastis (Eds.): HCII 2019, LNAI 11580, pp. 375–383, 2019.
https://doi.org/10.1007/978-3-030-22419-6_26

Keywords: Functional near infrared spectroscopy · Serious game · Virtual Reality

1 Introduction

Serious gaming modules have seen growing attention in medical training, especially due to their ability to provide learners with the possibility to learn in an interactive way. The advantages of interactive media over standard textbook lessons has been shown earlier [1]. Prior studies support the effectiveness of the game-based learning over the years [2, 3]. Various studies have also revealed that significant improvements in subject understanding, diligence, and motivation by using gaming technology [4]. Different display technologies like PC monitors, tablet PC's or VR (Virtual Reality) head mounted displays can be used in serious gaming modules. Due to its higher immersion level VR based serious gaming modules are nowadays widely used for teaching medical content. Because of the ability of VR to fully engage learners in the virtual environment, instructors intend to utilize VR-based approach in their teaching and learning activities [1, 5]. On the other hand, there is a need for a scientific-based approach to assess true effectiveness of these serious gaming modules specific to domain where they have been deployed for training. Functional near infrared spectroscopy (fNIRS) provides a non-invasive, validated, and practical method to study effectiveness of emerging training technologies and simulations in the field settings. fNIRS can detect changes in the concentration of oxygenated (oxyHb) and deoxygenated (deoxyHb) hemoglobin in the blood [6–8] and has been validated to quantify hemodynamic changes in response to expertise development in various field settings [6, 7, 9, 10]. Further, fNIRS is safe, noninvasive, relatively inexpensive and portable technology without constraints of other neuroimaging technologies, such as functional magnetic resonance imaging (fMRI) [7].

As there is still few research focusing on assessing the correlation between VR-based serious gaming performance and cognitive workload changes, here we studied this correlation with behavioral performance and fNIRS.

2 Methods

11 participants were included in this study. The participants consisted of 6 non-healthcare workers without prior experience of VR-gaming (Group 1), 5 healthcare workers with prior experience of VR-gaming (Group 2). After approval of the Ethical Committee of Acibadem Mehmet Ali Aydinlar University, the consent forms were signed by the participants of the study (Table 1).

Table 1. Distribution of the participants

		N
Group 1	Non-healthcare professional without prior VR experience	6
Group 2	Healthcare professional with prior VR experience	5

3DMedSim®Virtual Reality based BLS (Basic Life Support) serious gaming module which provides learners interactive learning of the basic life support algorithm in two different immersive, virtual environments. The software consists of three modules called as tutorial, seaside environment and subway station. All these modules are compatible with the ERC (European Resuscitation Council) 2015 Guideline, was used for study [11]. As the software is linked with its own LMS (Learning Management System), user scores are stored within this LMS after completing each module. The participants underwent the training sessions VR based serious gaming module and were expected to achieve a minimum score of 80 out of 100 points in order to become successful (Table 2).

Table 2. Scoring criteria of the VR based serious game

Criteria	Points
Checking consciousness	10
Head tilting	10
Checking breathing	10
Telling someone to call 911	10
Sending someone to fetch an AED (Automated External Defibrillator)	10
Controlling carotid pulse	10
Effective chest compression	10
Opening AED Device	10
Placement of AED pads	10
Defibrillation with AED	10
Total	**100**

At the first phase of the testing protocol a baseline recording with fNIRS device was performed for 10 s, while they were at the so called "Lobby" part of the serious game followed by a familiarization process of the game and hardware for 240 s. Then they were asked to complete the tutorial part of the game at the seaside environment. Afterwards a rest phase was given to the participants for 15 s. The participants were then asked to complete the test phase with the seaside environment followed by a resting time of 15 s. The last phases of the protocol consisted of a test with the subway environment and the last resting phase of 15 s. The participants were asked to take part at the same test protocol on the first day and on the 7th day of the study.

A continuous wave fNIRS system (fNIR Devices LLC, Potomac, MD) was used in this study to measure the hemodynamic response from the prefrontal cortex. The system is composed of three modules: a 16 channel sensor pad, a control box and a computer running data acquisition software. The sensor pad has 4 light emitting diodes (LED) as light sources and 10 detectors (Fig. 1). The light source and detector positioning on the sensor pad was designed to monitor anterior prefrontal cortex (PFC) underlying the forehead (Fig. 2) [8]. Two wavelengths of light (730 and 850 nm) from the LEDs were measured continuously at 500 ms intervals in 16 channels. The raw light intensity data were filtered using a low pass finite impulse hamming filter and sliding window motion artifact rejection tool of the signal processing and analysis software fNIRSoft® [12]. The filtered light intensity data were processed with the Modified Beer Lambert Law to calculate oxygenation values for each of the 16 channels. We have used the Oxy [oxygenation = oxyHb − deoxyH] and channel 3 which is known to be associated with working memory for this study [6]. The same methodology was also used in previous studies and proved that this is the area significant for training [13, 14].

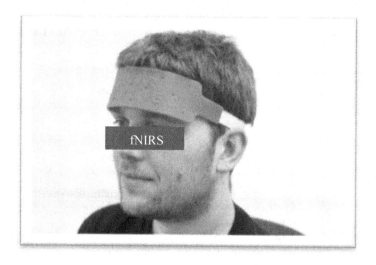

Fig. 1. A subject wearing fNIRS sensor used for this study.

Fig. 2. fNIRS measures were collected simultaneously with the VR headset.

Hypothesis of this study is to determine whether there is any decline in Oxy data as the participants learned and became familiar with VR & protocol. The data required from second group consisting of healthcare professionals with prior VR experience was used as reference, control data.

Statistical analysis was performed using the MedCalc Statistical Software version 12.7.7. For comparison of two non-normally distributed independent groups Mann Whitney U test was used and for comparison of two non-normally distributed dependent groups Wilcoxon Signed Rank test was implemented.

3   Results

The Oxy [oxygenation = oxyHb − deoxyHb] and channel 3, which are associated with working memory, were used to reveal the differences between the two groups in this study.

To investigate the training effect of the VR based serious gaming module, data acquired from Group 1 and Group 2 were compared. Besides fNIRS Oxy levels, data collected from the scoring system of the serious gaming module after each session were also compared. While a significant decline in the Oxy levels was observed in Group 1, a significant increase in BLS scores were detected after the first sessions in day 7. Collected fNIRS measures greatly indicated this change based on their brain responses (Table 3).

Table 3. Statistical analysis the two groups' Oxy levels and game scores (Wilcoxon test, *Mann-Whitney U test)

	Day	VR−, BLS−	VR+, BLS+	p*
		Mean ± SS	Mean ± SS	
OXY	1	0.76 ± 0.51	0.54 ± 0.28	0.397
	7	0.30 ± 0.21	0.81 ± 0.15	**<0.001**
p		**0.002**	**0.002**	
Scores	1	58.2 ± 2.48	94.5 ± 4.42	**<0.001**
	7	97.9 ± 0.88	95.6 ± 2.65	**0.035**
p		**<0.001**	0.453	

The results reveal that there is a statistically significant difference between Day 1 and Day 7 for the fNIRS (Oxy) measures within Group 1 ($p < 0.05$) (see Fig. 3), which validates our hypothesis for the training effect. One would expect lower oxygenation while their performance rises. Figure 4 depicts the behavioral performance score and shows statistically significant difference between day 1 and day 7 ($p < 0.05$).

However, no significant changes in the brain responses and BLS scores were shown in Group 2 (Figs. 5 and 6). This is also in line with the hypothesis that there should be no training effect, and thus no performance improvement for this expert group who has already prior task experience along with the VR experience.

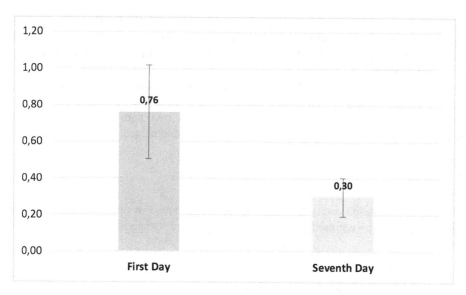

Fig. 3. Mean fNIRS Oxy levels of Group 1 on the first and seventh day of the study.

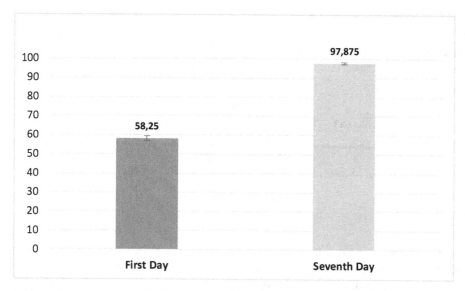

Fig. 4. Mean BLS scores of group 1 collected from the scoring system of the serious gaming module.

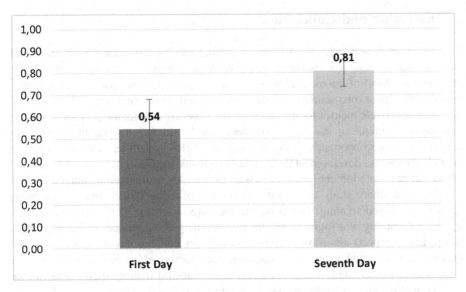

Fig. 5. Mean fNIRS Oxy levels of Group 2 on the first and seventh day of the study.

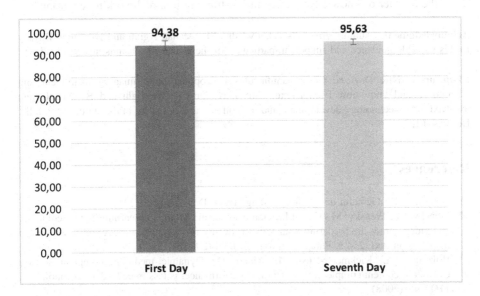

Fig. 6. Mean BLS scores of Group 2 (expert in task and VR training) collected from the scoring system of the serious gaming module.

4 Discussion and Conclusion

Both in research and industry, VR based serious gaming has been widely used and especially has growing attention in medical domain as an additional tool to train healthcare professionals. Most of the medical training systems, such as simulators or serious game modules have their own scoring systems, but the question is that whether these scoring systems can provide metrics for measures of true expertise development as a stand-alone assessment technique or should be complemented by another method to enhance the training and more importantly, better assessment. fNIRS system lends itself to monitor individual expertise development through hemodynamic measures from prefrontal cortex which can provide additional metric to assess the efficacy of the training approach.

This preliminary study investigated feasibility of using fNIRS, brain-based measures in VR-based training. Based on the measures from the PFC and performance scores, VR training was found to be effective and helped the trainees to learn the BLS (Basic Life Support) algorithm faster. Optical brain imaging also revealed its effectiveness as a potential metric for monitoring the learning effect in this study. Future study should include more participants and other standard training modules to further validate and develop brain-based metric to support the performance scores. The next study should not only include the further validation of the results, but also to investigate the transfer of knowledge gained in VR-training to a real-working environment.

Acknowledgment. Authors gratefully acknowledge Mr. Atahan Agrali and Dr. Engin Baysoy for his valuable guidance and support in particular for the fNIRS data acquisition and analyses.

Disclosure:. fNIR Devices, LLC manufactures the optical brain imaging instrument and licensed IP and know-how from Drexel University. Drs. K. Izzetoglu and B. Onaral were involved in the technology development and thus offered a minor share in the startup firm, fNIR Devices, LLC.

References

1. Deterding, S.: Gameful design for learning. Train. Dev. **67**(7), 3 (2013)
2. Prensky, M., Prensky, M., Gee, J.P.: Don't bother me Mom, I'm learning!: How computer and video games are preparing your kids for twenty-first century success and how you can help!, 1 edn, **xxi**, 254 s. Paragon House, St. Paul (2006)
3. Holzinger, A., Kickmeier-Rust, M., Albert, D.: Dynamic media in computer science education; content complexity and learning performance: is less more? Educ. Technol. Soc. **11**(1), 11 (2008)
4. Pellas, N., Fotaris, P., Kazanidis, I., Wells, D.: Augmenting the learning experience in primary and secondary school education: a systematic review of recent trends in augmented reality game based learning. Virtual Reality, p. 18 (2018)
5. Hanson, K., Shelton, B.E.: Design and development of virtual reality: analysis of challenges faced by educators. Educ. Technol. Soc. **11**(1), 14 (2008)
6. Ayaz, H., et al.: Optical brain monitoring for operator training and mental workload assessment. Neuroimage **59**(1), 36–47 (2012)

7. Bunce, S.C., et al.: Functional near-infrared spectroscopy. IEEE Eng. Med. Biol. Mag. **25**(4), 54–62 (2006)
8. Izzetoglu, M., et al.: Functional near-infrared neuroimaging. IEEE Trans. Neural Syst. Rehabil. Eng. **13**(2), 153–159 (2005)
9. Izzetoglu, K., Richards, D.: Human performance assessment: evaluation of wearable sensors for monitoring brain activity. In: Vidulich, M.A., Tsang, P.S. (eds.) Improving Aviation Performance through Applying Engineering Psychology: Advances in Aviation Psychology, 1 edn. CRC Press (2019)
10. Izzetoglu, K., Bunce, S., Onaral, B., Pourrezaei, K., Chance, B.: Functional optical brain imaging using near-infrared during cognitive tasks. Int. J. Hum. Comput. Interact. **17**(2), 211–227 (2004)
11. Perkins, G.D., et al.: European resuscitation council guidelines for resuscitation 2015: Section 2. adult basic life support and automated external defibrillation. Resuscitation **95**, 81–99 (2015)
12. Ayaz, H., et al.: Using MazeSuite and functional near infrared spectroscopy to study learning in spatial navigation. J. Vis. Exp. (56), e3443 (2011)
13. Sato, H., et al.: Correlation of within-individual fluctuation of depressed mood with prefrontal cortex activity during verbal working memory task: optical topography study. J. Biomed. Opt. **16**(12), 126007 (2011)
14. Sato, H., et al.: A NIRS-fMRI investigation of prefrontal cortex activity during a working memory task. Neuroimage **83**, 158–173 (2013)

Construction of Air Traffic Controller's Decision Network Using Error-Related Potential

Sim Kuan Goh[1], Ngoc Phu Tran[1], Duc-Thinh Pham[1], Sameer Alam[1(✉)], Kurtulus Izzetoglu[2], and Vu Duong[1]

[1] Air Traffic Management Research Institute,
School of Mechanical and Aerospace Engineering, Nanyang Technological University,
Singapore, Singapore
{skgoh,phutran,dtpham,sameeralam,vu.duong}@ntu.edu.sg
[2] School of Biomedical Engineering, Science and Health Systems, Drexel University,
Philadelphia, USA
ki25@drexel.edu

Abstract. Electroencephalography based brain computer interface has enabled communication of human's intention to a computer directly from the brain by decoding signatures that relay the intention information. Error-related potential has been adopted as a signature for natural communication and performance monitoring, among others. In this work, we investigate the use of error-related potential as an input channel to transfer human preference of a strategical advisory to a computer. Air traffic control task was used as a case study to make an empirical inquiry of error-related potential for higher level cognitive tasks (i.e. situation awareness in air traffic control tasks). The experimental task requires the subjects to monitor and assess air traffic scenarios presented on a simulated environment that provides conflict resolution advisories. The task is known to be highly mental demanding as it requires continuous situation awareness of the traffic. An interface and experimental protocol were developed for this experiment to validate that error-related potential can be used a new channel for preference. The implementation of the complete design was described together with the experimental evidence of error-related potential. According to the results, we found error-related potential that is in line with existing literature. We also discussed how the preliminary findings of this work can be used as an integral part of an intelligent conflict resolution advisory system that can learn from human preference and duplicate the decision making.

Keywords: Brain computer interface · Error-related potential · Decision network · Air traffic management · Air traffic control · Conflict detection and resolution

This research has been partially supported under Air Traffic Management Research Institute (NTU-CAAS) Grant No. M4062429.052.

D. D. Schmorrow and C. M. Fidopiastis (Eds.): HCII 2019, LNAI 11580, pp. 384–393, 2019.
https://doi.org/10.1007/978-3-030-22419-6_27

1   Introduction

Electroencephalography (EEG), a widely used brain-computer interface (BCI) that captures electrical brain activities, allows communication of human's intention directly from the human brain to computer system. To date, a number of brain patterns (e.g. P300 [21], motor imagery [20], error-related potential (ErrP) [17]) have been identified and utilized for BCI applications. Among these patterns, ErrP is naturally elicited when the brain observes an erroneous or unexpected behavior without explicit training or instructing participant to generate it [28]. Hence, ErrP has the advantage of facilitating natural and intuitive interaction between the brain and computer. A few pioneering work has employed ErrP in a human-machine setting. In [24], ErrP was monitored and decoded to correct robot mistakes. ErrP was also used to map human gesture to robot's action in an interactive setting [14]. Moreover, [28] used ErrP to select the heading of the vehicle in real-world driving task.

Existing BCI literature has been demonstrating the use of BCI to communicate instantaneous/operational command (i.e. left/right, start/stop) to control an external system on various tasks. Nonetheless, operational level BCI control is laborious, time-consuming and mentally demanding. Besides, even instances of similar tasks require human to generate a similar sequence of BCI commands repeatedly. However, there is little research work extends BCI to communicate tactical or strategical level commands, which can perform a set of sequential actions and generalize to solve similar tasks. Hence, we attempt to investigate the possibility of creating set of higher level instruction, decision network or preference using ErrP based BCI.

In air traffic management literature, BCI have been applied in air traffic management to monitor air traffic controller's (ATCO) performance. EEG was utilized to derive objective bio-marker of workload [2–4]. The findings are useful to ensure the situation awareness of ATCOs. Detector can then be developed to alert any sign of sleepiness or loss of vigilance. fNIRS was adopted to measure maturity and expertise of air traffic controllers [5] which will aid the training and selection process. While those work use BCI in a passive manner, our work has a very different objective which aims to use BCI to transfer preference actively.

In this work, we select one the most challenging task in air traffic control (i.e. conflict detection and resolution) to be an ErrP case study of higher level cognitive task. This task requires air traffic controller (ATCO) to maintain situation awareness of current and future air traffic condition to ensure a safe and efficient flow of every aircraft in a shared air space. Many researches and developments have been performed to develop assistance tools for ATCOs to reduce their workload whilst improving their performance. The pioneering researches relied on mathematical models of air-crafts, conflict scenarios, and airspace structure to compute conflict resolution strategy. An extensive model based approaches can be found in [15]. Recently, second-order cone programming [26], space-time Prism [9] approach, model predictive control (MPC) [13,27], surrounding traffic analysis [22], and large scale conflict resolution models for velocity maneuver [1] and 3D conflict resolution [16]. These mathematical models can hardly scale up for a big amount of aircraft and might fail to describe the complete dynamics of air traffic. Moreover, these automated tools were not fully trusted as most models behave like a black-box to ATCOs. Besides, the advisories might be very different from ATCO's expectation that leads to their low acceptance rate [11]. Hence, ATCOs have to remain taking active control in the management of air traffic.

While mathematical models showed their limitation in incorporating human preferences or strategies in their solution, artificial intelligence (AI) (e.g. deep learning and reinforcement learning (RL)) has achieved superhuman level in variety of strategical tasks (e.g. diagnosing a number of cancers, playing the game of Go and Atari games, etc.). Recently, the literature in behavior cloning [12] and inverse RL [8,18,19,23] have demonstrated machine abilities in mimicking the expert's behaviors from the demonstrations or even infer the the reward function of human strategies [7,10,25]. Following this line of research, RL can be adopted in air traffic control task to learn how ATCOs perform the conflict resolution.

In this study, we aims to integrate the advancement of BCI and RL. Our goal is to develop an BCI framework where a human can communicate and construct a goal-oriented sequential decision-making or preference command using ErrP signature. RL can be used as an engine to incorporate human's preference in learning model for conflict resolution. In this paper, we limit our work to fill the research gap on the empirical inquiry of error-related potential for higher level cognitive tasks (i.e. situation awareness in air traffic control tasks). In order to investigate our hypothesis that ErrP can be adopted in this air traffic monitoring task, we developed a simulated air traffic environment which can simulate different configuration of conflicts and visualize projected trajectories of traffic scenarios as well as advisory trajectory. The environment allows subject to monitor and assess the advisory naturally. We also design experimental protocol that facilitates the generation of ErrP that encapsulate ATCO's preference. The mapping between the ErrP and preference as well as how they can be useful will be discussed in succeeding parts of this paper.

The main contribution of this work includes (1) the study of ErrP for a higher level planning task (i.e. air traffic monitoring tasks) compared to previous work on instantaneous controlling tasks; (2) the development of simulated air traffic environment to emulate real-world traffic (3) the design of visual interfaces and experimental protocol for subjects to input their preferences using ErrP; (4) experimental evidence that our proposed protocol triggered brain signature that is consistent with existing ErrP literature.

The rest of this paper is structured as follows. Section 2 describes the methodology and how the whole BCI framework is designed and implemented to achieve the experimental goal of this work. It is followed by Sect. 3 that shows the experimental results of the ErrP generated by the subject. Finally, Sect. 4 discusses how the findings and experimental evidence can be used to adapt advisory tool.

2 Methodology

This is an exploratory study to investigate the extension of ErrP for higher level cognitive task. We implemented the BCI framework and design experimental protocol to validate the applicability of ErrP for a real-world problem in air traffic management. The experimental setup and paradigm are illustrated in Fig. 1.

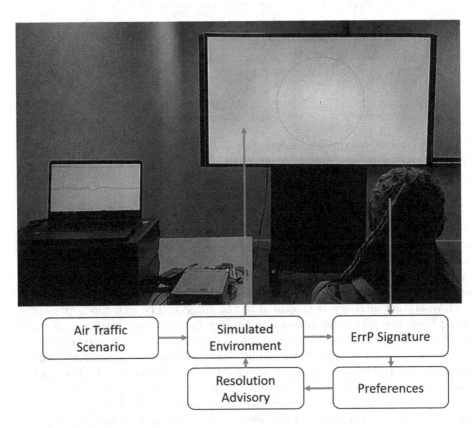

Fig. 1. The experimental setup and diagram of our BCI framework.

2.1 Simulated Environment and Scenario Generation

The simulated environment configures conflict scenarios with two air-craft as shown in Fig. 2. For simplicity, the airspace was assumed to a circular area. In the airspace presented to the subject, there is an ownship and an intruder together with their projected trajectories. We restrict the two air-crafts to fly as the same lateral speed in the environment. The conflict configuration can be characterized by the conflict angle and closest point of approach. Advisory tool can propose a heading change maneuver to resolve the conflict. An example is shown in Fig. 2.

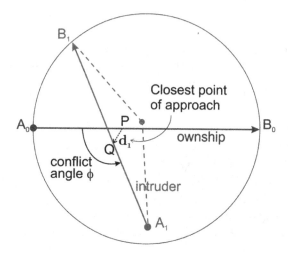

Fig. 2. An example of conflict configuration for two air-crafts. The dashed green line represents a random heading change maneuver. (Color figure online)

2.2 Experimental Protocol

The subject was seated in front of a screen displaying a variety of air traffic scenarios. The time-line of the experiment can be found in Fig. 3. A trial starts with a little dot in the middle of the screen with a audio cue. The color of the dot is hint of the quality of the solution showing next (green indicates good advisory while red indicates bad advisory). Subsequently, an air traffic scenario with resolution advisory is presented to the subject. The ownship is presented using black line while the intruder using gray line. On the ownship trajectory, there are 3 little circles. Subject monitor and assess the advisory trajectory of the ownship for 5 s. Next, the middle circle changes from unfilled to filled circle to cue the subject to get ready for the next visual event. The middle filled circle might either make a positional change in the same direction as the advisory trajectory or the opposite direction. Subject was told that the same direction signifies acceptance of the advisory and the opposite direction indicate rejection of the advisory. However, the direction is assigned randomly by design. We assume ErrP to be triggered when the assigned direction does not match with what the subject expects.

2.3 EEG Acquisition and Processing

When the subject was performing the task, EEG was acquired with a sampling frequency of 250 Hz using BrainAmp MR plus EEG device. We custom-made a circuit for EEG event marker using NI USB-6001. It facilitates our simulated environment to send triggers to the Brain Vision Recorder. In this study, we select and analyze the channel C2. Notch filter was applied to remove the line noise. The signal was then band-pass filtered between 1 and 5 Hz. During the

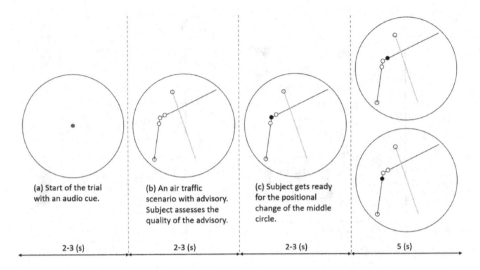

Fig. 3. Time-line of our experimental design. (Color figure online)

experiment, subject was instructed to prevent muscle movement that can induce artifact to the signal. The recorded signal is subsequently post-check to reject artifact trials. EEGLAB was used to extract the signal segments 0.5 s before and 1.5 s after the positional change on of the circle on the ownship trajectory.

3   Results

The event related potential of the experiment was shown in Fig. 4. Trials with positional changes that match subject's expectation are shown in Fig. 4 (left) and trials with positional changes that do not match subject's expectation are shown in Fig. 4 (right). By comparing the two figures, consistent event related potentials were observed for trials with positional changes that do not match subject's expectation. The difference between ErrP and Non ErrP was shown in Fig. 5. The pattern of the potential is similar to the ErrP reported in [6], a negative deflection followed by positive peak. Besides, we also observed longer ErrP latency compared to [6].

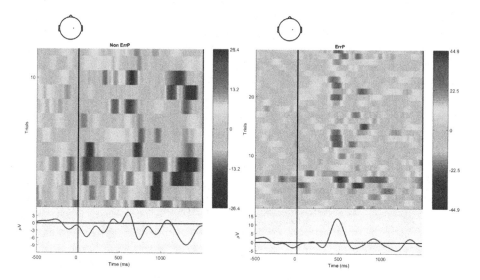

Fig. 4. Visualization of event related potential. (Left) Non ErrP. (Right) ErrP.

4   Discussion

The result of the experiment demonstrated that ErrP can be extended to conflict resolution in air traffic control, a higher level cognitive task. Based on the ErrP finding, we obtained experimental evidence that human preference can be encapsulated into the ErrP signature. Decoding of this signature can be used to adapt an intelligent agent to behave in the way human will trust. While the resolution advisory was hand designed in this experiment, it is possible implement the advisory model using inverse RL, where it can be proposing random maneuver initially and gaining air traffic control skill by iterative interaction with ATCOs.

In the succeeding parts of the paper, we will discuss how the ErrP obtained from our protocol can be used to adapt the advisory tool.

Table 1. Mapping between ErrP and preference.

	non ErrP	ErrP
Same direction	Accept	Reject
Opposite direction	Reject	Accept

4.1   Mapping Between ErrP and Preference

As ErrP is supposed to be triggered when the outcome of the positional change does not match the expectation. Hence, the expected positional change can be derived using the existence of ErrP and the outcome of the assigned positional change. The mapping between ErrP and preference (accept or reject) can be found in the Table 1.

4.2   ErrP as Building Block for Reinforcement Learning

The mapping obtained from Subsect. 4.1 can be used as a reinforcement signal to adapt advisory tool. One of the straightforward approach is to adopt reinforcement learning and define the reward function of reinforcement learning using ATCO's acceptability.

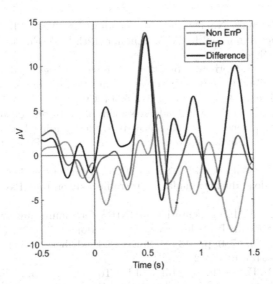

Fig. 5. The difference between ErrP and Non ErrP.

Let define the reward function $\mathcal{R}_0$ as follows:

$$\mathcal{R}_0(resolution) = \begin{cases} 1 & Preference = Accept \\ -1 & Preference = Reject \\ 0 & Preference = Not\,Available \end{cases} \qquad (1)$$

The reward function $\mathcal{R}_0$ can be used in addition to the environmental reward function $\mathcal{R}_1$ which access the quality of the resolution based on common criteria such as deviation, travel distance, travel time or fuel consumption. These rewards can be combined by weighted sum.

$$\mathcal{R}(resolution) = \omega_0 \mathcal{R}_0(resolution) + \omega_1 \mathcal{R}_1(resolution) \qquad (2)$$

4.3 Possible Future Work

As the focus of this work is to study ErrP for higher level cognitive task, several restriction was made to control and simplify the simulated environment. Relaxing the restriction and assumption part by part is important to improve the practicality of this work for solving real-world air traffic problem.

While the framework is designed for air traffic task, it is also applicable to amyotrophic lateral sclerosis (ALS) patients where they can construct their goal-oriented sequential decision making command.

References

1. Allignol, C., Barnier, N., Durand, N., Gondran, A., Wang, R.: Large scale 3D en-route conflict resolution. In: ATM Seminar, 12th USA/Europe Air Traffic Management R&D Seminar
2. Aricò, P., et al.: Adaptive automation triggered by eeg-based mental workload index: a passive brain-computer interface application in realistic air traffic control environment. Front. Hum. Neurosci. **10**, 539 (2016)
3. Ayaz, H., et al.: Cognitive workload assessment of air traffic controllers using optical brain imaging sensors. In: Advances in Understanding Human Performance: Neuroergonomics, Human Factors Design, and Special Populations, pp. 21–31 (2010)
4. Brookings, J.B., Wilson, G.F., Swain, C.R.: Psychophysiological responses to changes in workload during simulated air traffic control. Biol. Psychol. **42**(3), 361–377 (1996)
5. Bunce, S.C., et al.: Implementation of fNIRS for monitoring levels of expertise and mental workload. In: Schmorrow, D.D., Fidopiastis, C.M. (eds.) FAC 2011. LNCS (LNAI), vol. 6780, pp. 13–22. Springer, Heidelberg (2011). https://doi.org/10.1007/978-3-642-21852-1_2
6. Chavarriaga, R., Ferrez, P.W., Millán, J.R.: To Err is human: learning from error potentials in brain-computer interfaces. In: Wang, R., Shen, E., Gu, F. (eds.) Advances in Cognitive Neurodynamics ICCN 2007. Springer, Dordrecht (2008). https://doi.org/10.1007/978-1-4020-8387-7_134
7. Gao, Y., et al.: Reinforcement learning from imperfect demonstrations. arXiv preprint arXiv:1802.05313 (2018)
8. Hadfield-Menell, D., Russell, S.J., Abbeel, P., Dragan, A.: Cooperative inverse reinforcement learning. In: Advances in Neural Information Processing Systems, pp. 3909–3917 (2016)
9. Hao, S., Cheng, S., Zhang, Y.: A multi-aircraft conflict detection and resolution method for 4-dimensional trajectory-based operation. Chinese Journal of Aeronautics **31**(7), 1579–1593 (2018). http://www.sciencedirect.com/science/article/pii/S1000936118301705
10. Hester, T., et al.: Deep Q-learning from demonstrations. In: Thirty-Second AAAI Conference on Artificial Intelligence (2018)
11. Hilburn, B., Westin, C., Borst, C.: Will controllers accept a machine that thinks like they think? the role of strategic conformance in decision aiding automation. Air Traffic Control Q. **22**(2), 115–136 (2014)
12. Ho, J., Ermon, S.: Generative adversarial imitation learning. In: Advances in Neural Information Processing Systems, pp. 4565–4573 (2016)

13. Jilkov, V.P., Ledet, J.H., Li, X.R.: Multiple model method for aircraft conflict detection and resolution in intent and weather uncertainty. IEEE Transact. Aerosp. Electron. Syst. **55**, 1–1 (2018)
14. Kim, S.K., Kirchner, E.A., Stefes, A., Kirchner, F.: Intrinsic interactive reinforcement learning-using error-related potentials for real world human-robot interaction. Sci. Rep. **7**(1), 17562 (2017)
15. Kuchar, J.K., Yang, L.C.: A review of conflict detection and resolution modeling methods. IEEE Transact. Intell. Transp. Syst. **1**(4), 179–189 (2000)
16. Liu, Z., Cai, K., Zhu, X., Tang, Y.: Large scale aircraft conflict resolution based on location network. In: 2017 IEEE/AIAA 36th Digital Avionics Systems Conference (DASC), pp. 1–8 (2017)
17. Nieuwenhuis, S., Ridderinkhof, K.R., Blom, J., Band, G.P., Kok, A.: Error-related brain potentials are differentially related to awareness of response errors: evidence from an antisaccade task. Psychophysiology **38**(5), 752–760 (2001)
18. Odom, P., Natarajan, S.: Active advice seeking for inverse reinforcement learning. In: Proceedings of the 2016 International Conference on Autonomous Agents and Multiagent Systems, pp. 512–520. International Foundation for Autonomous Agents and Multiagent Systems (2016)
19. Pan, X., Shen, Y.: Human-interactive subgoal supervision for efficient inverse reinforcement learning. In: Proceedings of the 17th International Conference on Autonomous Agents and MultiAgent Systems, pp. 1380–1387. International Foundation for Autonomous Agents and Multiagent Systems (2018)
20. Pfurtscheller, G., Neuper, C.: Motor imagery and direct brain-computer communication. Proc. IEEE **89**(7), 1123–1134 (2001)
21. Polich, J.: Updating P300: an integrative theory of P3a and P3b. Clin. Neurophysiol. **118**(10), 2128–2148 (2007)
22. Radanovic, M., Eroles, M.A.P., Koca, T., Gonzalez, J.J.R.: Surrounding traffic complexity analysis for efficient and stable conflict resolution. Transp. Res. Part C Emerg. Technol. **95**, 105–124 (2018). http://www.sciencedirect.com/science/article/pii/S0968090X18302353
23. Ross, S., Gordon, G., Bagnell, D.: A reduction of imitation learning and structured prediction to no-regret online learning. In: Proceedings of the Fourteenth International Conference on Artificial Intelligence and Statistics, pp. 627–635 (2011)
24. Salazar-Gomez, A.F., DelPreto, J., Gil, S., Guenther, F.H., Rus, D.: Correcting robot mistakes in real time using EEG signals. In: 2017 IEEE International Conference on Robotics and Automation (ICRA), pp. 6570–6577. IEEE (2017)
25. Schaal, S.: Learning from demonstration. In: Advances in Neural Information Processing Systems, pp. 1040–1046 (1997)
26. Yang, Y., Zhang, J., Cai, K., Prandini, M.: Multi-aircraft conflict detection and resolution based on probabilistic reach sets. IEEE Transact. Control Syst. Technol. **25**(1), 309–316 (2017)
27. Yokoyama, N.: Decentralized conflict detection and resolution using intent-based probabilistic trajectory prediction. In: AIAA SciTech Forum, American Institute of Aeronautics and Astronautics (2018). https://doi.org/10.2514/6.2018-1857
28. Zhang, H., Chavarriaga, R., Khaliliardali, Z., Gheorghe, L., Iturrate, I., d R Millán, J.: EEG-based decoding of error-related brain activity in a real-world driving task. J. Neural Eng. **12**(6), 066028 (2015)

UAS Operator Workload Assessment During Search and Surveillance Tasks Through Simulated Fluctuations in Environmental Visibility

Jaime Kerr[1](✉), Pratusha Reddy[1], Shahar Kosti[2], and Kurtulus Izzetoglu[1]

[1] School of Biomedical Engineering, Science and Health Systems, Drexel University, Philadelphia, PA 19104, USA
jkk57@drexel.edu
[2] Simlat, Miamisburg, OH 45342, USA

Abstract. Unmanned aircraft system (UAS) sensor operators must maintain performance while tasked with multiple operations and objectives, yet are often subject to boredom and consequences of the prevalence-effect during area scanning and target identification tasks. Adapting training scenarios to accurately reflect real-world scenarios can help prepare sensor operators for their duty. Furthermore, integration of objective measures of cognitive workload and performance, through evaluation of functional near infrared spectroscopy (fNIRS) as a non-invasive measurement tool for monitor of higher-level cognitive functioning, can allow for quantitative assessment of human performance. This study sought to advance previous work regarding the assessment of cognitive and task performance in UAS sensor operators to evaluate expertise development and responsive changes in mental workload.

Keywords: Unmanned aircraft systems · Sensor operator · fNIRS · Cognitive workload

1 Introduction

Sensor operators are tasked with controlling unmanned aircraft systems (UAS) by maintaining flight systems and controlling visualization equipment onboard, typically in unpredictable and demanding environments regarding reconnaissance missions. Sensor operators face two main issues in the completion of the two aforementioned tasks, including: (1) high amounts of concurrent subtasks to complete, and (2) the nature of UAS ground control station (GCS) system operations removes operators from the physical environment, impairing their ability to assess contextual cues. As a consequence, the information-processing load and decision-making demands are increased for UAS operators, which in turn, impacts mission effectiveness, increases human error and safety concerns [1]. To ideally minimize human error among advancing levels of automation in

© Springer Nature Switzerland AG 2019
D. D. Schmorrow and C. M. Fidopiastis (Eds.): HCII 2019, LNAI 11580, pp. 394–406, 2019.
https://doi.org/10.1007/978-3-030-22419-6_28

UAS, development of operator-specific training protocols can aid in maximizing sensor operators' expertise development and enhance human performance [2].

Utilizing objective performance measures can support UAS operators by facilitating development and reinforcement of personalized training and real-time mental workload monitoring; thus, decreasing error and improving operator performance, with utility for training and in-field applications. UAS sensor operator roles can be arduous yet are often subject to boredom. Previous studies have analyzed lack of focus during detection tasks due to vigilance-related error and object prevalence, otherwise referred to as the prevalence effect, highlighting proportionality among detection miss-rates and target prevalence [3, 4]. However, such studies exhibit scope limited to assessments using subjective measures. Further examination regarding prevalence impacts on detection task performance have focused on subjective feature-dependent performance capabilities, including reported correlation with working memory capacity [5, 6].

Working memory (WM) is a cognitive system that temporarily stores limited information for processing availability with executive control domains; hence, WM is imperative for reasoning, decision-making and attentional control [7–9]. Although facets of theorized functional and mechanistic models of working memory are disputed, attentional control and cognitive load—individuals' capacity for targeting attention while attenuating unnecessary information, and the proportion of time or effort associated with task processing, respectively—are aspects involved in WM that are pertinent to learning [8]. Furthermore, adoption of appropriate cognitive resources taxed via highly realistic practice scenarios can aid in training attentional focus [2], and neurophysiological measures provide methods for monitoring higher-level cognitive functioning for assessing working memory and mental workload. To advance previous studies, we are examining changes in mental workload through neurophysiological data measures in response to the participants' completion of scanning and target identification tasks, using Simlat's C-STAR GCS training simulator [10]. Evidenced from prior studies, we hypothesized that increased level of task difficulty will result in mental workload increases. Task difficulty has been manipulated by administering realistic UAS flight operations under different daylight conditions and mental workload was assessed by behavioral performance measures and functional near-infrared spectroscopy (fNIRS).

1.1 Functional Near Infrared Spectroscopy

Emerging wearable functional brain activity monitoring technologies can help evaluate the cognitive status and capacities of sensor operators in GCS settings. Regarding brain activity monitoring of WM and cognitive load, functional magnetic resonance imaging (fMRI), electroencephalography (EEG) and fNIRS have been widely utilized. Various studies have discussed benefits and challenges of different monitoring technologies involved in measuring cognitive activity, particularly within the prefrontal cortex (PFC) and other associated brain areas correlated with WM assessment efficacy [2, 9, 11–15]. fNIRS and fMRI are frequently subject of such assessments, but fNIRS balances portability with easy-to-engineer design while providing relatively higher temporal resolution compared to fMRI [11]. Additionally, fNIRS is easy to deploy and its shorter calibration time makes it suitable for field settings.

fNIRS utilizes near-infrared light to monitor changes in hemodynamic responses, i.e., oxygenated and de-oxygenated hemoglobin from the PFC area. Changes in blood oxygenation are associated with neuronal activation, as neurons require oxygen to metabolize glucose for activation—oxygen is carried to the site in the form of oxy-hemoglobin (HbO) and converted to deoxyhemoglobin (HbR) once used. HbO and HbR absorb photons at different wavelengths of light, the magnitude of which is captured by source detectors at varying wavelengths within the optical window (i.e., 700 nm to 900 nm) [1, 14]. Relative concentrations of HbO and HbR can be quantified through the application of a modified Beer-Lambert law [1]. Previous studies have analyzed the application of fNIRS as a non-invasive measurement tool to evaluate key aspects of higher-level cognitive functioning, task performance, expertise development and cognitive workload through monitoring hemodynamic response from the prefrontal cortex (PFC) areas collated with behavioral measures [1, 2, 14–16].

2 Methods

2.1 Participants

To date, five participants between the ages of 18 and 42 participated in the IRB approved protocol, including 3 males and 2 females. All participants consented and compiled with inclusion criteria. Participants had no previous experience with flight simulators and verified fulfillment of inclusion and exclusion criteria by completing an Attention Deficiency questionnaire, Edinburgh Handedness survey (LQ > 0 acceptance for right-hand dominance) and video gaming experience survey.

2.2 Experimental Protocol

High-fidelity task representation within sensor operator training can assist in the trainee's adoption of appropriate cognitive styles to perform their role effectively [2]. Simlat's C-STAR GCS (Simlat Inc., Miamisburg, Ohio) simulator is a commercially available training simulation system for UAS pilots and utilized in previous proof of concept studies [1, 2, 16] (Figs. 1 and 2). Simlat's system collects behavioral data regarding the trainee's behavioral task performance by calculating evaluation metrics including scanned, not-scanned, and over-scan percent.

Participants with no previous UAS piloting experience completed three training sessions and two probative sessions, during which, they were required to navigate the UAS sensor over six sub-areas while engaged in route scanning and target identification tasks. For all five sessions the dimensions of each sub-area increased by fifteen percent consecutively along the flight path, and environmental variables for all tests were established from system sunlight and weather data for Mallorca (Majorca), Spain, on September 1, 2018 to administer different levels of task-load conditions. All simulated scenarios utilized a generic tactical unmanned vehicle (G-TAC UAV) system that maintained a fixed route, following the same scan areas and UAV initial state

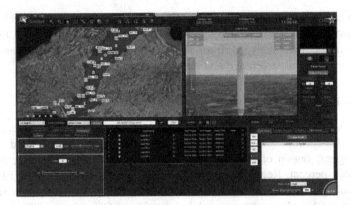

Fig. 1. Simlat C-STAR instructor screen

Fig. 2. Simlat C-STAR trainee screen

(Altitude: 2000 ft; Speed: 60 kts; Heading: 70°.). Differences among training sessions and probative sessions were established by manipulating the task load via simulated start time; three consecutive training sessions were set for a simulated time lapse from 11:00 to 11:30, while probative sessions one and two were set for simulated time lapses from 06:00 to 06:30 and 20:00 to 20:30, respectively.

Participants were instructed to complete two synchronous tasks: (1) operate the UAS sensor camera to scan along the designated flight path for each consecutive subarea, and (2) to identify and track at least one threat target within each scan area, marked by a red civilian bus. Participants were informed that a zoom angle lower than 15° and three-second target lock were classified as successful scans and target identification, respectively. The duration of each testing scenario was 24.12 min, and subjects were allotted breaks as requested.

Neurophysiological and behavioral performance data were simultaneously captured for elapsed time of participant sessions using a 16-channel fNIRS device and C-STAR integrated Performance Analysis & Evaluation Module (PANEL) logs, respectively. Further analysis was completed to evaluate cognitive performance during task completion, and potential effects of task-load changes in environmental visibility, using observed levels of oxygenated hemoglobin from participants' PFC area. Behavioral task performance was analyzed from system evaluations including scanned, not-scanned, over-scanned percentage and target detected indicator.

2.3 Behavioral Data Processing

The C-STAR PANEL module captured simulator performance metrics for each subject trial and area-specific scan task, including number of scans, field of view (FOV) polygons, region of interest (RoI) scanned percentage, RoI not-scanned, and RoI over-scan percent. Regarding the target identification tasks, the PANEL module captured trial and individual target identification task data, including trial start time, elapsed time, FOV polygon zoom level, target within FOV polygon (TRUE, FALSE) and confirmed target identification (Found = 1, Not Found = 0).

2.4 fNIRS Signal Processing

For the purpose of this study, we utilized a 16-channel fNIRS system with a sampling frequency of 2 Hz., consistent with previous studies [1, 2, 16] (see Fig. 3). fNIRS signals are susceptible to artifacts resulting from instrument noise, physiological noise and motion artifacts. Several signal pre-processing techniques were applied for artifact removal and to improve sensitivity and spatial specificity of brain activity measures [16].

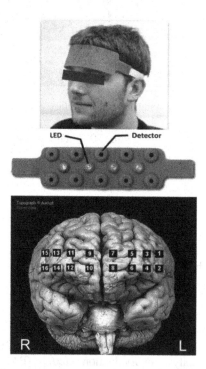

Fig. 3. 16-Channel fNIRS system, node and quadrant location diagram of PFC areas (quadrant 1: 1, 2, 3, 4) (quadrant 2: 5, 6, 7, 8) (quadrant 3: 9, 10, 11, 12) (quadrant 4: 13, 14, 15, 16)

Signal quality analysis and channel rejection were determined from saturation effects and high noise levels. Pre-processing was completed through application of a low pass filter with finite impulse response and cutoff frequency of 0.014 Hz for each channel, to tease instrumental noise, cardiac output, and respiration. To mediate low frequency drift, a linear detrending algorithm was applied. Additionally, due to DC shifts in amplitude for some channels, a temporal derivative distribution repair (TDDR) motion correction method was applied [17]. Following the completion of artifact removal and extraction of processed optical density data, a modified Beer-Lambert Law (MBLL) was applied for calculations of channel-specific changes in HbO and HbR. HbO and HbR measures were utilized for further derivation of channel-specific oxygenation (Hb-diff) and total hemoglobin (Hb_Tot). Finally, sampled outliers were categorized from real-time data as being greater than three standard deviations above the expected values and removed from subsequent analyses.

3 Results

Behavioral performance changes were most clearly determined from measurements for over-scan, rather than scanned, given that participants had no immediate indication of the RoI on their map screen. As an example, Fig. 4 shows the standard PANEL scan task evaluation: the RoI (blue) boundary indicates the region where the scanning task is assigned, but not shown to the operator. The polygon (orange) within field-of-view (FOV) shows a scan overstepping the RoI (blue) boundary, which is classified as over-scan. The polygon (green) encapsulated by the RoI is classified as a proper scan. Scan percentage is extrapolated as the intersection of the RoI area (blue) with the union of all FOV polygons during the elapsed time of the scan task. Alternatively, over-scan percentage was computed by subtracting the area of the RoI polygon from the union of all FOV polygons (black outline).

Provided the protocol design, the initial three training sessions were purposed to facilitate participants' sufficient familiarity of using the system under stable conditions regarding visibility. Upon fulfillment of these sessions, the goal was to study the effects of varying the task difficulty on mental workload, with the assumption all participants have no similar task familiarity and training time. Therefore, we focused on behavioral performance and neurophysiological measurement results of probative sessions only. Workload (i.e., probative) session one started under dark lighting conditions for scan-task one, followed by increasingly improved lighting conditions up to scan-task six— hence, we classified this session as decreasing in difficulty level with each consecutive scanning and target identification task. Figure 5 shows changes in participants' over-scan measures for scan-tasks one and six according to workload, noted as hard to easy, respectively. Figure 5 exhibits a general trend for subjects 1, 3, 6 and 11 of decreased performance associated with increasing workload conditions.

According to the hypothesis, workload conditions are positively related to oxygenation levels, meaning that as task difficulty increases, oxygenation levels should increase provided that the subjects stayed on task. fNIRS channels were grouped by quadrants (i.e., Q1 = Channels 1 to 4; Q2 = Channels 3 to 8; Q3 = Channels 9 to 12; Q4 = Channels 13 to 16) and Hb-diff levels were extracted and averaged per

Fig. 4. Simlat PANEL FOV polygon classification (Color figure online)

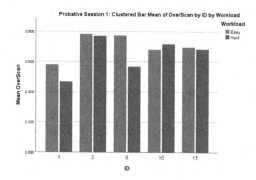

Fig. 5. Probative session 1: clustered means comparison of over-scan percentage sorted from easy workload conditions (blue) to hard (red) (Color figure online)

participant, workload condition and task. Figure 6 presents averages of oxygenation levels for each subject, for Q1 (a) to Q4 (d). Figure 6(a, b) Q1 and Q2 data present trends for subjects 1, 3, 6 and 11 consistent with our hypothesis of decreased oxygenation levels during the lower workload conditions; subject 10 did not present any significant interaction with the workload changes. Figure 6(c) shows Q3 trends among participants 1, 3, 10 and 11 consistent with our hypothesis, but subject 6 did not have apparent trends; whereas, Fig. 6(d) demonstrates that Q4 had hypothesis-consistent trends for subjects 6, 10 and 11, with no similar trend for 1 and higher oxygenation levels for lower workload conditions in 3.

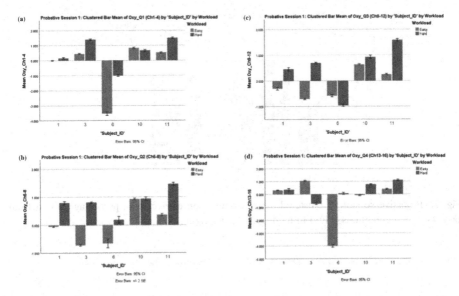

Fig. 6. Probative session 1 fNIRS quadrant Hb-diff Levels comparison between easy (blue) workload conditions and hard (red) for: (a) quadrant 1 (Ch 1, 2, 3, 4), (b) quadrant 2 (Ch 5, 6, 7, 8), (c) quadrant 3 (Ch 9, 10, 11, 12) and (d) quadrant 4 (Ch 13, 14, 15, 16) (Color figure online)

Consequently, for probative session two, participants' over-scan performance was expected to decrease from scan-task one to six, classified from easy to hard, respectively; over-scan performance trends identified in Fig. 7 are consistent with this prediction for subjects 1 and 6. However, improved scan performance was observed for subjects 3, 10 and 11. Figure 8(a–d) demonstrate Q1–Q4 consistency with our hypothesis of oxygenation level dynamics for subject 3. Figure 8(a) presents Q1 trends opposing our hypothesis, with lower levels of oxygenation for higher workload conditions in participants 1, 6, 10 and 11. Figure 8(b, c) show Q2 and Q3 consistency regarding opposing trends in 1, 10 and 11, while (b) presents no trend for 6 and (c) demonstrates an oxygenation level increase for 6. Fig. 8(d) demonstrates Q4 hypothesis consistent trends for subjects 6 and 10, yet opposing trends were observed in subjects 1 and 11.

To further examine the relationship of oxygenation level dynamics with workload conditions, the most difficult scan-task workload condition was compared to the least-difficult workload condition, taken as probative session two scan-task six and probative session one scan-task six, respectively. Difficulty level determinations were obtained from comparisons of visibility levels for probative scan-tasks to that of the training sessions, with the least difficult condition being similar to conditions of training tasks, and the most difficult being the task with the lowest lighting condition. Figure 9 demonstrates the comparison between over-scan measurements from the aforementioned *easiest* and *hardest* tasks. Fig. 9 indicates decreases in scan performance during the higher workload condition for subjects 1 and 6, and increases for 3, 10 and 11. Figure 10(a–c) present Q1–Q3 oxygenation level movements in line with our

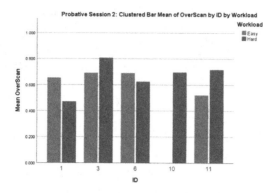

Fig. 7. Probative session 2: clustered means comparison of over-scan percentage sorted from easy workload conditions (blue) to hard (red) (Color figure online)

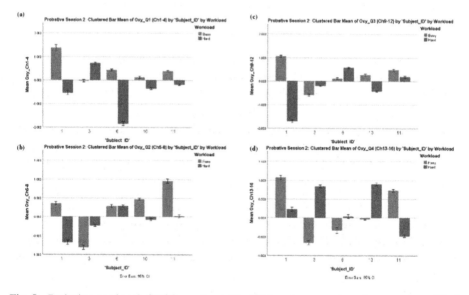

Fig. 8. Probative session 2 fNIRS quadrant Hb-diff Levels comparison between easy (blue) workload conditions and hard (red) for: (a) quadrant 1 (Ch 1, 2, 3, 4), (b) quadrant 2 (Ch 5, 6, 7, 8), (c) quadrant 3 (Ch 9, 10, 11, 12) and (d) quadrant 4 (Ch 13, 14, 15, 16) (Color figure online)

hypothesis for participants 3 and 6, with large disparity in Hb-diff levels between the hardest (red) and easiest(blue) workload conditions in 6, but conflicting Hb-diff level movements for subjects 1, 10 and 11—with significant difference between workload conditions in (b) and (c) for subject 1. Figure 10(d) demonstrates Hb-diff level movement in agreement with our hypothesis for subjects 6 and 10, no similar trend observed for 1, and opposing Hb-diff movement in 3 and 11.

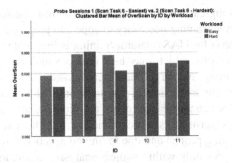

Fig. 9. Probative sessions over-scan comparison according to workload conditions: easiest (blue; probative session 1, scan-task 6), hardest (red; probative session 2, scan-task 6) (Color figure online)

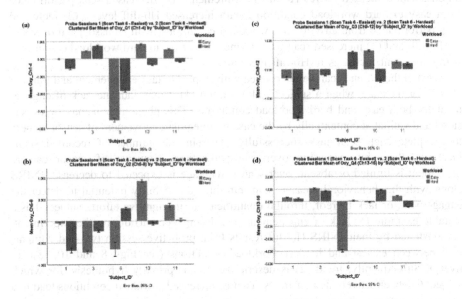

Fig. 10. Probative sessions fNIRS Hb-diff levels comparison between easiest (blue; probative session 1, scan-task 6) workload conditions, and hardest (red; probative session 2, scan-task 6) for: (a) quadrant 1 (Ch 1, 2, 3, 4), (b) quadrant 2 (Ch 5, 6, 7, 8), (c) quadrant 3 (Ch 9, 10, 11, 12) and (d) quadrant 4 (Ch 13, 14, 15, 16) (Color figure online)

4   Discussion

Sensor operators have a crucial role in the completion of mission objectives for UAS flight crew, particularly within the realm of reconnaissance missions. Increased information-processing load and decision-making demands require vast cognitive resource utility; however, such increases in attentional control and cognitive load have been associated with increases in human error, and thus escalating safety concerns.

Proper neurophysiological assessment with realistic training scenarios can facilitate personalized and adaptive training necessary to cope and prepare for such mental strain. Current evaluation criteria for UAS operator training is limited to subjective behavioral and performance evaluations, which may lack in-depth understanding of what factors may contribute to scanning and target-identification task performance. Use of neurophysiological measures for extraction of mental workload-associated parameters can provide objective performance measurements and contextual clues for evaluators to help sensor operators perform tasks more efficiently and effectively.

Since this is an ongoing project and due to limited number of participants, the scope of this manuscript includes only within-subject analysis. Evaluation of results from all subjects require understanding of engagement and training effect. There are four possible trends mainly observed between behavioral and fNIRS measurements in this preliminary study: (1) decreased scan performance from easy to hard workload conditions, with decreased fNIRS Hb-diff measurements; (2) decreased scan performance from easy to hard workload conditions, with increased Hb-diff levels; (3) increased scan performance from easy to hard workload conditions, with an associated decrease in Hb-diff; and (4) increased scan performance from easy to hard workload conditions, with concurrent increases in Hb-diff measures.

Each of these trends can be interpreted with respect to task engagement and training effect. For instance, when a subject follows trend (1) it may indicate lack of engagement for both easy and hard workload conditions. This phenomenon may be understood and validated contextually by the case with requisite movement of sensor camera to complete each scan task successfully, wherein the number of recorded scans increases, there by increasing over-scan percentage. However, if sensor camera movement is limited or absent, over-scan percentage is expected to decrease. fNIRS along with this behavioral measure can introduce a capability potential to detect the engagement, or lack thereof, through quantification of cognitive effort. Subject 1 is a suitable example for lack of engagement (i.e. disengagement) during the experiment, demonstrated by their fNIRS Hb-diff results from probative session two and comparison between easiest and hardest workload conditions (see Figs. 8 and 10, respectively). Subsequently, trend (2) is descriptive of our primary hypothesis, for which subject 6 data effectively demonstrates. That is, increased workload conditions lead to a rise in oxygenation changes acquired from the PFC region. This trend may suggest that if a subject is sufficiently trained on the task—which was achieved through completion of the three initial sessions with optimal visibility—they would likely have the ability to complete such tasks successfully with minimal cognitive effort. If the same trained participant continued active engagement, then administration of more difficult tasks would cause relative increases in cognitive load, revealing that user was able to stay on task and able to keep up with the challenge introduced. Subject data following trend (3) appears akin to trend (2) regarding easier workload conditions; however, improvement in behavioral over-scan percentage concurrent with decrease in brain activity assessed by Hb-diff measures. This preliminary finding suggests that the hardest workload conditions were too high for a given subject, during which (s)he was overloaded and failed to keep up with the task objectives, i.e., camera movement for route scanning and target identification. Finally, subjects outlining trend (4) exhibited proficient task engagement under high-workload conditions. Behavior of trend

(4) subjects are comparable to trend (3) under the easier workload conditions, yet their cognitive effort and behavioral performance rose with increased workload conditions.

The study reported here has ongoing recruitment and testing efforts; more data are being collected for further examination of correlations, trend classification and associated statistical significance. Data collection and analyses of the relationship between task engagement, task performance and correlational analysis with localized oxygenation changes in PFC shall be carried out across subjects in the following study. Although the sample population is limited, this study reported mainly within-subjects results with promising indication of trends and similar results reported in [1, 2, 14, 16].

Acknowledgements. The authors would like to thank Simlat, Inc for providing access to their UAS training simulator.

References

1. Richards, D., Izzetoglu, K., Shelton-Rayner, G.: UAV operator mental workload - a neurophysiological comparison of mental workload and vigilance. In: AIAA Modeling and Simulation Technologies Conference. American Institute of Aeronautics and Astronautics (2017)
2. Armstrong, J., Izzetoglu, K., Richards, D.: Using functional near infrared spectroscopy to assess cognitive performance of UAV sensor operators during route scanning. In: 11th International Joint Conference on Biomedical Engineering Systems and Technologies, vol. 3 (2018)
3. Wolfe, J.M., Horowitz, T.S., Kenner, N.M.: Rare items often missed in visual searches. Nature **435**, 439 (2005)
4. Wolfe, J.M., et al.: Low target prevalence is a stubborn source of errors in visual search tasks. J. Exp. Psychol. Gen. **136**(4), 623–638 (2007)
5. Schwark, J., Sandry, J., Dolgov, I.: Evidence for a positive relationship between working-memory capacity and detection of low-prevalence targets in visual search. Perception **42**(1), 112–114 (2013)
6. Peltier, C., Becker, M.W.: Individual differences predict low prevalence visual search performance. Cogn. Res. Princ. Implic. **2**(1), 5 (2017)
7. Chai, W.J., Abd Hamid, A.I., Abdullah, J.M.: Working memory from the psychological and neurosciences perspectives: a review. Front. Psychol. **9**, 401 (2018)
8. Cowan, N.: Working memory underpins cognitive development, learning, and education. Educ. Psychol. Rev. **26**(2), 197–223 (2014)
9. D'Esposito, M., Postle, B.R.: The cognitive neuroscience of working memory. Annu. Rev. Psychol. **66**(1), 115–142 (2015)
10. Products | Simlat, 04 March 2019. https://www.simlat.com/products
11. Ayaz, H., et al.: Continuous monitoring of brain dynamics with functional near infrared spectroscopy as a tool for neuroergonomic research: empirical examples and a technological development. Front. Hum. Neurosci. **7**, 871 (2013)
12. Berivanlou, N.H., Setarehdan, S.K., Noubari, H.A.: Quantifying mental workload of operators performing n-back working memory task: toward fNIRS based passive BCI system. In: 2016 23rd Iranian Conference on Biomedical Engineering and 2016 1st International Iranian Conference on Biomedical Engineering (ICBME) (2016)

13. Braver, T.S., et al.: A parametric study of prefrontal cortex involvement in human working memory. NeuroImage **5**(1), 49–62 (1997)
14. Hernandez-Meza, G., Slason, L., Ayaz, H., Craven, P., Oden, K., Izzetoglu, K.: Investigation of functional near infrared spectroscopy in evaluation of pilot expertise acquisition. In: Schmorrow, Dylan D., Fidopiastis, Cali M. (eds.) AC 2015. LNCS (LNAI), vol. 9183, pp. 232–243. Springer, Cham (2015). https://doi.org/10.1007/978-3-319-20816-9_23
15. Hill, A.P., Bohil, C.J.: Applications of optical neuroimaging in usability research. Ergon. Des. **24**(2), 4–9 (2016)
16. Reddy, P., Richards, D., Izzetoglu, K.: Cognitive performance assessment of UAS sensor operators via neurophysiological measures. In: 2nd International Neuroergonomics Conference (2018)
17. Fishburn, F.A., et al.: Temporal derivative distribution repair (TDDR): a motion correction method for fNIRS. NeuroImage **184**, 171–179 (2019)

Processing Racial Stereotypes in Virtual Reality: An Exploratory Study Using Functional Near-Infrared Spectroscopy (fNIRS)

Gyoung Kim, Noah Buntain, Leanne Hirshfield, Mark R. Costa, and T. Makana Chock[(✉)]

The S.I. Newhouse School of Public Communications, Syracuse University, Syracuse, USA
{gkim, nkbuntai, lmhirshf, mrcosta, tmchock}@syr.edu

Abstract. This within-subjects exploratory study examined users' (N = 13) neurological responses to a racially-charged VR experience. The goals of the study are (1) to test a new method of assessing neural activity while users are experiencing VR using non-invasive functional near-infrared spectroscopy (fNIRS) device and VR headset, and (2) to compare activation in areas of the pre-frontal cortex that have been found to be associated with prejudice and stereotyping (specifically the mPFC and lPFC) while participants are exposed to a racially-charged VR experience vs. a non-racially charged VR experience. There were no significant differences in mPFC activity between the two types of VR experiences suggesting no differences in empathy or "humanizing" for the characters. However, in the racially charged experience, significantly greater activation in the right lPFC was found which could indicate negative stereotype activation. In addition, significantly greater activation in the left lPFC occurred in the racially-charged VR experience which could indicate stereotype inhibition or regulation of negative stereotypes.

Keywords: Virtual reality · Brain measurement in social neuroscience · Racial prejudice

1 Introduction

Over the last few years, virtual reality (VR) technology has moved from laboratories to living rooms. In the process, the technology has branched into platforms that are fully immersive (VR), and those that use digital content to augment (AR) or interact with (MR) the real world. VR content has been created for journalism, health communications, education, entertainment, occupational training, and clinical psychology, among others.

VR has been proposed as tool that could reduce prejudice and stereotyping. For example, VR embodiment experiences have been used to temporarily decrease implicit

© Springer Nature Switzerland AG 2019
D. D. Schmorrow and C. M. Fidopiastis (Eds.): HCII 2019, LNAI 11580, pp. 407–417, 2019.
https://doi.org/10.1007/978-3-030-22419-6_29

racial and gender biases [1]. The immersive nature of VR experiences [2], however, and their potential for behavioral learning [3], could also potentially activate or reinforce negative stereotypes and prejudices. It is therefore important to develop a better understanding of users' reactions to racially-charged VR experiences.

This exploratory study examines users' neurological responses to a racially-charged VR experience. The goals of the study are first to test a new method of assessing neural activity while users are experiencing VR. We will be measuring neural activity using a modified non-invasive functional near-infrared spectroscopy (fNIRS) device and HTC Vive headset. Second, we will use this device to examine activation in areas of the prefrontal cortex that have been found to be associated with prejudice and stereotyping (specifically the mPFC and lPFC) while participants are exposed to a racially-charged VR experience.

2 Assessing Prejudice and Stereotyping

2.1 Developing a Measurement Device

Social neuroscience research has identified patterns of brain activation that are associated with specific types of social processes [4]. Much of the research in this area, however, has been limited to assessing responses to 2D visualizations.

Virtual reality offers researchers the potential for multi-sensory stimuli that more closely approximate real-world social situations [5]. VR headsets isolate viewers from their immediate environments and present visual and auditory input. Some VR experiences feature touch input in the form of force feedback in handheld controllers. Since these experiences are computer-generated, they benefit researchers by being flexible, adaptable, multimodal, repeatable, and more ecologically valid [6].

VR headsets, however, pose practical challenges for measuring the neural activations associated with cognitive processes. The two most popular methods of measuring brain activations are fMRI and EEG. EEG measures the electrical potential of large groups of neurons. While EEG has proved useful in establishing regions of interest, it has low spatial resolution [7] and produces a noise in signal by head movement. In addition, the straps that keep VR headsets such as the Oculus Rift and HTC Vive in place limit the placement of EEG sensors. Functional magnetic resonance imaging (fMRI) measures blood flow in the brain that correlates with neuronal activation. Widely used in neuroscience, it provides high spatial resolution [8]. However, the strong magnetic fields produced by the fMRI coils interfere with VR function [3], and head movement associated with users looking around a virtual environment make accurate tracking of regions of interest difficult to impossible in fMRI [6, 9].

Functional near-infrared spectroscopy (fNIRS) combines the spatial acuity of fMRI and the (relative) freedom of movement of EEG. The fNIRS device uses sensors that send near-infrared through the scalp to measure blood flow in the cerebral cortex. The sensors transmit and subsequently measure the reflected light, revealing changes in oxy- and deoxy-hemoglobin concentrations in the blood [10].

Despite these advantages, VR headsets are incompatible with the standard fNIRS sensor cap. The present study follows Seraglia et al. [11] in developing a custom-built fNIRS cap for use with VR, but updates the design to fit consumer-grade VR headsets [9].

2.2   Neural Activity in Prejudice and Stereotyping

Prejudice and stereotyping are interrelated but separate mental processes and are therefore associated with activation of different areas of the brain. Prejudice is an emotionally-laden attitude towards an individual or group and has both affective and motivational elements. Stereotyping involves the acquisition related to a group and the application of these concepts in judgments and related behaviors. A number of neural structures have been identified as playing a role in prejudice and stereotyping [12]. Although some of these structures (e.g. the amygdala) are not accessible using the fNIRS systems, other regions can be examined. Two key areas are the medial prefrontal cortex (mPFC) and lateral prefrontal cortex (lPFC).

The mPFC plays an important role in social cognition [12]. Numerous studies (see [21]) have determined that the mPFC is of key importance in the process of social mentalizing or making inferences or judgments about self and others. The mPFC is also involved in humanizing, empathizing and identifying with others [13–16]. The mPFC is therefore typically activated when people observe or think about others. A number of studies, however, have found that the mPFC is not recruited when people observe members of disparaged and despised groups [17–20]. The lack of activation in the mPFC may therefore indicate a lack of humanization or empathy for a despised out-group. This type of reaction, or lack of reaction, may reflect prejudice towards members of specific racial groups.

The mPFC has also been found to be engaged in the stereotyping process, although its precise role is unclear [12]. The process of inferring traits or characteristics about members of racial minority groups involves social mentalizing and thus activates the mPFC [21]. The mPFC is also activated in the process of making evaluative attitude judgments about those traits or characteristics [18, 22].

The lateral prefrontal cortex (lPFC) is involved in the processing of stereotypes. The left lPFC is recruited in the social reasoning process, drawing on working memory, and integrating information. Applying stereotypes drawn from working memory to social judgments and behaviors has been found to elicit activity in the left lPFC [20].

In contrast, the right lPFC has been found to play a role in stereotype regulation or inhibition. Numerous studies have found activation of the amygdala in response to exposure to images of other-race individuals. This automatic negative arousal response, however, can be regulated or suppressed by the right lPFC. This region is associated with behavioral control and has been found to co-vary with amygdala activation in experimental studies assessing implicit attitude bias. In individuals who are motivated to control these stereotype biases, the right lPFC is activated and serves as a form of inhibitory control on initial negatively arousing responses [12, 23].

2.3 Research Questions

As mPFC activation has been found in response to exposure to images or depictions of other people, we would expect to find similar types of activation in response to VR characters. However, responses to VR characters in a story that could activate negative prejudicial responses (an encounter between a White police officer and an African-American pedestrian) might not elicit mPFC activation to the same extent. Therefore, we propose the following research question:

> *RQ1: Are there significant differences in mPFC activation in response to a stereotype-inducing VR scenario (Police Encounter) compared to a neutral VR scenario (Holiday)?*

Given that the lPFC has been found to play a role in both stereotype activation and application (left lPFC) and in stereotype suppression (right lPFC), we will also examine the following:

> *RQ2(a): Are there significant differences in right lPFC activation in response to a stereotype-inducing VR scenario (Police Encounter) compared to a neutral VR scenario (Let Hawaii Happen)?*
> *RQ2(b): Are there significant differences in left lPFC activation in response to a stereotype-inducing VR scenario (Police Encounter) compared to a neutral VR scenario (Let Hawaii Happen)?*

3 Experiment

To examine these research questions, a within-subjects experiment was conducted comparing responses to a racially-charged VR experience and a non-racially charged experience. Activation in the mPFC and lPFC was examined using a modified fNIRS device.

3.1 Equipment

Functional Near-Infrared Spectroscopy. fNIRS data was collected using the Hitachi ETG-4000 functional near-infrared spectroscopy device to measure brain activity by calculating degrees of Oxy & De-oxy hemoglobin of the targeted area with a sampling of 10 Hz.

Using a VR headset in neuroscience research has been challenging because the headset mounts and headband interfere with the probes of the device. Recently, Kim and his colleagues [9] developed a special 3-D custom printed cap integrating brackets for the HMD and a sensor cap. In this exploratory study, however, we adapted two 3×3 fNIRS modules and attached these to the main band of the HMD using Velcro and multiple cable ties. See Fig. 1.

Fig. 1. fNIRS measurement setup with Head-Mounted Display: the custom probe design placed on participants was designed to cover the mPFC and lPFC brain regions.

Channel Configuration A total of 24 channels (2 3 × 3 optodes configuration) with 18 optodes were used in this study. Each optode works as either a detector or emitter. The blue circles in Fig. 2 indicate detectors and red circles indicate emitters.

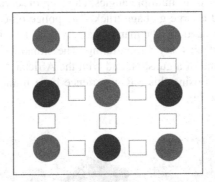

Fig. 2. Configuration of fNIRS optodes and channels (Color figure online)

Channels are measured in the areas between emitters and detectors. Using this combination, we setup 24 channels. Channel numbers have been rearranged to provide a concise report (Fig. 3).

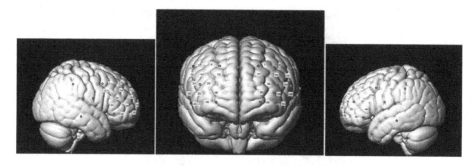

Fig. 3. Channel (*24 ch.*) setup for the experiment

3.2 Stimulus Materials

Study participants experienced two different VR short stories developed in Unity and compatible with the HTC Vive.

Racially-Charged VR Experience (Police Encounter)

A 3-minute short "story" was constructed by lab researchers using Unity. The VR story is set on a city street. During the first minute, the VR scene simply consists of a city street lined with parked cars. After the first minute, a police officer and an African American man have an encounter on the far side of the street. Although the participant may view this encounter from multiple angles, close-up observations of the encounter are blocked by cars and a large garbage truck. The police officer stops the pedestrian and asks to see his identification. The man repeatedly asks the officer to explain why he has been stopped and denies any wrong-doing. The situation escalates quickly. The scene ends with the sound of a gunshot, and with the African American man falling to the ground. Prior studies using this VR experience have found that it elicits racially charged stereotypes. See Fig. 4.

Fig. 4. A screenshot of Police scenario (VR)

Non-racially Charged VR Experience (Holiday in Hawaii)
A 3-minute segment of the VR film Let Hawaii Happen developed by Framestore VR Studio was used. In this experience, participants could navigate through a "holiday" experience set in Hawaii and populated by tourists relaxing amid beautiful scenery. See Fig. 5.

Fig. 5. A screenshot of let Hawaii happen (VR)

3.3 Protocol

Thirteen participants with no audiovisual impairments and no history of seizure disease took part in the experiment (6 female and 7 male). They also reported their race; 7 White, 3 Hispanic/Latino, and 3 Asian. All of participants were undergraduate students from a university in the Northeast of the United States. The experiment was administrated individually to participants in 30-minute sessions. Upon arrival in the laboratory, participants were asked to sign an informed consent form, seated at a computer and fitted with the fNIRS/VR headset (Fig. 6). They were then given a short trial session for training in the use of an HTC Vive and handheld controllers. After the trial session, they were asked to experience the Hawaii VR experience and then the Police VR experience.

Fig. 6. fNIRS and HMD setup for the experiment

4   Results

4.1   Exploratory Data Analysis

We used NIRS-SPM [24] to analyze the data we collected from ETG-4000. To clean the data, we used a Motion artifact Correction (MARA) functionality provided in the NIRS-SPM, and removed physiological noise at [0.12 0.35; 0.7 1.5]. In addition, hi-frequency noise was softened with Gaussian Temporal Smoothing.

4.2   Activation Area

Regarding *RQ1,* we found there were no notable differences in mPFC activation between the Police Encounter and Holiday condition. Channel 35, which measures a part of the mPFC, showed a higher level of activation in the Police Encounter experience then the Holiday in Hawaii experience. However, there were no other differences between experiences among all other channels in the mPFC region. Specifically, channels 36, 38, 39, 46, and 49 showed no difference in brain activation between experiences (Fig. 7).

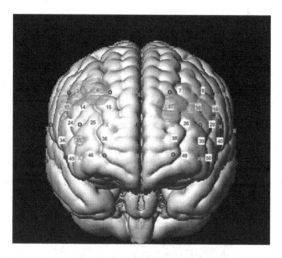

Fig. 7.   Activation area of the brain during Police Encounter scenario

Regarding *RQ2*(a), we found that right lPFC area showed a notably higher level of brain activation in the Police Encounter than in the Holiday in Hawaii experience.

We found that the part of participants' right lPFC area (channels 3, 4, 13) were highly activated while they are experiencing the Police Encounter scenario compared to activation during the Holiday in Hawaii experience. Figures 8 and 9 show the differences in participants' brain activity in the two conditions.

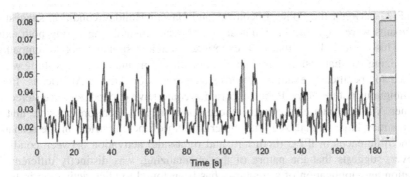

Fig. 8. Participants' HbO level of channel 13 in holiday in Hawaii condition (values in standard deviation)

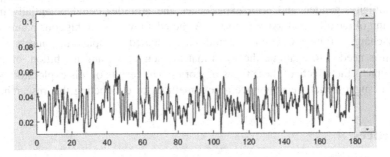

Fig. 9. Participants' HbO level of channel 13 in police encounter condition (values in standard deviation)

Regarding *RQ2*(b), there were also notable differences in left lPFC activation in response to a stereotype-inducing VR scenario (Police) compared to a neutral VR scenario (Holiday) In particular, channels 17, 18, 19 showed a higher brain activation level in the Police scenario than the Hawaii scenario.

5   Conclusion

The primary goal of this exploratory study was to determine if a modified fNIRS/VR headset could be used to identify changes in activity in the prefrontal cortex while participants were immersed in a VR experience. The ability to use fNIRS to examine areas of the brain that are particularly associated with social cognition provides researchers with the ability to not only study the effects of VR experiences in the brain, but provides a unique tool for examining social psychological constructs in an immersive environment.

In this exploratory study, we wanted to determine whether exposure to racially-charged VR environments elicited activation in areas of the PFC that have been found to be associated with prejudice and stereotyping. The lack of differences in activation in the mPFC between the Police Encounter and Hawaii Holiday condition suggest that participants were engaging in similar levels of social mentalizing during both experiences. This could indicate that both experiences elicited similar levels of empathy or "humanizing" or that both required similar amounts of mentalizing. This study was not intended to specifically examine negative stereotypes of African-Americans. Indeed, the ambiguous nature of the Police experience could have elicited negative stereotypes of either African-Americans, police officers, or both. It is therefore possible that participants may have empathized with either or both of the characters in the scenario.

The differences in lPFC (both left- and right-side) activation between conditions, however, suggests that the nature of that mentalizing, was distinctly different. The activation and application of stereotypes has been found to elicit activation in the left lPFC, while the activation in the right lPFC could indicate regulation or control of those stereotypes in interpreting and evaluation the VR experience.

There are a number of potential limitations to this exploratory study. Participants' race, ethnicity, gender, and prior experiences and attitudes could potentially impact processing of racially-charged VR stories. As noted above, the ambiguous nature of the racially-charged Police Encounter could have elicited complex responses. Further research is needed to examine the ways that the activation, and inhibition, of stereotypes impacts the effects of these types of stories. The results of this exploratory study, however, suggest that we now have a new method for conducting this research.

References

1. Schutte, N.S., Stilinović, E.J.: Facilitating empathy through virtual reality. Motiv. Emot. **41**, 708–712 (2017)
2. Shin, D., Biocca, F.: Exploring immersive experience in journalism. New Media Soc. (2018). https://doi.org/10.1177/1461444817733133
3. Hoffman, H.G., Garcia-Palacios, A., Carlin, A., Furness, T.A., Botella-Arbona, C.: Interfaces that heal: coupling real and virtual objects to treat spider phobia. Int. J. Hum. Comput. Interact. (2003). https://doi.org/10.1207/s15327590ijhc1602_08
4. Cacioppo, J.T., Berntson, G.G., Decety, J.: Social neuroscience and its relationship to social psychology. Soc. Cogn. (2011). https://doi.org/10.1521/soco.2010.28.6.675
5. Biocca, F., Delaney, B.: Immersive virtual reality technology. In: Communication in the Age of Virtual Reality, pp. 57–124 (1995)
6. Parsons, T.D., Gaggioli, A., Riva, G.: Virtual reality for research in social neuroscience. Brain Sci. **7**, (2017). https://doi.org/10.3390/brainsci7040042
7. Parasuraman, R., Rizzo, M.: Neuroergonomics: The Brain at Work. Oxford University Press, Oxford (2009)
8. Dimoka, A.: How to conduct a functional magnetic resonance (fMRI) study in social science research. MIS Q. **1**(36), 811–840 (2012)
9. Kim, G., Jeon, J., Biocca, F.: M.I.N.D. brain sensor caps: coupling precise brain imaging to virtual reality head-mounted displays. In: Schmorrow, D.D., Fidopiastis, C.M. (eds.) AC 2018. LNCS (LNAI), vol. 10915, pp. 120–130. Springer, Cham (2018). https://doi.org/10.1007/978-3-319-91470-1_11
10. Chance, B., et al.: A novel method for fast imaging of brain function, non-invasively, with light. Opt. Express **2**, 411–423 (1998). https://doi.org/10.1364/OE.2.000411

11. Seraglia, B., Gamberini, L., Priftis, K., Scatturin, P., Martinelli, M., Cutini, S.: An exploratory fNIRS study with immersive virtual reality: a new method for technical implementation. Front. Hum. Neurosci. **5**, (2011). https://doi.org/10.3389/fnhum.2011.00176

12. Amodio, D.M.: The neuroscience of prejudice and stereotyping. Nature Rev. Neurosci. **15** (10), 670 (2014)

13. Modinos, G., Ormel, J., Aleman, A.: Individual differences in dispositional mindfulness and brain activity involved in reappraisal of emotion. Soc. Cogn. Affect. Neurosci. (2010). https://doi.org/10.1093/scan/nsq006

14. Ochsner, K.N., Gross, J.J.: The cognitive control of emotion. Trends Cogn. Sci. **9**(5), 242–249 (2005)

15. Ochsner, K.N., et al.: For better or for worse: Neural systems supporting the cognitive down- and up-regulation of negative emotion. Neuroimage (2004). https://doi.org/10.1016/j.neuroimage.2004.06.030

16. Lee, K.H., Siegle, G.J.: Common and distinct brain networks underlying explicit emotional evaluation: a meta-analytic study. Soc. Cogn. Affect. Neurosci. (2012). https://doi.org/10.1093/scan/nsp001

17. Cikara, M., Farnsworth, R.A., Harris, L.T., Fiske, S.T.: On the wrong side of the trolley track: neural correlates of relative social valuation. Soc. Cogn. Affect. Neurosci. (2010). https://doi.org/10.1093/scan/nsq011

18. Cunningham, W.A., Johnson, M.K., Raye, C.L., Gatenby, J.C., Gore, J.C., Banaji, M.R.: Separable neural components in the processing of black and white faces. Psychol. Sci. (2004). https://doi.org/10.1111/j.0956-7976.2004.00760.x

19. Harris, L.T., Fiske, S.T.: Social groups that elicit disgust are differentially processed in mPFC. Soc. Cogn. Affect. Neurosci. (2007). https://doi.org/10.1093/scan/nsl037

20. Mitchell, J.P., Ames, D.L., Jenkins, A.C., Banaji, M.R.: Neural correlates of stereotype application. J. Cogn. Neurosci. (2009). https://doi.org/10.1162/jocn.2009.21033

21. Van Overwalle, F.: A dissociation between social mentalizing and general reasoning. Neuroimage **54**(2), 1589–1599 (2011)

22. Cunningham, W.A., Zelazo, P.D.: Attitudes and evaluations: a social cognitive neuroscience perspective. Trends Cogn. Sci. (2007). https://doi.org/10.1016/j.tics.2006.12.005

23. Forbes, C.E., Cox, C.L., Schmader, T., Ryan, L.: Negative stereotype activation alters interaction between neural correlates of arousal, inhibition and cognitive control. Soc. Cogn. Affect. Neurosci. (2012). https://doi.org/10.1093/scan/nsr052

24. Ye, J.C., Tak, S., Jang, K.E., Jung, J., Jang, J.: NIRS-SPM: statistical parametric mapping for near-infrared spectroscopy. Neuroimage **44**, 428–447 (2009). https://doi.org/10.1016/j.neuroimage.2008.08.036

An Enactive Perspective on Emotion: A Case Study on Monitoring Brainwaves

Vanessa Regina Margareth Lima Maike$^{(\boxtimes)}$ and M. Cecília C. Baranauskas

Institute of Computing, University of Campinas, Campinas, Brazil
{vanessa.maike,cecilia}@ic.unicamp.br

Abstract. In the growing field of ubiquitous computing research, there has been an understandable need to revisit the concept of a standard interface with goal-targeted conscious interaction. An enactive system, which draws on a phenomenological perspective, has as a core concept the dynamically coupling of mind and technology, where the interaction design is not goal-oriented, but driven by non-conscious control of the system. In this paper, we investigate the possibilities of the sensor measurements of an EEG device to in fact potentially contribute to the design of an enactive system. We then take the results of such exploration and look at them through the lens of the enactive approach to cognition and its perspective of emotion and cognition as intertwined. This perspective leads our discussion on how to bring the design of enactive systems closer to supporting, through interaction, the social and cultural construction of emotion.

Keywords: Human-Computer Interaction · MindWave · EEG · Universal Design

1 Introduction

Instead of making humans adapt to the computer world, ubiquitous computing, in essence, is about technology becoming invisible and blending into the human world [28]. The concept of Tangible User Interface (TUI) [13] extended this idea by proposing to transform digital information into concrete objects, which could be done with architectural elements (e.g. walls or doors), everyday objects (e.g. books or cards), or ambient conditions (e.g. sound, light or airflow). With the same intent but with a different approach is the concept of *enactive systems* [14], which rejects the idea of a goal-oriented and conscious interaction. Instead, in an enactive system, the person's body and spatial presence is the conduit that allows a non-conscious interaction with the system. The authors drew the *enactive* part from the concept of enaction proposed by Bruner [2], in the sense of "learning by doing", but it also resonates with what Varela et al. [26] called *enaction*. In particular, considering what are the frontiers of the body is important when talking about the design of enactive systems, and we take on the view of the

© Springer Nature Switzerland AG 2019
D. D. Schmorrow and C. M. Fidopiastis (Eds.): HCII 2019, LNAI 11580, pp. 418–435, 2019.
https://doi.org/10.1007/978-3-030-22419-6_30

Embodied Cognition (EC) theory, as it considers the cognitive system to be a network composed of the environment, the body and the brain [4].

Hence, in this paper we explore the possibilities brought by Brain-Computer Interface (BCI), in terms of non-conscious interaction in an enactive system, and analyzed through a lens based on phenomenology, such as that of enaction [2,26] and of Embodied Cognition [4]. As the name implies, BCI is the interaction between a person and a computer system using signals from the brain [16]. One way of providing BCI is to capture and record the electrical activity in the brain using electrodes attached to the surface of the head, a process called Electroencephalography (EEG). Until recently, EEG systems were restricted to hospital and laboratories, but now they are available to the general public through consumer-grade EEG devices [17]. Two examples of such technology are the Emotiv EPOC [6] and the Neurosky MindWave [19]. Both devices are capable of providing metrics on two emotional states: *attention* and *meditation*, i.e., how much a person is focused and how much she is relaxed. We can relate these metrics to the "arousal" and "pleasure" dimensions of the circumplex model of affect [21]. The values provided by the devices come from interpretations that their proprietary algorithms make of the person's brain waves. The availability of EEG devices, as well as the simple measures they can provide on a person's emotional state, make them an interesting option for using BCI in ubiquitous scenarios, or in enactive systems.

One major challenge that needs to be overcome by BCI technology is personalization [16]. This entails, for instance, adapting the system's algorithms to each person's individual brain waves, considering external factors such as possible distractions, or adapting to the person's mood on different occasions. Personalization might also be a desirable quality for Universal Design (UD), the approach to design that aims to make interactive products suitable for the widest possible range of users without requiring adaptations [5]. In a context that potentially tends to a variety of user characteristics and requirements – such as pervasive computing – it is crucial to provide usability and accessibility to all of them.

Such is the challenging scenario in which this work is situated. Therefore, in this paper, we will investigate if and how a consumer-grade EEG device, the Neurosky MindWave, can contribute to the design of an enactive system. Moreover, we wish such design to be informed by an enactive perspective, the theoretical basis from which the concept of enactive systems came. So, the paper is organized as follows: in Sect. 2 we present a literature review on BCI, in Sect. 3 we explain what is the enactive perspective, in Sect. 4 we present our case study with the MindWave, in Sect. 5 we discuss the results of the case study and its implications for the design of enactive systems; and in Sect. 6 we give our concluding remarks.

2   Emotion Captured Through EEG Devices

Literature has investigated gaming as a common application for research on EEG devices. For instance, [11] had four people play an audio-only horror game while wearing the Emotiv EPOC on their heads. The ambient sound of the game is meant to cause tension, as well as some of the goals players need to achieve, such as moving unarmed and evading enemies. The game was designed to have an equal number of moments of *calm* and *fear* (ten of each), since the author's goal is to test whether these states can be detected with the EEG device. After statistical analysis of the raw EEG data, the author found indications that it is possible to differentiate states of fear and calm, although more testing is needed to actually prove that. In addition, the author emphasizes that the electrical activity mapped by the EEG is unique for each individual, but some patterns emerged during the analysis.

Also in the gaming context, [12] used a simulation game to test whether the Neurosky MindWave can be used to detect the effects of *surprising* events on players. To do so, the authors made two versions of the game: one for control and another for experimental conditions. Twenty people played the game, ten for each version. Both versions had a moment for baseline recording – where players were asked to remain calm and inactive for five minutes – and a training phase, to teach the basic controls. The difference between the two versions was in the next phase, where players could either experience seven surprising events (experimental conditions) or regular gameplay, without surprises (control). Then, the final stage of the game is the same for both versions, with three surprising events. Results indicated it is possible to detect the effects of surprise using MindWave and that, furthermore, players from the experimental conditions group were more relaxed when they encountered the surprises on the final phase than the players from the control group.

Still in gaming context, [3] investigated if video game events can cause changes in player's emotions. They used the Emotiv EPOC in an experiment where twenty people played one of three different commercial games, each from a distinct genre: racing, shooting and pool. For each game, the authors established which kinds of events caused either *frustration* or *excitement*, the two emotions chosen for the study. The events were manually annotated by researchers, by watching video footage of participants playing the games. The authors used the Emotiv API, which measures emotion using a normalized value between 0 and 1. The authors converted this intensity into a time series, so that it would be possible to study its correlation with the game events. Hence, authors used linear regression, and found that (1) emotion peaks occurred about half a minute after game events, and (2) there is a strong correlation between game events and emotion peaks.

Also investigating how to apply BCI devices in games, [7] does it with emphasis on music and sounds. More specifically, the authors explore how to detect emotions elicited by certain sounds, to see if it would be possible to adapt a game's music according to the player's mental state. In this investigation, they compare the Emotiv EPOC and the Neurosky MindWave. They concluded that both devices are able to detect the four emotions needed for the experiment (*fear, joy, happiness* and *sadness*), despite the MindWave having less sensors. Furthermore, the participants reported they preferred MindWave because it felt more comfortable. The authors also performed an experiment to see if players can consciously create specific music notes using only a BCI device. At first, it was difficult for participants to reproduce notes by only listening to them. The solution authors found was to associate the note with an image and a gesture, which reduced the training time by half.

On a similar fashion, [8] developed a software that allows people to create drawings using the Neurosky MindWave. Artificial Intelligence (AI) algorithms interpret the brain signals, according to brain wave rhythms classifications, such as *arousal, anxiety* or *relaxation*. Twenty people experimented the software and, according to the authors, it gave them the opportunity to express their creativity in an unconscious way. After statistical and signal analysis, authors concluded that certain brain wave rhythms, as well as the levels of *attention* given by Mind-Wave, are only relevant for the creative process of people with arts education.

On the context of education and e-learning, [27] tests whether a person's levels of *attention* measured by the Neurosky MindWave change while watching a video and performing a task – counting how many times an event occurs in the video. The authors also test if a distraction within the video can have an effect on the levels of attention. The authors' final goal is to help improving performance assessment and evaluation for training videos, especially with students in remote locations. Results indicated that there was no significant difference in the levels of attention between participants who counted right and those who counted wrong. Furthermore, there was no significant difference in levels of attention between participants who saw the distraction and those who did not see it.

Finally, on the context of decision-making, [22] executed an experiment with ten participants where the Emotiv EPOC monitors their EEG while they perform a task. The authors' ultimate goal is to design a BCI system for decision-making. In the experiment, participants had to compare two sets of geometric forms, shown separately, and saying whether they were identical or not. They did this in two stages, each consisting of 56 comparisons. After each stage, participants answered a questionnaire about their feelings during the experiment. In the results analysis, authors did not find a relationship between the participants' self-reported perceptions and the Emotiv EPOC's readings of five possible emotions (*engagement, frustration, meditation, excitement*, and *long-term excitement*).

In summary, from the selected works we can notice a few trends in the domain of BCI and consumer-grade EEG devices. First, the applications we saw are still on an experimentation stage, and are all for individual use and in a controlled environment. Hence, the matters of a pervasive and personalized BCI have not been worked on yet. Second, most of the works performed some kind of statistical analysis on the EEG data. However, there is not a consensus on the statistical method, even among those that employed the same EEG device. Third, all works selected a few emotions to try to detect and classify in their experiments. This is an indication that emotion is being viewed as a type of information to be processed. In this sense, we can also see that there is not a consensus on the emotions that were selected; each study chose a different set.

These trends identified in the literature point to an open opportunity of investigation with regard to the design of ubiquitous systems using BCI. In this paper we take an approach that encourages a tight coupling between the system and the person using it, thus promoting pervasiveness. This approach does not treat emotion as just information, but instead views it as part of the whole cognitive process. In other words, such approach treats body, mind and computer system as a whole. We detail this approach in the next section.

3   Emotion Through the Lens of Enactive Approaches

An **enactive system**, as proposed by Kaipainen et al. [14], consists of a *"dynamic mind-technology embodiment"*, where the interaction is based on involvement of the body without a conscious control of the system, in contrast with the conventional interaction that is totally conscious and oriented by goals. The interface, then, can become implicit to the point of being directly linked to the person's physiological readings. In this case, Kaipainen et al. [14] relate the concept of *enactment* to the idea of *learning by doing*, proposed by Jerome Bruner [2].

Bruner's idea of learning through action comes from a differentiation of three experiences that happen in the learning process: the action-based (enactive), the image-based (iconic) and the language-based (symbolic). Such separation characterizes how higher-order cognition arises from joining the action of a task with its simple components [9]. This resonates with the idea that metaphoric concepts emerge from basic bodily experiences [10]. These views of the learning process are also compatible with the definition of *enaction* by Varela et al. [26]: *"In a nutshell, the enactive approach consists of two points: (1) perception consists in perceptually guided action and (2) cognitive structures emerge from the recurrent sensorimotor patterns that enable action to be perceptually guided"*. Hence, while perception is guided by action, cognitive structures – or higher-order cognition processes – are enacted, thus allowing the action to guide the perception.

This definition of enactive approach is a reflection of what Varela et al. [26] characterize as a shift in cognitive science; one that goes from seeing the world as independent and extrinsic, to viewing the world as inseparable from the processes of self-modification. Furthermore, this shift means looking at cognitive systems not in terms of input and output, but in terms of **operational closure**. According to the authors, *"A system that has operational closure is one in which the results of its processes are those processes themselves"*. Hence, such systems are autonomous in that they are defined by internal mechanisms of self-organization, not in a way that *represents* a detached world, but in a manner that *enacts* a domain that is inseparable from the embodied cognitive system.

Autonomy, however, cannot be defined exclusively by internal processes that recursively depend on each other. According to Thompson and Stapleton [25], an autonomous system – such as the human cognition – also has to regulate its interactions with the world, i.e., its network of internal processes needs to be **thermodynamically open**. Having this active regulation is what characterizes the adaptive autonomy that is necessary for **sense-making**, which, in turn, is the behavior the system adopts according to the *significance* and *value* that it gives to its current environment. Furthermore, such norms the system places on the outside world are not predetermined or fixed, but *enacted* by the system through its autonomy. Therefore, in the same way that the two points of the enactive approach described by Varela et al. [26] are interdependent, **autonomy** and **sense-making** also feed one another.

In essence, sense-making is the reasoning behind motivated action, which is a form of self-regulation, especially if it involves *affect*. Hence, the enactive approach sees that sense-making is as much about cognition as it is about *emotion* [25]. Moreover, in the same way that the cognitive system is not seen as simply input and output, emotion is not looked at as a type of *information*, to be transmitted back and forth from a person to a computer system. It is in this sense that Boehner et al. [1] propose an **interactional** approach to emotion, instead of an informational one.

The interactional approach *"sees emotions as culturally grounded, dynamically experienced, and to some degree constructed in action and interaction"* [1], which is a vision compatible with the enactive approach. Furthermore, in terms of computer systems, the interactional approach shifts the focus *"from helping computers to better understand human emotion to helping people to understand and experience their own emotions"*. In turn, this implies that computer systems designed with the interactional approach do not aim to guess the *correct* emotions people are feeling, but instead, their goal is to encourage individual or collective awareness and reflection on the emotions that were evoked during interaction. This way, feelings are not pre-existing facts, but something that develops with conversations and interactions, where an initially vague, ambiguous or even confusing sensation may consolidate into a meaning. Again, this is in accordance with the enactive approach and with Bruner's [2] idea of learning by doing.

In this sense, although Kaipainen et al. [14] relate their vision of an **enactive system** with Bruner's theory, the minimalist example they provide seems to be inclined towards the informational view of emotion. The enactive system they describe consists of sensors that make psycho-physiological readings, which, in turn, are interpreted by the computer to determine the user's emotional state from a possible set of emotions. Then, a computer-generated character changes its facial expression to match the user's interpreted emotion. Finally, this change should cause a reaction in the user, which would reflect on the psycho-physiological readings, closing a feedback loop that can be infinite. In terms of the enactive perspective we have presented so far, this example seems off due to how it treats emotion as information, but in a way it also can bring a person to have awareness and reflect upon her own emotions. Hence, looking at the enactive system in terms of **autonomy**, it has *operational closure* because of its internal feedback loop, but its internal processes are not *thermodynamically open*. In order for that to happen, they would have to somehow regulate their interactions with the outside world. One way of doing that would be to allow the meanings of emotions to emerge from interaction, instead of encoding them into specific patterns. For instance, Boehner et al. [1] present as an example of an interactional approach a system called "Affector" [23]. It consists of two video windows on each side of adjoining offices, each displaying real-time footage of the neighbor's office. The video, however, is distorted based on filters defined by the users according to what they feel is the affective mood of the office. In this example, the feedback loop between person and video represents the operational closure, while the distortion filters the user can apply to the video serve as self-regulation mechanisms, thus providing the thermodynamic openness and, consequently, *sense-making*.

Expanding this discussion to what we found in the previous section, we can see that, since most works focus on interpreting the EEG data, the trend in literature is also on the operational closure. Furthermore, since most systems we found were for individual use and on controlled environments, there is little room for sense-making, especially for the co-construction of meaning for the emotions that arise during the experiments. Bearing this in mind, in the next section we present our case study, where we take these experimental conditions found in literature as the starting point to our goal: an **enactive system** that follows the enactive perspective by providing both *autonomy* and *sense-making*.

4   Case Study

The object of our case study is the use of a consumer-grade EEG device in experimental conditions and with a single user at a time, following the trend found in literature. Our goal is to design an *enactive system* using the enactive perspective presented in the previous section. Therefore, we aim to see how far we can go with the EEG device as a starting point.

4.1   Technical Setup

The technical setup for our case study is twofold: the EEG device and the software which participants interacted with during the experiment.

EEG Device. In this study, we adopted a consumer-grade, non-invasive EEG device called MindWave, from Neurosky [19]. It is a brainwave sensing headset that has a single dry sensor the user places on the forehead. MindWave can communicate with the computer or a smartphone through Bluetooth, and can provide the following outputs: Attention value, Meditation value, brainwave band powers (e.g. delta, theta, alpha, beta, gamma), and raw EEG wave samples at 512 Hz. We chose to work with the two first outputs, Attention and Meditation. They are calculated by the device's proprietary algorithm, called *eSense*, which returns a value on a scale that goes from 0 to 100. According to the Neurosky developer documentation [18], the eSense scale has a meaning according to five different ranges, that indicate the current level of Attention or Meditation: from 1 to 20 it means a "strongly lowered"; from 20 to 40 it means "reduced"; from 40 to 60 it is "neutral" (baseline); from 60 to 80 it means "slightly elevated"; finally, from 80 to 100 it is "elevated". The meter value of 0 indicates the calculation is not being performed, probably due to poor reading of the signal. The scale with all this information is represented in Fig. 1.

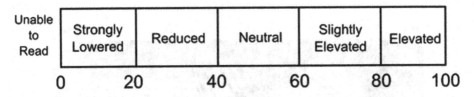

Fig. 1. Neurosky's eSense scale for both Mediation and Attention levels, based on developer documentation.

The developer documentation [18] also highlights how these ranges are relatively wide because the eSense algorithm has dynamic learning, so it sometimes adjusts to fluctuations that occur normally with EEG readings, and are particular to each person. Neurosky affirms this is what allows the device to work with a variety of personal and environmental conditions, maintaining reliable and accurate results. They also encourage developers to fine tune their use of the ranges according to the needs of the application; e.g. trigger an output only for values above 60.

On one hand, the eSense level of **Attention** indicates the magnitude of the person's mental focus, like the one that occurs during intense concentration. Factors that can bring it down are distractions, anxiety or wandering thoughts. On the other hand, the eSense level of **Meditation** corresponds to the *mental* calmness or relaxation, so simply relaxing the muscles of the body might not result in immediate rise in the Meditation level, although relaxing the body can help in relaxing the mind as well. In addition, closing one's eyes might be an effective method for increasing the Meditation level, since it turns off the mental activities that process images from the eyes. Factors that can lower the Meditation levels are the same that lower Attention levels, plus agitation and sensory stimulation.

Software: Quiz Game. The software consists of a quiz-like game, with a total of twelve Yes/No questions, taken from the appendix of the study of Sparrow et al. [24]. We took six questions the authors classified as easy (e.g. "*Are dinosaurs extinct?*"), and six that were considered hard (e.g. "*Do insects feel hunger?*"). The software was developed using the Scratch [20] programming language, because it was easy to integrate with the MindWave device, and it allowed us to program the software rather quickly.

The ultimate goal of the quiz is to detect whether relaxing and disturbing images can have an effect on the levels of Attention and Meditation captured by the MindWave. In this sense, the idea for the interface was to maintain the player's focus on the images, so other visual elements were kept to a minimum. In order to do that, the questions were read by a synthesized voice, and no text was displayed. The player only had three options of buttons: "Yes", "No", and a button to repeat the question. Figure 2 shows an example of the interface, displaying a relaxing image – the picture of a puppy.

Fig. 2. The minimalist interface of our quiz, showing a relaxing image.

The quiz is divided into three moments, each containing four questions. In the first moment, the player can only see the three buttons on a white background. After the player answers the fourth question, s/he enters the second moment, where each question has a different disturbing image as a background. Finally, after the player answers the eighth question, s/he goes into the third moment, where each question has a relaxing background.

The four **disturbing** images we chose were the following: the Napalm girl from the Vietnam war, three bare-chested starved children, a Somalian adolescent holding a rifle, and the explosion on the World Trade Center from the 9–11 plane crash. In turn, the **relaxing** images were these: a sleeping kitten, a puppy, reclining chairs in front of an ocean view, and a colorful sunny beach with a hammock attached to a palm tree.

4.2 Design of the Experiment

For every participant, the images always appear in the same order, although the order of the questions can change. As shown in Table 1, the twelve

Table 1. Questions from the quiz, with their corresponding answer, difficulty and set.

Set	Difficulty	Question	Answer
A	Easy	Are dinosaurs extinct?	Yes
A	Easy	Does 5 plus 7 equal 30?	No
A	Hard	Do insects feel hunger?	No
A	Hard	Is the average age of a human eyelash 150 days?	Yes
B	Easy	Are there 15 months in a year?	No
B	Easy	Is the formula for water H20?	Yes
B	Hard	Is a quince a fruit?	Yes
B	Hard	Is Krypton's atomic number 26?	No
C	Easy	Is a stop sign red in color?	Yes
C	Easy	Are there 24 hours in a day?	Yes
C	Hard	Do all countries have at least two colors in their flags?	No
C	Hard	Is myrmecophobia fear of ants?	Yes

questions are distributed between three sets: A, B and C, where each set contains two easy and two hard questions. The sets are used to organize the permutations that can be applied during the experiment. These permutations are the following: ABC, BCA and CAB. In other words, when the first permutation was active, the participant experienced the questions from group A with the white background, then the questions from group B with the disturbing images, and, finally, the questions from group C with the relaxing images. Within the groups, the order is never altered, i.e., no matter the permutation, the questions from group A always appear in the order shown in Table 1. The software has a configuration screen where the researcher can choose between the three permutations before the participant starts answering the quiz. This was made to add a bit of randomness to the order of the questions.

The experiment with our quiz and the EEG device MindWave was designed to be *within-group*, i.e., all participants experience the same conditions. The experiment was performed during a class of a 1-semester Human Factors course, and 16 students were present on the day of the experiment. In the classroom, one by one, students went to where the setup for the experiment was located: the MindWave device, a headphone, and a chair in front of the table with the laptop that was running the software. Before calling a participant, the researcher cleaned MindWave's forehead sensor, and selected one of the question permutations in the software. After the participant was called, the researcher helped with placing the headphones and the MindWave, which s/he wore throughout the entire quiz.

During the semester, the students were learning how to plan and execute formal experiments in the context of Human-Computer Interaction (HCI) [15], so this experience was presented to them as an example. Hence, instead of acting only as participants, students were also asked to act as observers after participating in the experiment, paying special attention to the body language of the current participant. Along with explanations about the workings of MindWave, this was the only instruction they received before the experiment started; details about the software were kept a secret, to maintain the surprise once they saw the images. In addition, the use of headphones was intended to keep the questions a secret as well, since they were only presented in audio format.

Another intentional design choice was allowing the player to answer only "Yes" or "No" in the quiz. This way, they have to guess, and cannot, for instance, skip a question. In addition, the software also does not provide feedback on whether the selected answer was right or wrong. This decision intended to minimize distractions.

After each participant completed the quiz, they were given a form with questions about the experiment, and also with a space for them to write their observations of other participants. The questions they had to answer were the following: (1) *Did you feel an impact seeing the disturbing images?*; (2) *Which image shocked you the most?*; (3) *Did you feel an effect seeing the relaxing images?*; (4) *Which image relaxed you the most?*.

At the end of the experiment, we also conducted a debriefing session to gather their oral impressions about the experiment. Therefore, we gathered both quantitative and qualitative data. **Quantitative data** consisted of the measures of the attention and meditation levels per second, gathered automatically by the software. **Qualitative data**, then, consisted of the answers from the forms, the ideas from the debriefing, and the written observations made by the students and by another researcher.

Finally, the **independent variables** of our experiment are the *difficulty of the questions* (easy or hard), and the *background during the quiz* (white, disturbing image, or relaxing image). In turn, our **dependent variables** are the *attention*, and *meditation* levels.

Furthermore, our null hypotheses are the following:

- **H0A:** There is no significant difference, in terms of *attention level*, between seeing a white background and seeing an image.
- **H0B:** There is no significant difference, in terms of *attention level*, between answering an easy question and answering a hard question.
- **H0C:** There is no significant difference, in terms of *meditation level*, between seeing a white background and seeing an image.
- **H0D:** There is no significant difference, in terms of *meditation level*, between answering an easy question and answering a hard question.

4.3   Quantitative Results

The first step was trying to reject the null hypotheses. To do so, we had to calculate the average levels of Attention and Meditation for each participant, so that we could then apply a T-Test. The calculated averages are shown in Table 2.

Table 2. Average levels of Attention (AT) and Meditation (MD) for each participant in the different types of images and question difficulties.

	WHITE		DIST.		RELAX.		EASY		HARD	
	AT	MD	AT	MD	AT	MD	AT	MD	AT	MD
P1	80	55	67	47	71	49	72	51	73	49
P2	75	46	54	44	57	40	64	33	61	51
P3	-	-	-	-	-	-	-	-	-	-
P4	51	44	46	32	56	53	56	45	48	43
P5	41	44	70	46	68	63	58	54	62	51
P6	67	49	64	56	65	50	74	46	59	56
P7	30	51	24	67	42	49	25	55	37	57
P8	67	60	73	79	75	70	70	72	73	68
P9	63	22	43	35	51	26	56	25	51	29
P10	42	40	64	36	74	47	66	57	56	25
P11	35	60	37	42	40	77	37	53	38	64
P12	47	80	56	81	58	67	53	81	53	73
P13	33	60	23	53	30	69	20	57	35	63
P14	28	42	50	73	34	57	36	54	37	57
P15	62	63	41	53	43	54	54	60	47	55
P16	89	63	75	79	58	61	82	64	71	69

It is important to note that there was some problem with the MindWave readings for participant P3, so such data could not be considered in the analysis. The next step, then, was trying to reject the null hypotheses related to the levels of Attention, H0A and H0B. The T-Test for comparison between the "White" and the "Disturbing" columns returned a P=0,73. Between "White" and "Relaxing", the result was P=0,86. Finally, for the "Disturbing" and "Relaxing" columns, the test returned P=0,35. Therefore, we cannot reject the null hypothesis H0A. Then, applying the T-Test to the samples from columns "Easy" and "Hard" returned a value of P=0,49. Hence, we also cannot reject null hypothesis H0B.

Lastly for our quantitative analysis, we tried to reject null hypotheses H0C and H0D, related to the levels of Meditation. Between columns "White" and "Disturbing", the test returned P=0,44. Comparing the samples from the "White" and the "Disturbing" columns, the result was P=0,18. Finally, between the "Disturbing" and "Relaxing" columns the T-Test returned P=0,86. Therefore, we cannot reject the null hypothesis H0C. The final T-Test, comparing the samples from columns "Easy" and "Hard", returned P=0,93, which also means we cannot reject the null hypothesis H0D.

Discussion of Quantitative Results. These quantitative results did not allow us to find statistically significant differences in the data collected from our experiment. This could be due to a number of factors, starting with the eSense algorithm. Since it is programmed to automatically adjust to fluctuations that occur in the EEG readings, such adjustment might not be, for instance, quick enough to adapt to sudden changes. During our experiment, the time participants spent on each question was relatively small: usually no more than five seconds. As reported by [3], emotion peaks can occur about half-minute after the event that triggered them. Therefore, it is possible that the EEG device was not able to detect in time the emotional reactions participants experienced, although these experiences in fact existed according to our qualitative data.

Another reason can be provided by looking at the works of [27] and [22]. The first found there was no significant difference between attention levels in people who performed a task wrong and those who performed it right. The other work reported finding no relation between the self-reported perceptions of emotions, and the EEG device's readings. These two works are examples of how the data from an EEG device might differ from the results we actually perceive.

4.4 Qualitative Results

First, we will look at the results from the post-experiment questionnaire. For the first question, "Did you feel an impact seeing the disturbing images?", of the sixteen participants, twelve answered they did feel an impact. Most reported they felt the image distracted them enough to cause difficulty in answering the question; some even highlighted how distracting it was the fact that the images were not related with the questions. Some participants also reported feelings of surprise from the sudden appearance of the images. From the four participants who said they were not affected by the images, one gave no explanation, two claimed the images were well-known, and one said once s/he realized the images had no relation with the questions, s/he stopped paying attention to them, staying focused on the questions.

For the second question, "Which image shocked you the most?", twelve of the sixteen participants reported they found the image of the starving children to be the most disturbing. Two recalled the 9–11 image, one mentioned the image of the adolescent holding a rifle, one mentioned the Vietnam girl, and one participant said none of the images was shocking.

For the third question, "Did you feel an effect seeing the relaxing images?", nine students said they did not feel an effect. Of the other seven participants, one said the relaxing image took her eyes away from the answer buttons, where they were to get away from the disturbing images. Another participant said she felt "peace and joy". One student said she perhaps felt relief, and that the images seemed less distracting than the disturbing ones, but maybe not relaxing. Another participant reported thinking "Wow, that's nice!", but then turned the focus back to the questions. Lastly, one participant said that the kitten made her smile a little.

For the last question, "Which image relaxed you the most?", nine participants reported not remembering any specific image. Interestingly, all but one of them remembered a specific disturbing image. Of the remaining seven participants, one said the puppy was the most relaxing image, three said it was the beach, and three said it was the kitten.

Regarding their observations of their colleagues' body language, there were interesting results. Despite receiving the same instructions, each participant had his/her own ways of interpreting their colleagues' gestures. On the one hand, some reported literal body language, like: moving fingers and feet, raise eyebrows, look up, move shoulders or head, intensity of blinks (quick, long or none), hand on chin, swallow, look away, dilated pupils, scratching, crossing legs, and beating on the table. On the other hand, there were observations associating direct meaning to their colleagues' expression: peaceful, "good expression", doubt, tension, discontentment, upset, nervous, uncomfortable, and indifferent. There were also some cases of a middle-ground, such as: "I don't know (eyes and mouth)", "whatever (shoulders)", "mocking laughter", and "signaling doubt with the lips".

Finally, on the debriefing session, participants gave good insights about the experiment. They pointed how the fact of knowing you are being observed is a possible bias; a few even admitted they tried to restrain their body language. Another bias could be of participants answering the questions quickly just to get over with the quiz as soon as possible. Regarding the questions, some said they had difficulty paying attention to the audio. They said the synthesized voice does not cause emotional interference, but its pronunciation can be confusing. Regarding body language, the students highlighted how they saw some people moved parts of their bodies when there were disturbing images, and how some participants tried to hide their reactions, for instance by putting their hand on their faces. They also recalled there were people who would look away from the screen to think. The students also felt that the relaxing images were easier to ignore, and a lot of them admitted the could not remember most of the images, or even of the questions from the quiz. Finally, they suggested improvements such as: changing the order of the images, giving a small pause between the questions, displaying the images on a larger screen to raise the impact, providing a more immersive atmosphere through lighting or sounds, displaying animated images, and making the "Yes" and "No" buttons appear with a delay, since their color is distracting.

Discussion of Qualitative Results. Georgiadis et al. [12] reported how people who encountered surprises earlier were more relaxed when they encountered later surprises than those who only experienced one surprising event. Our quantitative and qualitative data also corroborate this effect, since the disturbing images – which appeared first – were very striking for most participants, while the relaxing images – which came afterwards – were usually ignored or not easily remembered. Hence, during our experiment participants could have experienced some sort of numbness that prevented MindWave from detecting emotional reactions. Furthermore, the lack of correlation between self-reported emotions and the EEG readings found by Schuh & de Borba Campos [22] are very similar to ours, since our quantitative data did not provide insights that were present in our qualitative data; e.g., the impact the disturbing images had on the participant's concentration.

In fact, it is important to note how much richer the qualitative data was than the quantitative data. While the MindWave only measures levels of attention and meditation, the observations elicited a much wider variety of emotions, like peacefulness, doubt, tension, discontentment, and indifference. This is coherent with the interactional approach [1], which views emotion as much more than information. The way the participants interpreted each other's emotions, based only on body language, is a step towards emotion as a cultural, social and collaborative construction, like Boehner et al. talked about.

5   Discussion Towards an Enactive Scenario

Based on our analysis of the quantitative and qualitative results, we propose that, to follow the interactional approach, our quiz would have to harbor the kind of social meaning-making the students showed while observing each other. Watching another person play the quiz can lead to reflections on what that person might be feeling and what the images might be triggering for her, which, in turn, can lead to a self-reflection about one's own feelings when presented with the same experiences. Like the "Affector" example [23], our quiz could provide some sort of real-time output of how a player is feeling – like a video footage, or even the EEG reading – and allow other players to transform that output according to their own interpretations of it.

Providing a mechanism such as this would be a way to make our quiz *thermodynamically open*. For it to have *autonomy*, however, its internal processes would have to be recursively interdependent – which, in the current state, they are not. A way to do that would be to incorporate feedback loops, similar to what Kaipainen et al. [14] propose. On an individual level, we could make the quiz environment responsive to the player's EEG readings. For instance, if the readings indicate a high level of Meditation, an agitated music could play on the background, the ambient lighting could glow in warm colors, and the computer monitor could display disturbing or distracting images. If the Meditation levels went down, then calm music would play, ambient lights would glow in cold colors, and the displayed images would be comforting or relaxing. On a social level, we could make it so that it is not the current player's EEG that is affecting his environment, but someone else's. This way, players feed each other's environments, which could lead to co-construction of meaning if players are aware of whose emotions is affecting their environment. Again, a real-time video footage of the person, or some representation of her EEG data would suffice, as long as the interpretation of that data is left open-ended.

Such flexibility is important not only to allow sense-making to occur, but also because EEG readings are unique for each individual, as noted by Garner [11]. Therefore, if we are envisioning a pervasive system that responds to non-conscious control, it is beneficial to consider individual differences. In this sense, an approach like Universal Design is essential for a technology paradigm that needs to respond to the presence of different individuals in a seamless and unobtrusive way [5]. Considering EEG readings are so particular, it would be impossible to create one solution, based exclusively on them, that contemplates every user – a fact reinforced by how our quantitative analysis found no correlations. However, enactive systems, if designed with the enactive perspective we presented, have the potential to contemplate a wide variety of users, especially with the social component that emerged from our qualitative results. For instance, in our examples in which one person's emotional state affects another person's experience, as long as each one can develop their own sense-making of the other's situation, they are communicating with each other in an universal way. The ambient lights, the sounds, and the images, all embedded in the player's environment and making use of multimodality and multimedia, tend to a wide range of human abilities, skills and preferences.

6   Conclusion

In this paper we investigated how the MindWave EEG device can potentially contribute to the design of *enactive systems*, a concept of dynamic coupling between mind and technology. In our literature review, we saw how the use of EEG devices is still experimental, and meant for individual use in controlled environments. We also saw a focus on statistical analysis of the EEG data and on classification of emotions. We took these trends as a starting point to our case study, which involved an experiment that tested whether MindWave could detect emotional reactions from the participants in specified situations. Although our quantitative data did not allow us to make correlations between the experimental events and the EEG readings, our qualitative data showed to be quite rich. In particular, once we looked at it using the lens of the enactive perspective, we found significant contributions that could elevate our experimental setup to an enactive system. The concepts of *autonomy* and *sense-making* were crucial for this process, since they provided us with a scaffold to look at how the interactions with the system could be more pervasive and less goal-oriented.

In this sense, the social component emerged as an important factor not only for co-constructing emotions, but also for tackling the problem of personalization. Pervasive or ubiquitous computing needs to reach the widest possible range of users, without the need for special adaptations. Universal Design, then, is almost a necessity, and we believe enactive systems, with the enactive approach, are a viable path towards it.

Acknowledgments. This work is financially supported by the São Paulo Research Foundation (FAPESP) through grants #2015/16528-0 and #2015/24300-9, by Coordenação de Aperfeiçoamento de Pessoal de Nível Superior (CAPES) through grant #01-P-04554/2013 and by National Council for Scientific and Technological Development (CNPq) through grants #160911/2015-0 and #306272/2017-2.

References

1. Boehner, K., DePaula, R., Dourish, P., Sengers, P.: How emotion is made and measured. Int. J. Hum.-Comput. Stud. **65**(4), 275–291 (2007). https://doi.org/10.1016/j.ijhcs.2006.11.016

2. Bruner, J.: Toward a Theory of Instruction. W.W. Norton (1966). Books that live
3. Chen, D., James, J., Bao, F.S., Ling, C., Fan, T.: Relationship between video game events and player emotion based on EEG. In: Kurosu, M. (ed.) HCI 2016. LNCS, vol. 9733, pp. 377–384. Springer, Cham (2016). https://doi.org/10.1007/978-3-319-39513-5_35
4. van Dijk, J., van der Lugt, R., Hummels, C.: Beyond distributed representation: embodied cognition design supporting socio-sensorimotor couplings. In: Proceedings of the 8th International Conference on Tangible, Embedded and Embodied Interaction. TEI 2014, pp. 181–188. ACM, New York (2013). https://doi.org/10.1145/2540930.2540934
5. Emiliani, P.L., Stephanidis, C.: Universal access to ambient intelligence environments: opportunities and challenges for people with disabilities. IBM Syst. J. **44**(3), 605–619 (2005). https://doi.org/10.1147/sj.443.0605
6. Emotiv: EPOC. https://www.emotiv.com/epoc/. Accessed 20 June 2018
7. Folgieri, R., Bergomi, M.G., Castellani, S.: EEG-based brain-computer interface for emotional involvement in games through music. In: Lee, N. (ed.) Digital Da Vinci, pp. 205–236. Springer, New York (2014). https://doi.org/10.1007/978-1-4939-0536-2_9
8. Folgieri, R., Lucchiari, C., Granato, M., Grechi, D.: Brain, technology and creativity. BrainArt: A BCI-based entertainment tool to enact creativity and create drawing from cerebral rhythms. In: Lee, N. (ed.) Digital Da Vinci, pp. 65–97. Springer, New York (2014). https://doi.org/10.1007/978-1-4939-0965-0_4
9. Francis, K., Khan, S., Davis, B.: Enactivism, spatial reasoning and coding. Digital Experiences Math. Educ. **2**(1), 1–20 (2016). https://doi.org/10.1007/s40751-015-0010-4
10. Gallagher, S., Lindgren, R.: Enactive metaphors: learning through full-body engagement. Educ. Psychol. Rev. **27**(3), 391–404 (2015). https://doi.org/10.1007/s10648-015-9327-1
11. Garner, T.: Identifying habitual statistical features of EEG in response to fear-related stimuli in an audio-only computer video game. In: Proceedings of the 8th Audio Mostly Conference. AM 2013, pp. 14:1–14:6. ACM, New York (2013). https://doi.org/10.1145/2544114.2544129
12. Georgiadis, K., van Oostendorp, H., van der Pal, J.: EEG assessment of surprise effects in serious games. In: de De Gloria, A., Veltkamp, R. (eds.) GALA 2015. LNCS, vol. 9599, pp. 517–529. Springer, Cham (2016). https://doi.org/10.1007/978-3-319-40216-1_56
13. Ishii, H., Ullmer, B.: Tangible bits: towards seamless interfaces between people, bits and atoms. In: Proceedings of the ACM SIGCHI Conference on Human factors in computing systems, pp. 234–241. ACM (1997)
14. Kaipainen, M., et al.: Enactive systems and enactive media: embodied human-machine coupling beyond interfaces. Leonardo **44**(5), 433–438 (2011). https://doi.org/10.1162/LEON_a_00244
15. Lazar, J., Feng, J.H., Hochheiser, H.: Research Methods in Human-Computer Interaction. Wiley Publishing (2010)
16. Lightbody, G., Galway, L., McCullagh, P.: The brain computer interface: barriers to becoming pervasive. In: Holzinger, A., Ziefle, M., Röcker, C. (eds.) Pervasive Health. HIS, pp. 101–129. Springer, London (2014). https://doi.org/10.1007/978-1-4471-6413-5_5
17. Maskeliunas, R., Damasevicius, R., Martisius, I., Vasiljevas, M.: Consumer-grade eeg devices: are they usable for control tasks? PeerJ **4**, e1746 (2016)

18. Neurosky: Developer Documentation. http://developer.neurosky.com/docs/doku. php?id=esenses_tm. Accessed 20 June 2018
19. Neurosky: MindWave. http://neurosky.com/biosensors/eeg-sensor/biosensors/. Accessed 20 June 2018
20. Resnick, M., et al.: Scratch: programming for all. Commun. ACM **52**(11), 60–67 (2009). https://doi.org/10.1145/1592761.1592779
21. Russell, J.A.: A circumplex model of affect. J. Pers. Soc. Psychol. **39**(6), 1161 (1980)
22. Schuh, Â., de Borba Campos, M.: Design of a decision-making task for a collaborative brain-computer interface system based on emotiv EEG. In: Harris, D. (ed.) Engineering Psychology and Cognitive Ergonomics: Cognition and Design, pp. 115–132. Springer, Cham (2017). https://doi.org/10.1007/978-3-319-58475-1_9
23. Sengers, P., Boehner, K., Mateas, M., Gay, G.: The disenchantment of affect. Pers. Ubiquitous Comput. **12**(5), 347–358 (2008). https://doi.org/10.1007/s00779-007-0161-4
24. Sparrow, B., Liu, J., Wegner, D.M.: Google effects on memory: cognitive consequences of having information at our fingertips. Science **333**(6043), 776–778 (2011). https://doi.org/10.1126/science.1207745
25. Thompson, E., Stapleton, M.: Making sense of sense-making: reflections on enactive and extended mind theories. Topoi **28**(1), 23–30 (2009). https://doi.org/10.1007/s11245-008-9043-2
26. Varela, F.J., Thompson, E., Rosch, E.: The Embodied Mind: Cognitive Science and Human Experience. MIT press (2017)
27. Wang, C., Cesar, P., Geelhoed, E.: An invisible gorilla: is it a matter of focus of attention? In: Huet, B., Ngo, C.-W., Tang, J., Zhou, Z.-H., Hauptmann, A.G., Yan, S. (eds.) PCM 2013. LNCS, vol. 8294, pp. 318–326. Springer, Cham (2013). https://doi.org/10.1007/978-3-319-03731-8_30
28. Weiser, M.: The computer for the 21st century. Sci. Am. **265**(3), 94–104 (1991)

An Immersive Brain Painting: The Effects of Brain Painting in a Virtual Reality Environment

Willie McClinton$^{(\boxtimes)}$, Sarah Garcia$^{(\boxtimes)}$, and Marvin Andujar$^{(\boxtimes)}$

University of South Florida, Tampa, FL 33620, USA
{wmcclinton, sarahgarcia}@mail.usf.edu,
andujar1@usf.edu

Abstract. Brain Painting is a brain-computer interface (BCI) application that allows users to paint on a virtual canvas without requiring physical movement. Brain Painting has shown to improve the Quality of Life of patients with Amyotrophic lateral sclerosis (ALS), by giving the patients ways of expressing themselves and affecting society through art exhibitions. Although there is currently no known cure for ALS, through such outlets, we can help mitigate the physical and psychological impairments. Therefore, this paper discusses a study where users paint with their brain with a BCI in an immersed Virtual environment, also known as Brain Painting. This study evaluates the cognitive workload, presence, and changes in affect through standardized questionnaires. Also, to explore the validity for users to express themselves creatively can lead to a more positive experience using the application in an immersed environment, in comparison to non-immersive.

Keywords: Brain-computer interface · Brain Painting ·
Electroencephalography (EEG) · P300 · Event-related potentials (ERP) ·
Virtual Reality (VR)

1 Introduction

The interaction between human's brains and machines is an emerging area in computer science. Bringing aspects of Human-centered computing into this area, especially where Brain-Computer Interfaces (BCI) is used as a mechanism for control, adds a hands-free dimension to computing. The research presented in this paper is the first study ever demonstrating able-bodied users painting with their brains in an immersive Virtual Reality (VR) environment as a medium for creative expression. The immersive brain painting environment allows users the ability to paint in a virtual canvas without requiring physical movement [1–3].

The art of creative expression is considered to be until this day a purely human ability and skill. Art has taken many forms like sculpting and painting. We are introducing a new mechanism for interactive method of creating art using only the Brain in an immerse environment. Brain Painting has shown to improve the quality of life (QOL) of patients with ALS, by giving the patients ways of expressing themselves and affecting society through art exhibitions [1]. Although there is currently no known

© Springer Nature Switzerland AG 2019
D. D. Schmorrow and C. M. Fidopiastis (Eds.): HCII 2019, LNAI 11580, pp. 436–445, 2019.
https://doi.org/10.1007/978-3-030-22419-6_31

cure for ALS, through such outlets, we can help mitigate the physical and psychological impairments of those living with ALS.

In order to understand if our novel VR Brain Painting application elicits a more positive experience, we tested our system on participants without ALS. This allowed us to validate, as well as find improvements on our system before running trials directly with the ALS community. Although further research is needed to fully validate the effect of immersion during the healthy patient trials, these trials accelerate the research process, due to the larger participant pool, and lead to more insights into aspects of user experience that both populations share. If the user experience and its usability is good, it is expected that the participants will be able to express themselves naturally through their brain paintings. Since the focus in BCI research has mostly been on the reliability of applications, no standardized methods to assess the user experience for BCIs exist at the moment. Therefore, in this study we use standardized questionnaires adapted in Affective Computing and Human-Computer Interaction research to collect data about our participant's experience. During this study, we show that Brain Painting in an immersive virtual environment can help provide an experience that makes the users feel more "present" [6] while they express themselves creatively and this leads to a more positive user experience. We do this by comparing their emotional state given by the Positive and Negative Affect Schedule (PANAS) [4], which gives a score for positive and negative affect between 10 and 50 points, before and after using traditional Brain Painting and VR Brain Painting. We also measure their cognitive workload with the NASA Task Load Index (TLX) [5] and the felt presence through the Presence Questionnaire [6] to determine how present they felt while using the application. We believe that the immersive VR Brain Painting will allow the user to feel more present using the application and have a better experience measured by the PANAS survey.

2   Related Work

The world's first Brain–computer interfaces (BCIs) that enabled creative expression in paralyzed patients were first introduced by a group at the University of Tübingen [1]. They investigated the efficacy and user-friendliness of P300-based Brain Painting, which is an application developed to paint pictures using brain activity only. The application used two screens for the painter: one screen displayed the P300 matrix while another screen showed the painting canvas. The standard P300 speller matrix, proposed by Donchin et al., was adapted to contain symbols indicating different colors, objects, grid sizes, object sizes, transparency, zoom and cursor movement. They showed the usability of P300s in painting applications and the qualitative results of peoples' experiences. Their user study with three ALS and 10 healthy participants got an average accuracy of over 70% and bit transfer rate of 4.41 bits/min during the brain painting trials, which reflects the accuracies and transfer rates in modern P300 applications. The surveys given in the study measured each participants motivation and mood prior to the study, but no comparative measure before and after were given to measure the benefits of using the device.

This study was followed by an extended version of the first Brain Painting with a much greater selection and longer study [3]. They conducted a more extensive user study of 2 ALS patients during 27 home use days over 3.5 months to demonstrate how P300-based Brain Painting could be integrated into their everyday life to promote their QOL. Since their results are based on only 2 end-users, they could not be generalized to the population of potential users, but their achievement of high satisfaction with their participants demonstrates the benefit of adopting user-centered design in BCI development. This was a long-term study and they used the QUEST2.0 [7] to along with a self-reported satisfaction score between 1 and 10 to assess the satisfaction and usability of the device over the course of the study. As our experiment is measuring the direct effects of one session of using Brain Painting, we decided to use a before and after PANAS surveys over self-reported satisfaction scores and since our study is over healthy participants the devices are not considered assistive, so we left out the QUEST2.0 survey. Although Brain Painting in Virtual Reality has not been explored yet, there have been many other attempts to bring BCI into virtual reality through conventional methods like Motor Imagery, P300, SSVEP, and even Hybrid P300/SSVEP with much success [8–11]. They have been used in applications for navigation [11], object control [10], and even movement [9] due to recent technological advancement in VR headsets and the design of BCI devices. Surprisingly, none of the new advancements have been utilized in the Brain Painting space, but in this research, we will explore the first glimpse of P300 Brain Painting in Virtual Reality.

Our approach is different to past studies in 2 significant ways: (1) this research focuses on the direct effects of Brain Painting on a person's affective state and (2) is the first evaluation of Brain Painting in an immersive VR environment with a VR headset.

3   Methodology

3.1   Study Design

A within-subject study was performed to evaluate the participant's experiences according to their changes in their affective state during the use of Brain Painting, and to test if the new environment affected the participants' measured cognitive workload (NASA TLX) or felt presence (Presence Questionnaire), along with their experience.

A total of eight participants took part in this study. They were all students at the University of South Florida, Tampa, Florida, USA with an age range from 18–31. All of them mentioned that they have used a BCI before. This information was recorded by the pre-experiment survey which asks questions regarding demographics, experience with brain-computer interface devices, and asked for a pre-experiment PANAS survey to measure the current affective state of the participant before the study. This PANAS survey is used to compare the change at the end of both the immersive and non-immersive task. The participants were also asked to fill out surveys about their experience during the task, which included: the post-task PANAS survey, NASA TLX, and Presence Questionnaire.

Study Procedure
Each study session lasted approximately two hours and consisted of 8 parts: (1) pre-experiment surveys, (2) setup and training session of the first application, (3) testing of the first application, (4) post-survey of the first application (5) training session of the

second application, (6) testing of the second application, (7) post-survey of the second application, and finally, (8) post-experiment surveys.

After the pre-experiment surveys, the next part of the study consisted of a mounting session where the participant was asked to mount the G.Tec Nautilus device and Oculus Rift. After mounting the devices, they were asked to train the P300 by focusing on randomly flashing symbols on the screen to generate the EEG data which we used to train a Linear Discriminant Analysis (LDA) classifier. After a series of 10, 12 row flashes (chosen based on previous studies, with training time in mind [10]), the times when the elements that contained the chosen symbols were flashed, P300 event-related potentials (ERP) are elicited in the user. The event-related potentials (ERPs0 were then trained to be detected by the LDA classifier, which then transfers them into commands for the paint utility. After the training was completed (approximately 30 min), the participant went through a short familiarization session in which they are asked to make any five selections from the painting interface using the P300 stimuli. After the pre-testing phase, the participant was given a session of 10 more commands to use the application. Their task was to use the Brain Painting application to draw whatever they wanted. This process to gauge the effects of using the application in an uncontrolled recreational setting is similar to other studies where the session is done with the ALS participants at home [1, 2]. Each participant tested both the immersive and non-immersive version of the application and they were split equally into two groups: one group used the non-immersive application first, then the immersive. The other used the immersive application first, then the non-immersive. This was done to reduce bias in the results. At the end of each task the participants were asked to fill out surveys about their experience.

3.2 System Design

For our brain-painting control interface, we use 16 dry-electrode channels with the G Tec Nautilus electroencephalography (EEG) device. We chose this device because it is lightweight, wireless, does not require electrode gel, and can collect high quality EEG data. These features allow for easy integration with the Oculus Rift head mounted display, which is used to project the canvas and brush to the user while painting. Electrodes were placed at positions Cz, CPz, P1, P3, P5, P7, Pz, P2, P4, P6, P8, PO3, PO7, POz, PO6, and PO4 according to the 10–20 international system. These positions are located on the parietal lobe which is the most prominent for P300 [12].

For this study, two different applications were made (Fig. 1). One of applications has the non-immersive interface, which consists of the paint canvas and symbol matrix of visual stimuli on the computer monitor, without a VR headset. The other has an immersive interface which also consists of the paint canvas and symbol matrix of visual stimuli, but these are now displayed within Oculus Rift VR Headset that the user is wearing.

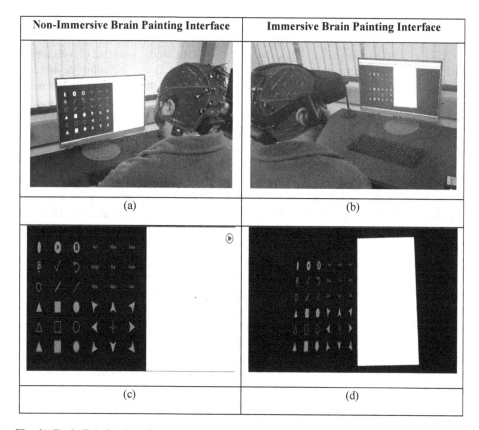

Fig. 1. Brain Painting Interface – A user wearing the G. Tec Nautilus EEG headset performing the 2 tasks. On the left side, the non-immersive interface (a) and symbol matrix of visual stimuli (c) are shown. On the right side, the immersive interface (b) and symbol matrix of visual stimuli (d) are shown with the user wearing an Oculus Rift VR Headset.

User Interface

The P300-based brain-computer interface for abstract painting in an immersive VR environment was developed using the Unity Game Engine. The immersive environment is divided into two components comprising the total field of view seen on the Oculus Rift VR headset. The first component is used to incorporate the visual stimulus. This is done by displaying a 6-by-6 grid containing symbols that represent actions in a painting utility (i.e. movement, changing color, or switching brushes). The second component consists of a canvas where the painting takes place. A cursor, placed on the canvas, responds to a selected option from the Visual Stimulus panel. These include moving in ten directions (up, down, left, right, and all diagonals) in order to draw lines connecting the cursor's current position to its next position on the canvas. Other features of the application allow for printing shapes on the canvas, changing the color of the lines and shapes, and size of the brush (Fig. 2).

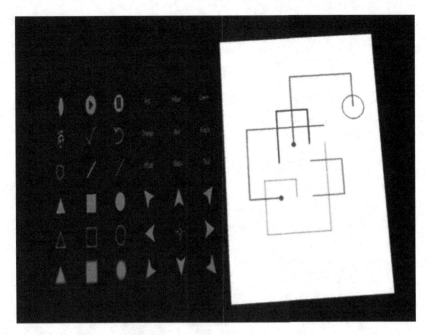

Fig. 2. Immersive P300 Brain Painting User Interface designed for Oculus Rift

Control Interface

The control interface of our application was achieved via the use of OpenVibe, a software platform used to design, develop, and test brain-computer interfaces. We used OpenVibe as a mechanism to acquire and classify EEG data coming from the G.Tec Nautilus, as well as an application driver to manage the synchronization between our user interface in Unity and data acquisition from the G.Tec. As a base for the acquisition, training and the online task of our application, we used the built-in P300 Speller within OpenVibe. The application was extended by creating a virtual environment in Unity on top of the existing OpenVibe application and established a communication link between our user interface and OpenVibe via the LabStreamingLayer (LSL).

Figure 3 is the OpenVibe online task scenario. Identical to the filtering process done when training the LDA classifier. The online task reads and filters data coming from the acquisition client using channels 1-16 (in the 1–20 Hz frequency band with a fourth-order digital Butterworth filter). The data is then decimated by a factor of five and cut into signal segments, averaging over the epochs. After the data is preprocessed, it is passed into the classifier trained from the acquisition step. The result of the classification is passed into Unity along with the interface variables via the LSL Blocks. The interface variables include information about the state of the P300 Speller. Our user interface in Unity, using the LSL4Unity library (https://github.com/xfleckx/LSL4Unity), receives the LSL data and then mirrors the instance of the P300 Speller on OpenVibe.

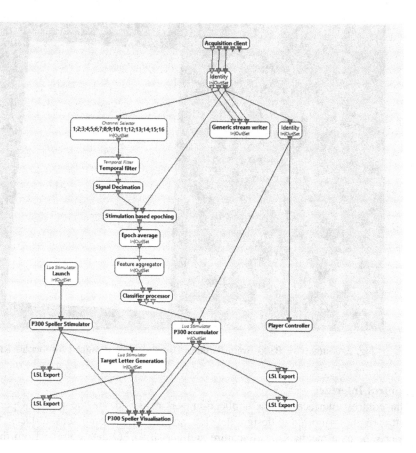

Fig. 3. OpenVibe Online Task Scenario – This contains the coding blocks used to gather and classify data during the 3D Brain painting online task. (The template is similar to: http://openvibe.inria.fr/openvibe-p300-speller/).

4   Results and Discussion

Data was collected by a total of 8 participants. Therefore, individual data will be reported descriptively (Tables 1, 2, and 3). Our post-task PANAS (Table 1) showed that 7 of the 8 participants had a higher positive score after using the immersive Brain Painting interface over the non-immersive interface and 6 of the 8 had higher or equal negative score when using the non-immersive Brain Painting interface. The pre-PANAS score was higher than the post-PANAS. One of the reasons the why the score decreased after the tasks might be because of fatigue. The training and testing time took over an hour for both tasks and this could lead to boredom and irritation. Also having a small sample size can lead to the large variation in the results.

Table 1. Average PANAS affect scores after immersive and non-immersive tasks.

Task	Mean positive score	Mean negative score	Difference in positive score	Difference in negative score
Pre-experiment	34.88 (SE 1.99)	15.38 (SE 2.30)	0.00	0.00
Non-immersive	30.50 (SE 2.33)	15.50 (SE 2.69)	-4.38	0.125
Immersive	32.00 (SE 2.83)	15.38 (SE 2.17)	-2.88	0.00
p-value	0.44	0.94	0.44	0.94

The NASA TLX (Table 2) showed a mean measured workload of 49.25 (Standard Error (SE) 5.84) and 50.5 (SE 4.90) over all participants performing the task in non-immersive and immersive Brain Painting, respectively. In contrast to the difference seen in the positive affect of the immersive environment measured in the PANAS surveys, the workload showed to be the same between tasks. The similarity between the measured cognitive workload of the tasks can be attributed to the alike interfaces and mechanisms for selecting a command.

Table 2. Average Score for NASA TLX after immersive and non-immersive tasks.

Task	Non-immersive	Immersive	p-value
Mental demand	13.25 (SE 1.93)	14.00 (SE 1.56)	0.62
Physical demand	4.38 (SE 1.85)	5.38 (SE 1.32)	0.51
Temporal demand	9.63 (SE 1.75)	8.88 (SE 1.06)	0.69
Performance	4.00 (SE 1.07)	4.13 (SE 1.04)	0.92
Effort	12.88 (SE 1.85)	12.25 (SE 1.79)	0.53
Frustration	5.13 (SE 1.65)	5.875 (SE 1.893)	0.46
Total	49.25 (SE 5.84)	50.5 (SE 4.90)	0.78

Our post-task results for the categories measured in the Presence Questionnaire were slightly higher in the immersive task than the non-immersive task in almost all categories (Table 3). Also, six out of the eight participants said they would use the brain painting device recreationally, showing their enthusiasm towards the novel system.

Table 3. Average score for presence questionnaire after immersive and non-immersive tasks.

Task	Non-immersive	Immersive	p-value
Realism	23.00 (SE 2.83)	23.13 (SE 2.11)	0.95
Possibility to act	12.88 (SE 1.74)	14.50 (SE 1.86)	0.16
Quality of interface	15.63 (SE 1.57)	15.50 (SE 1.32)	0.90
Possibility to examine	10.50 (SE 1.52)	10.63 (SE 1.19)	0.93
Self-evaluation of performance	7.13 (SE 1.42)	8.00 (SE 1.00)	0.40
Total	69.13 (SE 6.17)	71.75 (SE 5.15)	0.32

5 Conclusion and Future Work

This paper discussed a novel method to allow users to paint with their brain in an immersive virtual environment and its early user testing. This paper is meant to be the first step towards the study of immersive environments in brain painting applications from an HCI perspective. This work was done through surveys that measure the affective state of participants before and after the task, measure cognitive workload and felt presence when evaluating the experiences of the participants. We wanted to see what aspects of the participants experience improved with the immersive interface. The proposed system did not seem to improve the measured cognitive workload. This is probably due to the complex control aspects of the interfaces. Nevertheless, users felt more present during the use immersive environment than in the non-immersive interface. Although this is only the pilot work, this application has great potential to improve the quality of life of patients with ALS and could give a new mechanism for the physically impaired to express themselves without physical movement.

References

1. Münßinger, J.I., et al.: Brain painting: first evaluation of a new brain–computer interface application with ALS-patients and healthy volunteers. Front. Neurosci. **4**. 182 (2010). https://doi.org/10.3389/fnins.2010.00182
2. Zickler, C., Halder, S., Kleih, S.C., Herbert, C., Kübler, A.: Brain painting: usability testing according to the user-centered design in end users with severe motor paralysis. Artif. Intell. Med. **59**(2), 99–110 (2013). https://doi.org/10.1016/j.artmed.2013.08.003
3. Botrel, L., Holz, E., Kübler, A.: Brain painting V2: evaluation of P300-based brain-computer interface for creative expression by an end-user following the user-centered design. Brain-Comput. Interfaces **2**(2–3), 135–149 (2015). https://doi.org/10.1080/2326263x.2015.1100038
4. Watson, D., Clark, L.A., Tellegen, A.: Development and validation of brief measures of positive and negative affect: the PANAS scales. J. Pers. Soc. Psychol. **54**(6), 1063 (1988)
5. Hart, S.G., Staveland, L.E.: Development of NASA-TLX (Task Load Index): results of empirical and theoretical research. Adv. Psychol. Hum. Ment. Workload **52**, 139–183 (1988). https://doi.org/10.1016/s0166-4115(08)62386-9
6. Witmer, B.G., Singer, M.J.: Measuring presence in virtual environments: a presence questionnaire. Presence: Teleoper. Virtual Environ. **7**(3), 225–240 (1998). https://doi.org/10.1162/105474698565686
7. Demers, L., Weiss-Lambrou, R., Ska, B.: Development of the quebec user evaluation of satisfaction with assistive technology (QUEST). Assist. Technol. **8**(1), 3–13 (1996). https://doi.org/10.1080/10400435.1996.10132268
8. Koo, B., Choi, S.: SSVEP response on oculus rift. In: The 3rd International Winter Conference on Brain-Computer Interface (2015). https://doi.org/10.1109/iww-bci.2015.7073028
9. Koo, B., Lee, H., Nam, Y., Choi, S.: Immersive BCI with SSVEP in VR head-mounted display. In: 2015 37th Annual International Conference of the IEEE Engineering in Medicine and Biology Society (EMBC) (2015). https://doi.org/10.1109/embc.2015.7318558

10. Chun, J., Bae, B., Jo, S.: BCI based hybrid interface for 3D object control in virtual reality. In: 2016 4th International Winter Conference on Brain-Computer Interface (BCI) (2016). https://doi.org/10.1109/iww-bci.2016.7457461
11. Bevilacqua, V., et al.: A novel BCI-SSVEP based approach for control of walking in virtual environment using a convolutional neural network. In: 2014 International Joint Conference on Neural Networks (IJCNN) (2014). https://doi.org/10.1109/ijcnn.2014.6889955
12. Manyakov, N.V., Chumerin, N., Combaz, A., Hulle, M.M.: Comparison of classification methods for P300 brain-computer interface on disabled subjects. Comput. Intell. Neurosci. **2011**, 1–12 (2011). https://doi.org/10.1155/2011/519868

Predicting Java Computer Programming Task Difficulty Levels Using EEG for Educational Environments

Ramaswamy Palaniappan[1]([⊠]) [iD], Aruna Duraisingam[1],
Nithyakalyani Chinnaiah[1,2] [iD], and Murugappan Murugappan[3] [iD]

[1] Data Science Research Group, School of Computing, University of Kent,
Canterbury, UK
r.palani@kent.ac.uk, ad543@kentforlife.net,
chinnaiahn@mgsg.kent.sch.uk
[2] Mayfield Grammar School, Gravesend, Kent, UK
[3] Kuwait College of Science and Technology, Doha, Kuwait
m.murugappan@kcst.edu.kw

Abstract. Understanding how difficult a learning task is for a person allows teaching material to be appropriately designed to suit the person, especially for programming material. A first step for this would be to predict on the task difficulty level. While this is possible through subjective questionnaire, it could lead to misleading outcome and it would be better to do this by tapping the actual thought process in the brain while the subject is performing the task, which can be done using electroencephalogram. We set out on this objective and show that it is possible to predict easy and difficult levels of mental tasks when subjects are attempting to solve Java programming problems. Using a proposed confidence threshold, we obtained a classification performance of 87.05% thereby showing that it is possible to use brain data to determine the teaching material difficulty level which will be useful in educational environments.

Keywords: Confidence threshold · Education · EEG · Java ·
Mental task level · NASA TLX · Programming task

1 Introduction

Gerjets et al. [1] describes optimum learning conditions as providing learning at the appropriate level and pace for the learner. To be able to tailor the teaching material, it is first imperative to decide on the difficulty level of material as perceived by the learner. But assessment measures such as obtaining correct vs incorrect responses in the exams may not be a good indication to gauge the understanding of the students. Often, learners could miss-assume their level of understanding and therefore causing incorrect tailoring of the material, pace etc.

Therefore, it becomes necessary to utilise measures that can correctly predict the difficulty level. Subjective and dual-task procedures can be used for this purpose but likely to interrupt the subjects in-between the experiments and it may annoy them though it could produce less noisy data and provide promising results [2].

© Springer Nature Switzerland AG 2019
D. D. Schmorrow and C. M. Fidopiastis (Eds.): HCII 2019, LNAI 11580, pp. 446–460, 2019.
https://doi.org/10.1007/978-3-030-22419-6_32

Electroencephalogram (EEG) is a suitable approach for unobstructive and continuous measure of the task level difficulty [3] as it can measure the brain's response to the learning material presented and therefore offers a direct measure on the task difficulty level (TDL). Furthermore, EEG is non-invasive, portable and relatively cheap when compared to other measures of brain activity such as functional magnetic resonance imaging (fMRI).

Klimesch [4] has proposed using event related desynchronisation (ERD) feature extracted from the EEG as a measure of task difficulty level. ERD measures the extent to which neuron populations no longer oscillate synchronously to process the given task [5]. Band energies in specific EEG bands such as delta, alpha and beta in frontal areas of the brain have also been used to predict the memory load [6–8]. Here, we set out to use more channels to cover more areas of the brain and additionally combine inter-hemispheric asymmetry (ASR) features [9] as additional measure of cognitive load. We also use subjective measurement with NASA TLX index [10].

The band energies, ERD and ASR are used individually and in combination with six different classifiers: Quadratic Discriminant Analysis (QDA), Support Vector Machine (SVM), Naïve Bayes (NB), k-Nearest Neighbour (KNN), neural network (NN) and random forest decision tree (TREE) to classify the programming mental task into either the easy or the difficult levels. We also employ a confidence approach to further increase the prediction performance. Java programming language was used here as it is popular in Computer Science programmes throughout the world but any coding language could have been used instead.

2 Methodology

2.1 Experimental Paradigm

Nine subjects were recruited from a pool of postgraduate students from School of Computing, University of Kent, who had at least six months of Java experience or have taken Java programming module as a part of their postgraduate course. Out of nine subjects, seven were males and two females. Subjects age ranged between 20 and 37 years (mean = 26±3.74). However, data from two male subjects could not be used as they did not complete a baseline task which was necessary to compute the ERD features (discussed later).

Ethical approval was obtained from University of Kent Sciences Research Ethics Committee and subjects signed a voluntary consent form and were paid £15 each. The subjects were briefed on the tasks and the experiment was designed such that the subjects would be able to understand the given program and perform the code execution mentally. Subjects have to give the final output of the program code as an answer and this method was chosen to avoid inductive bias. All codes were written in Java programming language and initially, a total of 20 Java programs were developed into three categories (spatial relation, visual object grouping, mathematical execution), each from two different TDL (easy, difficult). From this, six Java programs as deemed easy and difficult by questionnaire responders were selected (three for easy and three for difficult categories).

The easy and difficult TDL were pre-determined using questionaire responses from 15 subjects who were not involved in the EEG data collection. The volunteers (age: 28.8±4.63, 9 males and 6 females, all non-related to University of Kent, who have sufficient experience in Java (currently working or proficient in Java - mean experience of 30.53±3.56 months). This good Java experience ensures correct 'ground truth' of choosing different task difficulty level; there was no statistical difference in age range for these and the volunteers for EEG based study. These subjects completed a questionnaire on time-spent and task difficulty level rating for each task. The difficulty level ranged from 1 to 10 (where 1 is very easy task, 10 is impossible to solve mentally). Only those questionnaire with correct answers to the questions were considered. The different task categories to be solved were:

- Spatial relation tasks that test subject's spatial reasoning skills like visualising shape of objects mentally. For example, visualising two rectangle objects mentally using parameters of x-axis and y-axis coordinates, width and height and to solve whether the two rectangles overlaps or not.
- Visual object grouping tasks that utilises subject's working memory to recall the swapped, mapped or sorted shape of objects group correctly. For example, given a number of shape objects mapped to variables and grouped in an array in different order, subject has to map the object variable name with the shape objects correctly and output those objects in order.
- Mathematical execution where the subject had to perform arithmetic calculations mentally. For example, subject has to compute the mean of an array of integers.

Prior to performing the tasks, subjects were asked to relax for one minute (EEG was also collected during this time as baseline). Table 1 shows the GUI steps in collecting the EEG data. Steps 3 and 4 will repeat until all six programs are shown (in random order). Figure 1 shows an example of the task screen.

This GUI not only serves as a front-end, but also communicates with the EEG collection device via COM port (emulated serial port) by sending different markers values for different user activities such as relax and task execution states. Table 2 gives the details of marker type and the sent value to EEG device during the experiment. This information can be used to segment the EEG into the different tasks.

Subjects were demonstrated the working of the GUI and were asked to perform the practise tasks in order to familiarise with the tool. Before the experiment was started, subjects sat comfortably. They were discouraged to make physical movement (example avoid blinking where possible, excessive swallowing or any hand gestures etc during the task experiment and to focus on the presented task while solving the program code. Figures 2, 3 and 4 show examples of the tested Java codes.

Table 1. GUI sequence for the experiment

Screen	Description
1. Welcome	Displays general information on the research
2. Instruction	Displays instructions to be followed by the subjects during the experiment
3. Relax	During this screen, the subjects will relax or complete the NASA TLX feedback for the tasks they solved (no time limit)
4. Task	This screen contains the Java codes to solve and a text box for the subject to input their responses
5. Thank you	This screen contains thank you message and indicates the end of the experiment

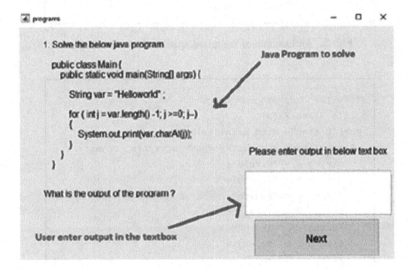

Fig. 1. Task screen.

Table 2. Marker values sent by GUI to EEG device

Marker type	Value
Start of task	1
End of task	2
Start of relax state (to indicate baseline recording)	3

```
Solve the Java program below. Note: Rectangle object
is defined as drawRectangle (x-axis, y-axis, width,
height)

  class MyCanvas extends JComponent {
     public void paint(Graphics g) {
        g.drawRect (2, 2, 10, 10);
        g.drawRect (5, 5, 10, 10);
     }
  }

Will two rectangles overlap?
```

Fig. 2. An example of the tested spatial relation Java code.

```
Solve the Java program below
public class Main {
 public static void main(String[] args) {
  ArrayList<Object> array = new ArrayList<Object> ();
    String a = "Circle";
    String b = "Triangle";
    String c = "Square";
    String d = "Triangle";
    String e= "Circle";
    String f = "Triangle";
    array.add(b);
    array.add(a);
    array.add(d);
    array.add(e);

    for (int i = 1; i < 4; i++) {
      System.out.print(array.get(i) + " ");
    }
 }
}

Last three shape objects are?
```

Fig. 3. An example of the tested visual object grouping Java code.

```
Solve the Java program below
public class Main {
 public static void main(String[] args) {
   int [] vars = {4, 8, 10, 12, 16, 10, 18, 2, 3, 5};
   int value = 0;
   int count = 0;

   for(int i = 0; i< vars.length; i++){
     value += vars[i];
     count++;
   }
   int temp = value/count;
   System.out.println(temp);
 }
}
What is the output of the program?
```

Fig. 4. An example of the tested mathematical Java code.

2.2 NASA TLX Survey

After solving each task, the subjects were instructed to fill a paper based NASA TLX rating sheet based on their perception on task difficulty level. NASA TLX index is a six dimensional subjective measurement method developed NASA to measure cognitive loads [10]. The six dimensional sub scales are mental demand, physical demand, temporal demand, performance, effort and frustration level. The workload is evaluated in two procedures for each task: first, subjects have to give their perspective in a sub-scale rating range from 0–100 (divided into 20 equal intervals) and second is the sub-scale weights created by forming 15 possible pair from six dimensional elements and subjects choose the most important dimension or factor contributing to the workload.

Here, after marking the six dimension ratings, the subject were instructed to circle the most important dimension that contributes to the task which is given in pairs as mentioned above. The overall Weighted Workload Score (WWS) is computed from the subjects rating and weight that contribute to the cognitive workload. This procedure is similar to the usage of NASA TLX Index form in the study done by Fritz et al. [11].

2.3 EEG Data

The EEG data was obtained from Emotiv Epoc 14 channels (configuration as shown in Fig. 5) wireless EEG device sampled at 128 Hz. During the experiment, the signal strength was continually checked and adjusted to ensure all the electrodes had good contact with the scalp through the use of saline solution.

The EEG data was segmented to one second lengths. Elliptic IIR filters were used to filter the segmented EEG signals in delta (1–4) Hz, theta (4–8 Hz), alpha (8–12 Hz), beta (12–30 Hz), gamma (30–50 Hz) bands [12] and feature extraction techniques were performed on these segments. Eighty such segments were obtained for each category and as such there were 480 patterns from six tasks altogether (easy and difficult tasks from three categories).

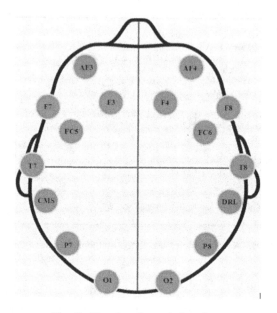

Fig. 5. Emotive electrode locations

2.4   EEG Analysis

ERD was computed by band pass filtering the EEG signal within the specified frequency band and percentage band power change was computed between the relaxed state and task execution state using (1):

$$ERD_b = (BE_r - BEtask_b)/BE_r \qquad (1)$$

where band energy during resting was computed using

$$BE_r = \sum_{i=1}^{n} (x - \bar{x})^2 \qquad (2)$$

and band energy during task using

$$BEtask_b = \sum_{i=1}^{n} (x - \bar{x})^2 \qquad (3)$$

with x is EEG data from each channel with length n from either the rest or task execution state and $\bar{x}$ is the mean of each channel. Given 14 channels and 5 bands, there were 70 ERD features for each one second EEG.

The ASR of each spectral band was computed using (2) and as in [9]:

$$ASR_b = \left(BE_{left} - BE_{right}\right) / \left(BE_{left} - BE_{right}\right) \tag{4}$$

where ASR is the asymmetric ratio between left and right hemispheres, BE_{left} is the spectral energy from one channel in left hemisphere (computed using (3)) and BE_{right} is spectral energy from opposite channel in right hemisphere. Since there were 14 channels (7 on each hemisphere) and 5 spectral bands, ASR resulted in a total of 35 features.

In addition, band energies (EN) for each channel in the five bands were also computed using (3) giving 70 features. Finally, all the available features were combined giving all feature (AF) set of 175 features.

2.5   Classification

These features were used by six different classifiers: QDA, SVM, NB, KNN, NN and TREE. For KNN, Euclidean distance was used whereas for QDA, the covariance matrices could vary among classes. TREE approach used an ensemble of 100 decision trees. For NN, the two output layer nodes values were set as either [1 0] or [0 1] with 10 hidden units (size chosen randomly) and trained using Matlab's *trainlm*. For the rest, classifier default parameters as available in Matlab's *fitcsvm*, *fitcnb*, *fitcensemble*, *fitcdicsr*, *patternnet* and *fitcknn* were used [13]. The two easy and difficult TDL were predicted with randomly split 40 fold cross validation.

Classifier Confidence. The classifier confidence (CC) approach used here worked by computing the output of the classifier for the test data. From the results, it was found that NN gave the best performance for most subjects, so only the output of this classifier was used. Also, all the features gave the best performance for majority of the subjects, so these features were used. The two classifier outputs for each test pattern were checked and the predicted class was seen as confident only if the two outputs differed by at least 0.1. With perfect classification, the best outputs would differ by 1 since one output would have a value of 1 and the other value of 0. Hence, having a 10% threshold value of 0.1 is appropriate though this value would need to be experimented in future to obtain the best threshold. It should be noted that some data will be discarded where classification outputs have lower confidence than the threshold. Figure 6 shows the flow of the experimental design.

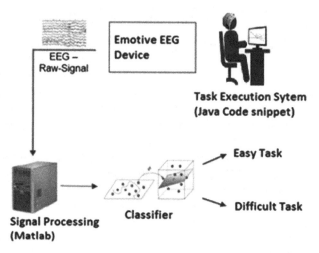

Fig. 6. Experimental flow design

3 Results and Discussion

Figure 7 shows the overall WWS from NASA TLX for the different task difficulty levels. Non-parametric Kruskal-Wallis test (as normality was not assumed) showed that there is significant difference between TDL ($p < 0.01$). Comparing each sub-scale (refer to Table 3), there were significant differences (using sign rank tests, $p < 0.01$) between TDL for mental demand, temporal demand, frustration and effort. Performance and physical demand did not indicate any difference. The latter is not surprising since there is no physical effort required in the tasks, though it is somewhat surprising there was no difference in performance measure. This clearly indicates the necessity of utilising measures such as EEG as subjects are unable to differentiate different levels of performance required to complete the tasks.

Kruskal-Wallis test showed that there is a statistically significant difference between easy and difficult tasks of EEG features ($p < 0.05$). Table 4 shows the classification results for EN, ERD, ASR and combined features for the five different classifiers for subject 1.

Similarly, Tables 5, 6, 7, 8, 9 and 10 show the results for rest of the subjects. To decide on the best classifier, all the features were combined and statistical test revealed significant difference between the classifier performances (p < 0.05). The mean rank comparison showed that NN classifier gave the best performance. It also gave the best performance for five out of seven subjects.

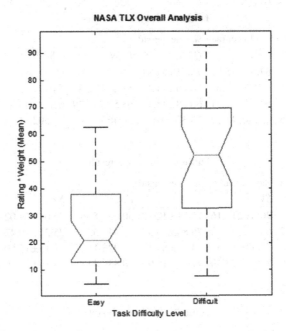

Fig. 7. Boxplot of overall NASA TLX index mean weighted workload for different task difficulty levels.

Table 3. NASA TLX – subscale

Sub scale	Mean (tasks)		p value
	Easy	Difficult	
Mental demand	7.52	15.36	7.1e–5
Physical demand	0.14	0.30	5.9e–1
Temporal demand	5.82	10.41	3.7e–3
Performance	4.84	6.23	3.4e–1
Effort	7.27	13.00	5.5e–3
Frustration	0.80	4.68	7.3e–3

Table 4. Subject 1 results

Classification method and accuracy % (mean±std)

	TREE	SVM	QDA	KNN	NB	NN
EN	69.17±12.26	73.54±14.48	73.33±11.66	63.54±14.94	69.17±12.26	72.86±10.95
ERD	67.50±11.75	73.75±10.77	73.33±13.63	59.79±15.89	67.50±11.75	70.00±13.84
ASR	68.12±11.31	68.33±13.10	65.00±8.48	62.50±12.66	68.12±11.31	73.75±6.77
AF	70.42±14.24	71.25±12.65	75.21±13.14	61.46±13.30	70.42±14.24	**82.50±6.57**

Table 5. Subject 2 results

Classification method and accuracy % (mean±std)

	TREE	SVM	QDA	KNN	NB	NN
EN	63.33±13.84	67.08±15.09	76.25±14.81	51.67±5.06	63.33±13.84	68.33±7.78
ERD	64.17±15.58	66.87±14.68	76.46±11.15	51.46±4.58	64.17±15.58	65.42±12.16
ASR	67.71±13.76	67.92±12.31	66.88±15.15	63.54±10.79	67.71±13.76	74.58±10.57
AF	72.50±11.35	73.54±11.77	74.37±9.69	51.46±3.72	72.50±11.35	**77.92±7.88**

Table 6. Subject 3 results

Classification method and accuracy % (mean±std)

	TREE	SVM	QDA	KNN	NB	NN
EN	76.04±14.02	70.63±11.16	75.21±12.30	50.83±5.91	76.04±14.02	72.62±11.36
ERD	75.21±10.42	68.96±12.66	75.83±13.58	51.04±5.72	75.21±10.42	65.83±13.18
ASR	73.13±15.50	72.50±14.52	74.58±8.43	67.92±11.72	73.13±15.50	77.08±5.13
AF	73.13±12.59	74.79±13.80	77.92±10.93	50.83±4.92	73.13±12.59	**79.58±6.33**

Table 7. Subject 4 results

Classification method and accuracy % (mean±std)

	TREE	SVM	QDA	KNN	NB	NN
EN	73.75±12.02	75.21±12.59	75.42±13.07	51.46±4.95	73.75±12.02	73.54±12.07
ERD	72.29±11.23	75.00±12.52	74.38±15.14	51.67±4.70	72.29±11.23	70.63±14.12
ASR	69.17±13.63	67.92±12.31	75.83±12.63	64.79±12.30	69.17±13.63	73.13±12.59
AF	75.63±11.07	75.21±15.27	**78.54±11.62**	52.50±5.72	75.63±11.07	73.75±7.88

Table 8. Subject 5 results

Classification method and accuracy % (mean±std)

	TREE	SVM	QDA	KNN	NB	NN
EN	70.63±12.94	72.50±14.40	74.79±12.87	52.29±5.66	70.63±12.94	75.48±9.89
ERD	71.04±10.67	71.46±11.77	74.17±11.75	52.92±6.69	71.04±10.67	72.50±11.51
ASR	68.54±12.87	68.33±13.10	71.67±13.84	67.29±14.42	68.54±12.87	77.08±7.09
AF	75.21±13.41	71.25±14.85	76.04±11.51	53.33±5.60	75.21±13.41	**80.42±6.77**

Table 9. Subject 6 results

Classification method and accuracy % (mean±std)

	TREE	SVM	QDA	KNN	NB	NN
EN	66.25±14.12	76.25±12.16	82.71±9.69	56.46±7.66	66.25±14.12	72.08±13.68
ERD	67.08±15.33	75.83±11.45	82.92±9.61	56.67±8.89	67.08±15.33	72.71±12.94
ASR	73.33±15.58	68.54±14.31	77.71±11.38	62.29±13.21	73.33±15.58	71.67±6.33
AF	72.29±12.14	73.96±13.24	**83.54±10.92**	58.33±11.79	72.29±12.14	73.75±6.77

Table 10. Subject 7 results

Classification method and accuracy % (mean±std)						
	TREE	*SVM*	*QDA*	*KNN*	*NB*	*NN*
EN	70.63±12.08	74.58±14.12	73.75±12.88	55.21±6.45	70.63±12.08	74.52±12.77
ERD	67.29±11.54	75.63±12.43	73.54±10.66	55.42±7.20	67.29±11.54	73.96±14.15
ASR	70.63±11.32	70.00±11.45	71.04±12.66	68.75±14.09	70.63±11.32	75.83±5.98
AF	69.79±12.33	74.37±12.14	73.33±12.83	55.83±9.66	69.79±12.33	**83.75±7.39**

Next, using NN classification results (as it gave the best significant performance overall), significant difference was obtained in the classification accuracies between the different feature extraction approaches, $H(3) = 26.33$, $p = 8.12e-6$. Comparing the mean rank values (EN: 581.03, ERD: 576.88, ASR: 478.44, AF:605.66) showed that ERD features had the highest discriminatory information to separate the two mental tasks with combination of all features giving the best results. Using all the features also gave the best accuracy for six out of seven subjects with the remaining subject having ERD giving the best accuracy.

Using CC approach revealed further improvement in the classification performance. As NN gave the best performance, it was decided to use this classifier with the best performing all feature combination. Figure 8 shows the performance for the seven subjects and it can be seen that performances were higher when CC was used. The improvements were statistically significant for all subjects (sign rank test, $p < 0.05$) except subject 6. This is as expected since only the classification outputs that have slightly more confident predictions are being used (the experiment revealed that about 10% of patterns were dropped).

Table 11 shows the average response time (i.e. the time taken to complete the tasks). It can be seen that the difficult tasks take longer to complete as compared to easy tasks as expected.

This research was limited by significant noise in experiment design procedure with some subjects verbalising, flicking pens, nodding etc. Eye blinks occurred in the EEG data as shown in Fig. 9 (example shown for one subject but similar artifacts were

Table 11. Average completion time (secs) for different task levels.

Subject	Average time (in seconds)	
	Easy	*Difficult*
1	35.43	87.94
2	92.74	180.08
3	38.47	97.46
4	35.78	210.12
5	31.51	75.45
6	93.60	174.22
7	88.34	143.59
Mean	59.41	138.41

458   R. Palaniappan et al.

observed for other subjects too). While these could have been removed in the pre-processing stage (for example using independent component analysis), we chose not to in order to simulate actual classroom settings where it will be difficult to force students to adhere to strict no-movement instruction.

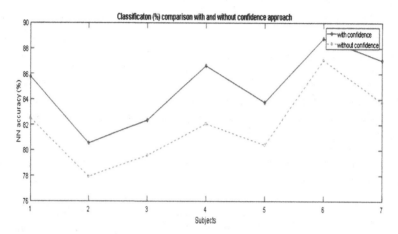

Fig. 8. Classification (%) comparing the improvement with confidence approach (blue: with confidence, red: without confidence). (Color figure online)

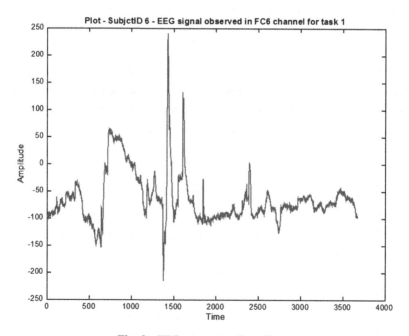

Fig. 9. EEG segment with artifacts.

4 Conclusion

Both NASA TLX and task completion time showed significant differences between TDL. NASA TLX has been used as a non-physiological measure to discriminate different cognitive load for different programming language [14]. However, based on the lack of statistical difference in the performance measure in TLX sub scale, we can infer that it is difficult for subjects to estimate the TDL, hence showing the necessity to have measures that directly measure the ability.

In this report, we have shown that it is possible to differentiate the task difficulty of Java programming code using EEG signals. Though the subject pool is small and the performance needs improvement for real-life implementation, there is sufficient promise in the method to be studied further. The combination of proposed ASR with ERD and EN features improves the classification performance and among the tested classifiers, NN gave the best performance. The use of CC approach further improved the performance to give a maximum accuracy of 87.05%. It is possible that with proper feature selection and tuning of classifier parameters could further improve the accuracy.

In conclusion, the findings here will hopefully pave the way for future research studies on tailoring learning material with appropriate level of difficulty, which will be especially useful for those with independent learning plans.

References

1. Gerjets, P., Walter, C., Rosenstiel, W., Bogdan, M., Zander, T.O.: Cognitive state monitoring and the design of adaptive instruction in digital environments: lessons learned from cognitive workload assessment using a passive brain-computer interface approach. Front. Neurosci. **8**, 1–21 (2014)
2. Mihalcaa, L., Salden, R.J.C.M., Corbalan, G., Paas, F., Miclea, M.: Effectiveness of cognitive-load based adaptive instruction in genetics education. Comput. Hum. Behav. **27**, 82–88 (2011)
3. Lee, J.C., Tan, D.S.: Using a low-cost electroencephalograph for task classification in HCI research. In: Proceedings of the 19th ACM Symposium on User Interface Software and Technology, pp. 81–90 (2006)
4. Klimesch, W.: EEG alpha and theta oscillations reflect cognitive and memory performance: a review and analysis. Brain Res. Rev. **29**(2–3), 169–195 (1999)
5. Crk, I., Kluthe, T., Stefik, A.: Understanding programming expertise: an empirical study of phasic brain wave changes. ACM Trans. Comput.-Hum. Interact. **23**(1), 2 (2015)
6. Zarjam, P., Epps, J., Chen, F.: Characterizing working memory load using EEG delta activity. In: Proceedings of 19th European Signal Processing Conference, pp. 1554–1558 (2011)
7. Jensen, O., Tesche, C.D.: Frontal theta activity in humans increases with memory load in a working memory task. Eur. J. Neurosci. **15**(8), 1395–1399 (2002)
8. Klimesch, W.: Alpha-band oscillations, attention, and controlled access to stored information. Trends Cogn. Sci. **16**(12), 606–617 (2012)
9. Duraisingam, A., Palaniappan, R., Samraj, A.: Cognitive task difficulty analysis using EEG and data mining. In: IEEE Conference on Emerging Devices and Smart Systems, Salem, India (2017)

10. Hart, S.G., Staveland, L.E.: Development of NASA-TLX (Task Load Index): results of empirical and theoretical research. Adv. Psychol. **52**, 139–183 (1988)
11. Fritz, T., Begel, A., Müller, S.C., Yigit-Elliott, S., Züger, M.: Using psycho-physiological measures to assess task difficulty in software development. In: ACM Proceedings of the 36th International Conference on Software Engineering, pp. 402–413 (2014)
12. Palaniappan, R.: Utilizing gamma band spectral power to improve mental task based brain computer interface design. IEEE Trans. Neural Syst. Rehabil. Eng. **14**(3), 299–303 (2006)
13. Mathworks, Statistics and Machine Learning Toolbox (2018)
14. Yousoof, M., Sapiyan, M.: Optimizing instruction for learning computer programming – a novel approach. In: Intan, R., Chi, C.-H., Palit, H.N., Santoso, L.W. (eds.) ICSIIT 2015. CCIS, vol. 516, pp. 128–139. Springer, Heidelberg (2015). https://doi.org/10.1007/978-3-662-46742-8_12

Towards Hybrid Multimodal Brain Computer Interface for Robotic Arm Command

Cristian-Cezar Postelnicu[(✉)], Florin Girbacia,
Gheorghe-Daniel Voinea, and Razvan Boboc

Department of Automotive and Transport Engineering,
Transilvania University of Brașov, Brașov, Romania
cristian-cezar.postelnicu@unitbv.ro

Abstract. A hybrid brain-computer interface (BCI) is a system that combines multiple biopotentials or different types of devices with a typical BCI system to enhance the interaction paradigms and various functional parameters. In this paper we present the initial development of a hybrid BCI system based on steady state evoked potentials (SSVEP), eye tracking and hand gestures, used to command a robotic arm for manipulation tasks. The research aims to develop a robust system that will allow users to manipulate objects by means of natural gestures and biopotentials. Two flickering boxes with different frequencies (7.5 Hz and 10 Hz) were used to induce the SSVEP for the selection of target objects, while eight channels were used to record the electroencephalographic (EEG) signals from user's scalp. Following the selection, the users were able to manipulate the objects from the workspace by using the Leap Motion controller to send commands to a Jaco robotic arm.

Keywords: Hybrid brain computer interface · SSVEP · Robotic arm

1 Introduction

Typical brain-computer interface (BCI) systems allow people to communicate without performing movements and speaking. This category of systems are mainly designed for severely disabled persons to provide them an alternative communication and control channel. However, the research community has recently started developing BCIs also for persons with less severe disabilities or even for clinically healthy users. Nowadays is a high demand for natural user interfaces and even if BCIs may seem difficult to use in the beginning, by using them for a long term period the performances and acceptance rates keep increasing for the majority of users.

Most of the BCIs are based on the signals recorded from the electroencephalogram (EEG). The most significant results have been obtained by using one of the following signals: event-related potential (ERPs) P300, steady-state visual evoked potential (SSVEP) and event-related desynchronization/synchronization (ERD/ERS) [1–4]. ERD-based BCIs require subjects in general to imagine natural movements of hands and/or feet and translating them in commands to control devices like mobile robots, wheelchairs and orthoses, to control cursors on a computer monitor or to navigate through virtual environments and games [5–8]. Their main advantages are related to the

© Springer Nature Switzerland AG 2019
D. D. Schmorrow and C. M. Fidopiastis (Eds.): HCII 2019, LNAI 11580, pp. 461–470, 2019.
https://doi.org/10.1007/978-3-030-22419-6_33

short amount of time required to learn to control such a BCI and the high values obtained for classification rates [5]. In general this type of BCIs are limited to very few commands that can be generated, while the information transfer rate (ITR) falls dramatically if more than two commands are used [9].

The P300 potential is a positive amplitude in the EEG signals that appears around 300 ms after the user perceives a target visual, auditory or somatosensory stimulus. The P300-based BCI systems requires the users to silently count the target event among a large set of such stimuli [10–13]. Initially, these BCIs have been mainly developed for communication applications called spellers [12]. They are based on the presentation of visual flashing items (letters, digits, punctuation marks and various typical keyboard commands) on a computer monitor, thus allowing the users to generate a command from a large set of items. Multiple presentation paradigms have been proposed so far by researchers, like: row-column, single-character, checkerboard, lateral-single character or half checkerboard [12–17]. This category of BCIs offers both high ITRs and classification accuracy rates, and just a couple of minutes are required for users to learn how to use such a system. Furthermore, the mentioned presentation paradigms have been successfully used for various types of applications such as smart homes control, Internet browsing, environment control and even for drawing [18–23].

SSVEP-based BCIs allow also a large number of choices, but it is limited by the number of frequencies that can be analyzed in the recorded EEG signals. SSVEP has been used in studies that involve low-level cognitive processes in the brain because it is an intrinsic neuronal response which is mostly independent of higher-level cognitive processes [23]. The number of SSVEP-based applications is increasing every year due to its robustness, high ITR, a relatively simple system configuration and a very short training time [3, 24–27]. Also, the SSVEP response of the brain is straightforward to model and interpret. In order to have a usable BCI application, correct decisions should be made by using short signal segments, with intervals ranging from 0.5 s to 4 s, while an interval of 2 s is proved to achieve very high classification accuracy rates [27].

Latest research activities have started focusing on various combinations of EEG-based signals, and also on combinations between eye tracking technologies, electrooculography (EOG) and electromyography (EMG) with typical BCIs [2, 4, 16]. This new category of BCIs, namely hybrid BCIs (hBCIs), have already been validated to provide increased ITRs and higher classification rates [16]. Such hybridization solutions can offer the users multiple input methods by working in a sequential or consecutive paradigms, or the various signals used can represent redundant checks to maximize the classification rates and maximize ITRs.

The present study aims to develop a hybrid multimodal interface based on SSVEP, eye tracking technologies and user's hand gesture identified by the Leap Motion controller [28], to remotely provide commands to a Jaco robotic arm [29] to manipulate objects in a specific workspace. The proposed system is mainly developed for clinically healthy users who are supposed to use it in the future on a daily basis. For the initial validation phase presented in this paper the user was placed in the same room with the robotic arm in order to facilitate the users' capability to learn how to specifically use the hybrid multimodal interface.

2 Materials and Methods

2.1 Interaction Paradigm

Increasing and diversification of the information related to commands that can be send to robotic systems presents limitations that can be translated into considerable cognitive efforts for the operator. Most often the limitations appear at the training stage when the operator has to make judgments on the commands used for a particular task, identify and select the right ones. Obviously, as a certain routine is established, the cognitive effort decreases, but also for this stage is important to organize and improve the interaction modalities.

A common approach to reduce cognitive effort is the use of multimodal interfaces. These can allow the distribution of task information to several communication channels when the cognitive effort increases, therefore allowing the user to distribute and manage the cognitive effort. Instead of using various joystick commands and consuming an extra cognitive effort, the multimodal interface allows the use of perceptual and mental systems at a normal level, so the user can focus on the task that needs to be solved.

The use of biopotentials in combination with other human-machine communication channels, for interaction with a robotic system, offers a possible solution to the problems of the current interfaces. This approach allows different ways to accomplish a robotic task when required. The use of biopotentials in the robot interaction paradigms is gaining more and more interest in the research community. Dedicated interfaces improve user performance through specific interaction methods. Developing new interaction methods, acquiring and automating them can minimize multiple processes in the user's brain, thus reducing his cognitive effort.

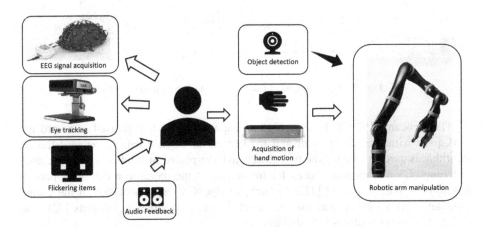

Fig. 1. System architecture of hybrid multimodal interface

The proposed hybrid multimodal interface is capable of adapting to the user's intentions and behavior, in dynamically changing environments. It combines latest BCIs technology and concepts to provide a natural interaction between humans and the robotic systems, which is unconstrained and robust. Thus, the interface provides to the users the means to command a robotic arm for manipulation tasks. The users are placed in front of a computer monitor where they are able to see the workspace where the robotic arm is placed and to select the objects in the workspace (see Fig. 1). The workspace scene is displayed on the computer's monitor by acquiring the video stream from a Logitech C920 webcam. Automatically the system identifies the moments when the user is gazing at the monitor by using an eye tracking system, Tobii X120 [30]. Thus, the system enables the selection of an object by superimposing a set of flashing blocks on two different frequencies (7.5 Hz and 10 Hz) over the objects found in scene (see Fig. 1). The flashing blocks trigger SSVEP signals in the recorded EEG signals, while the classification of EEG data will determine which object was selected by the user (see Fig. 2). An object selection will trigger an audio feedback to the user and will automatically stop the flashing blocks on the computer's monitor.

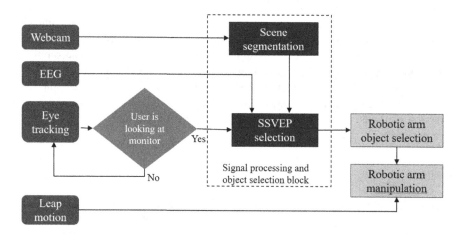

Fig. 2. Flow diagram of the hybrid multimodal interface

The webcam (1920 × 1080 resolution, captured at 30 fps) is used to record the workspace with the robotic arm and the objects to be selected. An edge-detection algorithm is used to identify the two objects that are placed in front of the robotic arm. The Emgu CV library was chosen for image processing, because it can use OpenCV functions in .NET language [31]. As such, an OpenCV implementation of the Canny edge detection algorithm was used to detect the edges of the target objects [32]. The Canny Algorithm contains 5 basic steps:

(1) apply a Gaussian filter to eliminate the noise in the image;
(2) search for image intensity gradients;
(3) apply a non-maximum suppression to get rid of the false response to edge detection;

(4)   a double threshold is applied to determine the potential edges;

(5)   the hysteresis route is followed and the edges are finally detected by suppressing the lower intensity and not connected to the most pronounced ones.

For this initial study, the objects are placed in predefined positions in the workspace. After the user selects a specific object, the robotic arm is automatically positioned near the object waiting for manipulation commands. For the command of the Jaco robotic system the Leap Motion controller was used because, based on [33], it allows accurate tracking of hand movements. The Leap Motion controller can detect the position, orientation and velocity of hands along with the position, orientation and velocity of each component finger [34]. The user can send translation (X axis) and rotation (roll, pitch, yaw angles) commands to the Jaco robotic arm by moving the right hand. Grasping and releasing an object is controlled by closing and opening the right hand. The implementation of robotic arm commands using natural user interaction is based on the functions available in the Leap Motion SDK. The actual position and orientation of the right hand is acquired continuously from the Leap Motion controller and compared with the previous one. If the difference is higher than a predefined threshold, the movement command is sent to the Jaco robotic arm.

2.2   Experimental Design and Procedure

The purpose of this initial study is to familiarize the users with the proposed hybrid interface, without aiming to define a qualitative or quantitative evaluation. Most of the BCIs related studies require users to test a specific system after a very short training period. The entire study we aim to conduct assumes that users will know exactly how to interact with the developed system. Thus, we asked three users to participate in a series of practice sessions for a period of three weeks (one session per week).

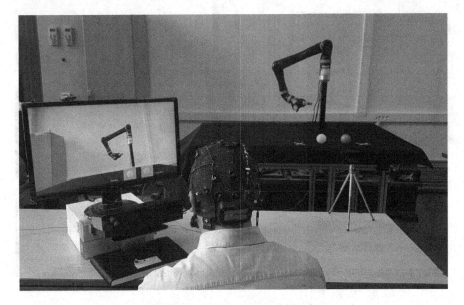

Fig. 3. SSVEP-based object selection

The experimental design assumes three phases related to familiarization and calibration procedures followed by a practice session during which the users are required to perform a set of selection and manipulation tasks.

The familiarization procedure allows the participants to use the Leap Motion controller to command the Jaco robotic arm to grasp and manipulate the two objects placed in the workspace until they feel confident in using the interface.

The calibration procedure is related to the eye tracking system calibration and the SSVEP classifier training dataset. The eye tracking calibration is performed by successively gazing at a series of 5 points displayed on the computer's monitor. The Tobii SDK functionalities were used for both calibration procedure and the continuous tracking of users' gaze. The calibration procedure can be repeated if the calibration results are not satisfactory. The SSVEP calibration is described in Sect. 2.4.

During the practice session the subjects are requested to perform a set of 10 manipulation task, 5 for each object placed in the workspace. The subjects are requested to alternatively select (see Fig. 3), grasp and move (see Fig. 4) the two yellow and blue objects from their initial positions to the other two predefined positions marked in the workspace. After the subject successfully managed to move each object in the workspace, an experiment observer repositions the two objects to their initial locations. Between each two consecutive trials is a pause of 12 s during which the robotic arm is automatically moving to the default position (see Fig. 3).

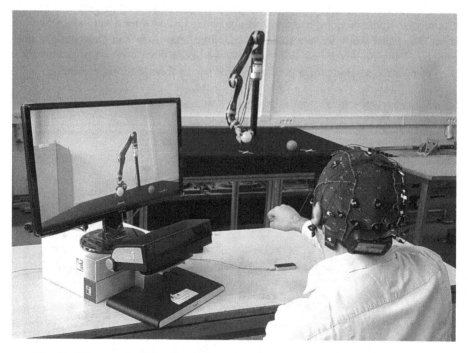

Fig. 4. Object manipulation based on Leap Motion controller and Jaco robotic arm

2.3 Participants

Data was collected from three healthy subjects (mean age = 22.33 years, SD = 2.05, range = 20–25 years), students of the Transilvania University of Brasov, with no prior experience with BCIs. The subjects were informed about the purpose of the study, were given a written consent form to sign and did not receive a financial reward for their participation.

2.4 EEG Data Processing

Subjects were placed on a seat in front of a 24 inches monitor (resolution: 1920 x 1080 pixels, vertical refresh rate: 60 Hz) at a distance of around 60 cm. The EEG signals were recorded by a 16-channels g.Nautilus amplifier channels with a 250 Hz sampling rate, and were further bandpass filtered between 0.5–30 Hz and also a notch filter at 50 Hz was applied. Standard g. SAHARA dry active electrodes were used to acquire the data from the surface of the scalp. Data were recorded by using eight electrodes from the occipital region of the brain, with the ground electrode placed on the Fpz and referenced to the right earlobe. The electrodes were attached at locations O1, Oz, O2, PO7, PO3, POz, PO4 and PO8.

For the calibration phase the subjects were asked to perform a set of trials by gazing at a set of flashing boxes on the computer's monitor. For each trial they were requested to focus at a specific flashing box for 10 s after which the trial was over and a pause of 5 s was allowed before next trial. The stimulation boxes were oscillating on two frequencies 7.5 Hz (left stimulus) and 10 Hz (right stimulus). The users were requested to alternatively gaze at the two flashing stimuli resulting in total 8 trials. The resulting data was used to train the classifier which was next used to classify the EEG during the on-line test.

During the online practice session the subjects can select one of the two objects in the workspace based on the SSVEP. The classification of recorded EEG signals is based on the minimum energy combination (MEC) which is a method that overcomes the limitations of the bipolar and Laplacian approaches and can use an arbitrary number of electrodes [35]. MEC implies finding the best combinations of electrode signals that can cancel as much of the noise and interferences as possible, thus optimizing the signal-to-noise ratio. The algorithm used the Levinson AR Model with epochs of 3 s consisting of 750 samples. The obtained features were entered into a linear discriminant analysis classifier for pattern classification.

3 Discussion

The present study aims to develop a hybrid multimodal interface to command a Jaco robotic arm by using biopotentials and hand gestures. The initial phase of the research is presented in this paper. Three users were asked to participate for three weeks to a couple of evaluation session during which we aimed to identify if the proposed hybrid interaction paradigm is appropriate for tasks involving objects selection based on SSVEP and manipulation with a robotic arm by means of hand gestures.

One of our findings during this validation phase are related to the calibration of the eye tracking system for which the distance from the computer's monitor influenced the calibration results and sometimes it was required to repeat this step for 2 or 3 times.

Related to the SSVEP-based selection all users were able to correctly select the target object, thus achieving 100% accuracy.

During the tests performed in the first week two of the users had some issues in correctly positioning the object in the requested area. Four times they released the object before reaching the target area and three times they moved the robotic arm too far from the target area. For the next sessions after familiarizing more with the Leap Motion controller they managed to correctly position all the time the objects in the requested area.

4 Conclusions

Overall, the proposed hybrid multimodal BCI system for robotic arm manipulation task presents promising initial testing results. The users were able to correctly select all the time the target object after the successful calibration for eye tracking system and for SSVEP-based classifier. Also, the interaction based on the Leap Motion controller was found to be reliable and easy to master even if in the beginning for a couple of trials the users didn't correctly position the object to the target area (83 successful trials out of 90).

The future work is focused on the integration within the interface of selection possibilities for more than two objects and also on the object's manipulation possibilities in the 3D space. Also, we aim to realize a comprehensive evaluation of the proposed interface for performance parameters, i.e. ITR, time-to-complete of a specific task, SSVEP classification accuracy, ergonomics assessment.

Acknowledgments. This work was supported by a grant of the Ministry of National Education and Scientific Research, RDI Programme for Space Technology and Advanced Research - STAR, project number 566.

References

1. Wolpaw, J.R., Birbaumer, N., McFarland, D.J., Pfurtscheller, G., Vaughan, T.M.: Brain–computer interfaces for communication and control. Clin. Neurophysiol. **113**(6), 767–791 (2002)
2. Allison, B.Z., et al.: Toward smarter BCIs: extending BCIs through hybridization and intelligent control. J. Neural Eng. **9**(1), 013001 (2012)
3. Guger, C., et al.: Poor performance in SSVEP BCIs: are worse subjects just slower?. In: 2012 Annual International Conference of the IEEE Engineering in Medicine and Biology Society, pp. 3833–3836 (2012)
4. Pfurtscheller, G., et al.: The hybrid BCI. Front. Neurosci. **4**(30) (2010)
5. Guger, C., Edlinger, G., Harkam, W., Niedermayer, I., Pfurtscheller, G.: How many people are able to operate an EEG-based brain-computer interface (BCI)? IEEE Trans. Neural Syst. Rehabil. Eng. **11**(2), 145–147 (2003)

6. Millan, J.R., Mourino, J.: Asynchronous BCI and local neural classifiers: an overview of the adaptive brain interface project. IEEE Trans. Neural Syst. Rehabil. Eng. **11**(2), 159–161 (2003)
7. Scherer, R., et al.: Brain–computer interfacing: more than the sum of its parts. Soft. Comput. **17**(2), 317–331 (2012)
8. Guger, C., Allison, B.Z., Edlinger, G.: Brain-Computer Interface Research: A State-of-the-Art Summary. Springer, Heidelberg (2013)
9. Edlinger, G., Allison, B.Z., Guger, C.: How many people can use a BCI system? In: Kansaku, K., Cohen, L., Birbaumer, N. (eds.) Clinical Systems Neuroscience. Springer, Tokyo (2014). https://doi.org/10.1007/978-4-431-55037-2_3
10. Guger, C., et al.: How many people are able to control a P300-based brain-computer interface (BCI)? Neurosci. Lett. **462**(1), 94–98 (2009)
11. Donchin, E., Spencer, K.M., Wijesinghe, R.: The mental prosthesis: assessing the speed of a p300-based brain-computer interface. IEEE Trans. Rehabil. Eng. **8**(2), 174–179 (2000)
12. Farwell, L.A., Donchin, E.: Talking off the top of your head: toward a mental prosthesis utilizing event-related brain potentials. Electroencephalogr. Clin. Neurophysiol. **70**(6), 510–523 (1988)
13. Guan, C., Thulasidas, M., Wu, J.: High performance P300 speller for brain–computer interface. In: IEEE International Workshop on Biomedical Circuits and Systems, S3/5/INV - S3/ pp. 13–16 (2004)
14. Pires, G., Nunes, U., Castelo-Branco, M.: Comparison of a rowcolumn speller vs. a novel lateral single-character speller: assessment of BCI for severe motor disabled patients. Clin. Neurophysiol. **123**(6), 1168–1181 (2012)
15. Townsend, G.T., et al.: A novel P300-based brain-computer interface stimulus presentation paradigm: moving beyond rows and columns. Clin. Neurophysiol. **121**(7), 1109–1120 (2010)
16. Postelnicu, C.-C., Talaba, D.: P300-based brain-neuronal computer interaction for spelling applications. IEEE Trans. Biomed. Eng. **60**(2), 534–543 (2012)
17. Fazel-Rezai, R., Allison, B.Z., Guger, C., Sellers, E.W., Kleih, S.C., Kubler, A.: P300 brain computer interface: current challenges and emerging trends. Front. Neurosci. **5**, 14 (2012)
18. Edlinger, G., Krausz, G., Groenegress, C., Holzner, C., Guger, C., Slater, M.: Brain-computer interfaces for virtual environment control. In: Lim, C.T., Goh, J.C.H. (eds.) 13th International Conference on Biomedical Engineering IFMBE Proceedings, vol. 23. Springer, Heidelberg (2009)
19. Sirvent Blasco, J.L., Ianez, E., Ubeda, A., Azorin, J.M.: Visual evoked potential-based brain-machine interface applications to assist disabled people. Expert Syst. Appl. **39**(9), 7908–7918 (2012)
20. Botrel, L., Holz, E.M., Kubler, A.: Brain painting V2: evaluation of P300-based brain computer interface for creative expression by an end-user following the user-centered design. Brain-Comput. Interfaces **2**(2–3), 135–149 (2015)
21. Edlinger, G., Holzner, C., Guger, C.: A hybrid brain-computer interface for smart home control. In: Jacko, Julie A. (ed.) HCI 2011. LNCS, vol. 6762, pp. 417–426. Springer, Heidelberg (2011). https://doi.org/10.1007/978-3-642-21605-3_46
22. Postelnicu, C.-C., Covaci, A., Panfir, A.N., Talaba, D.: Evaluation of a P300-based interface for smart home control. In: Camarinha-Matos, L.M., Shahamatnia, E., Nunes, G. (eds.) DoCEIS 2012. IAICT, vol. 372, pp. 179–186. Springer, Heidelberg (2012). https://doi.org/10.1007/978-3-642-28255-3_20
23. Nunez, P.L., Srinivasan, R.: Electric fields of the brain: the neurophysics of EEG. Oxford University Press, Oxford (2006)

24. Sozer, A.T., Fidan, C.B.: Novel spatial filter for SSVEP-based BCI: a generated reference filter approach. Comput. Biol. Med. **96**, 98–105 (2018)
25. Volosyak, I., Gembler, F., Stawicki, P.: Age-related differences in SSVEP-based BCI performance. Neurocomputing **250**, 57–64 (2017)
26. Guger, C., et al.: How many people could use an SSVEP BCI? Front. Neurosci. **6**, 169 (2012)
27. Erkan, E., Akbaba, M.: A study on performance increasing in SSVEP based BCI application. Eng. Sci. Technol. Int. J. **21**(3), 421–427 (2018)
28. Leap Motion. https://www.leapmotion.com/. Accessed 10 Jan 2019
29. JACO robotic arm. https://www.kinovarobotics.com/en/products/assistive-technologies/kinova-jaco-assistive-robotic-arm. Accessed 10 Jan 2019
30. Tobii X120. https://www.tobiipro.com/. Accessed 10 Jan 2019
31. Shi, S.: Emgu CV Essentials. Packt Publishing Ltd. (2013)
32. Canny, J.: A computational approach to edge detection. IEEE Trans. Pattern Anal. Mach. Intell. PAMI **8**(6), 679–698 (1986)
33. Weichert, F., Bachmann, D., Rudak, B., Fisseler, D.: Analysis of the accuracy and robustness of the leap motion controller. Sensors **13**(5), 6380–6393 (2013)
34. Girbacia, F., Girbacia, T., Butnariu, S.: Design review of CAD models using a NUI leap motion sensor. J. Ind. Design Eng. Graph. **10** (2015)
35. Friman, O., Volosyak, I., Graser, A.: Multiple channel detection of steady-state visual evoked potentials for brain-computer interfaces. IEEE Trans. Biomed. Eng. **54**(4), 742–750 (2007)

Interpolation, a Model for Sound Representation Based on BCI

Hector Fabio Torres-Cardona[1,2]([envelope]), Catalina Aguirre-Grisales[1,3,4]([envelope]),
Victor Hugo Castro-Londoño[1]([envelope]), and Jose Luis Rodriguez-Sotelo[1,5]([envelope])

[1] Universidad de Caldas, Manizales, Colombia
hector.torres_c@ucaldas.edu.co, victorhcastrolondon@gmail.com
[2] Department of Music, Universidad de Caldas, Manizales, Colombia
[3] Universidad Autónoma de Manizales, Manizales, Colombia
catalina.aguirreg@autonoma.edu.co
[4] Electronic Engineering Department, Universidad del Quindío, Manizales, Colombia
caaguirre@uniquindio.edu.co
[5] Electronic Engineering Department, Universidad Autónoma de Manizales,
Manizales, Colombia
jlrodriguez@autonoma.edu.co
http://www.ucaldas.edu.co/

Abstract. Brain state control has been well established in the area of
Brain-computer interfaces over the last decades in which the active applications allow controlling external devices consciously. The purpose of this
study was to develop a real-time graphical sound representation system based on an interaction design that allows navigating through the
motor imagery cognitive task in a bidimensional plane. This representation was developed using the OpenBCI EEG acquisition system in order
to record the necessary information which was sent and processed in
Max/MSP software. The system operates under a metaphorical Graphical User Interface (GUI) programmed in Processing. The system was
tested through an experiment under controlled conditions in which six
professional musicians participated. From the experimental results, it
was found that all participants achieved different control levels associated to their static and dynamic response with an average of 26.73% and
73.27% respectively.

Keywords: Brain-computer interface · Electroencephalogram ·
Motor imagery · Neurofeedback · Sound synthesis

1 Introduction

Brain-computer interface (BCI) is a system that records the brain activity (Electroencephalogram - EEG) to communicate the brain in real time with external
devices without relying on peripherical nerves or muscles [1–3]. Even though
BCI is a non-invasive and mature technology, nowadays its applications are
considered as an emerging research field focused on improving brain-machine

© Springer Nature Switzerland AG 2019
D. D. Schmorrow and C. M. Fidopiastis (Eds.): HCII 2019, LNAI 11580, pp. 471–483, 2019.
https://doi.org/10.1007/978-3-030-22419-6_34

communication [4,5]. For its features, the BCI system has become an important tool for neurofeedback applications [6,7] that allows users to be aware of changes in their brain activity and react to those changes. All these mechanisms, used in the field of brain activity, can be translated into data, and be processed to the synthesis of sound or metaphorization of musical structures [8–11], allowing the user to navigate between limited control, automation, and creation, introducing the concept of Brain-Computer Music Interface (BCMI).

The earliest example of transforming the EEG signal to sound appears in the literature shortly after the invention of EEG. In 1934, Adrian and Matthews [12] were the first researchers who monitored their own EEG with sound through the replication of the earliest EEG descriptions of the posterior dominant rhythm (PDR). However, the first brain composition was performed in 1965 by the composer and experimental musician Alvin Lucier controlling percussion instruments via strength of EEG PDR [11,13]. Following Lucier's experience, David Rosenboom created a music piece in 1970 in which EEG signals of several participants were processed through individual electronic circuits, generating visual and auditory performance for the Automation House in New York [14]. These early BCI music performances were based on EEG signals sonification. Nevertheless, in 1997 Rosenboom introduced an attention performance driving music by detecting several EEG parameters [15]. Few years later Eduardo Miranda started to create music using EEG signals processing, allowing the composers to modulate the tempo and dynamics of the music thus giving place to the BCMI concept [11,16]. After the first work of Miranda, several researchers like Dan Wu [17,18], Ian Daly [10], Thomas Deuel [9], and Mei Lin Chen [19] among others, proposed different BCMI compositions and sound representation systems using diverse EEG patterns and processing techniques.

In this context, a real-time graphical sound representation based on metaphorical interaction design was proposed, which allows navigating through the motor imagery cognitive task in a bidimensional plane. This representation was developed using the OpenBCI EEG acquisition system [20] to generate a pipeline information processing based on data packets transmission over Open Sound Control (OSC) [21] from BCI to Max/MSP [22] and a metaphorical Graphical User Interface (GUI) programmed in Processing [23]. This system presents a neurofeedback interaction through a graphical and sonorous output, the first related to sound representation based on image processing, and the second related to the sound synthesis convolution as a result of motor imagery user interaction. In the python codification stage, a storage and data normalization were carried out to obtain spectral energy data of brain waves in the frequency domain. Then data packing and transmission were performed via OSC protocol to Max/MSP, where all brain data were mapped to a floating point. Here, a statistical discrete processing allowed the classification and parameterization of data into sound representation. This metaphorization was generated through amplitude modulation (AM), frequency modulation (FM), granular and subtractive sound synthesis and their convolution, setting up the graphical sound representation by quadraphonic signals and 3D navigation. The navigation is

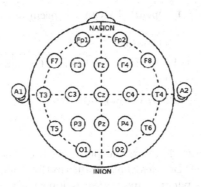

Fig. 1. 10–20 electrodes location

associated aurally with synthesis and sonification in order to make the visual synchronism with space control conscience. Therefore, the color background and the image pointer change in an interpolated way in the 3D GUI. The sound and visual responses evoke stimulation changes, allowing the user to make decisions and execute variations of his conscious brain activity.

The layout of the paper is as follows: Sect. 2 presents a detailed description of all materials and protocol used; Sect. 3 introduces the results and their discussion; and finally, Sect. 4 offers the conclusions of the research.

2   Materials and Methods

2.1   Brain-Computer Music Interface

Brain-Computer Music Interface (BCMI) is a BCI used specifically in Music and Sound Representation applications. The BCMI is an acquisition and processing system that takes EEG signals measure from the user's brain electrical activity in order to generate commands in real time. The BCMI provides a communication channel between the brain and an external sonorous softwares or devices [2, 7, 10, 11]. The BCMI system allows the users to learn how to regulate consciously their brain activity in order to control a specific external music application [11, 24].

2.2   Interpolation BCMI Prototype Design

Python Patch for the OpenBCI. The prototype proposed uses the *OpenBCI* EEG acquisition system in which two Cyton and Daisy boards for the acquisition, amplification, signal processing and data transmission were used. Each board acquires 8 EEG channels, using the international standard 10–20 for the location of the electrodes (Fig. 1).

The OpenBCI is an open source and presents the hardware data software capabilities described in Table 1:

For Interpolation of the BCMI prototype development, an adjustment of Hurtado [25] was made. This patch creates a graphical interface for EEG data

Table 1. OpenBCI technical specifications

OpenBCI Cyton specification	
Data format	32 bytes, 24 for 8 EEG and 6 auxiliary channels
Sampling rate	250 Hz
Scale factor	$0.02235\,\mu V$ from 1–24

representation in the time domain, the percentage of energy in the frequency domain of brain rhythms. Besides that, brain rhythm calculation through FFT is made and then its data packing and transmission to Max/MSP is accomplished via OSC.

Max/MSP Patch. The interpolation system is associated with a motor imagery conscious task which fluctuates between 13 and 15 Hz, mixing with alpha waves from occipital brain activity, related at the same time with μ brain rhythm. In their interaction, the electrochemical brain signals are translated into data and the processes are related to sound synthesis. This system allows the user to navigate with control, automation and creation restrictions. For this purpose, a sound graphical representation in real-time, which implies a basic training and pipeline communication between OpenBCI, Python, Max/MSP, and Processing is proposed.

In the prototype formulated, a parallel process of active and passive brain states is applied to control a GUI interface and the parameter variation of AM, FM, granular, and subtractive sound synthesis schemes, as well as the parallel between the active or conscious control process through motor imagery and the passive or unconscious control process through brain rhythms energy variation. The Max/MSP programming is focused on discrete data treatment, synthesis parameterization, and sound representation, and the processing programming is centered on visual representation (GUI).

The general programming of Max/MSP is divided into the following steps:

1. Connection system between Python and OpenBCI, which is in charge of analyzing the incoming brain information in frequency domain and scaling it to execute the audio control.
2. Visualization of brain rhythms variation.
3. Cursor variation and its position randomness which is modified by high alpha and low gamma rhythm variation, and its discrete classification results in a motor imagery task.
4. Position of the cursor into the navigation interface and synthesis components selection.
5. Sound synthesis of each Navigation GUI extreme distributed as follows: left, granular synthesis; up, FM synthesis; right, AM synthesis; down, subtractive synthesis.

6. OSC communication with Processing to send and receive relevant information about the position and data related with the volume and the audio reproduction.
7. Reverb subpatch for all the synthesis signals.
8. Output audio wave visualization and compressor stage which avoids overload in audio output.

Sound Synthesis Process. The sound synthesis process is divided into amplitude and frequency modulation synthesis, granular synthesis and subtractive synthesis. These syntheses were inspired by the brain signals behavior based on the hypothesis that the brain rhythms can change in function of a specific brain feature as a signal modulation.

Amplitude Modulation (AM) Synthesis. The amplitude modulation is based on the ring modulation described in the Max/MSP cycling 74 Tutorial Part 2 and 3 [26]. Here, the carrier and modulator signals are multiplied sample by sample, resulting in a signal in which the amplitude of the carrier changes following the amplitude modulator signal, while the frequency of the carrier remains the same. This modulation process is described in Eq. 1, where, f_m is the modulator frequency, f_a is the frequency for the amplitude variation, f_c is the carrier frequency and k is a constant.

$$V_{AM}(t) = (0,25k\sin(2\pi f_m t)) + \sin(2\pi f_a t)\sin(2\pi f_c t) \tag{1}$$

Frequency Modulation (FM) Synthesis. The frequency modulation encodes the information in a carrier signal of constant amplitude in a proportion of the instantaneous frequency of the modulator signal. The FM syntheses used multiplied the carrier signal with the harmonic relation F_m/F_c. The FM modulation allows the emulation of instruments and the generation of sounds with a complex spectral response. The behavior of the FM synthesis is described in Eqs. 2 and 3
    For $0 < t \le t_a$

$$V_{FM}(t) = \frac{t_{mi}}{4}\sin\left(f_c + f_{mo}t_{mi}\left[f_{mo}t_{mi} - floor\left(\frac{1}{2} + f_{mo}t_{mi}\right)\right]\right) \tag{2}$$

where f_c is the carrier frequency, f_h is the harmonic frequency and f_{mo} is the modulation frequency given by $f_{mo} = f_c\sin(2\pi f_h t_{mi})$
    For $t \ge t_a$

$$V_{FM}(t) = \frac{t_a}{4}\sin\left(f_c + 2f_c t_a\sin(2\pi f_h t_a)\left(f_m mot_a - floor\left(\frac{1}{2} + f_{mo}t_a\right)\right)\right) \tag{3}$$

Granular Synthesis. The granular synthesis consists of the wave conversion into wave grains. It is based on small acoustic events generation called acoustic grains with a temporary duration less than 50 ms. The granular synthesis applied consists of creating a simple waveform in a signal segment. Following, the signal is multiplied by a saw-tooth signal to granulate it, and then an envelope signal is applied with a phase shift to avoid unwanted spikes. This synthesis is described in Eq. 4.

$$r_1(t) = \begin{cases} 0, \ t < 0 \\ t, \ 0 \leq t < t_t \\ t_a \ t \geq t_t \end{cases}$$

$$\delta\left(\frac{3}{4}t\right) = \begin{cases} 0, \quad t \neq 0 \\ \frac{3}{4}t_a, \ t = 0 \end{cases}$$

$$P(t) = \frac{1}{2}(1 + saw(t)) = \frac{1}{2} + \left[f_g t - floor\left(\frac{1}{2} + f_g t\right)\right]$$

$$V_{GN} = 50\sin(2\pi f_0 t)P(2\pi f_g t)r_1(t)\frac{1}{2}\left[1 + \sin\left(P(2\pi f_g t) + \delta\left(\frac{3}{4}t_t\right)\right)\right] \qquad (4)$$

Where, f_0 is the original carrier frequency, f_g is the grain frequency, t_a is the amplitude of t, and t_t is the slope of t.

Subtractive Synthesis. The subtractive synthesis consists of a selective band pass filter to eliminate specific harmonics of the signal. The bandwidth of the filter is determined by the delta rhythm bandwidth, and the central frequency of the filter is calculated according to the highest spectral energy of gamma rhythm.

Processing Programming. Processing is an open-access software based on JavaScript used in visual arts. The proposed system uses it for making a real-time visual representation of the BCI through a GUI that includes a cursor and a brain illustration. The cursor might move into five regions according to the brain activity responses addressed by the auditory synthesis and the sonification process, allowing the users to improve the visual synchronism with the brain control commands.

The processing allows sending information packets via OSC where they are verified and compared to generate new parameters of the 3D brain illustration. Figure 2 shows the five cursor regions. Each region corresponds to a specific element. The horizontal location represents the speaker fount sound (left or right), the vertical position is the volume intensity (High or Low), and the middle region is silence. Every time the cursor moves between regions both the GUI background and the cursor change their color in an interpolated way.

Neurofeedback. The prototype proposes a neurofeedback model through four spatialized speakers to reproduce three-dimensional panels according to the GUI navigation. The user is located in the center of the speakers in front to a screen

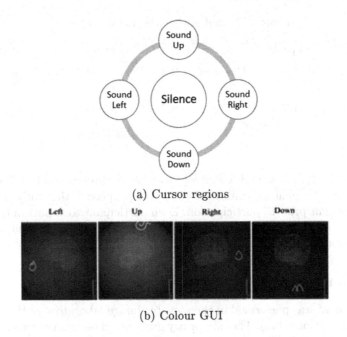

(a) Cursor regions

(b) Colour GUI

Fig. 2. GUI response according to the cursor location

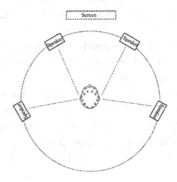

Fig. 3. User experiment location

(Fig. 3). Here, the user is able to do the actions shown in Table 2. In a controlled environment, the user may be able to move and transform the cursor at will, as well as to stop the movement through a relaxation state, allowing the system to return to a neutral state.

2.3 Participants

Six healthy musicians who self-reported normal hearing participated in the study. This group was divided into two subgroups, an experimental group with

Table 2. Neurofeedback controlled process

IF	THEN	IF	THEN
Push (P)	Move (M) Change (c)	Relax (R)	Stop (S) Remain (R_m)
R	SR_m	P	M
P^n	MC^x	SP	$SM + R_m$
R^n	$SM + R_m Y$	P	MC

three blind participants, and a control group with three normal vision participants. Each participant was informed about the purpose of the study and signed informed consent prior to participation. Each participant took part in the experiment three times at an interval of one week. Participants were monetary compensated for their time.

2.4 Experiment Setup

The experiment was performed in the movable image laboratory of the Universidad de Caldas (Colombia). The laboratory guaranteed an quiet with acoustically isolated environment. The experimental equipment used in this experiment was an OpenBCI for the EEG signal acquisition, a screen for the GUI visualization, two cameras one for the facial recording of the participants and the second one for the overall experiment recording. Likewise, it was necessary to use two computers, one for the signal acquisition and pattern recognition process, and the second one for the Max/MSP processing, and a quadraphonic audio system for the sonorous immersion (Figs. 4 and 5).

Previous to the experiment it was necessary to perform a musical and interface control training. The procedure for each experiment had the following steps: 5 min of relaxing state; 5 min for placing the ECG acquisition system; 2 min to be in a neutral state; and 2 min for control imagery experiment. After each experiment, the participants answered a self-assessment

Fig. 4. Experimental scene

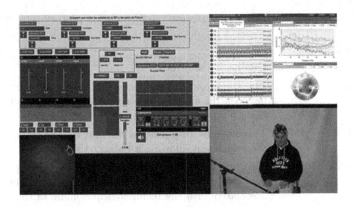

Fig. 5. Final prototype

test focused on sound level perception, integration difficulty level, mental control level, cognitive strategies, memory, sensations, neuro-feedback cognitive awareness, experience, and movement imagination level among others.

3   Results

For the interpolation prototype, the analysis was made through the brain rhythms variation in function of the motor imagery control of all participants. These levels were measured using static an dynamic levels. The general results (Table 3) show that the high dynamic responses of the participants can be associated with the motor imagery learning process, allowing them to control the system step by step. The static level reflects the low interaction between four control stages of the prototype. As the user starts to navigate in a conscious way into the system, the static level decreases and the dynamic level increases. These measure levels also allow inferring if the participants got a reproduction, integration or neurofeedback level.

According to Table 3, the participant I1 had a low control of the interpolation system in which only 50% of the experiment time was possible to control the motor imagery system assigning a low reproduction level. In the case of the participants I2, I3, C1 and C3, the high dynamic level responses showed an integration process within the sound generation in function of the system navigation. Finally, the highest dynamic response of the participant C2, exhibited a complete control of the sound generation mechanism related to the interactivity between the conscious navigation and the desired sound and achieved a neurofeedback level. Likewise, it was possible to see that in the sound moments produced by the visual control of the interpolation system, the occipital and

Table 3. Self-assessment results, interpolation prototype

Interpolation			
User	Static	Dynamic	Qualitative analysis
I1	40.77%	59.23%	The static and dynamic value represent the permanence between the subtractive and AM synthesis which implies a basic motor imagery control
I2	38.56%	61.44%	The dynamic responses of the user reflects a control state the movement allowing a balanced sonification
I3	18.62%	81.38%	The high dynamics reflect the result of training of the participant. On the other hand, this result might be associated with the technological domain of the environment allowing a good musical performance
C1	20.84%	79.16%	The high dynamics presented by the participant show the capacity to balance the sound and the movement imagination relationship where the movement directly depends on the sound pauses made intentionally
C2	8.95%	91.05%	The high dynamics of the participant demonstrates his upper motor imagery control level and his correlation with sound generation allowing him to move in the entire sound plane. In this case, the participant is performing a conscious feedback process
C3	32.65%	67.35%	The static and dynamic response of the participant shows the motor imagery control difficulty presented at the beginning of the experiment. However, once it was controlled, it remains in the same state
Mean	26.73%	73.27%	

parietal regions showed several changes in their activity levels (Fig. 6), confirming the visual and auditory synthesis related to these brain regions. Also, the EEG average level showed a synchronized response in the temporal lobes in the research participants I1, I2 and I3, while the control participants C1, C2 and C3 showed a desynchronized response.

Similarly, in the EEG activity recorded in the control participants, two activity patterns stood out on the O1, O2, C3 and C4 channels associated with the occipital and central brain regions, respectively. These EEG characteristics were generated through the sound control using the visual motor imagery interface. However, these signal patterns did not appear in the research participants' data due to their visual limitations. The sound control process of the research participants was carried out through auditory navigation, because of the training process performed before the experimental phase.

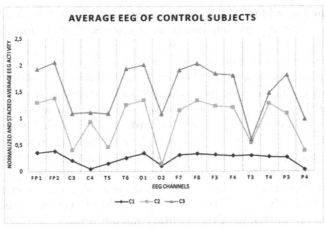

(a) Average EEG of control subjects

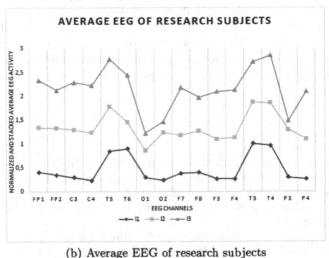

(b) Average EEG of research subjects

Fig. 6. Average response of the sixteen EEG channels for all experiment participants.

4   Conclusion

The interpolation system was proposed as a sonorous navigation model based on motor imagery. This system used a BCI interface as a way of communication between the brain and a technological external device. Here, the sound generation was proposed through the convolution of four sonorous syntheses: AM, FM, granular and subtractive. To evaluate the sound system operation, an experiment was made under controlled conditions, where six professional musicians participated. The participants were divided into two sub groups, the first one was formed by three blind participants (research participants); and the second one was formed by three participants without any visual limitation (control participants).

Into the system evaluation, it was found that all participants achieved different control levels associated to their static and dynamic response, as a consequence of the training process, the musical experience and the degree of commitment. During the experiment process, it was observed that all the participants with normal vision controlled the system through the graphic navigation interface based on motor imagery, where a high brain activity in the central and occipital regions was evidenced. On the other hand, it was found that the blind participants controlled the system through an auditory processing as a perception tool of the navigation interface, allowing them to generate sound changes intentionally. This kind of control showed a brain activation in the temporal regions, which are in charge of the auditory human process. During the experiment, all the participants expressed feeling a physical movement sensation and speed of thought accompanied by a floating sensation.

Finally, with the results obtained, the hypothesis that EEG signals acquired during the experimental process can have signals deflections in a latency between 8 to 15 ms is proposed, which according to the literature [27] can be related as early auditory evoked potentials, as a response to the sonorous generation of the system.

References

1. Al-Nafjan, A., Hosny, M., Al-Ohali, Y., Al-Wabil, A.: Review and classification of emotion recognition based on EEG brain-computer interface system research: a systematic review. Appl. Sci. **7**(12), 1239 (2017)
2. Wolpaw, J., Wolpaw, E.W.: Brain-Computer Interfaces: Principles and Practice. OUP, USA (2012)
3. Pinegger, A., Hiebel, H., Wriessnegger, S.C., Müller-Putz, G.R.: Composing only by thought: novel application of the p300 brain-computer interface. PloS One **12**(9), e0181584 (2017)
4. Tan, D., Nijholt, A.: Brain-computer interfaces and human-computer interaction. In: Tan, D., Nijholt, A. (eds.) Brain-Computer Interfaces. Human-Computer Interaction Series, pp. 3–19. Springer, London (2010). https://doi.org/10.1007/978-1-84996-272-8_1
5. Panetta, K.: 5 trends emerge in the gartner hype cycle for emerging technologies. https://www.gartner.com/smarterwithgartner/5-trends-emerge-in-gartner-hype-cycle-for-emerging-technologies-2018/. Accessed 16 Aug 2018
6. Basmajian, J.V.: Biofeedback: Principles and Practice for Clinicians. Williams & Wilkins (1979)
7. Fedotchev, I., Parin, S., Polevaya, S., Velikova, S.: Brain-computer interface and neurofeedback technologies: current state, problems and clinical prospects. **9**(1) (2017). (eng)
8. Zbikowski, L.M.: The Cambridge Handbook of Metaphor and Thought. Metaphor and music, pp. 502–524 (2008)

9. Deuel, T.A., Pampin, J., Sundstrom, J., Darvas, F.: The encephalophone: a novel musical biofeedback device using conscious control of electroencephalogram (EEG). Front. Hum. Neurosci. **11**, 213 (2017)

10. Daly, I., et al.: Affective brain-computer music interfacing. J. Neural Eng. **13**(4), 046022 (2016)

11. Miranda, E.R., Castet, J.: Guide to Brain-Computer Music Interfacing. Springer, London (2014). https://doi.org/10.1007/978-1-4471-6584-2

12. Adrian, E.D., Matthews, B.H.: The berger rhythm: potential changes from the occipital lobes in man. Brain **57**(4), 355–385 (1934)

13. Lucier, A.: Music for solo performer (1965). https://www.youtube.com/watch?v=bIPU2ynqy2Y

14. Rosenboom, D.: A Model for Detection and Analysis of Information Processing Modalities in the Nervous System Through an Adaptive, Interactive, Computerized, Electronic Music Instrument. Ann Arbor, MI, Michigan Publishing, University of Michigan Library (1975)

15. Rosenboom, D.: Extended musical interface with the human nervous system (1997)

16. Miranda, E.R., Boskamp, B.: Steering generative rules with the EEG: an approach to brain-computer music interfacing. Proc. Sound Music Comput. **5** (2005)

17. Wu, D., Li, C.Y., Yao, D.Z.: Scale-free music of the brain. PloS One **4**(6), e5915 (2009)

18. Wu, D., Li, C., Yao, D.: Scale-free brain quartet: artistic filtering of multi-channel brainwave music. PloS One **8**(5), e64046 (2013)

19. Chen, M.L., Yao, L., Jiang, N.: Music imagery for brain-computer interface control. In: Schmorrow, D.D., Fidopiastis, C.M. (eds.) AC 2017. LNCS (LNAI), vol. 10285, pp. 293–300. Springer, Cham (2017). https://doi.org/10.1007/978-3-319-58625-0_21

20. OpenBCI: Open source brain-computer interface (2017). https://openbci.com/

21. Wright, M., Freed, A., et al.: Open soundcontrol: a new protocol for communicating with sound synthesizers. In: ICMC (1997)

22. Puckette, M., Zicarelli, D., et al.: Max/msp. Cycling **74**, 1990–2006 (1990)

23. Reas, C., Fry, B.: Processing: programming for the media arts. AI Soc. **20**(4), 526–538 (2006)

24. Miranda, E.R.: Brain-computer music interface for composition and performance. Int. J. Disabil. Hum. Dev. **5**(2), 119–126 (2006)

25. Hurtado, J.V.: python-brain-rhythms (2015). https://github.com/JuanaV/python-brain-rhythms

26. Place, T.: Advanced max: Ffts. https://cycling74.com/tutorials/advanced-max-ffts-part-2

27. Picton, T.W.: Human Auditory Evoked Potentials. Plural Publishing (2010)

Wavelet Packet Entropy Analysis of Resting State Electroencephalogram in Sleep Deprived Mental Fatigue State

Yanjing Wang[1,2], Zhongqi Liu[1,2], Qianxiang Zhou[1,2(✉)],
and Xuewei Chen[3]

[1] Key Laboratory for Biomechanics and Mechanobiology of the Ministry
of Education, School of Biological Science and Medical Engineering,
Beihang University, Beijing 100191, China
vmove_yjwang@sina.com, {liuzhongqi,zqxg}@buaa.edu.cn
[2] Beijing Advanced Innovation Centre for Biomedical Engineering,
Beihang University, Beijing 102402, China
[3] Tianjin Institute of Environmental and Operational Medicine,
Tianjin 300000, China
chenxueweill@sina.com

Abstract. In order to explore the characteristics of the complexity of resting electroencephalogram (EEG) in mental fatigue state after sleep deprivation, 36 healthy subjects were recruited to participate in the 30 h complete sleep deprivation test, and the resting state electroencephalogram (EEG) in open and closed eyes were collected, and the entropy analysis of the resting state electroencephalogram (EEG) signal alpha was carried out before and after sleep deprivation. The results showed that compared with resting state before sleep deprivation, the closed eye alpha 1 band wavelet packet entropy decreased significantly after sleep deprivation, and the wavelet packet entropy of alpha 1 band in eye opening did not change significantly after sleep deprivation. The results showed that the entropy of the alpha 1 band wavelet packet entropy decreased obviously after sleep deprivation. It was an important feature of the electroencephalogram changes under the condition of increasing sleep pressure.

Keywords: Sleep deprivation · Mental fatigue · Resting state ·
Electroencephalogram · Wavelet packet entropy

1 Introduction

With the acceleration of work pace and changes in lifestyle habits, sleep deprivation has become extremely common in the modern society [1, 2]. Sleep deprivation refers to a forced reduction in the number of hours of sleep due to a particular (some) reason(s). Sleep deprivation can be classified as follows: full sleep deprivation and partial sleep deprivation according to the amount of sleep deprived, acute sleep deprivation and chronic sleep deprivation according to the nature of the condition, or rapid eye movement sleep deprivation and non-rapid eye movement sleep deprivation (or slow wave sleep deprivation) according to the stage of sleep that is deprived. The influence

© Springer Nature Switzerland AG 2019
D. D. Schmorrow and C. M. Fidopiastis (Eds.): HCII 2019, LNAI 11580, pp. 484–494, 2019.
https://doi.org/10.1007/978-3-030-22419-6_35

of sleep deprivation on human physiological (e.g., biochemical indicators, brain electrical activity, cardiac electrical activity, etc.) and psychological (memory, emotions, and cognitive states) functions and the mechanisms by which it occurs have remained important research directions in the fields of medicine, brain science, neuropsychiatry, psychology and aerospace medicine. The investigation of the influence of sleep deprivation on human electrophysiological and psychological functions and the elucidation of the mechanisms underlying its occurrence will contribute to the formulation of active and effective countermeasures and minimization of adverse effects. Studies have shown that individuals experience an increase in the degree of sleepiness and a decrease in the cognitive function after sleep deprivation, which have an important impact on job performance [3].

Further damage of the cognitive function due to an increase in the degree of fatigue not only results in a decrease in work efficiency, but may also lead to accidents due to human error during operations [4]. Therefore, an in-depth exploration of the effects of sleep deprivation on the physiological and psychological functions of the body is of great social significance.

Medical imaging techniques (computed tomography (CT), magnetic resonance imaging (MRI), etc.) can provide information on the brain structure and spatial localization, however, the drawback of such techniques is poor temporal resolution. For neuroscience research, electrophysiological signals can capture more details of changes. An electroencephalogram (EEG) can reproduce the information-processing process in the brain on the millisecond scale with sufficient temporal resolution. It can be used to study functional responses under external stimuli as well as to reflect spontaneous brain activities during resting state. As an important indicator in the monitoring of the sleep-wake state in organisms, EEG can reflect changes in the neurological function of the brain, and is an ideal research tool that plays an important role in indicating changes in the physiological and psychological functions of the body [5]. Early EEG research relied on manual analysis and judgment, and was thus heavily affected by human factors. The EEG power spectrum, a method developed in recent years, can be used for the quantitative study of electrophysiological changes, however, reports of its application in studies on sleep deprivation have been relatively rare. Previous studies have found that the brain electrical activity of sleep-deprived individuals showed trends of increasing low-frequency components and decreasing high-frequency components [6], suggesting that the body requires sleep to repair mental functions that have been reduced or damaged.

Brain electrical activity is a complex time series. The nonlinear dynamical method is a new method that can analyze the pattern of neural activity at the overall level, and can provide more brain electrical activity information compared with conventional time, frequency, and time-frequency analysis methods. In this method, the human brain is regarded as a complex and nonlinear system. Wavelet transform and wavelet packet transform are characterized by the joint analysis of signals of non-stationary time series in both the frequency and time domains. A large number of studies based on brain electrical activity have applied wavelet transform in the multiresolution analysis and transient feature extraction of EEG signals of sleep-deprived subjects, and numerous results have been obtained. The wavelet-packet entropy method was developed based on wavelet transform and wavelet packet transform. It incorporates the advantages of both transforms, further examines the overall characteristics of brain electrical activity from a system perspective, and can characterize the complexity of brain electrical activity sequences. The wavelet-packet entropy method was adopted in this study, as it can reflect the degree of disorder of multi-frequency component signals, provide dynamic characteristics of signals, and function as an indicator to evaluate the complexity of EEG signals [7]. However, studies on the influence of full sleep deprivation on physiological functions of the body from the perspective of wavelet packet entropy in different frequency bands are still relatively rare. The present study aimed to analyze the influence of sleep deprivation on awake resting state brain electrical activity of individuals from the perspective of wavelet packet entropy by comparing the changes in wavelet packet entropy of the frequency bands ($\alpha 1$ bands) closely related to the waking function in the awake resting brain electrical activity of individuals before and after sleep deprivation.

2   Materials and Methods

2.1   Subjects

Thirty-six healthy university students voluntarily participated in this experiment (age: 18–28 years, mean age: 23.5 years). All subjects were right-handed, of normal intelligence, did not have a history of neurological or psychiatric diseases, and maintained regular 8-hour sleep periods for two weeks before participating in the experiment. During the experimental period, subjects abstained from alcohol, coffee, tea and other stimulant beverages and did not engage in intense physical activities. The subjects provided informed consent prior to the experiment, and received appropriate compensation at the end of the experiment.

2.2   Experimental Procedure

Subjects reported to the laboratory one day before the actual experiment. After one night of sleep monitoring, baseline EEG measurements were performed at 7 am on the following day. Subsequently, the subjects, accompanied by the chief investigator, underwent full sleep deprivation. During the sleep deprivation period, subjects could watch movies and play poker, but had to stay awake all times. At 13:00 on Day 3, subjects received a second EEG measurement and reported whether they remained awake before and after the data were recorded (Fig. 1).

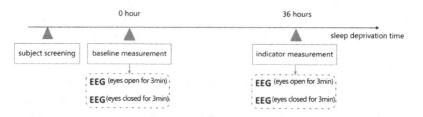

Fig. 1. A schematic map for measuring sleep deprivation effects

2.3 EEG Acquisition

Scalp cleaning was performed on subjects prior to data collection, then each subject was seated in a relaxed state on a fixed experimental chair. The chief investigator explained the precautions of the experiment to the subjects in detail, and requested the subjects to relax as much as possible and avoid recalling events that could cause relatively strong stimulations. Skin resistance was reduced with conductive paste to ensure that the scalp resistance of the subjects was less than 5 kΩ during data acquisition. After the start of the experiment, continuous EEG data were collected when the subjects' eyes were closed and open for 3 min, respectively.

A Neuroscan EEG recorder was used for EEG acquisition. The EEGs of subjects were amplified and converted using a SynAmps2 amplifier and recorded with Scan 4.3 software. The international 10-20 positioning system was used as the standard for electrode placement [8], and the mastoids on the left and right sides were used as reference electrodes. A 32-lead electroencephalogram of each subject in resting state was recorded, and horizontal electrooculography (HEOG) and vertical electrooculography (VEOG) were recorded simultaneously for horizontal and vertical eye movements, respectively. In the present study, EEG signals during the sleep deprivation experiment were collected using the electrodes placed at frontopolar (FP), inferior frontal (IF), frontal (F), temporal (T), central (C), posterior temporal (PT), parietal (P) and occipital (O) areas, and the corresponding electrode positions are shown in Table 1. The electrodes were placed on the head according to the international 10-20 positioning system developed by the International Federation of Societies for Electroencephalography and Clinical Neurophysiology, and the mastoid electrodes M1 and M2 were used as reference electrodes (as shown in Fig. 2). EEG signal acquisition software was used to collect the EEG signals of the subjects. The acquisition frequency (A/D), amplification factor (gain), and filter band-pass were 1000 Hz, 1000, and 0.05–100 Hz, respectively. Continuous acquisition was performed in the alternating current (AC) mode, and a 50 Hz notch filter was applied to remove power line interference simultaneously during data acquisition. After the EEG cap was connected with the amplifier, conductive paste was injected into each electrode hole. Prior to the injection of conductive paste, electrodes in the software interface were displayed in red (i.e., a resistance of greater than 100 KΩ). During the process of injecting the conductive paste into each electrode, the electrode color gradually turned yellow-green and

Table 1. The 10-20 electrode name matching table for electrode lead system

Area	Abbreviation	Electrodes
Frontopolar	FP	Fp1, Fp2
Inferior frontal	IF	F7, F8
Frontal	F	F3, Fz, F4
Temporal	T	T7, T8
Central	C	C3, C4, Cz
Posterior temporal	PT	P7, P8
Parietal	P	P3, P4, Pz
Occipital	O	O1, O2, Oz

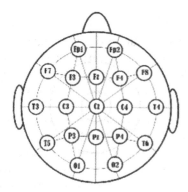

Fig. 2. An international 10-20 system diagram of electroencephalogram acquisition electrode placement

eventually turned green. The experimental requirement was satisfied when scalp resistance was less than 10 KΩ.

2.4   EEG Data Preprocessing

The collected EEG data were preprocessed using MATLAB and EEGLAB. For each subject, the resting EEG data at 30–90 s (including eyes open and eyes closed) were selected, and EEGLAB was used to remove artifacts such as electrooculogram (EOG), electromyography (EMG), and ECG. Subsequently, independent component analysis (ICA) was performed and a finite impulse response (FIR) filter was used to filter the EEG data. The preprocessed EEG data was then analyzed by wavelet packet entropy.

The main frequency range of EEG signals is 0.5–30 Hz, and the frequency of EMG is generally above 100 Hz. Therefore, by directly filtering out frequency components exceeding 100 Hz, EMG artifacts can be removed without causing substantial effects on the EEG signal. During the signal acquisition process, slow voltage drifts caused by movements of the electrodes and slow fluctuations caused by sweating can interfere with the baseline voltage of the EEG signal. Therefore, slow voltage drift artifacts can be removed by directly filtering out frequency components below 0.5 Hz. In addition, the 50 Hz power line interference can also be directly filtered. In view of the afore-mentioned points, an infinite impulse response (IIR) filter was used to perform 0.5–35 Hz (48 dB/oct) band-pass filtering and 50 Hz notch filtering to remove EMG and slow voltage drift artifact artifacts as well as power line interference. Besides these interferences, EOG artifacts were the main source of noise. In this study, the ICA method was used to remove EOG artifacts. Figure 7 shows the EEG signals before and after the removal of EOG artifacts. It can be seen that the ICA method could effectively remove the EOG artifacts and provide a good basis for subsequent analysis (Fig. 3).

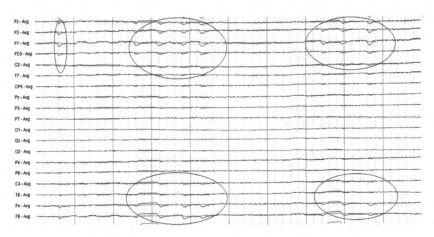

Fig. 3. Contrast map before and after removal of eye electrical artifacts

2.5 Wavelet Packet Entropy Analysis

Wavelet packet transform is an effective method for the analysis of non-stationary processes, and it provides more detailed signal decomposition. According to the characteristics of the signal to be analyzed, the corresponding frequency band is selected to match the signal spectrum in a self-adaptive manner, thus improving the time-frequency resolution.

The Shannon information entropy can measure the information contained in various probability distributions. It is mainly used to examine the degree of uncertainty of the sequence and estimate the complexity of random signals [9]. The wavelet packet entropy can be obtained by combining the energy distribution of wavelet-packet decomposition coefficient with the information entropy [13].

$$WEP = -\sum P_l \ln[P_l] \qquad (1.1)$$

The wavelet packet was used to finely decompose the signal into various frequency bands. By decomposing the sleep-deprivation EEG signal with seven layers of wavelet packets, the signal could be divided into 128 frequency bands. Since the acquisition frequency of the EEG data in the experiment was 1024 Hz, the frequency range of the signals was 0–512 Hz [10]. The seven-layer wavelet packet decomposition of the frequency band range corresponding to each node is shown in Table 2. From the table, it can be seen that the frequency band corresponding to the α1 band was the (7, 2) node.

Table 2. Corresponding values of different frequency bands of decomposition signals

Node	Frequency band range	Node	Frequency band range
(7,0)	0–4 Hz	(7,8)	32–36 Hz
(7,1)	4–8 Hz	(7,9)	36–40 Hz
(7,2)	8–12 Hz	(7,10)	40–44 Hz
(7,3)	12–16 Hz	(7,11)	44–48 Hz
(7,4)	16–20 Hz	(7,12)	48–52 Hz
(7,5)	20–24 Hz	(7,13)	52–56 Hz
(7,6)	24–28 Hz	(7,14)	56–60 Hz
(7,7)	28–32 Hz	(7,15)	60–64 Hz

2.6 Wavelet Packet Entropy Analysis

Statistical analysis of EEG wavelet-packet entropy data before and after sleep deprivation was performed with the SPSS 18.0 software package, and statistical testing was conducted using repeated-measures analysis of variance (ANOVA).

3 Results

3.1 Effect of Sleep Deprivation on the α1 Band Wavelet-Packet Entropy in Resting EEG

The F3, FZ, F4, C3, CZ, C4, P3, PZ, and P4 lead data were selected to compare the α1 band wavelet-packet entropy values before and after sleep deprivation. The results showed that compared with the values before sleep deprivation, the α1 band wavelet-packet entropy during the eye-closed state significantly decreased after sleep deprivation ($F(1,34) = 7.473$, $P = 0.010$), and there was no significant change in α1 band oscillation strength with eyes open ($F(1,34) = 0.026$, $P = 0.872$). The results are shown in Figs. 4 and 5.

Under conditions of sleep deprivation, the body needs to maintain normal waking functions. Previous studies have shown that the maintenance of normal waking function is related to the default mode network of the brain [11]. Functional magnetic resonance imaging studies have shown that after full sleep deprivation, the functional connectivity within the individual's default mode network as well as that between the default mode network and the executive control functional network declined [12]. In the present study, through the analysis of changes in the EEG power spectrum in different bands, it was found that changes also occurred in the power spectrum in the EEG network after sleep deprivation. There was a significant shift in the EEG intensity spectrum after sleep deprivation. Changes in the functions of the brain's default mode network after sleep deprivation may be an important contributing reason for the increase in slow-wave energy in the EEG.

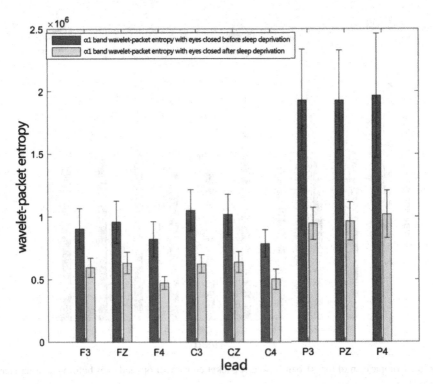

Fig. 4. Comparison of the α1 band wavelet packet entropy of closed eyes before and after sleep deprivation

According to the theory of energy transfer during sleep, energy transfer can occur within the body when an individual experiences long-term acute sleep loss [13]. This energy transfer is related to the continuously increasing sleep pressure of the body. With prolonged sleep deprivation, sleep pressure continues to increase [14], and the need for sleep increases constantly as well. An increase in the degree of fatigue leads to an enhanced inhibitory effect on the brain. A shift of the EEG power spectrum of the brain to slow waves may also be a manifestation of constant sleep stress or need for sleep.

By comparing changes in the EEG power spectrum under eyes-open and eye-closed conditions, it was found that the eyes-open and eyes-closed conditions had substantial effects on the characteristics of changes in EEG power. For the α1 band, the oscillation strength was significantly decreased with eyes closed. The above results suggest that the eyes-open and eyes-closed conditions after sleep deprivation have a great influence on the wavelet packet entropy of EEG waves, and it is necessary to consider the eyes-open and eyes-closed conditions as a factor during resting-state EEG analysis.

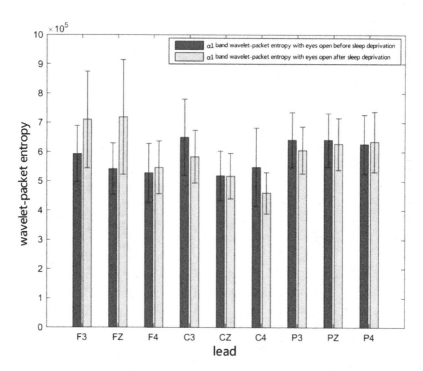

Fig. 5. Comparison of the α1 band wavelet packet entropy of opened eyes before and after sleep deprivation

4 Conclusion

After sleep deprivation, the α1 band wavelet-packet entropy with eyes closed was significantly decreased, which is an important feature of the EEG changes under increasing sleep pressure. As the present study only analyzed changes in the EEG power spectrum, further studies should focus on the different EEG networks and the pattern of changes in the effective connectivity of the brain, in order to provide evidence for an in-depth understanding of the impact of sleep loss on the physiological and psychological functions of the body.

References

1. Ning, L., Yan, W., Xiyu, L., et al.: Research on effect of sleep deprivation on cognitive brain function. J. Biomed. Eng. **25**(5), 1197–1200 (2008)
2. Fugui, W., Yongcong, S., Jianlin, Q., et al.: Effects of total sleep deprivation on executive function in young men. Chin. Ment. Health J. **24**(7), 541–545 (2010)
3. Chen, L., Kang, L., Tao, C.: Effects of acute sleep deprivation on male military soldiers cognitive funtion. Chin. Health Psychol. J. **25**(2), 195–199 (2017)
4. Chunhua, W., Shuqing, W., Ying, L.: Influence of sleep deprivation on nurses' mood. Chin. Nurs. Res. **20**(29), 2658–2659 (2006)

5. Wenbin, L.Y.Z., Qingtao, R., et al.: Study of whole night polysomnography in patients with chronic fatigue syndrome. J. Neurol. Neurorehabil. **6**(1), 44–46 (2009)
6. Mingshi, W., Jin, L., Qiang, Z., et al.: Effects of sleep deprivation on brain cognition and EEG complexity. J. Tianjin Univ. **38**(4), 343–346 (2005)
7. Yuliang, X., Weixing, H., Xiaoping, C., et al.: Application of wavelet transform and wavelet entropy in the characteristic study of sleep electroencephalogram signal change. Chin. J. Clin. Rehabil. **10**(25), 118–120 (2006)
8. Bing, Q.: American clinical neurophysiology society (5) standard electrode position naming guide. J. Epileptol. Electroneurophysiol. **20**(6), 377–378 (2011)
9. Ling, T., Guang, C., Wen, L.: Information entropy and error entropy of measurement data. J. Univ. Electron. Sci. Technol. China **36**(5), 935–937 (2007)
10. Yanjing, W., Xiaoyan, Q., Peng, L., et al.: Classification of motor imagery task based on wavelet packet entropy and support vector machines. Chin. J. Sci. Instrum. **31**(12), 2729–2735 (2010)
11. De Havas, J.A., Parimal, S., Soon, C.S., et al.: Sleep deprivation reduces default mode network connectivity and anti-correlation during rest and task performance. Neuroimage **59**(2), 1745 (2012)
12. Sämann, P.G., Tully, C., Spoormaker, V.I., et al.: Increased sleep pressure reduces resting state functional connectivity. Magn. Reson. Mater. Phys. Biol. Med. **23**(5–6), 375–389 (2010)
13. Schmidt, M.H.: The energy allocation function of sleep: a unifying theory of sleep, torpor, and continuous wakefulness. Neurosci. Biobehav. Rev. **47**, 122 (2014)
14. Maric, A., Lustenberger, C., Werth, E., et al.: Intraindividual increase of homeostatic sleep pressure across acute and chronic sleep loss: a high-density EEG study. Sleep **40**(9) (2017)
15. Cajochen, C., Blatter, K., Wallach, D., Belgica, P.: Circadian and sleep-wake dependent impact on neurobehavioral function. Psychol. Belgica **44**(1), 59–80 (2011)
16. Yang, C.M., Wu, C.H.: The situational fatigue scale: a different approach to measuring fatigue. Qual. Life Res. **14**(5), 1357–1362 (2005)
17. Belz, S.M., Robinson, G.S., Casali, J.G.: Temporal separation and self-rating of alertness as indicators of driver fatigue in commercial motor vehicle operators. Hum. Factors **46**(1), 154–169 (2004)
18. Moller, H.J., Kayumov, L., Bulmash, E.L., Nhan, J., Shapiro, C.M.: Simulator performance, microsleep episodes, and subjective sleepiness: normative data using convergent methodologies to assess driver drowsiness. J. Psychosom. Res. **61**(3), 335–342 (2006)
19. Schmidt, E.A., Schrauf, M., Simon, M., Fritzsche, M., Buchner, A., Kincses, W.E.: Drivers' mis-judgement of vigilance state during prolonged monotonous daytime driving. Accid.: Anal. Prev. **41**(5), 1087–1093 (2009)
20. Russo, M., Thomas, M., Thorne, D., Sing, H., Redmond, D.: Oculomotor impairment during chronic partial sleep deprivation. Clin. Neurophysiol. **114**(4), 723–736 (2003)
21. Stern, J.A., Boyer, D., Schroeder, D.: Blink rate: a possible measure of fatigue. Hum. Factors **36**(2), 285–297 (1994)
22. Liu, J., Zhang, C., Zheng, C.: EEG-based estimation of mental fatigue by using KPCA–HMM and complexity parameters. Biomed. Sig. Process. Control **5**(2), 124–130 (2010)
23. Egelund, N.: Spectral analysis of heart rate variability as an indicator of driver fatigue. Ergonomics **25**(7), 663–672 (1982)
24. Li, Z., Jiao, K., Chen, M., Wang, C.: Effect of magnitopuncture on sympathetic and parasympathetic nerve activities in healthy drivers-assessment by energy spectrum analysis of heart rate variability. Eur. J. Appl. Physiol. **88**, 401–410 (2003)
25. Lal, S.K., Craig, A.: A critical review of the psychophysiology of driver fatigue. Biol. Psychol. **55**(3), 173–194 (2001)

26. Lal, S.K., Craig, A.: Electroencephalography activity associated with driver fatigue: implications for a fatigue countermeasure device. J. Psychophysiol. **15**(1), 183–189 (2001)
27. Lal, S.K., Craig, A.: Driver fatigue: electroencephalography and psychological assessment. J. Psychophysiol. **39**(3), 313–321 (2002)
28. Waard, D.D., Brookhuis, K.A.: Assessing driver status: a demonstration experiment on the road. Accid. Anal. Prev. **23**(4), 297–307 (1991)
29. Jap, B., Fischer, P., Fischer, P., Bekiaris, E.: Using EEG spectral components to assess algorithms for detecting fatigue. Expert Syst. Appl. **36**(2), 2352–2359 (2009)
30. Kar, S., Bhagat, M., Routrary, A.: EEG signal analysis for the assessment and quantification of driver's fatigue. Transp. Res. Part F Traffic Psychol. Behav. **13**(5), 297–306 (2017)
31. Jung, T., Makeig, S., Stensmo, M., Sejnowski, T.J.: Estimating alertness from the EEG energy spectrum. IEEE Trans. Biomed. Eng. **44**(1), 60–69 (1997)
32. Lal, S.K., Craig, A.: Electroencephalography activity associated with driver fatigue: implications for a fatigue countermeasure device. J. Psychophysiol. **15**(13), 183–189 (2001)
33. Mao, Z., Chu, X., Yan, X., Wu, C.: Advances of fatigue detecting technology for drivers. China Saf. Sci. J. **15**(3), 108–112 (2005)
34. Schier, M.A.: Changes in EEG alpha energy during simulated driving: a demonstration. Int. J. psycho physiol. **37**(2), 155–162 (2000)
35. Wilson, G.F., Swain, C.R., Ullsperger, P.: EEG energy changes during a multiple level memory retention task. Int. J. Psychophysiol. **32**(2), 107–118 (1999)
36. Qin, S., Zhong, J.: Extraction of features in EEG signals with the non-stationary signal analysis technology. In: IEEE Engineering in Medicine and Biology Society, 1–4 September, San Francisco (2004)
37. Daubechies, I.: The wavelet transform, time-frequency localization and signal analysis. J. Renew. Sustain. Energy **36**(5), 961–1005 (2015)
38. Mallat, S.: A wavelet tour of signal processing, vol. 31. Academic Press, San Diego (1999)
39. Belyavin, A., Wright, N.A.: Changes in electrical activity of the brain with vigilance. Electroencephalogr. Clin. Neurophysiol. **66**(2), 137–144 (1987)
40. Torsvall, L., Akerstedt, T.: Sleepiness on the job: continuously measured EEG changes in train drivers. Electroencephalogr. Clin. Neurophysiol. **66**(6), 502–511 (1987)
41. Lal, S.K., Craig, A., Brood, P., Kirkup, L., Nguyen, H.: Development of an algorithm for EEG-based drive fatigue countermeasure. J. Saf. Res. **34**(3), 321–328 (2003)
42. Hong, J.E., Min, K.C., Kim, S.H.: Electroencephalographic study of drowsiness in simulated driving with sleep deprivation. Int. J. Ind. Ergon. **35**(4), 307–320 (2005)

Augmented Learning

Holographic and Related Technologies for Medical Simulation

Christine Allen[1]([⊠]), Sasha Willis[1], Claudia Hernandez[1],
Andrew Wismer[1], Brian Goldiez[1], Grace Teo[1],
Lauren Reinerman-Jones[1], Mark Mazzeo[2], and Matthew Hackett[2]

[1] University of Central Florida, Institute for Simulation and Training (UCF IST),
Orlando, USA
callen@ist.ucf.edu
[2] Combat Capabilities Development Command, Orlando, USA

Abstract. Holographic technologies allow for direct three-dimensional (3D) imaging without the need for special glasses or headwear. Holographic imaging ranges from static (i.e., unchanging) toward dynamic (i.e., changing) presentations. Since dynamic holographic products are in their developmental infancy, this study utilized static holographic images to predict future needs and preferences for dynamic holography. Using a single anatomical model, five static holograms were created for subjective evaluation from respondents. Four major research questions addressed the aim of this study, to determine the impact of color, hogel size, polygon density, and directional resolution on user preferences and perceived image quality of holograms within the medical field.

Data collection took place at Orlando Regional Medical Center, a part of Orlando Health from November 2017–February 2018. A total of 32 medical educators and providers viewed the static holograms, answering a series of questions related to each hologram. Overall perceptions/preferences were reported. Trends suggest that participants preferred the color over monochrome hologram, even when both are of the same image quality. The highest polygon density, 3.3 M (3.3 million polygons) was rated as rendering significantly higher image quality than lower polygon densities. Furthermore, there is a potential interaction between hogel size (like a pixel) and directional resolution (angular rays in each hogel). This study provides useful technical recommendations for future development of static and dynamic holograms as a possible alternative to current 3D visualization mechanisms in the medical domain.

Keywords: Hologram · Three-dimensional (3D) visualization · Training

1 Introduction

The human body exists in three dimensions. It follows that technology should be able to display it as such. However, students often learn anatomy through two-dimensional (2D) pictures and must integrate these pictures into a three-dimensional (3D) mental model [1–3]. One drawback to using 2D images in anatomical education is that multiple pictures or diagrams (also known as "key views") are needed to fully represent a 3D structure [2]. A second limitation of using 2D images is that they lack the depth

D. D. Schmorrow and C. M. Fidopiastis (Eds.): HCII 2019, LNAI 11580, pp. 497–516, 2019.
https://doi.org/10.1007/978-3-030-22419-6_36

cues needed to accurately and completely depict the spatial relationships among and within anatomical structures [4, 5].

Static holography can supplement digital technologies that are new to anatomical education classrooms, such as virtual and augmented reality systems. While virtual reality has recently garnered a considerable amount of attention, the technology is also not without its own disadvantages. Stereopsis, used in virtual reality headsets which allows for depth perception within the head-mounted display, presents a separate image to each eye and is the current standard for user-centered medical simulation [6–8]. While virtual and augmented reality systems can be effective tools for displaying 3D relationships, there are several negative usability effects on users with sustained use (e.g., visual fatigue and muscle fatigue in the neck), making alternative technologies for medical simulation more attractive [6, 7, 9].

A study by Rizzolo [10] describes specific benefits that holography can provide for medical curricula. These benefits include a clearer understanding of how 2D images can be reconstructed into a depiction of a 3D structure and improved visualization of 3D structures and relationships. Additionally, static holograms are reusable, such that students can take turns viewing the images individually or in groups, and the plates can be stored away for future use [11].

This paper outlines an overview of holograms and their associated technical components such as color, contrast, hogel size, polygon density, and directional resolution. We also evaluate whether differences in color, hogel size, polygon density, and directional resolution affect a viewer's ability to identify anatomical structures. We expect these outcomes reported will guide developers to produce images at or above a level of minimal usability, therefore, providing a recommended path for further technical development of both static and dynamic holography.

1.1 Holographic Displays Described

Before a discussion can take place on the classification of holograms used in this research, it is important to describe a holographic display. Bruckheimer [12] defines holography as "a method for creating an exact visual representation of an object in three physical dimensions using light" (p. 1). Holographic displays exhibit a remarkable advantage over other types of 3D displays. They are autostereoscopic, meaning users can view the displayed images in three dimensions without glasses or other supplementary eyewear, and have multiple viewpoints [7, 13, 14]. However, autostereoscopy affords a more comfortable experience as compared to various other methods of 3D visualization, such as virtual reality headsets, even when an external light source is needed [7]. Autostereoscopy may also decrease the potential for motion-related sickness [15].

In addition to being autostereoscopic, holographic displays can present several depth cues that 2D images cannot. Some of these depth cues are due to changes in how an image looks from different viewing angles and are referred to as parallax. When the effect of parallax is present, objects that are closer to the viewer move more quickly than objects that are further from the viewer [16]. Movement or motion parallax refers to the phenomenon that contributes to the perception of slightly different images when the viewer changes his or her position [13]. It is the form of parallax that is most important for the static holographic images used in this research. Movement parallax happens over time and with motion, so the displayed image is viewed from multiple viewpoints and provides information about the relative distances of objects in a visual scene [13, 16].

A second noticeable depth cue that is an important feature in holographic imagery is occlusion. Occlusion is the visual effect experienced when an object is blocked from view by another. The degree of occlusion changes when parts of the image come into view as the viewer changes position [7, 8, 13]. For instance, a building that appears to be blocking a tree is perceived as being in front of the trees. From another angle, the tree may be blocking the building and would then appear as being in front of the building. This cue becomes more pronounced as objects are displayed at increasingly larger distances apart [8].

In some types of holographic displays, a point light source (e.g., a flashlight) is needed to recreate the perception of the 3D image that was first generated during the printing process [17, 18]. In such displays, ambient light largely impedes the capability of viewing the printed image. Changing the location and/or angle of the direct light source when viewing a hologram can also affect both occlusion and motion parallax, depending on the type of hologram that is being viewed. In this way, motion parallax and occlusion operate together to produce multiple viewpoints of the same visual scene and can even help improve performance during interactive tasks involving reaching in depth [16].

1.2 Features of Holograms

There are several technical aspects of a static hologram that can be manipulated including color, contrast, hogel size, polygon density, and directional resolution. Generally, each of these static holography features can inform the development of dynamic hologram displays. Using the holograms created by Zebra Imaging Inc. (now known as HoloTech Switzerland AG) we can directly evaluate the usability of a 3D display system by adjusting several technological aspects while simultaneously allowing for various depth cues to be observable within the display.

Color. Monochrome refers to images that are printed in varying in shades of green rather than grey. Monochrome prints are best for models that lack texture, a limitation for medical imaging purposes where texture is of considerable importance [1, 18]. The coloring is manipulated during the image generation process, prior to printing. In either case, color and monochrome holographic images are displayed on a black background to distinguish the image from the empty space surrounding it.

Polygon Density. Polygon density is also altered entirely during the digital image generation process. The layout of polygons in a representation of a 3D image is sometimes referred to as a mesh [19]. Printed hologram images may contain millions (notated as M) of polygons, dependent on the count chosen during image generation. A higher polygon density generally corresponds to a sharper image but requires more processing power and is limited by the capabilities of the computer program used during image generation, the physical printing process, and the material holograms are printed on [18]. Additionally, missing or "sparkling" black hogels may appear in a holographic print if polygons become too large (i.e., low polygon density) and interfere with the hogel pitch.

Hogel Size. Hogel size refers to the dimensions of the holographic elements that comprise the image. Hogels are the holographic equivalent or version of pixels [20, 21]. The holograms are set to contain hogel sizes in millimeters (mm) or fractional parts of millimeters (e.g., 0.50 mm). Decreasing hogel size significantly impacts the cost required to generate and print a holographic image [22].

Directional Resolution. Directional resolution in static holography signifies the number of angular rays present in each hogel and does not exist for 2D displays [23]. In other words, it is the range of viewing angles from which hogel may be seen [23, 24]. For instance, the directional resolution of 128 × 128 has a total of 16,384 angular rays per hogel. It is important to determine if directional resolution influences user preferences and perceived image quality, as it measures the extent to which parallax is present in the displayed image [23].

1.3   Use of Static Holograms to Inform Dynamic Hologram Technology

Static holograms are studied in this research effort because they can be studied directly, whereas dynamic holographic displays are not as prevalent in use and lack in technical maturity. By implementing static holograms in the present experiment, we gain a higher understanding of the features that should be prioritized when evolving a realistic and useful dynamic holographic display for users. Following this, developers can focus on advancing the most important features, while maintaining a conservative expense and resource approach to the areas of holography that are not as important to users. For example, if it is found that the directional resolution of an image is critical in accurately perceiving an image, whereas color does not affect image perception, developers can allocate more resources into further improving directional resolution and distribute less resources towards improving or implementing color capabilities. By studying the impact of different criteria on end-user preferences (i.e., healthcare professionals) and the visualization materials they use, we seek to gain a higher understanding of the baseline features needed to create realistic and useful static and/or dynamic holographic displays for healthcare training purposes.

2   Methodology

2.1   Participants

Study participants consisted of 32 volunteers and included nurses (31%), medical students (25%), nursing students (12.5%), Emergency Medical Technicians (12.5%), medical residents (6%), physicians (6%), one nurse practitioner (3%) and one medical assistant (3%). Fifty-six percent of participants were females (n = 18), 43% worked in the emergency department, and the average age across all participants was 32 years. All participants had either normal or corrected-to-normal vision and none of the participants noted color vision deficiency (i.e., color blindness). Experience levels ranged from 1 to 20 years, with an average of 6.28 years. When asked about previous training experiences, participants indicated that they were primarily exposed to cadavers, deceased animals, simulations, and lectures, while 3D movies, virtual reality, and computer software were not commonly used during their anatomy training.

2.2   Study Material: Holograms

The static holograms shown to participants during this study were generated by a multi-step process. First, a real-world model was obtained from the Combat Capabilities Development Command which features a simulated bilateral leg amputation (as seen in Fig. 1 left). This model was then scanned using the Eva Professional Handheld Scanner by Artec 3D. The resulting images were meshed in the Artec Studio 11 software program. Further image manipulation was conducted using the software program MAYA12 from Autodesk [25]. Finally, the computer images were converted into static holograms using a multi-step lamination process and printed onto photopolymer plates. Figure 2 (right) was generated digitally from the Multiple Amputation Trauma Trainer MATT^{TM} Simulator and translated across each of the 18 channels.

Fig. 1. (Left) Full MATT^{TM} Simulator with bilateral leg amputations; blurred for graphic content

In total, five holograms were created by Zebra Imaging Inc., each hologram exhibited full parallax, thus making the 3D image visible from every overhead viewing angle. Each channel (or side of the hologram containing a 3D image) displayed a single image. Four of the holographic plates (Holograms 1, 2, 4, and 5) displayed one image on all four possible channels. One holographic plate (Hologram 3) compared just two channels. In all, 18 total channels were evaluated for perceived image quality, usability, and usefulness.

By using multiple iterations of the same image, the researchers were able to evaluate user perceptions across the four technical features of interest (e.g., color, polygon density, hogel size, and directional resolution). Please refer to Fig. 3 for an overview of the specific features that varied in each hologram.

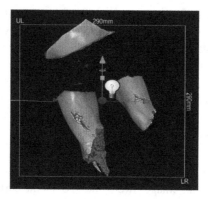

Fig. 2. (Right) Computer-generated 3D model of the MATT^{TM} Simulator with bilateral leg amputations

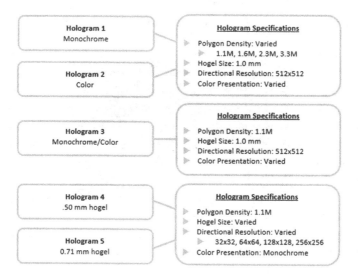

Fig. 3. Holograms overview

2.3 Research Questions

The five holograms were created by combining different levels of the four hologram features (i.e., color, polygon density, hogel size, directional resolution) in a way that allowed comparisons between different holograms to address the research questions (Table 1).

2.4 Study Procedures

All study materials were approved by the University of Central Florida's Institutional Review Board. After giving their consent for study participation, participants were

Table 1. Research questions summary

Questions	Hologram(s) compared
(A) **What is the optimal perceived image quality when** <u>**polygon density**</u> **is varied?** (Participant presented with holograms H1 and H2 in turn. Each of these holograms was rotated to show the adjacent channels that varied on polygon densities. Participant compared the image quality between adjacent channels)	(A1) **Monochrome** with **varying polygon density 1.1 M, 1.6 M, 2.3 M, 3.3 M (H1)** (A2) **Color** with **varying polygon density 1.1 M, 1.6 M, 2.3 M, 3.3 M (H2)**
(B) **Do users prefer** <u>**color or monochrome presentation of holograms?**</u> (Participant presented with H3, a hologram that was rotated to show its color and monochrome channels. Participant was asked which channel they preferred)	(B1) **Color vs monochrome (H3)**
(C) **What is the optimal perceived image quality when** <u>**directional resolution**</u> **is varied?** (Participant presented with holograms H4 and H5 in turn. Each of these holograms was rotated to show adjacent channels that varied on directional resolutions. Participant compared the image quality between adjacent channels)	(C1) Monochrome with **hogel size .50 mm,** and **varying directional resolutions 32 × 32, 64 × 64, 128 × 128, and 256 × 256** rays per hogel (H4) (C2) Monochrome with **hogel size .71 mm,** and **varying directional resolution 32 × 32, 64 × 64, 128 × 128, 256 × 256** rays per hogel (H5)
(D) **What is the optimal perceived image quality when** <u>**hogel size**</u> **is varied?** (Three holograms, H1, H4, and H5, that differed on hogel size were laid out from left to right before the participant. Participant compared the image quality of the holograms)	(D1) Monochrome with **hogel size 1.0 mm (H1)** placed on **Left** (D2) Monochrome with **hogel size .50 mm (H4)** placed in **Center** (D3) Monochrome **hogel size .71 mm (H5)** placed on **Right**

assigned unique participant identifiers which were created to ensure participant confidentiality and anonymity was protected (using https://www.randomizer.org). These IDs assisted with the hologram random sequencing order. Next, a demographics questionnaire and restrictions survey were administered. Based on answers to the restriction's questionnaire, the experimenter verified that there was no history of photosensitive seizures, color vision deficiency, or a physical or psychological sensitivity toward viewing anatomical images; two participants were dismissal from the study. Additionally, this study employed short questionnaires before, during, and after the hologram viewing task. All surveys were created through Qualtrics, a trusted third-party survey platform and administered via a tablet or paper copy.

During the study, a LED flashlight was handed to the participant for viewing and the room lights were turned off. A green light was used for monochrome holograms and white light was used for colored holograms [17, 26]. While standing, participants were instructed to hold the flashlight at ear level, aligned with their line of sight for optimal viewing. To facilitate the rotation and minimize potential damage to the surface, each hologram was placed on a rotating platform on the floor (except for holograms used for Research Question D), and a second experimenter rotated the platform when participants indicated they were ready to see the next image. (Note: interrater reliability was confirmed across experimenters to ensure conditions and rotation methods were consistent across participants.) Participants stood approximately eight inches away from the holograms. The participants were not allowed to view previous channels again, and each channel was only viewed for five - ten seconds to encourage participants to respond with their initial instincts.

Each participant randomly viewed all five holograms and were exposed to all the research questions listed in Table 1. For Research Questions A and C, participants were asked to indicate if they noticed a difference in image quality between adjacent channels following each of the four 90-degree rotations. The final comparison always compared the two most extreme differences (e.g., highest versus lowest polygon density or directional resolution). Question B asked about image quality regarding Hologram 3's two channels after a 180-degree rotation. Research Question D also asked about image quality in relation to the three holograms presented and participants were told to step laterally in front of each hologram they were viewing and made judgements about perceived image quality. For each comparison of adjacent channels, if the participant indicated a difference in image quality, they were asked if they perceived the quality had increased or decreased from the previous side (or from the previous hologram in the case of Research Question D). Additionally, respondents were asked if they could identify any anatomical structures (as seen in Figs. 1 and 2) in the holographic images and rate their ability to distinguish detail in the hologram, overall, between the choices of easy, moderate, and difficult.

3 Results

This study presents an evaluation of five static holograms with varying technical features, as explored through four major research questions. Evaluation included the impact changes to polygon density (Research Question A), color (Research Question B), directional resolution (Research Question C), and hogel size (Research Question D) had on perceptions of holographic image quality. Participants completed questionnaires before and after viewing the five holograms. Additionally, the questions answered by participants during hologram viewing provided insight on how the end user's perceptions of image quality are affected by the possible feature changes mentioned above. The results of this study are intended to guide future hologram technology features.

A within-subjects design was implemented, and research questions were assessed through a series of binomial tests. Chi-square tests were also conducted, when appropriate, and described in-text. An initial binomial test was conducted for each evaluation of a perceived image quality change during hologram viewing ("Do you see a difference in image quality: yes or no?"). This was performed to examine whether there was a significant difference in perceived image quality change from one hologram channel (or hologram, in the case of Research Question D) compared to the next. A second binomial test was conducted on responses regarding the direction of change (increase or decrease, if a change in image quality was perceived) to evaluate if one direction of change was perceived significantly more often than the other direction, for each evaluation. The binomial tests were run with the assumption (i.e., "expectation") that the proportions would be equivalent. Therefore, when comparing the two groups, it was expected to be 50-50. A significant result means one group responded significantly above chance.

All analyses were conducted using JASP v. 9 [27] and accompanying Bayes factors (BFs) are provided. Bayes factors provide an estimate of the likelihood of differences, such that BFs less than 1 provide evidence for the null hypothesis (i.e., no difference), while BFs 3 and larger provide anecdotal (1–3), substantial (3–10), or greater evidence (10+) for the alternative hypothesis [28]. When applicable, binomial tests are also reported to evaluate any potential of order effects that may exist from participants being exposed to the highest or lowest level of a factor (i.e., polygon density, color, directional resolution, or hogel size) first.

3.1 Research Question a on Polygon Density: What Is Optimal Perceived Image Quality When *Polygon Density* Is Varied?

To test Research Questions A1 and A2, the influence of polygon density changes on perceived image quality was evaluated within both a monochrome and a color hologram, respectively. As described in the Methodology section, both Holograms 1 and 2 used 1.0 mm hogel sizes and a directional resolution of 512×512 rays per hogel. Each hologram had four channels, with each channel containing one of the following four polygon densities: 1.0 M, 1.6 M, 2.3 M, and 3.3 M, where M = million. Since hologram rotation order was approximately counterbalanced (either the lowest to highest or the highest to lowest), responses by hologram channel viewing order are presented first to provide a sense of potential order effects. Next, the results of the binomial tests are presented.

Monochrome (Research Question A1). First, polygon density was considered within a monochrome hologram (Hologram 1). Order effects were evaluated by chi-square tests using the Yates correction due to small sample sizes. There were no significant order effects with respect to hologram channel viewing order (i.e., lowest or highest polygon density viewed first) in the monochrome hologram (all p's > .265; all BF's < 1.21). Binomial tests revealed a significantly larger number of participants than chance perceived differences in image quality between polygon densities 1.1 M and 1.6 M ($p = .050$; $BF = 2.02$), 1.6 M and 2.3 M ($p = .002$; $BF = 38.67$), and 3.3 M and 1.1 M ($p = .020$; $BF = 4.64$; see Table 2). Among the participants who did perceive a difference in image quality viewing the 1.1 M and 1.6 M channels, there was a significant difference in the proportion of participants who saw an "increase" in image quality $\chi^2(1) = 11.83$, $p < .001$, $BF = 1210.65$, especially when viewed from low to high. In this comparison, 100% of the participants reported the 1.6 M channel as yielding higher image quality, see Table 3.

Color (Research Question A2). Polygon density was also evaluated within a color hologram (Hologram 2). Order effects were again evaluated by chi-square tests using the Yates correction due to small sample sizes. There were no significant order effects

Table 2. Proportion of participants *perceiving a difference in image quality* between channels of different polygon densities in a monochrome hologram

Polygon density comparison	Hologram channel order	Proportion of participants	
		No difference	Difference
1.1 M vs. 1.6 M	Lowest first (n = 9)	0.11	0.89
	Highest first (*n* = 23)	0.39	0.61
	Total (*n* = 32)	**0.31***	**0.69***
1.6 M vs. 2.3 M	Lowest first (n = 9)	0.33	0.66
	Highest first (*n* = 23)	0.17	0.83
	Total (*n* = 32)	**0.22†**	**0.78†**
2.3 M vs. 3.3 M	Lowest first (n = 9)	0.44	0.56
	Highest first (*n* = 23)	0.48	0.52
	Total (*n* = 32)	**0.47**	**0.53**
3.3 M vs. 1.1 M	Lowest first (n = 9)	0.11	0.89
	Highest first (*n* = 23)	0.35	0.65
	Total (*n* = 32)	**0.39‡**	**0.61‡**

Note. Typographical symbols superscripts denote significant differences in perception of image quality change.

Table 3. Proportion of participants *perceiving an increase or decrease in image quality* between channels of different polygon densities in a monochrome hologram

Polygon density comparison	Hologram channel order	Proportion of participants	
		1.1 M better	1.6 M better
1.1 M vs. 1.6 M**	Lowest first (*n* = 8)	0	1
	Highest first (*n* = 14)	0.86	0.14
	Total (*n* = 22)	**0.55**	**0.45**
		1.6 M better	2.3 M better
1.6 M vs. 2.3 M	Lowest first (*n* = 6)	0.33	0.67
	Highest first (*n* = 19)	0.79	0.21
	Total (*n* = 23)	**0.68**	**0.32**
		2.3 M better	3.3 M better
2.3 M vs. 3.3 M	Lowest first (*n* = 5)	0.60	0.40
	Highest first (*n* = 12)	0.67	0.33
	Total (*n* = 17)	**0.65**	**0.35**
		3.3 M better	1.1 M better
3.3 M vs. 1.1 M	Lowest first (*n* = 8)	0.50	0.50
	Highest first (*n* = 15)	0.40	0.60
	Total (*n* = 23)	**0.43**	**0.57**

**Denotes that an order effect was observed for this comparison.

in the proportion of participants perceiving a difference in image quality in any of the four evaluations between the two viewing orders (all p's > .204; all BF's < 1.42). However, there was a significant difference in the proportion of participants who saw a directional "increase" in image quality $\chi^2(1) = 16.97$, $p < .001$, $BF = 35530$. Specifically, when viewed from low to high, 100% of participants reported the 1.6 M channel to be an increase in image quality over the 1.1 M channel. Conversely, when viewed from high to low, 100% of participants reported the 1.1 M channel to be an increase in image quality over the 1.6 M channel (see Table 5). The remaining three comparisons (e.g., 1.6 M vs. 2.3 M) showed no significant order effects (all p's > 0.080; all BF's < 4.87).

Table 4. Proportion of participants *perceiving a difference in image quality* between channels of different polygon densities in a color hologram

Polygon density comparison	Hologram order	Proportion of participants	
		No difference	Difference
1.1 M vs. 1.6 M	Lowest first ($n = 18$)	0.28	0.72
	Highest first ($n = 14$)	0.43	0.57
	Total ($n = 32$)	**0.34**	**0.66**
1.6 M vs. 2.3 M	Lowest first ($n = 18$)	0.22	0.78
	Highest first ($n = 14$)	0.50	0.50
	Total ($n = 32$)	**0.34**	**0.66**
2.3 M vs. 3.3 M	Lowest first ($n = 18$)	0.28	0.72
	Highest first ($n = 14$)	0.29	0.71
	Total ($n = 32$)	**0.28***	**0.72***
3.3 M vs. 1.1 M	Lowest first ($n = 18$)	0.28	0.72
	Highest first ($n = 14$)	0.21	0.79
	Total ($n = 32$)	**0.25†**	**0.75†**

Note. Typographical symbols superscripts denote significant differences in perception of image quality change.

Binomial tests revealed a significantly larger number of participants than chance perceived differences in image quality between polygon densities of 2.3 M and 3.3 M ($p = .020$; $BF = 4.64$), and between 3.3 M and 1.1 M ($p = .007$; $BF = 12.37$; see Table 4). Within participants who did perceive a difference in image quality, there was a significantly higher proportion of participants who rated the 3.3 M channel (75%) as higher in image quality than the 1.1 M channel ($p = .023$; $BF = 4.99$; see Table 5).

3.2   Research Question B on Color: Do Users Prefer Color or Monochrome Presentation of Holograms?

As detailed in the methodology, Hologram 3 contained two opposing channels showing the same image: one in color and one in monochrome. This hologram used 1.1 M polygon density, 1.0 mm hogel size, and a directional resolution of 512 × 512 rays per

Table 5. Proportion of participants *Perceiving an Increase or Decrease in Image Quality* between channels of different polygon densities in a color hologram

Polygon density comparison	Hologram order	Proportion of participants	
		1.1 M better	1.6 M better
1.1 M vs. 1.6 M**	Lowest First (*n* = 13)	0	1
	Highest First (*n* = 8)	1	0
	Total (*n* = 21)	**0.38**	**0.62**
		1.6 M better	2.3 M better
1.6 M vs. 2.3 M	Lowest First (*n* = 14)	0.21	0.79
	Highest First (*n* = 7)	0.71	0.29
	Total (*n* = 21)	**0.38**	**0.62**
		2.3 M better	3.3 M better
2.3 M vs. 3.3 M	Lowest First (*n* = 13)	0.31	0.69
	Highest First (*n* = 10)	0.30	0.70
	Total (*n* = 23)	**0.30**	**0.70**
		3.3 M better	1.1 M better
3.3 M vs. 1.1 M	Lowest First (*n* = 13)	0.62	0.38
	Highest First (*n* = 11)	0.91	0.09
	Total (*n* = 24)	**0.75‡**	**0.25‡**

Note. Typographical symbols superscripts denote significant differences in perception of image quality change.
**Denotes that an order effect was observed for this comparison.

hogel. Participants responded by stating whether they saw a difference in image quality between the two channels, the direction of the difference (increase or decrease), and any preference for one channel or the other. These responses were analyzed in the same way as the earlier analyses.

An analysis of order effects revealed that, regardless of whether the monochrome or color channel was viewed first, no order effects were present regarding a difference in perceived image quality ($p = 1.000$; $BF = 0.30$), direction of change ($p = .445$; $BF = 0.90$), or channel preference ($p = 1.000$; $BF = 0.41$).

Difference in Perceived Image Quality. Significantly more participants than chance (90.6%) reported seeing a difference in image quality between the color and monochrome channels, $p < .001$, $BF = 26240.03$.

Direction of Change. While most participants saw a difference in image quality between the two channels, there was no significant difference between the proportion of participants who saw either the color or monochrome channel as being of higher image quality than the other, $p = .711$, $BF = 0.26$. Thus, there was no consensus as to whether the color or monochrome channel was perceived to have higher image quality.

Channel Preference. When asked to provide the channel (i.e., color or monochrome) they preferred, more participants preferred the color hologram (*n* = 19) to the monochrome hologram ($n = 11$); however, this difference was not statistically significant ($p = .200$).

3.3   Research Question C on Directional Resolution: What Is Optimal Perceived Image Quality When Directional Resolution Is Varied?

Directional resolution was varied on two different four-channel holograms: one with a hogel size of 0.50 mm (Research Question C1; Hologram 4) and one with a hogel size of 0.71 mm (Research Question C2; Hologram 5). Both holograms were monochrome and used a 1.1 M polygon density. The four channels of directional resolution include 32×32, 64×64, 128×128, and 256×256 rays per hogel. Separate analyses were conducted for each set of comparisons within a hologram of a set hogel size.

Hogel Size 0.50 mm (Research Question C1). A set of analyses were conducted for comparisons of directional resolution on the hologram with hogel size 0.50 mm (Hologram 4). There were no order effects for a perceived difference in image quality dependent on which channel was viewed first (all p's $> .591$; all BF's < 0.55). For direction of change (i.e., an increase or decrease), a chi-square test revealed that there was a significant order effect for the comparison of 128×128 versus 256×256, $\chi^2(1) = 3.86$, $p = .049$, $BF = 7.36$, such that participants who viewed the highest resolution first were more likely to see 128×128 as an improvement over 256×256, whereas the opposite was more likely (i.e., 256×256 rays per hogel was seen as an increase in image quality over 128×128 rays per hogel) for participants who viewed the channels from lowest to highest resolution. No other order effects for direction of change were found (all p's > 0.479; all BF's < 0.95) for the other three comparisons (e.g., 32×32 vs. 64×64 rays per hogel).

There were significantly more participants who saw a difference between each pair of channels for all four comparisons than chance (all p's $< .021$; all BF's > 4.63; see Table 6). However, there was only a significant effect for the direction of change for the 256×256 versus 32×32 rays per hogel directional resolution comparison, with more participants responding 256×256 was better ($n = 21$) than 32×32 ($n = 5$), $p = .002$, $BF = 37.79$ (see Table 7). The other comparisons were not significantly different in terms of direction of change (all p's > 0.211; all BF's < 0.72), suggesting that there was no common perception of which resolution was higher in terms of image quality for intermediate steps in resolution.

Hogel Size 0.71 mm (Research Question C2). A corresponding set of analyses were conducted for comparisons of directional resolution on the hologram with hogel size 0.71 mm (Hologram 5). There was a significant order effect, with respect to hologram channel viewing order for the 32×32 vs 64×64 rays per hogel comparison, $\chi^2(1) = 10.24$, $p = .001$, $BF = 380.09$. In this case, significantly more people than chance perceived a difference in image quality between directional resolutions of 32×32 and 64×64 with a higher proportion of participants ($n = 15/15$) seeing a difference in image quality between the two resolutions when viewed from lowest to highest than when viewed highest to lowest ($n = 7/17$). There were no other significant

Table 6. Proportion of participants *perceiving a difference in image quality* between channels of different directional resolutions in a 0.50 mm hogel size hologram

Directional resolution comparison	Hologram order	Proportion of participants	
		No difference	Difference
32 × 32 vs. 64 × 64	Lowest first (*n* = 11)	0.36	0.64
	Highest first (*n* = 21)	0.24	0.76
	Total (*n* = 32)	**0.28***	**0.72***
64 × 64 vs. 128 × 128	Lowest first (*n* = 11)	0.18	0.82
	Highest first (*n* = 21)	0.33	0.67
	Total (*n* = 32)	**0.28†**	**0.72†**
128 × 128 vs. 256 × 256	Lowest first (*n* = 11)	0.36	0.64
	Highest first (*n* = 21)	0.24	0.76
	Total (*n* = 32)	**0.28‡**	**0.72‡**
256 × 256 vs. 32 × 32	Lowest first (*n* = 11)	0.09	0.91
	Highest first (*n* = 21)	0.24	0.76
	Total (*n* = 32)	**0.19§**	**0.81§**

Note. Typographical symbols superscripts denote significant differences in perception of image quality change.

order effects for the other three differences in image quality comparisons (e.g., 64 × 64 vs. 128 × 128; all *p*'s > .305; all *BF*'s < 0.93). There was also a significant order effect for the 32 × 32 vs 64 × 64 direction of change comparison, $\chi^2(1) = 4.99$, *p* = .026, *BF* = 13.90, with considerably more participants who viewed the channels in order from lowest to highest seeing the 64 × 64 rays per hogel image as higher quality than the 32 × 32 (*n* = 13/15), compared to participants who viewed the channels from highest to lowest (*n* = 2/7). No other significant order effects for direction of change judgments were found (all *p*'s > .477; all *BF*'s < 0.78).

There were significant differences in the proportion of participants who saw a difference in image quality between all sets of comparisons (all *p*'s < .051; all *BF*'s > 2.01), with most people noticing a difference in image quality in each case (see Table 8).

Specifically, there were significant differences in the proportion of participants who saw the 128x128 channel as higher image quality than the 64 × 64 channel, *p* = .023, *BF* = 4.99, and between the 256 × 256 and 32 × 32 channels, *p* = .002, *BF* = 59.38, with a higher proportion of responses indicating an increase in image quality for the higher resolution channel in each case (see Table 9).

3.4   Research Question D on Hogel Size: What Is Optimal Perceived Image Quality When *Hogel Size* Is Varied?

As detailed in the Methodology, the effect of hogel size on perceived image quality was evaluated by comparing three channels of varying hogel sizes across three separate

Table 7. Proportion of participants *perceiving an increase or decrease in image quality* between channels of different directional resolutions in a 0.50 mm hogel size hologram

Directional resolution comparison	Hologram order	Proportion of participants	
		32 × 32 better	64 × 64 better
32 × 32 vs. 64 × 64	Lowest first (n = 7)	0.43	0.57
	Highest first (n = 16)	0.69	0.31
	Total (n = 23)	**0.61**	**0.39**
		64 × 64 better	128 × 128 better
64 × 64 vs. 128 × 128	Lowest first (n = 9)	0.33	0.67
	Highest first (n = 14)	0.36	0.64
	Total (n = 23)	**0.35**	**0.65**
		128 × 128 better	256 × 256 better
128 × 128 vs. 256 × 256**	Lowest first (n = 7)	0.71	0.29
	Highest first (n = 16)	0.19	0.81
	Total (n = 23)	**0.35**	**0.65**
		256 × 256 better	32 × 32 better
256 × 256 vs. 32 × 32	Lowest first (n = 10)	0.80	0.20
	Highest first (n = 16)	0.81	0.19
	Total (n = 26)	**0.81‖**	**0.19‖**

Note. Typographical symbols superscripts denote significant differences in perception of image quality change.

**Denotes that an order effect was observed for this comparison.

holograms all monochrome and 1.1 M polygon density (only one channel per hologram was shown). Hogel size ranged from 0.50 mm (Hologram 4), to 0.71 mm (Hologram 5), to 1.0 mm (Hologram 1). Due to hologram processing limitations, directional resolution also varied, with 512 × 512 rays per hogel only on the 1.0 mm hogel size hologram, and 256 × 256 rays per hogel on the 0.50 mm and 0.71 mm holograms. For all three comparisons (0.50 mm vs. 0.71 mm, 0.71 mm vs 1.0 mm, and 0.5 mm vs 1.0 mm), most participants perceived a difference in image quality (all p's < .001; all BF's > 26240.00). There was no significant difference in the proportion of participants who saw either 1.0 mm or 0.5 mm holograms as higher quality than the other, $p = .856$, $BF = 0.24$. However, a majority of participants perceived the 0.71 mm hologram as higher quality than the 0.50 mm hologram ($p < .001$; $BF = 753.47$) as well as the 0.71 mm hologram as higher quality than the 1.0 mm hologram ($p = .050$; $BF = 2.02$).

Table 8. Proportion of participants *perceiving a difference in image quality* between channels of different directional resolutions in a 0.71 mm hogel size hologram

Directional resolution comparison	Hologram order	Proportion of participants	
		No difference	Difference
32 × 32 vs. 64 × 64**	Lowest First (*n* = 15)	0.00	1.00
	Highest First (n = 17)	0.59	0.41
	Total (*n* = 32)	**0.31***	**0.69***
64 × 64 vs. 128 × 128	Lowest First (*n* = 15)	0.13	0.87
	Highest First (*n* = 27)	0.35	0.65
	Total (*n* = 32)	**0.25†**	**0.75†**
128 × 128 vs. 256 × 256	Lowest First (*n* = 15)	0.20	0.80
	Highest First (*n* = 17)	0.18	0.82
	Total (*n* = 32)	**0.19‡**	**0.81‡**
256 × 256 vs. 32 × 32	Lowest First (*n* = 15)	0.20	0.80
	Highest First (*n* = 17)	0.12	0.88
	Total (*n* = 32)	**0.16§**	**0.84§**

Note. Typographical symbols superscripts denote significant differences in perception of image quality change.

**Denotes that an order effect was observe for this comparison.

4 Discussion

Key takeaways from the results are included, along with a summary of technical parameters (Table 10) concluding the following trends:

1. When evaluating a variety of polygon densities (1.1 M to 3.3 M) 3.3 M polygon density was rated as significantly higher image quality than the lowest extreme polygon density (1.1 M) for color holograms, with no clear direction of polygon density impact within monochrome holograms.
2. Color presentations did not show higher perceptions of image quality over monochrome presentations, but participants did indicate directly that they preferred color over monochrome holographic images.
3. When addressing direction resolution, 256 × 256 rays per hogel directional resolution with either 0.71 mm or 0.50 mm hogel size was noted as being of significantly higher image quality than directional resolutions of 32 × 32 rays per hogel. Within the 0.71 mm hogel hologram, 128 × 128 rays per hogel was rated as significantly higher in image quality than 64 × 64, with a similar trend observed for the 0.50 mm hogel hologram.
4. When assessing hogel size, a hogel size of 0.71 mm was rated as being of significantly higher quality over hogel sizes of 0.50 mm and 1.0 mm. However, hogel size may interact with the directional resolution.

It should be noted that each technical parameter comes with factors such as cost, performance expectation, and user preference. This study is about placing the correct

Table 9. Proportion of participants *perceiving an increase or decrease in image quality* between channels of different directional resolutions in a 0.71 mm hogel size hologram

Directional resolution comparison	Hologram Order	Proportion of participants	
		32 × 32 better	64 × 64 better
32 × 32 vs. 64 × 64**	Lowest first (*n* = 15)	0.13	0.87
	Highest first (*n* = 7)	0.71	0.29
	Total (*n* = 22)	**0.32**	**0.68**
		64 × 64 better	128 × 128 better
64 × 64 vs. 128 × 128	Lowest first (*n* = 13)	0.09	0.91
	Highest first (*n* = 11)	0.36	0.64
	Total (*n* = 24)	**0.25‖**	**0.75‖**
		128 × 128 better	256 × 256 better
128 × 128 vs. 256 × 256	Lowest first (*n* = 12)	0.42	0.58
	Highest first (*n* = 14)	0.21	0.79
	Total (*n* = 26)	**0.31**	**0.69**
		256 × 256 better	32 × 32 better
256 × 256 vs. 32 × 32	Lowest first (*n* = 12)	0.83	0.17
	Highest first (*n* = 15)	0.80	0.20
	Total (*n* = 27)	**0.82¶**	**0.18¶**

Note. Typographical symbols superscripts denote significant differences in perception of image quality change.

**Denotes that an order effect was observed for this comparison

Table 10. Summary of suggested technical parameters

Polygon density (1.1 M, 1.6 M, 2.3 M, 3.3 M)	Color vs. Monochrome	Directional resolution (32 × 32, 64 × 64, 128 × 128, 256 × 256, 512 × 512)	Hogel size (0.5 mm, 0.71 mm, 1.0 mm)
3.3 M	Color preferred, but monochrome may be sufficient	128 × 128, or 256 × 256 (512 × 512 possible with color holograms)	0.71 mm

usage of technology in combination with other factors. Additionally, some of these factors may interact with each other in ways that negatively affect user preferences. Therefore, the suggestions in Table 10 are subject to change based on the interaction effects and user population for future studies.

5   Limitations and Challenges

A limitation of the study is the relatively small sample size. With an increased sample size, counterbalancing would have been more equal across hologram viewing orders. Additionally, the researchers were limited on the amount of time that could be dedicated for each data collection session (i.e., 15 min per participant) which prevented more in-depth questions to be explored with participants and further comparisons between channels (e.g., 64×64 vs 256×256 rays per hogel). Finally, the study did not account for individual differences in acuity and quality of vision in participants, including differences in depth perception. These may have accounted for why some participants saw increases in image quality when others saw decreases for the same comparisons.

Many of the comparisons, specifically for hogel size and resolution changes, were evaluated using monochrome holograms. Future studies could evaluate if study results generalized to color holograms.

The comparison between the three hogel sizes had a difference in the directional resolution. Due to hologram processing limitations, directional resolution varied, with 512×512 rays per hogel only on the 1.0 mm hogel size hologram, and 256×256 rays per hogel on the 0.50 mm and 0.71 mm holograms. This created a challenge to analyze the results equally across the hogel variations. We were able to extrapolate outcomes from other hologram comparisons but cannot confirm the precision of the results. While it is not possible to print a single hologram with channels varying in hogel size (thereby keeping directional resolution constant within just one hologram), additional holograms with equal directional resolutions would have been preferred.

Finally, the holograms were produced by Zebra Imaging Inc., which is no longer in business. The first set of holograms produced by Zebra Imaging Inc. were substandard to our initial requirements. A second set of prints was ordered to correct the issues between the directional resolution channels. Future studies will need to employ the use of holograms from a different company since it is unclear if holograms with the same parameters printed by other companies would be different.

6   Conclusions and Future Work

A systematic approach is used to determine the ability of users to detect differences in static holograms based on initial reactions to viewing and comparing several 3D images. These evaluations are pertinent to the design of future autostereoscopic static and dynamic hologram displays. The methodology described herein provides an easy-to-execute approach to quickly ascertain the image characteristics users find most important and inform the development community of those characteristics. Theoretical implications will contribute to understanding the limitations of the human visual system to detect differences in new forms of visual media. Applications of the results will help to inform guidelines for development of the next generation of holographic displays and how tradeoffs among various display factors affect usability of the holograms.

Future work should continue to address the interaction of technical factors such as the combination of polygon density of 3.3 M, color versus monochrome, direction resolution of 128×128 versus 256×256 versus 512×512 rays per hogel (for color), and hogels of 0.50 mm versus 0.71 mm, as well as the implications on user preference and cost. Furthermore, a comparison of these factors across different fields of study or domains will allow a more comprehensive and generalizable evaluation of the effect of varying technical factors.

Acknowledgements. This research was sponsored by the U.S. Army Research Laboratory, currently the Combat Capabilities Development Command (CCDC) and was accomplished under Cooperative Agreement Number W911NF-15-2-0011. The views and conclusions contained in this document are those of the author's and should not be interpreted as representing the official policies, either expressed or implied of CCDC or the U.S. Government. The U.S. Government is authorized to reproduce and distribute reprints for government purposes notwithstanding any copywrite notation herein.

References

1. Codd, A.M., Choudhury, B.: Virtual reality anatomy: is it comparable with traditional methods in the teaching of human forearm musculoskeletal anatomy? Anat. Sci. Educ. **4**(3), 119–125 (2011)
2. Estevez, M.E., Lindgren, K.A., Bergethon, P.R.: A novel three-dimensional tool for teaching human neuroanatomy. Anat. Sci. Educ. **3**(6), 309–317 (2010)
3. Preece, D., Williams, S.B., Lam, R., Weller, R.: "Let's get physical": advantages of a physical model over 3D computer models and textbooks in learning imaging anatomy. Anat. Sci. Educ. **6**(4), 216–224 (2013)
4. Ballantyne, L.: Comparing 2D and 3D imaging. J. Vis. Commun. Med. **34**(3), 138–141 (2011)
5. Heylings, D.J.A.: Anatomy 1999–2000: the curriculum, who teaches it and how? Med. Educ. **36**(8), 702–710 (2002)
6. Goldiez, B., Abich, J., Carter, A., Hackett, M.: Perceived image quality for autostereoscopic holograms in healthcare training. In: Medical Imaging 2017: Image Perception, Observer Performance, and Technology Assessment, vol. 10136, p. 1013615. International Society for Optics and Photonics (2017)
7. Held, R.T., Hui, T.T.: Special report: a guide to stereoscopic 3D displays in medicine. Acad. Radiol. **18**, 1035–1048 (2011). https://doi.org/10.1016/j.acra.2011.04.005
8. Yang, L., Dong, H., Alelaiwi, A., Saddik, A.E.: See in 3D: state of the art of 3D display technologies. Multimed. Tools Appl. **74**(19), 1–35 (2015). https://doi.org/10.1007/s11042-015-2981-y
9. Knight, J.F., et al.: (2006) Assessing the wearability of wearable computers. In: 10th IEEE International Symposium on Wearable Computers, pp. 75–82 (2006)
10. Rizzolo, L.J., et al.: Design principles for developing an efficient clinical anatomy course. Teacher **8**(2), 142–151 (2006)
11. Shaked, N.T., Katz, B., Rosen, J.: Review of three-dimensional holographic imaging by multiple-viewpoint-projection based methods. Appl. Opt. **48**(34), H120–H136 (2009)
12. Bruckheimer, E., et al.: Computer-generated real-time digital holography: first time use in clinical medical imaging. Eur. Heart J. Cardiovasc. Imaging **17**(7) (2016) https://doi.org/10.1093/ehjci/jew087
13. Dodgson, N.A.: Autostereoscopic 3D displays. Computer **38**(8), 31–36 (2005)
14. Geng, J.: Three-dimensional display technologies. Advances in optics and photonics **5**(4), 456–535 (2013)
15. Tay, S., et al.: An updatable holographic three-dimensional display. Nature **451**(7179), 694–698 (2008). https://doi.org/10.1038/nature06596

16. Aras, R., Shen, Y., Noor, A.: Quantitative assessment of the effectiveness of using display techniques with a haptic device for manipulating 3D objects in virtual environments. Adv. Eng. Softw. 7643–7647 (2014). https://doi.org/10.1016/j.advengsoft.2014.05.009

17. Gibson, T.: Into the deep: 3-D holographic technology provides detailed human intelligence. Soldiers Mag. **8**, 6 (2011)

18. Zebra Imaging Inc. (n.d.) Frequently asked questions. https://www.zebraimaging.com/faq-1

19. Friedman, T., Michalski, M., Goodman, T.R., Brown, J.E.: 3D printing from diagnostic images: a radiologist's primer with an emphasis on musculoskeletal imaging—putting the 3D printing of pathology into the hands of every physician. Skeletal Radiol. **45**(3), 307–321 (2016)

20. Bourzac, K.: A step toward holographic videoconferencing. MIT Technol. Rev. (2010). https://www.technologyreview.com/s/421540/a-step-toward-holographic-videoconferencing/

21. Tsutsumi, N., Kinashi, K., Tada, K., Fukuzawa, K.: Fully updatable three-dimensional holographic stereogram display device based on organic monolithic compound. Opt. Soc. Am. (2013). https://www.osapublishing.org/DirectPDFAccess/97796550-D0A2-5319-D003AF8783595D82_260338/oe-21-17-19880.pdf?da=1&id=260338&seq=0&mobile=no

22. Scullion, C.: Personal communication, 28 June 2017

23. Kovács, P.T., Bregović, R., Boev, A., Barsi, A., Gotchev, A.: Quantifying spatial and angular resolution of light-field 3-D displays. IEEE J. Sel. Topics Signal Process. **11**(7), 1213–1222 (2017)

24. Aron, J.L., Patz, J. (eds.): Ecosystem Change and Public Health: A Global Perspective. JHU Press, London (2001)

25. Artec3D: Artec Studio 11 [computer software]. Artec USA, Palo Alto (2016)

26. Newswanger, C.: Personal communication, optimal viewing conditions, 4 April 2017

27. JASP Team: JASP (Version 0.9) [Computer software] (2018)

28. Jeffreys, H.: Theory of Probability, 3rd edn. Oxford University Press, Oxford, UK (1961)

Nature Inspired Scenes for Guided Mindfulness Training: Presence, Perceived Restorativeness and Meditation Depth

Mark R. Costa[1]([✉]), Dessa Bergen-Cico[1], Trevor Grant[1],
Rocio Herrero[2], Jessica Navarro[3], Rachel Razza[1], and Qiu Wang[1]

[1] Syracuse University, Syracuse, New York, USA
mrcosta@syr.edu
[2] Universidad Jamie I, Castellón de la Plana, Spain
[3] Universidad de Valencia, Valencia, Spain

Abstract. Practicing mindfulness-based stress reeducation and other contemplative practices generates a number of health and human performance benefits. However, limited access to qualified training and practice support, as well as poor practice environments, makes it difficult to sustain the habits necessary to develop the attentional regulation skills needed to benefit from mindfulness. In this paper, we report on our research, which focuses on developing immersive environments to support mindfulness-based stress reduction practices. We specifically look at how the design of a virtual environment can foster a restorative experience, if that restorative experience is associated with the depth of the meditation session, and if there are associations between presence and the depth of the meditation session and the restorative properties of the virtual experience. Results show there are significant relationships between the three core concepts, suggesting future work is needed to determine if there are causal relationships exist between the presence, meditation depth, and perceived restorativeness. Understanding how the design of virtual environments may facilitate mindfulness and other contemplative practices has implications for promoting the use of the practices in a variety of contexts.

Keywords: Mindfulness · Virtual reality · Mental health · Presence · Meditation

1 Introduction

There is growing empirical evidence demonstrating the efficacy of contemplative practices and mindfulness-based interventions for treating and managing physical and mental health problems; as well as improving attentional regulation [33]. One of the most widely studied programs is mindfulness-based stress reduction (MBSR) which helps veterans with post-traumatic stress disorder [7]; patients with chronic pain [30]; as well as job stress and interpersonal relationships [3, 29].

Although we know that training in contemplative practices can benefit individuals in various professions, we find that accessibility to qualified trainers, time commitments associated with attending guided practices, environmental distractors, and lack of social support can make sustained engagement with training and practice difficult. The

© Springer Nature Switzerland AG 2019
D. D. Schmorrow and C. M. Fidopiastis (Eds.): HCII 2019, LNAI 11580, pp. 517–532, 2019.
https://doi.org/10.1007/978-3-030-22419-6_37

effectiveness of mindfulness and other contemplative practices are dependent on quality of instruction and group instruction is the prevailing format of delivery. However, access to qualified instructors is limited and many people often find that access to groups is limited due to time and travel constraints, coupled with limited class offerings. Veterans and active duty military who are juggling multiple priorities face extra time constraints and are often in isolated environments where it is difficult to find other individuals to practice with.

Our ongoing research focuses on the question of whether immersive technologies can increase access to mindfulness and other contemplative practices. We identified three areas to focus where immersive technologies can help. First, we are exploring whether immersive environments like virtual and mixed reality (VR, MR) can be designed to provide a comfortable and engaging space to practice mindfulness. Second, whether technology-based solutions can adapt the to the learner to optimize the skill acquisition process. For novice meditators that may be a secluded and quiet virtual forest with basic guided meditations, while more advanced practitioners may want specific feedback on aspects of their practice or to be exposed to more challenging scenarios where attention is harder to maintain. Second, whether virtual environments can support the social component of practice, including group meditation, to mitigate geographical challenges and practice such that geographically distributed users can experience the social interactions necessary to learn contemplative skills like MBSR.

This paper reports on the first question related to whether virtual environments can serve as a comfortable and engaging space to practice mindfulness. Our current research focuses on providing basic instruction in awareness of breath (AOB) and sitting meditation in virtual nature environments. We are specifically working on fine-tuning virtual environments to assist people who may have difficult with meditation practice due to personal barriers and environmental distractions that tax limited attentional resources. This report presents our research process related to creating an optimized environment for people to learn mindfulness-based meditation. The specific research questions guiding this paper are – Can virtual nature immersive environments have restorative properties? and Do the virtual nature immersive environments improve the quality of a meditation session?

2 Theoretical Framework and Research Questions

Our teams' research focuses on developing interventions to help high stress populations manage their stress effectively through the use of complementary health practices. In particular, we focus on questions related to the efficacy of mindfulness-based practices for self-regulation of emotions and behaviors, based on the mindfulness-based stress reduction MBSR curriculum [22].

Mindfulness and other contemplative approaches require regular practice for sustained change. However, many people find engaging in regular practice difficult, as noted by Grow and colleagues who found daily practice drops off significantly [17]. For those of us who are interested in developing tools to make mindfulness more widely available, one of the core challenges we face is making is easier for people to maintain a consistent practice schedule because consistent and frequent practice accelerates the skill acquisition process [13].

One of the fundamental skills associated with contemplative practices is attentional regulation. The ability to engage in advanced techniques is predicated upon the ability to monitor and direct one's attention. Some of the early training novices receive in mindfulness-based intervention is on breath awareness and focus, which helps the participant learn to monitor their attention and redirect it as necessary. Part of the mindfulness practice is to set aside time each day to work on managing attention. In some sense, there is a feedback loop between attentional regulation and quality of mindfulness practice. Attentional regulation makes it easier to practice mindfulness, but practicing mindfulness improves attentional regulation. As other researchers have noted, many people who need or want to train their attentional regulation skills have difficulty focusing, which makes mindfulness practice more difficult [2, 12, 26].

There are several factors that make it difficult to direct one's attention. There are external factors, such as noise and light pollution, and internal factors, such as stress, that make it difficult to focus on mindfulness practices; this is especially true for novices [26]. When we say something acts as a drain on attentional resources, we are adopting the position that attention is a limited resource, and chronic stress and/or intense focus will deplete that resource [25]. Therefore, if someone is chronically stressed or they are distressed, they have difficulty focusing. Mindfulness can help mitigate the negative consequences of stress, but one needs to be able to focus in order to practice mindfulness. If stress has depleted a person's attentional resources, they cannot tune out distractors to focus on the practice. In situations where there is a confluence of factors, such as chronic stress and distracting environments, serving to impede practice, we can leverage aspects of technology to mitigate those impediments. Thus, we specifically look at whether virtual reality environments can help mitigate the effects of chronic stress on attention fatigue through simulation of natural environments, and whether a controlled virtual environment increases the quality of meditation sessions because of its ability to mitigate the effects external stimuli.

With respect to directed attention fatigue, attention restoration theory [18, 23] posits that attention is a limited resource that is frequently depleted through stress and over taxation due to excessive stimulation and cognitively demanding environments. Research has shown that natural environments have inherently restorative properties, if they cultivate a sense of being away, extent, compatibility, and soft fascination [23]. In fact, there is a larger movement of research on biophilic design principles, which focuses on the ways in which the integration of nature into our built environments affects us physiologically and psychologically [31]. These design features include:

Extent is the quality of environments that leaves you totally immersed. An environment that satisfies the extent quality has both coherence and scope such that it constitutes a "whole other world" [23].

Compatibility broadly refers to the match between the person and the environment. Kaplan argues there are two types of compatibility. The first is information compatibility, where a person understands exactly what he or she can do in that environment with little effortful direction of attention towards decision making [24]. The second type of compatibility is motivational compatibility, where there is a match between what the environment affords and what the person wants to do.

Fascination can either be hard fascination, where your attention is held easily by a fascinating activity (like a game), or soft fascination, where your attention is held by a less stimulating activity. Our work focuses more on soft fascination, where we create virtual environments to gently relax one's attention to ease them into the practice of meditation [23].

Being away refers to the sense of being separated from usual concerns and the environment those concerns arise from [23].

If an environment meets several of these criteria, it can increase the perceived restorativeness of the environment [18]. A restorative environment promotes the recovery of mental resources and well-being [9, 18]. Lymeus et al. [27] argue that a restorative environment refreshes the mental capacity of individuals and can thus make a meditation session more productive, particularly for novice practitioners. **Our first research question is** – what is the relationship between perceived restorativeness of an environment and the quality of the meditation session? Two questions that would help us answer that question are – (1) Do any of the factors associated with **perceived restorativeness** have a stronger relationship with the **quality of the meditation** session than the others?, and (2), Is it possible that certain factors are necessary, but not sufficient for quality meditation sessions?

Any time we design a virtual environment to have restorative properties, we need to identify ways in which we can maximize the user's sense of extent, compatibility, fascination, and being away with the virtual environment. Before we can meet any of those criteria, the user needs to feel present in the virtual environment. The concept of physical presence, or the sense of "being there", best captures what we need to optimize [32]. However, the exact relationship between presence and perceived restorativeness of the environment has yet to be determined; therefore, we are not proposing a causal model in this paper.

In the past, we could argue that a sense of presence is required before the media produced the desired psychological effects. However, recent research has shown that certain elements of a virtual environment might induce higher levels of presence if the element or stimuli has a significant relationship to emotions outside of the virtual environment. For example, the presence of phobia inducing stimuli in a VE increased the subjects' sense of presence [14]. It may be possible that there is a positive, causal relationship between compatibility and presence or that several variables related to the design of the VE mediate the relationship between presence and the factors of perceived restorativeness. Based on the uncertain relationships between the constructs of presence and perceived restorativeness, we identified our **second research question** - What is the relationship between **physical presence** and the **perceived restorativeness** of an environment? Sub-questions related to that question are - Do any of the factors associated with perceived restorativeness have a stronger relationship with the presence than the others?

3   Methods and Measures

3.1   Testbed and Environment

For this study, we built a custom virtual reality-based meditation app to test the relationships between concepts outlined in the theoretical framework section above. The meditation area was designed with specific features that other researchers have found to be critical to creating the sense of nature that is required for restorativeness. More specifically, the meditation area was an open forest scene next to a waterfall and stream with running water. We have trees and grass swaying gently in the background. Additionally, we incorporated bird song and scenic views into the virtual environment. We also set the environmental conditions to a bright, sunny day with clouds periodically passing overhead (Fig. 1). Part of the experiment included a 10-min guided meditation, which was recorded by a member of the research team who is a certified Mindfulness-Based Stress Reduction coach; the track was edited by local sound engineers to improve the sound quality of the track.

Fig. 1. A nature-inspired theme with running water and birdsong

The MIND Lab has a suite of sensors and data collection computers supporting the virtual environment. We displayed the virtual environment on an HTC Vive while the subjects were seated in an office chair. The virtual environment was configured to send signals to a suite of sensors when certain interactions or events happened in the virtual environment (see below). The suite of sensors includes a BOPAC system with galvanic skin response (tech specs), and ECG (tech specs), and a 52-channel Hitachi functional near infrared spectroscopy system (fNIRS) (tech specs) to capture participants' neural hemodynamics. Signals from the virtual environment to the sensors were time locked via a custom Arduino setup, which allows us to compare data from different sensors that sample at different frequencies.

3.2   Protocol

Adult subjects were recruited from Syracuse University and the surrounding Syracuse, New York community. Exclusion criteria included epilepsy and advanced neurodegenerative diseases (e.g. Parkinson's Disease). Subjects were invited into the lab, where they were greeted and given a brief explanation of the experiment. After listening to the lab manager's explanation, the individuals were placed in front of a standard computer and given access to a digital version of the informed consent document. If the participants agreed to the informed consent, the digital survey would proceed on to the pre-test measures; if the subjects declined to give consent, they would be thanked for their time and be allowed to leave. No subjects declined to participate. At the end of the pre-test surveys, the survey system displayed a thank you message along with information on the next steps. The lab manager explained how the virtual reality headset worked and what was going to happen over the next ten minutes. The manager used the following text:

In a moment we will place the headphones on your head and outfit the sensor [fNIRS] over the headset. Once that is set up, we will begin the meditation track. Please follow the instructions of the audio track as best as possible. If you feel your mind wander, pull the trigger on the control [technician demonstrates which trigger to pull]. Once you are done, let the technician know and they will help you remove the headset.

After the participants acknowledged that they understood the next steps, they were connected to the physiological sensors and the headset and headphones were placed on their heads and adjusted for comfort. We asked the subjects to indicate when they noticed their mind wandering so we incorporate feedback in the future to bring attention to the wandering mind; that feedback system will be built on a classified that identifies, from the physiological data, when a person's mind is wandering.

The mindfulness track incorporated elements of awareness of breath The narrator also leveraged stimuli in the virtual environment as objects of focus for the track, including the swaying grass, falling water, water running down the stream, and passing clouds. The audio track began one minute and thirty seconds after the participants entered the virtual environment. We delayed the onset of the track because we noticed, even among members of the research staff, that people who entered the virtual environment needed a few moments to get settled in the space and to look around.

When the track was complete, the technician removed the fNIRS cap, headphones, and virtual reality headset from the subjects. The technician then removed the remaining sensors and asked the subjects to complete the post-test measures via an online survey. After the subjects completed the survey, they were thanked for their time and informed them that the experiment was complete.

3.3   Measures

Pre-test measures included a basic demographic survey, which included self-reported measures of meditation experience (style of meditation and estimated hours practiced), whether the subject had ever attended a meditation retreat, and veteran status. We asked subjects to provide self-reported measures of total practice and experience with retreats per [15] because both total practice time and time spent at retreats has been found to affect a person's overall mindfulness and the level of physiological changes associated with long-term meditation. Participants also completed the following self-reported measures:

Five Facets of Mindfulness Questionnaire (FFMQ; [4]) [Pre-test]. FFMQ is a 39-item, five factor instrument designed to capture one's mindful disposition towards daily life. The five factors are: observing, describing, acting with awareness, non-judging of inner experience, and non-reactivity to inner experience.

Positive and Negative Affect Scale – Short Form (PANAS-SF [34]). [Pre- and Post- test] We used a modified version of the PANAS-SF, worded to capture the strength of their current positive and negative moods. Subjects completed both a pre- and post- test PANAS to capture changes in mood due to the mindfulness session.

Perceived Stress Scale (PSS [11], 1983) [Pre-test]. PSS consists of ten items grouped into two dimensions – positively and negatively worded questions related to how one appraises the stressfulness of situations in their life.

Perceived Restorativeness Scale (PRS [18]) [Post-test] PRS is a four-factor, 26-item scale intended to capture properties of natural environments that are expected to facilitate emotional and attentional rejuvenation.

SUS Presence Questionnaire (SUSPQ, [18]) [Post-test] SUSPQ has six questions, focused on three presence indicators: sense of being there, extent to which the virtual environment becomes more "real" than reality, and the extent to which the virtual environment is thought of as a place visited [6].

Meditation Depth Questionnaire (MDQ [28]) [Post-test]. MDQ was Developed to measure the depth or intensity of a meditation experience while simultaneously being agnostic to the school of meditation. The thirty-item scale has five factors – hinderances, relaxation, personal self, transpersonal qualities, and transpersonal self.

The following physiological instruments were used to capture data during the experiment:

Electrodermal Activity (EDA) for arousal [5]. BIOPAC MP-150 system, data transmitted wirelessly with BIONOMADIX, data collected with BIOPAC's Acknowledge Software. The sampling rate was set at 2000 Hz. Raw data from the module pair is bandlimited from DC to 10 Hz for both channels. Two sensors were placed on the participant's non-dominant hand.

Electrocardiogram (ECG) for heart rate variability for arousal [5]. The module pair has a fixed gain rate of 2,0000Db, with bandlimits from 1 Hz to 35 Hz. Two sensors were placed on the participant's collar bone and a ground electrode on the right hip using the BIOPAC MP 150.

Respiration Sensor (RSP) for breath rate. Raw RSP data from the module pair is bandlimited from DC to 10 Hz, to provide for the measurement of relatively static respiratory conditions, such as cessation of breathing, up to the extremely rapid respiratory effort variations (up to 600 breaths per minute) associated with coughing or sneezing. An elastic band with a pressure sensor was wrapped around participant's chest so that the pressure sensor was located at the inferior section of the participant's sternum using the BIOPAC MP 150.

Function Near Infrared Spectroscopy (fNIRS) for hemodynamic patterns of the brain for assessing cognitive states [19, 21]. Captured with a Hitachi ETG-4000, sampling oxygenated and deoxygenated hemoglobin at 10 Hz. Probes were placed in 2, 3 × 5 arrangements above both the right and left frontal cortex, which resulted in 24 channels of data.

4 Analysis

This paper focuses on the associations between mindfulness disposition via the Five Facets of Mindfulness Questionnaire, sense of presence experienced by participants in the virtual environment via the SUS Presence Questionnaire, perceived restorativeness of the virtual environment via the Perceived Restorativeness Scale, and quality of the mindfulness session via the Meditation Depth Questionnaire. We will report on the associational relationship among the other measures in subsequent publications.

4.1 Descriptive Statistics and Measure Assessment

Twelve subjects participated in the study (Female = 5), ranging in age from 20–54 (M = 32.92, SD = 11.22). Five of the subjects had a master's degree or greater, five either had no degree or some college but no degree, and the remaining had college degrees.

Six of the subjects practiced meditation in the past, covering a variety of styles, including focused attention meditation, yoga meditation, body scan meditation, and transcendental meditation. Of the six subjects who have practiced meditation in the past, three participated in retreats. The least experienced practitioner had about one hour of meditation experience, the most around 10,000 h (M = 110 h, SD = 3996).

Table 1 summarizes the descriptive statistics for the four main variables. Results for tests of normality in the last column ($p > .05$) indicate the data adhere to a normal distribution (Fig. 2). The Presence Scale is calculated differently in that the final score is a count of all instances where a participant rated an item a 6- or a 7- on the Likert Scale [6].

Table 1. Descriptive statistics and normality tests for main variables

Measure	Mean	CI (95%) Lower/upper	S.D.	Shapiro-Wilk	p
Five facets mindfulness	3.370	3.165/3.574	.323	.974	.947
Perceived restorativeness	3.435	2.966/3.905	.739	.966	.865
Meditation depth	2.694	2.253/3.135	.694	.951	.649
Presence	1.167	.632/1.70	.835		

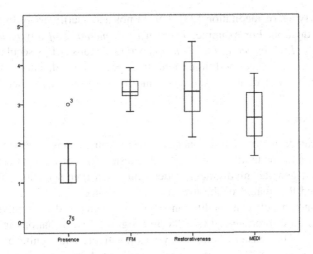

Fig. 2. Boxplots for main variables

We attempted to conduct confirmatory factor analysis on the variables with multiple scales (Five Facets of Mindfulness, Perceived Restorativeness, and Meditation Depth Questionnaire), but the sample size was too small to produce results. The (sub) scales were also tested for internal consistency; all but two scored in the acceptable to good range ($0.7 < \alpha < 0.9$), except for Presence scale and the *Describe* subscale from the Five Facets of Mindfulness questionnaire (see Table 2).

Table 2. Reliability analysis

Scale	Subscale	No. items	Cronbach's α
Five facets of mindfulness	Observe	7	.806
	Describe	8	**.698**
	Act	8	.891
	Non-judgmental	8	.883
	Non-reactive	7	.812
Perceived restorativeness	Away	5	.883
	Fascination	8	.871
	Coherence	4	.707
	Compatibility	9	.838
Meditation depth questionnaire	Hinderances	6	.756
	Relaxation	3	.893
	Transpersonal qualities	8	.799
	Transpersonal self	6	.810
	Personal self	7	.857

Another area where we expected to see divergent patterns is in the meditation depth results. Some of the factors and their associated items were written to ostensibly written to apply to all types of meditation [28], but do not necessarily apply to mindfulness-based stress reduction. For example, *There was no meaning of any meditation techniques any more, I felt myself at one with everything*, because of its secular bent. In the future, we expect to collect additional data to support more detailed analysis of the relationship between meditation experience and reported meditation depth.

4.2 Correlations

Two-tailed Pearson's correlation analyses were run among presence, perceived restorativeness, and meditation depth. We also tested for correlations between meditation depth and general mindfulness to determine if a predisposition towards mindfulness impacted the quality of the meditation experience.

The relationship between meditation depth and perceived restorativeness of the experience was more pronounced (see the five significant correlations in the last column of Table 3). The sense of being away was positively and significantly correlated with hinderances ($r = .842$, $p = .001$). As a sense of Away increased, Hinderances, such as *I found it difficult to relax*, and *There was a constantly change of thoughts in my mind*, decreased. Positive correlations are shown in Table 3 because the values were reverse coded for Hinderances. The sense of being Away was also positively correlated with Relaxation ($r = .847$, $p = .001$), Transpersonal Qualities ($r = .690$, $p = .013$), and General Meditation Depth ($r = .767$, $p = .004$). Relaxation also had significant correlations with Fascination ($r = .580$, $p = .048$) and Compatibility ($r = .672$, $p = .017$). Compatibility was significantly correlated with Transpersonal Qualities ($r = .700$, $p = .011$) and General Meditation Depth ($r = .680$, $p = .020$). Overall Perceived Restorativeness had positive correlations with all factors of Meditation Depth except Transpersonal Self.

Table 3. Relationships between meditation depth and perceived restorativeness of the virtual environment, along with factors

			Perceived restorativeness				
			Away	Fascination	Coherence	Compatibility	Perceived restorativeness
Meditation depth	Hinderances	r	**.842**	.308	.370	.573	**.610**
		p	**.001**	.330	.236	.052	**.035**
	Relaxation	r	**.847**	**.580**	.569	**.672**	**.768**
		p	**.001**	**.048**	.053	**.017**	**.004**
	Personal self	r	.533	.484	.505	.496	**.576**
		p	.075	.111	.094	.101	**.050**
	Transpersonal qualities	r	**.690**	.361	.450	**.700**	**.629**
		p	**.013**	.249	.142	**.011**	**.028**
	Transpersonal self	r	.351	.219	.402	.384	.385
		p	.263	.494	.196	.218	.216
	General meditation depth	r	**.767**	.475	.544	**.658**	**.702**
		p	**.004**	.119	.068	**.020**	**.011**

In Table 3, several $p < 0.10$, indicating the marginal significance due to the impact of small sample size. Looking at Tables 4 and 5, there are several significant or marginally significant correlations between Presence and factors of Perceived Restorativeness or Meditation Depth. Presence was significantly associated with Compatibility ($r = .596$, $p = .041$), Transpersonal Self ($r = .765$, $p = .004$), Personal Self ($r = .679$, $p = .015$), and General Meditation Depth ($r = .673$, $p = .017$).

Table 4. Correlations between presence and perceived restorativeness and its factors

		Perceived restorativeness				
		Away	Fascination	Coherence	Compatibility	Perceived restorativeness
Presence	r	.404	.467	.506	**.596**	.554
	p	.193	.126	.093	**.041**	.061

Table 5. Correlations between presence and meditation depth and its factors

		Meditation depth					
		Transpersonal self	Hinderances	Relaxation	Personal self	Transpersonal qualities	General meditation depth
Presence	r	**.765**	.493	.433	**.679**	.493	**.673**
	p	**.004**	.103	.160	**.015**	.103	**.017**

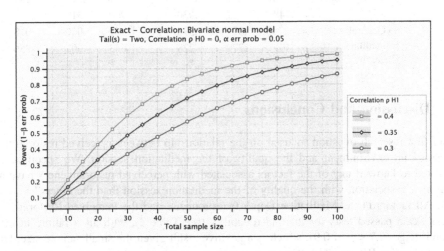

Fig. 3. Post-hoc empirical power analysis for sample sizes across three correlations

One area where we expected to see stronger correlations was between general mindfulness, as measured by the Five Facets of Mindfulness, and meditation depth, as measured by the Meditation Depth Questionnaire (Table 6). Only Acting with Awareness showed significant correlation relationships with Personal Self ($r = .619$, $p = .032$) and Transpersonal Self ($r = .623$, $p = .030$). The small correlation (i.e., $r < .30$) [10] between meditation depth and mindfulness disposition could be because so few skilled meditators were in the group. The empirical power analysis (Fig. 3) shows that a larger sample size (e.g., N $\geq$ 85) is needed to reach a significant relationship. Alternatively, the non-meditators may have rated their mindfulness disposition inconsistently because they do not fully understand the experience they are trying to rate. Having novices use the Meditation Depth Questionnaire may be problematic because the questionnaire was designed for, and validated on, expert meditators to assess the depth of their meditation sessions [28].

Table 6. Relationships between mindfulness and meditation depth, include factors

			Five facets of mindfulness (FFM)					
			Observe	Describe	Acting with awareness	Non-judging	Non-reactivity	FFM
Meditation depth	Hindrances	r	−.278	.167	−.007	−.230	.033	−.168
		p	.381	.604	.984	.471	.920	.601
	Relaxation	r	−.301	−.045	.145	−.202	−.200	−.258
		p	.341	.890	.653	.530	.534	.419
	Personal self	r	−.213	−.002	**.619***	.071	−.003	.200
		p	.506	.995	**.032**	.826	.992	.533
	Transpersonal qualities	r	.005	.221	.267	−.265	.160	.121
		p	.988	.491	.401	.405	.620	.709
	Transpersonal self	r	−.298	.141	**.623***	−.041	.315	.284
		p	.348	.662	**.030**	.900	.319	.371
	General meditation depth	r	−.263	.091	.376	−.150	.040	.018
		p	.409	.779	.228	.643	.902	.956

5   Discussion and Conclusions

Our first research question focused on the relationship between perceived restorativeness of an environment and the quality of the meditation session. We specifically wanted to know if any of the factors associated with perceived restorativeness have a stronger association with the quality of the meditation session than the others?

All factors in the Meditation Depth Questionnaire and the Perceived Restorativeness Scale passed tests for internal reliability (see Table 2), with all α falling in the good range (0.7 < α < 0.9), which is a positive result given the small sample size. Of the factors in Perceived Restorativeness, *Away*, or the sense of being separate from problems and spaces they arise in, had the strongest correlation with aspects of meditation depth. Despite participants rating the environment as being a place separate from the participants' usual concerns, there was no correlation between the sense of presence in the environment and the sense of being away. There is no clear reason for this disconnect. These findings may be reflective of the limitations of the instruments used and/or the need for a larger sample size in future research.

In addition to the factor *Away*, *Compatibility* had a significant correlation with meditation depth, which has face validity. If the participants were not comfortable with the virtual environment, it would not be a relaxing experience and the lack of comfort might interfere with the quality of the meditation session.

The fact that Restorativeness had significant correlations with all factors of *Meditation Depth*, except *Transpersonal Self*, suggests that the act of meditation may be restorative. Experienced meditators may be able to conjure up the feeling of being away, without the need to travel (virtually or physically). For those participants who were not as skilled in meditation, the sense of being away (*Away*), in an environment that supported gentle fascination (*Fascination*), and was compatible with their schemas of what constitutes a relaxing environment (*Compatibility*), was relaxing and restorative.

The finding in this paper related to meditation depth mirror the findings in [26], which reported greater depth of meditation along all factors in the virtual environment with neurofeedback than the virtual environment alone, which in turn had greater levels of depth than a standard computer screen. The extreme variance in time spent meditating makes it difficult to use the self-report data to explore in greater detail the relationship between meditation experience, meditation depth, and perceived restorativeness of the environment. Our current work focused on classifying mindful experiences [20] may provide more objective ways to tease out the complex relationship between meditation experience, the role of practice environment, the quality of a meditation session, and the restorative property of the environment.

Our second research question focused on the relationship between physical presence and the perceived restorativeness of an environment. We were specifically interested in whether any of the factors associated with perceived restorativeness have a stronger association with the presence than the others. Only *Compatibility* and *Coherence* showed significant correlations to presence. This is expected because incoherence, or a mismatch between what is expected of an environment and what is perceived, can break the perceptual illusion that presence is founded on. We suggest that compatibility is important for presence because incompatibility creates a barrier to accepting the environment and being in the present moment of the experience; although the exact conceptual linkage would need to be further explored. Presence was also marginally correlated with General Perceived Restorativeness, which would indicate that being in the virtual environment had some effects associated with restorative environments on the participants. We would need to further tease out whether the restorativeness of the environment helped improve the meditation depth of the session.

We do see significant correlations between transpersonal self, personal self, and general meditation depth, and presence. We attribute these results to participants' abilities to reach transpersonal and personal self-states. Their ability to do so may enable them to feel more present, regardless of the context, which may support greater meditation depth. It is noteworthy that, a sense of being in the space was correlated with overall meditation depth, which is encouraging because it would indicate the environment facilitated meditation.

Finally, we explored whether general mindfulness had any correlation with meditation depth. Results show little support for the effects of disposition towards mindfulness, as measured by the FFMQ, and the quality of the meditation session. Only acting with awareness showed any correlation with personal self and transpersonal self.

Brown and Ryan [8] argue that present-centered attention, or acting with awareness, is foundational to mindfulness; therefore, we would expect participants who report greater awareness of self and of transpersonal experiences to have higher levels of the basic skill of acting with awareness.

Similar to previous studies [1, 2, 12, 16], the results from this study indicate that a nature inspired virtual environment may has some restorative or relaxing properties and that the restorative effects may translate to higher quality or deeper mindfulness sessions. Researchers and practitioners who are promoting mindfulness as a possible intervention may want to consider paying attention to the physical space they are teaching in, as it will have an impact on the perceived depth of the session.

Major limitations to this study include a small sample size, lack of longitudinal analysis assessing whether the virtual practice environment supports sustained practice, and reliance on self-report measures to assess depth of the meditation experience. Our future work will integrate the use of sensors to assess the quality of the meditation session as well as deploy the virtual reality app to test whether the environment promotes sustained engagement in the practice.

References

1. Andersen, T., et al.: A preliminary study of users' experiences of meditation in virtual reality. In: 2017 IEEE Virtual Reality (VR), pp. 343–344 (2017). https://doi.org/10.1109/VR.2017.7892317
2. Allison, P., et al.: Relaxation with immersive natural scenes presented using virtual reality. Aerosp. Med. Hum. Perform. 88(6), 520–526 (2017). https://doi.org/10.3357/AMHP.4747.2017
3. Anderson, V.L., Levinson, E.M., Barker, W., Kiewra, K.R.: The effects of meditation on teacher perceived occupational stress, state and trait anxiety, and burnout. School Psychol. Q. 14(1), 3–25 (1999). https://doi.org/10.1037/h0088995
4. Baer, R.A., Smith, G.T., Hopkins, J., Krietemeyer, J., Toney, L.: Using self-report assessment methods to explore facets of mindfulness, using self-report assessment methods to explore facets of mindfulness. Assessment 13(1), 27–45 (2006). https://doi.org/10.1177/1073191105283504
5. Bandara, D., Song, S., Hirshfield, L., Velipasalar, S.: A more complete picture of emotion using electrocardiogram and electrodermal activity to complement cognitive data. In: Schmorrow, Dylan D.D., Fidopiastis, Cali M.M. (eds.) AC 2016. LNCS (LNAI), vol. 9743, pp. 287–298. Springer, Cham (2016). https://doi.org/10.1007/978-3-319-39955-3_27
6. Van Baren, J., W.A IJsselsteijn, J., Deliverable 5. measuring presence: a guide to current measurement approaches (2004)
7. Bergen-Cico, D., Possemato, K., Pigeon, W.: Reductions in cortisol associated with primary care brief mindfulness program for veterans with PTSD. Med. Care 52, S25 (2014). https://doi.org/10.1097/MLR.0000000000000224
8. Brown, K.W., Ryan, R.M.: The benefits of being present: mindfulness and its role in psychological well-being. J. Pers. Soc. Psychol. 84(4), 822 (2003)
9. Carrus, G., et al.: a different way to stay in touch with 'urban nature': the perceived restorative qualities of botanical gardens. Front. Psychol. 8 (2017). https://doi.org/10.3389/fpsyg.2017.00914
10. Cohen, J.: Statistical Power Analysis for the Behavioral Sciences, 2nd edn. Routledge, New York (1988)

11. Cohen, S., Kamarck, T., Mermelstein, R.: A global measure of perceived stress. J. Health Soc. Behav. **24**(4), 385–396 (1983). https://doi.org/10.2307/2136404

12. De Kort, Y., Meijnders, A.L., Sponselee, A.A.G., IJsselsteijn, W.A.: What's wrong with virtual trees Restoring from stress in a mediated environment. J. Environ. Psychol. **26**(4), 309–320 (2006)

13. Ericsson, K.A., Charness, N.: Expert performance: its structure and acquisition. Am. Psychol. **49**(8), 725 (1994)

14. Felnhofer, A., Hlavacs, H., Beutl, L., Kryspin-Exner, I., Kothgassner, O.D.: Physical presence, social presence, and anxiety in participants with social anxiety disorder during virtual cue exposure. Cyberpsychol. Behav. Soc. Netw. **22**(1), 46–50 (2018). https://doi.org/10.1089/cyber.2018.0221

15. Golem, D., Davidson, R.J.: Altered Traits: Science Reveals How Meditation Changes Your Mind, Brain, and Bod. Avery, New York (2017)

16. Gromala, D., Tong, X., Choo, A., Karamnejad, M., Shaw, C.D.: The virtual meditative walk: virtual reality therapy for chronic pain management. In: Proceedings of the 33rd Annual ACM Conference on Human Factors in Computing Systems (CHI 2015), pp. 521–524. https://doi.org/10.1145/2702123.2702344

17. Grow, J.C., Collins, S.E., Harrop, E.N., Marlatt, G.A.: Enactment of home practice following mindfulness-based relapse prevention and its association with substance-use outcomes. Addict. Behav. **40**, 16–20 (2015)

18. Hartig, T., Korpela, K., Evans, G.W., Gärling, T.: A measure of restorative quality in environments. Scand. Hous. Planning Res. **14**(4), 175–194 (1997). https://doi.org/10.1080/02815739708730435

19. Hincks, S.W., et al.: Entropic brain-computer interfaces-using fNIRS and EEG to measure attentional states in a bayesian framework. In: PhyCS, pp. 23–34 (2017)

20. Hirshfield, L., Bergen-Cico, D., Costa, M.R., Jacob, R.J.K., Hincks, S., Russell, M.: Measuring the neural correlates of mindfulness with functional near infrared spectroscopy. In: Empirical Studies of Contemplative Practices. Nova (2019)

21. Hirshfield, L., Williams, T., Sommer, N., Grant, T., Gursoy, S.V.: Workload-driven modulation of mixed-reality robot-human communication. In: Proceedings of the Workshop on Modeling Cognitive Processes from Multimodal Data, p. 3 (2018)

22. Kabat-Zinn, J., Hanh, T.N.: Full catastrophe living: using the wisdom of your body and mind to face stress, pain, and illness. Delta (2009)

23. Kaplan, S.: The restorative benefits of nature: toward an integrative framework. J. Environ. Psychol. **15**(3), 169–182 (1995). https://doi.org/10.1016/0272-4944(95)90001-2

24. Kaplan, S.: Meditation, restoration, and the management of mental fatigue. Environ. Behav. **33**(4), 480–506 (2001). https://doi.org/10.1177/00139160121973106

25. Kaplan, S., Berman, M.G.: Directed attention as a common resource for executive functioning and self-regulation. Perspect. Psychol. Sci. **5**(1), 43–57 (2010). https://doi.org/10.1177/1745691609356784

26. Kosunen, I., Salminen, M., Järvelä, S., Ruonala, A., Ravaja, N., Jacucci, G.: RelaWorld: neuroadaptive and immersive virtual reality meditation system. In: Proceedings of the 21st International Conference on Intelligent User Interfaces (IUI 2016), pp. 208–217. https://doi.org/10.1145/2856767.2856796

27. Lymeus, F., Lundgren, T., Hartig, T.: Attentional effort of beginning mindfulness training is offset with practice directed toward images of natural scenery. Environ. Behav. **49**(5), 536–559 (2017). https://doi.org/10.1177/0013916516657390

28. Piron, H.: The meditation depth index (MEDI) and the meditation depth questionnaire (MEDEQ). J. cand Meditation Res. **1**(1), 69–92 (2001)

29. Roeser, R.W., et al.: Mindfulness training and reductions in teacher stress and burnout: Results from two randomized, waitlist-control field trials. J. Educ. Psychol. **105**(3), 787 (2013)

30. Rosenzweig, S., Greeson, J.M., Reibel, D.K., Green, J.S., Jasser, S.A., Beasley, D.: Mindfulness-based stress reduction for chronic pain conditions: variation in treatment outcomes and role of home meditation practice. J. Psychosom. Res. **68**(1), 29–36 (2010). https://doi.org/10.1016/j.jpsychores.2009.03.010

31. Ryan, C.O., Browning, W.D., Clancy, J.O., Andrews, S.L., Kallianpurkar, N.B.: Biophilic design patterns: emerging nature-based parameters for health and well-being in the built environment. Int. J. Architect. Res.: ArchNet-IJAR **8**(2), 62–76 (2014). https://doi.org/10. 26687/archnet-ijar.v8i2.436

32. Slater, M., Usoh, M., Steed, A.: Depth of presence in virtual environments. Presence: Teleoperators Virtual Environ. **3**(2), 130–144 (1994). https://doi.org/10.1162/pres.1994.3.2. 130

33. Tang, Y.-Y., et al.: Short-term meditation training improves attention and self-regulation. Proc. Nat. Acad. Sci. **104**(43), 17152–17156 (2007)

34. Thompson, E.R.: Development and validation of an internationally reliable short-form of the positive and negative affect schedule (PANAS). J. Cross-Cultural Psychol. **38**, 2 (2007)

Calculating Cognitive Augmentation – A Case Study

Ron Fulbright[(✉)]

University of South Carolina Upstate, Spartanburg, SC, USA
rfulbright@uscupstate.edu

Abstract. We are entering an era in which humans will increasingly work in partnership and collaboration with artificially intelligent entities. For millennia, tools have augmented human physical and mental performance but in the coming era of cognitive systems, human cognitive performance will be augmented. We are only just now beginning to define the fundamental concepts and metrics to describe, characterize, and measure augmented and collaborative cognition. In this paper, the results of a cognitive augmentation experiment are discussed and we calculate the increase in *cognitive accuracy* and *cognitive precision*. In the case study, cognitively augmented problem solvers show an increase of 74% in cognitive accuracy—the ability to synthesize desired answers —and a 27% increase in cognitive precision—the ability to synthesize only desired answers. We offer a formal treatment of the case study results and propose cognitive accuracy and cognitive precision as standard metrics to describe and measure human cognitive augmentation.

Keywords: Cognitive augmentation · Cognitive accuracy ·
Cognitive precision · Information theory · Representational information ·
Cognitive work · Cognitive systems · Cognitive computing

1 Introduction

With recent advances in artificial intelligence (AI) and cognitive systems (cogs), we are at the beginning of a new era in human history in which humans will work in partnership with artificial entities capable of performing high-level cognition rivaling or surpassing human cognition. The new era will see human cognitive performance augmented by working with such artificial entities—*human cognitive augmentation*. Needed is a way to measure how cognitively augmented a human is by virtue of working with artificial entities. We propose two new metrics: *cognitive accuracy* and *cognitive precision*. Cognitive accuracy measures the ability to produce a desired result. Cognitive precision measures the ability to produce only the desired result and not any undesired results.

We can envision the situation as shown in Fig. 1. Together, a human and an artificial entity work together, forming a virtual entity, with the goal of performing a cognitive task—the transformation of information in an input form to a desired output form. We can compare this situation to a human performing all cognition alone without the help of an artificial entity. The human involved in collaboration with the cog is

© Springer Nature Switzerland AG 2019
D. D. Schmorrow and C. M. Fidopiastis (Eds.): HCII 2019, LNAI 11580, pp. 533–545, 2019.
https://doi.org/10.1007/978-3-030-22419-6_38

cognitively augmented and therefore able to perform at a higher level than the human working alone. However, the question is how can we measure the degree of cognitive augmentation?

The result of any cognitive process can be either the desired result (or close to it) or an undesired result. We define *cognitive accuracy* (C_A) as the propensity to produce the desired result. We define *cognitive precision* (C_P) as the propensity to not produce something other than the desired result. Note, these are not necessarily equivalent to "correct" and "incorrect" results. Often, the result of cognitive processing cannot be labeled as correct or incorrect. For example, asking a person what things in life are important to them will generate a number of answers. It is not possible to determine if one of those answers is correct and the rest incorrect. However, we can identify a particular answer as being the one we desire. Once we have chosen the target, we can calculate accuracy and precision of any set of answers relative to the target.

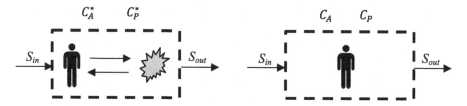

Fig. 1. In the coming era, human cognitive performance will be enhanced by partnering with artificially intelligent entities. The human/artificial ensemble will achieve a higher cognitive accuracy and cognitive precision than a human working alone.

This paper presents the results of a case study in which students were given an innovation problem and asked to synthesize as many solutions as they could think of in a period of time. Some students received expert advice in the form of suggested concepts pertaining to a class of preferred solutions. Other students received no assistance. Students receiving the expert advice represent cognitively augmented humans and students receiving no assistance represent non-augmented humans. Results show cognitively augmented students achieved a 74% increase in cognitive accuracy and a 27% increase in cognitive precision.

2 Previous Work

The idea of enhancing human cognitive ability with artificial systems is not new. In the 1640s, mathematician Blaise Pascal created a mechanical calculator called the Pascaline [1]. Thousands of years before this, mechanical devices such as the abacus aided basic arithmetic operations. Throughout history, humans have created thousands of such devices. Using these devices, a human is able to perform mathematical calculations difficult or impossible for an unaided human. These devices augment human mental performance, but the human still does all the thinking.

In the 1840s, Ada Lovelace was among the first to envision a machine performing a human task—musical composition [2, 3]. Lovelace imagined the machine composing the music, not a machine enhancing a human's ability to compose music. However, ideas like this were a century before their time. In the 1940s, Bush envisioned a system called the Memex and discussed how employing associative linking could enhance a human's ability to store and retrieve information [4]. Similar to the above-mentioned calculating devices, the Memex made the human more efficient but did not actually do any of the thinking on its own.

In 1950, Turing discussed whether or not machines themselves could think and offered the "imitation game" as a way to decide if a machine is exhibiting intelligence [5]. Since coining the phrase artificial intelligence (AI) in 1955, several generations of researchers have sought to create an artificial system capable of human-like intelligence [6]. Also in the 1950s, Ashby coined the term *intelligence amplification* maintaining human intelligence could be synthetically enhanced by increasing the human's ability to make appropriate selections on a persistent basis [7]. But here again, the idea is the human does all of the thinking. The synthetic aids just make the human more efficient.

In the early 1960s, Engelbart and Licklider envisioned human/computer symbiosis. Licklider imagined humans and computers becoming mutually interdependent, each complementing the other [9]. However, Licklider envisioned the artificial aids merely assisting with the preparation work leading up to the actual thinking which the human would do. Engelbart's H-LAM/T framework described the human as a part in a multicomponent human/computer system allowing human and artificial systems to work together to perform problem-solving tasks [8]. Through the work of Engelbart's Augmentation Research Center, and other groups in the 1950s and 1960s, many of the devices we take for granted today were invented as "augmentation" tools including: the mouse, interactive graphical displays, keyboards, trackballs, WYSIWYG software, email, word processing, and the Internet. However, while making it easier for the human to think and perform, none actually do any of the thinking themselves. In essence, these are just modern versions of devices aiding human mental activity.

Recently, researchers have discussed entities capable of performing cognition on their own. One branch of AI has sought to develop semi-autonomous *intelligent software agents* to act on behalf of a user or other program [10, 11]. These agents are designed to interact as if they were human but also perform on their own without supervision from the human user. The concept of the agent—a self-contained, interactive and concurrently executing object, possessing internal state and communication capability—can be traced to the Hewitt's Actor model [12]. Software agents act autonomously and only occasionally communicate with the human user. However, the field of *human-autonomy teaming* has studied real-time interaction between humans and artificial systems. One active area involves military applications. Constraints must be met though because combat requires systems to respond rapidly and efficiently while attaining mission objectives [13]. One goal of this research is an Autonomous Squad Member where a human squad member, either in a military or law enforcement setting, is assisted by an autonomous agent in mission environments [14]. NASA has researched the idea of having fewer human operators on long space flights by using artificial intelligence [15, 16].

However, these are all highly-specialized applications. What about the average person? Thirty years ago, Apple, Inc. envisioned an intelligent assistant called the Knowledge Navigator [17]. The Knowledge Navigator was an artificial executive assistant capable of natural language understanding, independent knowledge gathering and processing, and high-level reasoning and task execution. The Knowledge Navigator was envisioned as a colleague anyone could work with. The Knowledge Navigator concept was well ahead of its time, however, some of its features are seen in today's voice-controlled "digital assistants" such as Siri, Cortana, and Amazon Echo.

More recently, [18] described companion cognitive systems as software collaborators helping their users work through complex arguments, automatically retrieving relevant precedents, providing cautions and counter-indications as well as supporting evidence. Companions assimilate new information, generate and maintain scenarios and predictions, and continually adapt and learn about the domains they are working in, their users, and themselves. Companions, operate in a limited domain however achieve expert-level performance in that domain, often exceeding that of a human expert.

[19] challenged the cognitive systems research community to develop a synthetic entertainer, a synthetic attorney, and a synthetic politician as a way to drive future research in integrated cognitive systems. The vision here is to develop a virtual human. We maintain the goal should be not to create a virtual human capable of being an entertainer, an attorney, or a politician, but rather create a cognitive system capable of expert-level performance in entertainment, a different cognitive system capable of exhibiting expert performance in a subfield of law, and a cognitive system capable of expert politicking.

[41–45] foresees the creation a cognitive system for virtually any human endeavor and maintains the democratization of expertise will change the way we live, work, and play over the next several decades much like the computer and Internet have changed our lives over the last few decades.

A significant step toward this vision occurred in 2011, when a cognitive computing system built by IBM, called Watson, defeated two of the most successful human *Jeopardy!* champions of all time [20]. Watson received clues in written natural language and gave answers in natural spoken language. Watson's answers were the result of searching and deeply reasoning about millions of pieces of information and aggregation of partial results with confidence ratios. Watson was not programmed to play *Jeopardy!* Instead, Watson was programmed to *learn* how to play *Jeopardy!* which it did in many training games with live human players before the match [21, 22]. Watson *practiced* and achieved expert-level performance within the narrow domain of playing *Jeopardy!* Watson represents a new kind of computer system called a cognitive system [3, 23].

Instead of replacing humans, cognitive systems seek to act as partners with and alongside humans. John Kelly, Senior Vice President and Director of Research at IBM describes the coming revolution in cognitive augmentation as follows [24]:

> *The goal isn't to... replace human thinking with machine thinking. Rather...humans and machines will collaborate to produce better results – each bringing their own superior skills to the partnership. The machines will be more rational and analytic – and, of course, possess encyclopedic memories and tremendous computational abilities. People will provide judgment, intuition, empathy, a moral compass and human creativity."*

Cognitive systems, whether or not they are human or artificial, process and transform information. To measure and characterize cognition then, we must consider how to measure information and quantify information processing. Most information content metrics devised so far key off of the structure of the information being processed. In 1948, Claude Shannon developed the basis for what has become known as *information theory* [25–27]. Shannon, like Hartley before him, equates order/disorder and information content. In the 1960's, Solomonoff, Chaitin, Kolmogorov and others developed the concept of *algorithmic information theory* (Kolmogorov-Chaitin complexity) as a measure of information [28–31]. The algorithmic information content, *I*, of a string of symbols, *w*, is defined as the size of the minimal program running on the universal Turing machine generating the string. This measure of information concerns the complexity of a data structure as measured by the amount of effort required to produce it. A string with regular patterns can be "compressed" and produced with fewer steps than a string of random symbols which requires a verbatim listing symbol by symbol. Like the entropic measures described above, this description equates order/disorder to information content, although in a different manner by focusing on the computational resources required. In 1990, Stonier suggested an exponential relationship between entropy, *S*, and information [32–34]. Stonier also maintained information content is dependent on the *structure* present and uses Shannon's entropy to provide the measure of that structure.

Vigo has recently proposed a new kind of information theory, *generalized representational information theory* (GRIT) [35, 36]. Key to GRIT is how humans learn concepts from information. Empirical evidence shows human concept extraction is based on the detection of patterns—invariance in the information. Vigo's *generalized invariance structure theory* (GIST) maintains it is easier to extract a concept from information with less variance (more similarity between elements) than it is from information with a more variance (less similarity between elements). A key measure of information is then the *structural complexity*—a different measure of structure than Shannon entropy. Values calculated with GIST formulae agree with empirical evidence from human trials.

Fulbright has proposed several metrics based on GRIT and GIST for describing cognitive processing [37–39]. Cognitive work, is an accounting of all changes in structural complexity caused by a transformation of information. Cognitive work is a measure of the total effort expended in the execution of a cognitive process. When humans (H) and artificial entities, or cogs, (C) work together, each are responsible for some amount of change (cognitive gain (G)) and each expend a certain amount of cognitive work (W):

$$W^* = W_H + W_C \qquad G^* = G_H + G_C \qquad (1)$$

Given that we can calculate the individual cognitive contributions, it is natural to compare their efforts. In fact, doing so yields the *augmentation factor*, A^+:

$$A_W^+ = \frac{W_C}{W_H} \qquad A_G^+ = \frac{G_C}{G_H} \qquad (2)$$

Note humans working alone without the aid of artificial entities are not augmented at all and have an $A^+ = 0$. If humans are performing more cognitive work than artificial entities, $A^+ < 1$. This is the world in which we have been living so far. However, when cogs start performing more cognitive work than humans, $A^+ > 1$ with no upward bound. That is the coming cognitive systems era. Fulbright defines other efficiency metrics by comparing cognitive gain and cognitive work to each other and to other parameters such as time, t, and energy, E.

However, all of these metrics focus on the microscopic features of the information itself—structure. As such, they are based on quantities difficult, if not possible, to measure and calculate. This renders these metrics interesting conceptually but useless in a practical sense. In this paper, we discuss two macroscopic metrics: *cognitive accuracy* and *cognitive precision*. Instead of focusing on characteristics of the information, these two metrics focus on the results of the cognitive processing making them easy to calculate and useful across multiple domains.

3   Cognitive Accuracy and Cognitive Precision

To define the notions of *cognitive accuracy* and *cognitive precision*, we first model the students in the class with a general information machine (GIM) [40]. A GIM is a stochastic Turing machine accepting information as an input and producing information as an output. In a traditional Turing machine, rules specify symbol transitions dictating a deterministic transformation of an input to a specific output. However, the

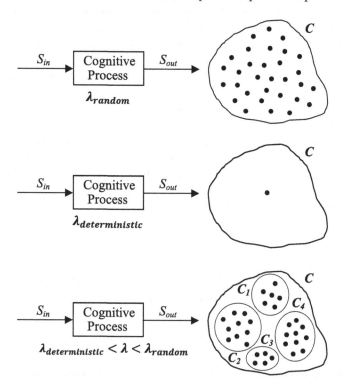

Fig. 2. Cognitive processing modeled as stochastic manipulation of information.

rules in a GIM are stochastic in nature. Each transition rule is associated with a probability rather than being a deterministic certainty. Therefore, for a given input, the GIM's output may vary with each run. Over a number of runs, given the same input, a set of outputs (C) is created. The pattern of outputs in C is determined by the randomness of the probabilities within the GIM and denoted as λ as shown in Fig. 2.

If the GIM is truly random, the outputs in C are evenly distributed with the average probability of each output, c, being $1/|C|$ where $|C|$ is the cardinality of C or simply the number of different outputs. If the GIM is truly deterministic (such as that of a traditional Turing machine), one and only one output will be generated 100% of the time. Of course, the probability of that output is 100% and the probability of any other possible output is zero. If, however, the randomness of the GIM is an intermediate value a pattern of outputs will emerge. Some of these outputs will be very similar to other outputs and can be grouped together into subsets of C. The probability of a subset can be calculated by comparing the cardinality of the subset with the cardinality of C.

Critical is the distribution pattern of C. If we choose one of the subsets in C as the preferred type of output we can characterize any distribution pattern based on the ideas of *accuracy* and *precision* as shown in Fig. 3. Accuracy involves the propensity to hit the preferred subset. Precision involves the propensity to hit *only* the preferred subset. The goal, of course, is for every output to fall within the preferred subset (upper right quadrant). This represents high accuracy and high precision. It is possible for outputs to be very similar to each other (forming a tight cluster) but not falling within the preferred subset (lower right quadrant). This represents high precision but low accuracy.

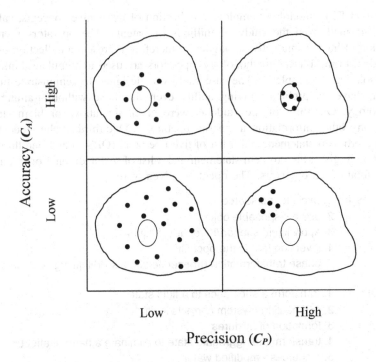

Fig. 3. Precision and accuracy relative to a target subset.

Outputs centered on the preferred subset but not tightly clustered (upper left quadrant) represents high accuracy but low precision. Outputs with low accuracy and low precision (lower left quadrant) have only accidental relationship to the preferred subset.

Since the outputs in the model are the result of cognitive processing, we call these two measures *cognitive accuracy* (C_A) and *cognitive precision* (C_P).

4 The Case Study

For the case study, an innovation problem was given to a classroom of students registered into the INFO 307: Systematic Innovation course at the University of South Carolina Upstate. INFO 307 teaches students an innovation methodology called I-TRIZ. This test was done on the first day of two consecutive semesters before any instruction took place. This was done so as to not bias any results with the effect of learning the methodology. One semester consisted of 25 students and the second semester consisted of 21 students. Students were given ten minutes to write down as many solutions as they could think of to the following problem:

> **Problem:** *Skeet shooting is a recreational and competitive activity where participants, using shotguns, attempt to break clay targets mechanically flung into the air from fixed stations at high speed from a variety of angles. The problem is the shattered skeet litter the grass field harming the grass. As you know, grass needs sun, water, and nutrients to be healthy. The skeet fragments prevent water and sun from reaching the grass and diminish the healthy nutrients in the soil after they eventually dissolve.*

The I-TRIZ methodology employs a collection of innovative concepts, called *operators*, gleaned from the study of millions of patents. The operators represent a distillation of human innovative thought and therefore represent a collection of expert knowledge. In the I-TRIZ methodology, operators are used to stimulate thinking and inspire solutions to problems. The case study was designed to demonstrate the effect operators have on problem solving ability even for those without training in the methodology. One-third of the students were given the above problem statement without any other information at all (no operators). One-third of the students were given the problem statement and a list of five operators (OPS 1) and one-third of the students were given the problem statement and a list of five additional operators (OPS 2) for a total of ten operators. The operators given were:

OPS 1 = 1. exclude the source
2. use a disposable object
3. apply liquid support/Introduce a liquid
4. inversion (apply the opposite)
5. phase transformation (freeze/melt/boil; solid/liquid/gas/plasma)

OPS 2 = 1. transform a substance to a fluid state
2. self-healing (system corrects itself)
3. formation of mixtures
4. transform the aggregate state to eliminate a harmful effect
5. resources - modified water

These operators were chosen for a reason. Based on previous experience, solutions to this problem generally fall into three distinct categories:

F:→ modifying the field/cleaning up the field
T:→ modifying the target (skeet)
G:→ modifying the gun or bullet.

Solutions in the field category (F) include: covering the field with a tarp or net, various ways of cleaning up the field, and relocating to a location without a grass field. Solutions in the target category (T) include: using targets made of biodegradable material, using targets made of fertilizer and other nutrients good for the grass, and enhancing the rapid dissolvability of the targets. Solutions in the gun category (G) include: using a different kind of gun or bullet, and using a laser or electromagnetic gun/target.

For the case study, the modifying the target (T) category of solutions was chosen as the preferred type of solution because solutions in the other two categories are considered obvious solutions (the first ones most people think of off the top of their head). Operators in OPS1 and OPS2 shown above were chosen with the intent of driving student thinking toward solutions in the T category.

The results of the case study are shown in Fig. 4. A total of 128 solutions were created by the students. Of course, there were many duplicative solutions. In all, 11 different solutions were synthesized by the class over the two semesters. Each of the 128 solutions turned in by the students were classified into one of the three solution categories and the simple percentage for each category was calculated.

Students receiving only the problem statement and no operators produced solutions in the preferred subset just over one quarter of the time for an accuracy of, $C_A = 27\%$. Students receiving five operators (OPS 1) produced solutions in the preferred subset with an accuracy of $C_A = 40\%$. Students receiving ten operators (OPS 1 + OPS 2) achieved an accuracy of $C_A = 47\%$. Therefore, the operators increased the students' cognitive accuracy by $(40\% - 27\%)/27\% = 48\%$ (for OPS1) and $(47\% - 27\%)/27\% = 74\%$ (for OPS 1 + OPS 2).

Non-augmented students tended to focus on the modifying the field (F) type of solution (61% of the solutions). Together with the modifying the gun/bullet (G) type of solution (12%), non-augmented students therefore missed the preferred subset 73% of the time. However, students augmented with the operators created far fewer solutions outside of the preferred subset: $52\% + 8\% = 60\%$ (for OPS 1) and $43\% + 10\% = 53\%$ (for OPS 1 + OPS 2). Therefore, augmented students achieved a reduction of the miss-rate $(60\% - 73\%)/73\% = -18\%$ (for OPS 1) and $(53\% - 73\%)/73\% = -27\%$ (for OPS 1 + OPS 2) representing an *increase in cognitive precision*.

Students were cognitively augmented by the operators. Even without formal training in the IPS methodology, students were able to create solutions of the preferred (non-obvious) type far in excess of students solving the problem without operators.

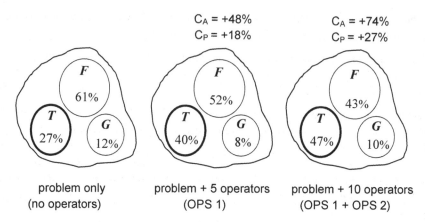

Fig. 4. The cognitive augmentation effect of operators on problem solving is demonstrated relative to the preferred type of solution *T*. Cognitive accuracy (C_A) was increased by 48% and 74% respectively. Cognitive precision (C_P) was increased by 18% and 27% respectively.

5   Conclusion

In the coming era of cognitive systems, cognition will be the result of human collaboration with artificially intelligent entities (cogs). Cogs will think on their own and offer expert advice to the human. The human will consider the expert advice and use it to enhance the quality of his or her own thinking. Together, the biological/artificial collaboration will yield a cognitive product greater than each of the entities could have produced on their own. Humans will be *cognitively augmented.* We have proposed two new metrics to describe the result of such cognitive augmentation. *Cognitive accuracy* is the ability of the human/artificial ensemble to create the desired output. *Cognitive precision* is the ability of the ensemble to create nothing but the desired output.

The results of a case study were presented. In the case study, student problem solving was augmented by considering problem solution concepts called operators. The cognitive accuracy of the cognitively augmented students was increased by as much as 74% meaning they were significantly more likely to create solutions of the preferred type. Furthermore, the cognitive precision of the group was increased by 27% meaning the students were less likely to create solutions not of the preferred type.

We believe the metrics of cognitive accuracy and cognitive precision are two ideas applicable to any study of cognition, biological or artificial. Since these ideas rely only on the results of cognitive processing and do not rely on the structure of the information involved nor on any particulars of how the cognition is performed these metrics can be used to compare and contrast results across multiple domains.

References

1. Chapman, S.: Blaise Pascal (1623–1662) Tercentenary of the calculating machine. Nature **150**, 508–509 (1942)
2. Hooper, R.: Ada Lovelace: my brain is more than merely mortal (2012). New Scientist https://www.newscientist.com/article/dn22385-ada-lovelace-my-brain-is-more-than-merely-mortal. Accessed Nov 2015
3. Isaacson, W.: The Innovators: How a Group of Hackers, Geniuses, and Geeks Created the Digital Revolution. Simon & Schuster, New York (2014)
4. Bush, V.: As we may think. Atlantic **176**, 101–108 (1945)
5. Turing, A.: Computing machinery and intelligence. Mind **LIX**(236), 433 (1950)
6. McCarthy, J., Minsky, M., Rochester, N., Shannon, C.: A proposal for the dartmouth summer research project on artificial intelligence (1955). Stanford University http://www-formal.stanford.edu/jmc/history/dartmouth/dartmouth.html. Accessed Nov 2015
7. Ashby, W.R.: An Introduction to Cybernetics. Chapman and Hall, London (1956)
8. Engelbart, D.C.: Augmenting Human Intellect: A Conceptual Framework, Summary Report AFOSR-3233, p. 1962. Stanford Research Institute, Menlo Park (1962)
9. Licklider, J.C.R.: Man-Computer Symbiosis. IRE Trans. Hum. Factors Electron. **1**, 4–11 (1960)
10. Nwana, H.: Software agents: an overview (1996). https://teaching.shu.ac.uk/aces/rh1/elearning/multiagents/introduction/nwana.pdf. Accessed 9 Oct 2018
11. Schermer, B.W.: Software agents, surveillance, and the right to privacy: a legislative framework for agent-enabled surveillance. Leiden (2007). https://openaccess.leidenuniv.nl/bitstream/handle/1887/11951/Thesis.pdf. Accessed 9 Oct 2018
12. Hewitt, C., Bishop, P., Steiger, R.: A universal modular actor formalism for artificial intelligence (1973). WorryDream http://worrydream.com/refs/Hewitt-ActorModel.pdf. Accessed 9 Oct 2018
13. Barnes, M., Chen, J.Y.C., Hill, S.: Humans and autonomy: implications of shared decision-making for military operations (2017). Army Research Laboratory http://www.arl.army.mil/arlreports/2017/ARLTR-7919.pdf. Accessed 6 Oct 2018
14. Chen, J.Y.C., Stowers, K., Barnes, M.J., Selkowitz, A.R., Lakhamni, S.G.: Human-autonomy teaming and agent transparency. In: HRI 2017 Proceedings of the Companion of the 2017 ACM/IEEE International Conference on Human-Robot Interaction. Association for Computing Machinery, March 2017
15. Shivley, R.J., Bandt, S.L., Lachter, J., Matessa, M., Sadller, G., Battise, V.: Application of human autonomy teaming (HAT) patterns to reduce crew operations (RCO) (2016). National Aeronautics and Space Administration https://ntrs.nasa.gov/archive/nasa/casi.ntrs.nasa.gov/20160006634.pdf. Accessed 17 Sept 2018
16. Shivley, R.J., et al.: Why human autonomy teaming (2018). ResearchGate https://www.researchgate.net/publication/318182279_Why_Human-Autonomy_Teaming. Accessed 9 Oct 2018
17. Apple (1987). Knowledge Navigator. YouTube https://www.youtube.com/watch?v=JIE8xk6Rl1w. Accessed Apr 2016
18. Forbus, K., Hinrichs, T.: Companion cognitive systems: a step toward human-level AI. AI Mag. **27**(2) (2006)
19. Langley, P.: Three challenges for research on integrated cognitive systems. In: Proceedings of the Second Annual Conference on Advances in Cognitive Systems (2013)

20. Jackson, J.: IBM watson vanquishes human Jeopardy Foes (2011). PC World http://www.pc world.com/article/219893/ibm_watson_vanquishes_human_jeopardy_foes.html. Accessed May 2015
21. Ferrucci, D.A.: Introduction to "This is Watson". IBM J. Res. Dev. **56**(3/4) (2012)
22. Ferrucci, D., et al.: Building watson: an overview of the DeepQA project. AI Mag. **31**(3), 59–79 (2010)
23. Wladawsky-Berger, I.: The era of augmented cognition (2013). Wall Street J.: CIO Reporthttp://blogs.wsj.com/cio/2013/06/28/the-era-of-augmented-cognition/. Accessed May 2015
24. Kelly, J.E., Hamm, S.: Smart Machines: IBMs Watson and the Era of Cognitive Computing. Columbia Business School Publishing, New York (2013)
25. Hartley, R.V.L.: Transmission of information. Bell Syst. Tech. J. **7**(1928), 535–563 (1928)
26. Shannon, C.E.: A mathematical theory of communication. Bell Syst. Tech. J. **27**(3), 379–423 (1948)
27. Weaver, W., Shannon, C.E.: The Mathematical Theory of Communication. University of Illinois Press, Urbana (1949)
28. Chaitin, G.J.: On the length of programs for computing finite binary sequences. J. Assoc. Comput. Mach. **13**(1966), 547–569 (1966)
29. Chaitin, G.J.: Algorithmic information theory. IBM J. Res. Dev. **21**, 350–359 (1977)
30. Kolmogorov, A.N.: Three approaches to the quantitative definition of information. Probl. Inf. Transm. **1**(1965), 1–17 (1965)
31. Solomonoff, R.J.: A formal theory of inductive inference. Inf. Control **7**, 1–22 (1964)
32. Stonier, T.: Information and the Internal Structure of Universe. Springer, London (1990)
33. Stonier, T.: Beyond Information: The Natural History of Intelligence. Springer, London (1992). https://doi.org/10.1007/978-1-4471-1835-0
34. Stonier, T.: Information and Meaning: An Evolutionary Perspective. Springer, Berlin (1997). https://doi.org/10.1007/978-1-4471-0977-8
35. Vigo, R.: Complexity over uncertainty in generalized representational information theory (GRIT). Information **4**, 1–30 (2013)
36. Vigo, R.: Mathematical Principles of Human Conceptual Behavior. Psychology Press, New York (2015). ISBN 978-0-415-71436-5
37. Fulbright, R.: How personal cognitive augmentation will lead to the democratization of expertise. In: Fourth Annual Conference on Advances in Cognitive Systems, Evanston, IL, June 2016. http://www.cogsys.org/posters/2016. Accessed Jan 2017
38. Fulbright, R.: Cognitive augmentation metrics using representational information theory. In: Schmorrow, Dylan D., Fidopiastis, Cali M. (eds.) AC 2017. LNCS (LNAI), vol. 10285, pp. 36–55. Springer, Cham (2017). https://doi.org/10.1007/978-3-319-58625-0_3
39. Fulbright, R.: On measuring cognition and cognitive augmentation. In: Yamamoto, S., Mori, H. (eds.) HIMI 2018. LNCS, vol. 10905, pp. 494–507. Springer, Cham (2018). https://doi.org/10.1007/978-3-319-92046-7_41
40. Fulbright, R.: Information domain modeling of emergent systems. Technical report CSCE 2002–2014, May 2002. Department of Computer Science and Engineering, University of South Carolina, Columbia, SC (2002)
41. Fulbright, R.: On measuring cognition and cognitive augmentation. In: HCI International 2018, Las Vegas, July 2018
42. Fulbright, R.: Cognitive augmentation metrics using representational information theory. In: HCI International 2017, Vancouver, July 2017

43. Fulbright, R.: ASCUE 2067: how we will attend posthumously. In: Proceedings of the 2017 Association of Small Computer Users in Education (ASCUE) Conference, June 2017
44. Fulbright, R.: How personal cognitive augmentation will lead to the democratization of expertise. In: The Fourth Annual Conference on Advances in Cognitive Systems, June 2016
45. Fulbright, R.: The cogs are coming: the coming revolution of cognitive computing. In: Proceedings of the 2016 Association of Small Computer Users in Education (ASCUE) Conference, June 2016

Designing an Interactive Device to Slow Progression of Alzheimer's Disease

Ting-Ya Huang$^{(\boxtimes)}$, Hsi-Jen Chen, and Fong-Gong Wu

Department of Industrial Design, National Cheng Kung University,
No. 1, University Road, Tainan City 701, Taiwan
sum84olc4@gmail.com

Abstract. Inactive sedentary working styles result in many people exercising insufficiently, leading to poor blood circulation and decreased blood flow of the brain. Consequently, amyloid beta (Aβ) accumulates in the brain, impairing nerve conduction and potentially causing nervous necrosis, both of which cause Alzheimer's disease (AD). The early symptoms of Alzheimer's disease include inability to recognize orientation in space and memory decline, along with amyotrophic Hippocampus, and low Neurotrophic content and Cerebral blood flow. Brain-derived neurotrophic factor (BDNF) is a crucial component for brain nerve growth, abundant BDNF alleviates nervous necrosis. An environment with BDNF secretion is directly proportional to cerebral blood flow. Research has indicated that engaging in puzzle games and finger motor exercises can mitigate and ameliorate Alzheimer's disease. Performing tasks with the fingers make extensive use of the brain's motor and somatosensory cortices, thus promoting activity and increased blood flow in the brain to generate more BDNF facilitating brain neuron growth. The goal of this research is to increase cerebral circulation, through patient-oriented activity, to elucidate which activities increase cerebral circulation, and sustained local motion enhancing muscle contraction that increases blood flow velocity. These three game guidelines are devised to improve spatial cognition and increase cerebral circulation.

Keywords: Alzheimer disease (AD) · Cerebral blood circulation · Brain-derived neurotrophic factor (BDNF) · Spatial cognition

1 Introduction

1.1 Background

According to Alzheimer's Disease International (ADI) data from 2017, it is estimated that on average every 4 s one person will suffer Dementia onset, with Alzheimer's disease (AD) accounting for the majority of Dementia sufferers (about 60–80%) [1].

Contemporary eating habits combined with long-term sedentary work patterns along with slowed metabolism in the middle-aged and elderly and lack of exercise, lead to increasing prevalence of suffering from Alzheimer's disease. As cerebral disease is often complicated and considered dangerous by patients, and cannot be easily detected, or when detected there is already irreversible nerve damage; therefore, cerebral health is often ignored. Especially in the senior years, when the body's metabolism is

© Springer Nature Switzerland AG 2019
D. D. Schmorrow and C. M. Fidopiastis (Eds.): HCII 2019, LNAI 11580, pp. 546–562, 2019.
https://doi.org/10.1007/978-3-030-22419-6_39

gradually slowing, the lack of exercise, resulting in poor blood circulation, affecting and resulting in slow cerebral blood flow and vasoactivity, causes the brain to aggregate unnecessary substances, and finally results in pathological cerebral states and disease. The brain is one of the most important human organs, and maintaining cerebral health and an active lifestyle are keys to preventing senescence and enjoying longevity in the aging society [2, 3].

In today's mainstream allopathic medicine, treatments and prevention methods are divided into pharmacological and non-pharmacological treatment approaches, pharmacological treatment aims to increase Acetylcholinesterase inhibitors to reduce cerebral hypoxia, while for improving patient status through invasive methods, the price of medicine is not only expensive, but patient willingness to take medicine for a long time is also not high [4]. Non-pharmacological preventive treatment includes regular exercise, reminiscence therapy and music therapy, through speech communication or improvement of emotions, and thus slowing progressive degeneration [5, 6]. This reduces senior citizens' needs for pharmacological treatment, hence, it is extremely important to improve the prevention of neurodegenerative conditions among senior citizens through performing simple activities.

This study aims to promote people's activities in a simple and convenient way, to explore the etiology of Alzheimer's disease, to analyze the causes of cognitive impairment, to consider from which point of view we can effectively prevent AD, enhance senior citizens' willingness for engaging in activities, and promote cerebral vascular circulation, increasing brain-derived neurotrophic factor (BDNF), analyzing how the holistic approach accelerates cerebral vascular circulation and brain activation, and consider human factors designs to develop a suitable human-computer interaction designed device to slow down the increasing rate of the AD population, as well as reduce the future mental, economic and social pressures on primary and secondary caregivers.

1.2 Research Purposes

This study requires senior citizen interaction including the needs of senior citizens and their activities in device development and research. The main purpose of this study is to develop a set of interactive devices with elderly preferences and suitability by understanding the causes of cognitive function and memory degeneration caused by AD.

During the process of the research, this study will accompany senior citizens to deploy several smart games on the market and explore the relevant literature on improving AD. The activities and difficulties can be used as evaluation experiments for future device design. Test results and literature review organization will specifically summarize the useful parameters of the senior citizens in the smart game activities that can be used as a reference for the development of wearable evaluation devices in the future.

The main purposes of this study are summarized in these 3 following points: (1) Analysis of the relationship between puzzle/smart games on the market and prevention of AD. (2) Exploring the etiologies influencing cognitive function and neurodegeneration as the theoretical basis for the design of the evaluation device. (3) Focusing on the design of interactive devices and evaluation devices for promoting cerebral vascularity and non-sedentary activities.

It is hoped that this study will elucidate the differences in the games for reducing memory degeneration and cognitive training, elicit the learnability of the products and needs for improvement, and try to integrate the medical operations of cognitive impairment and memory degeneration as referents in future design.

2   Related Work

2.1   Alzheimer's Disease

In the initial phase of AD, the temporal lobe's entorhinal cortex and hippocampus nerve cells are damaged resulting in inability to form new memory. With the aggregation of amyloid beta (Aβ) and it's slowly spreading throughout the brain, the affected brain areas extend wider from the hippocampus to the temporal lobe to cause dysphasia and further influence the frontal lobe, affecting cognition, logic and judgment for decision-making ability. In the late phase of AD, there will occur cerebral atrophy and neurodegeneration, and brain functions will be affected as the patient loses ability for independent living [7].

Harm of Aβ to nerve cells:

(1) Etiology of Brain Aβ:

Aβ will aggregate into a senile plaque, over 10 to 20 years, Tau protein will be coagulated in the nerves, causing neurofibrosis, blocking neuro processes or links, and resulting in neural twisting, causing nerve cell atrophy and death [8, 9].

(2) Tau Neurofibrillary Tangles:

Aβ oligomers attached on the nerve cell membrane may activate Phosphorylase kinase, causing Tau proteins excessive phosphorylation and leaving the microtubules, while the microtubules without Tau protein stability contribution leads to disintegration, aggregation into neurofibrillary tangles, with cell body substances unable to be transported to the axon distal end, resulting in Neuronal synapse atrophy and degeneration [10].

(3) Generate Reactive Oxygen and Free Radicals:

Aβ binds to the scavenger receptor of cerebrovascular epithelial cells and produces reactive oxygen via NADPH oxidase [11], which reduces brain blood flow, promotes Aβ aggregation, and forms cerebral amyloid angiopathy (CAA). Aβ also reduces the efficiency of mitochondrion to produce ATP, and increases Reactive oxygen generation [12].

(4) Causes Cerebrovascular Constriction:

The mechanism eliminating Aβ is the blood-brain barrier (BBB) and if it is blocked, there will be increasing Aβ aggregation in the blood vessels, inducing cerebrovascular constriction and reducing brain blood flow [13] (Fig. 1).

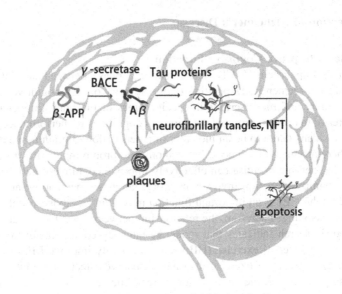

Fig. 1. Formation process of Alzheimer's disease

2.2 Importance of Cerebral Vascular Blood Circulation to Prevent Alzheimer's Disease

The brain is the most metabolically active organ in the human body. Because the energy reserves of the brain are scarce, it is extremely important to continuously provide nutrient energy to the brain. The brain must rely on circulation to maintain the supply of nutrients and oxygen to effectively remove dysfunction from metabolic waste. AD is a disease that may cause neurovascular dysfunction and aging to form neurodegeneration [14]. Cerebrovascular injury promotes aggregation of AD Aβ in the brain, and Aβ aggregation leads to BBB blockage in the brain of those with Alzheimer's disease. AD promotes inflammation and cytotoxicity, so cerebrovascular destruction can cause damage to the blood-brain barrier, causing the neurotoxic cycle to directly trigger neuronal damage [15]. Vascular dysfunction also affects Aβ clearance, leading to elevated Aβ concentration in the brain [16].

In AD patients, the most important way to discharge Aβ is the blood-brain barrier. If the elimination rate is too slow, Aβ concentration will be deposited in the brain micro vascularity, which will cause the cerebral blood vessels to lose elasticity and damage the brain blood flow adjusting its function and forming a vicious neuropathological circle [17, 18]. Therefore, the discharge of Aβ reduces the concentration of the pathological or toxic substances, when the blood circulation is enhanced. The interaction between the aggregation and blood circulation to reduce Aβ is a non-negligible factor in the design of this product [13].

2.3 Prevention of Alzheimer's Disease

Exercise Helps Improve Cognitive Function

Mild cognitive impairment (MCI) is a transitional phase between normal cognitive function and mild Dementia. Depending on the severity of AD, cognitive impairment will become more and more serious in specific areas of the brain (hippocampus and prefrontal areas). In order to promote the region of cognitive task performance, cognitive degeneration is considered to be an indicator of deterioration of brain tissue and brain's function. There are many references that demonstrate promoting cardiovascular health and performing aerobic exercise can effectively improve cognitive function [19, 20].

Holistic cognitive function involves memory, language, perception, and executive function (EF), where EF is an important function in working memory, problem solving, and complex reasoning. Therefore, the relationship between executive function and AD is considered. Research indicates that moderate-intensity continuous exercise (MCE), and high-intensity interval exercise (HIIE) can effectively improve EF. In addition to aerobic exercise, resistance exercise also has a positive effect on cognitive function. Scholars also propose that the combination of resistance exercise and aerobic exercise is more than aerobic exercise alone or resistance exercise has a greater influence on cognitive function [21]. Aerobic exercise can improve several aspects of executive function (EF), especially inhibitory control (IC). Inhibitory control, as a cognitive process, involves the prefrontal cortex (PFC), and Caudate nucleus participation control with subthalamic nucleus. Inhibitory control (IC) can improve neural activities in the brain as induced by aerobic exercise and resistance exercise [22].

Exercise Promotes BDNF Secretion

Brain-derived neurotrophic factor (BDNF) acts on the Central Nervous System (CNS) and the Peripheral Nervous System (PNS) in the neuron, helping the neuron to grow and derive new neurons and synapses, particularly focusing on the hippocampus and cerebral cortex, so learning and memory are extremely important [23]. Many surveys show that people with higher education or those who need to work with the brain have lower rates of dementia, and environmental stimulation, learning or exercise require the brain's neurological action [24]. The aforementioned stimulation can promote nerve health by increasing the secretion and function of BDNF. If nerve cells increase or there are more synapse links between cells, when a synapse is blocked by Aβ aggregation, it can transmit information via other synapses. When the brain's blood flow is reduced, it will affect BDNF generation, and cause brain damage [25].

Exercise is the most effective method to reduce the risk of AD. The induction of aerobic exercise and resistance exercise can promote improved cognition, and also promote BDNF. The literature repletely demonstrates that physical exercise delays cognitive decline. A non-pharmaceutical intervention to stave off neurodegeneration, neuroprotective factors are not only associated with brain plasticity and neurobehavioral efficacy, but also have a great relationship with angiogenesis [21]. The reason for this is that exercise will increase the activity of neplysin, which decomposes Aβ. In addition to the goal of preventing AD Aβ aggregation, nerve cells can be made into BDNF to achieve neuroprotective effects [25, 26].

Finger Activities Stimulate the Brain

Neurosurgeon Wilder Penfield plotted a homunculus, which shows that the five fingers and the palm account for about one-third of the total neurotopology affected by the exercise area, and the somatosensory cortex constitutes about a quarter of the total. The Motor Cortex area is responsible for performing body movements. The function of the Somatosensory Cortex area is for feeling, so about one-third of the brain's field is used to control the fingers and hands, especially the fingertips which have many nerve cells connected in series [27]. Some studies have pointed out that finger stimulation specified tasks can increase blood flow in the brain and also find that in the finger touch recognition task, the sympathetic nerve will induce blood vessels constriction, resulting in changes in the fingertip blood flow [28, 29].

The three main functions of the nerve are sensory function, synthesis and command function and exercise function. The peripheral nerve contains sensory nerves and exercise nerves. When we feel an object through the sense of touch, we will receive the message through the sensory nerves and transmit it to the central nervous system (CNS). The exercise nerves complete a series of actions, so the finger is at the end of the peripheral sensory nerve. The starting point for transmitting the message is also the place where the nerve distribution is the mostly dense. In the task of tactile perception, it is equivalent to performing indirect brain nerve activities [30] (Fig. 2).

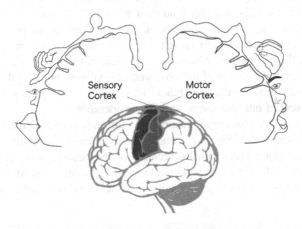

Fig. 2. Homunculus neurotopology

2.4 Prevention Device Design Guidelines and Principles for Alzheimer's Disease

Holistically organizing the above-mentioned studies on the causes and prevention of AD allow for discrimination in the following three directions:

- Exercise can promote brain's secretion of BDNF, training cognitive function to activate the brain
- Finger activities and tactile movements combine the operation of sensory nerves, exercise nerves and the brain to increase cerebral blood flow.

- Increase in cerebral blood flow can promote BDNF production and reduce Aβ aggregation.

This study used observation, literature review and invited experts in various fields to carry out a focus group, as well as considering how to advance cerebral blood circulation, types of senior citizen activities and market trend to discuss the design direction and development.

3   Research Method

The literature points out that brain's blood circulation is one of the key factors in preventing AD. This study clarifies the problem and establishes the goal of innovative design.

3.1   Participant Observation

This study focuses on the training of senior citizens for preventing AD; therefore, we observe the type of elderly activities in daily life at nursing homes, and learn about the interests of the elderly and their participation willingness for various activities, among which Dementia and people with mobility problems accounted for the majority. By interacting with the elderly to understand their lifestyles and interests, the participants' wishes and appropriateness were determined by observing the status of the activities, and asking the local caregivers to design activities and objectives of the activities, in order to obtain the most realistic scenarios, by maintaining the most natural context, reducing the interaction effects of the observed person, and reducing the preconceived views of the individual by the researcher. This study incorporates the life style and activities as observed into the future design considerations.

Problems found by observation are divided into five major items: (1) it is impossible to synchronize multiple tasks, such as simultaneous stepping and clapping. (2) When performing exercise, the correctness of the posture is not clear. (3) Trying new things is difficult. (4) High participation rate for things related to familiarity or reminiscence memory. (5) Low participation in difficult and complex activities (Fig. 3).

Fig. 3. Observation process

3.2 Focus Group

After carrying out the observation and literature review, the apparent problems and potential solutions were elicited and discussed by the Focus Group team. In order to facilitate the senior citizens' ease in getting started, we began by analyzing hand activities in daily life. The most commonly used actions are rotation, pressing and holding. Therefore, these three actions are included in the gestures of the game. As for spatial ability training, the space position training was based on the relative position of the azimuth positioning, distance estimation and mental rotation. Then we designed guidelines to think about the game modes of play.

The key points and principles are as follows: (1) The hand muscles' activities are promoted to facilitate the brain's blood circulation, and the gripping gestures are paired with the space training game mode suitable for synergetic participation, and the play pattern updates frequently. Putting an object into the correct matching position activates the memory to train the spatial relationship ability and the spatial sensory ability. (2) According to the research, the activity of the thumb effectively enhances brain blood circulation, and the button is used to make the thumb perform simple activities, such as in the mental arithmetic game mode. (3) Training for spatial positioning ability is most complex since the rotating gesture is the most complicated action, so the psychological rotation of spatial visual ability and the training of spatial organization ability are more suitable.

Features of future design: After discussion by the focus group, the training modes are divided into: the conjunction of spatial relationship ability and the psychological rotation of spatial visual ability to train the space ability game to promote hippocampus activities to promote the circulation of that brain area. The partial exercise mode: Combine the rotation and pressing of the hand to increase the frequency of the whole hand activity to promote brain blood circulation. For the main design goals, this study seeks to indirectly stimulate the brain with directly affected brain and local movements, to maximize the positive effect on brain blood circulation (Fig. 4).

Fig. 4. Focus group process

3.3 Design of Evaluation Device

This study primarily aims to increase the amount of hand activities to promote brain blood circulation as the main design content. Before the game content design, an evaluation device is needed to evaluate the finger activities and blood flow status. This study is designed with the Arduino board design, and a set of wearable gloves as the evaluation device. The device set includes Arduino, a bending sensor, and a pulse sensor module.

Arduino bending sensing is mainly used to detect the number of fingers bending. Through 10 bending sensors, the frequency of 10 fingers used in the game can be evaluated. The main structure can be divided into the information receiving end, and USB is connected to the computer to receive signals. As the computer software data is processed, the bending sensor will increase the resistance due to the bending, and the data will be converted into the number of instances of bending after the data is collected. The pulse sensor is clamped to the earlobe, and the pulse is a non-invasive detector. The Photoplethysmography (PPG) instrument itself is based on the calculation of the relative change of the blood flow using the reflection of light to calculate the pulse rate by Near Infrared Spectroscopy (NIRS). This study mainly uses the change of the blood flow state as the main indicator, so the process of estimating the pulse rate is required, and the blood flow is presented as a relative change waveform (Fig. 5).

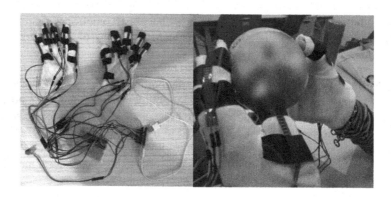

Fig. 5. Glove evaluation device

3.4   Pretest Experiment

In this study, the purpose of the pretest experiment is to test and evaluate whether the glove can effectively detect the number of activities of the finger and the blood flow value, and establish the relationship between the amount of activity and the brain's blood flow.

Experiment mode: (1) Using the current puzzle/smart game products (Tangram and Wisdom Ball) as the test game, use the Study design evaluation glove to calculate the finger activities frequency and blood flow status during the game. (2) For the blood flow part, measure the neck blood flow with a sophisticated Doppler Bi-mode Blood Flow Monitor, compare the measured data of the precision instruments on the market with the sensing ability of this study's wearable device, and detect it. The measured finger activity and blood flow data were statistically analyzed and explore the relationship between hand activity and blood flow.

Experiment Arrangement

In this experiment, we evaluated the effect of the wearable glove design in an objective way. We evaluate the performance of the device through the game of the puzzle, Wisdom Ball and Tangram. There were 5 testees all of whom are female, in the age range of 24-30 years old with no major illnesses and physical disabilities that may affect the outcome. Participants in the experiment process must not perform any other exercise. Testees wear a designed wearable device. The set up includes a table, a chair, and a laptop. The experiment environment allows participants to stay relaxed and calm during the test (Fig. 6).

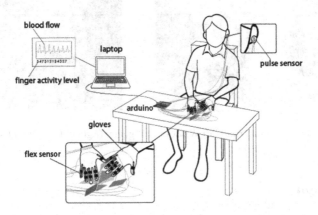

Fig. 6. Experiment arrangement

Experiment Procedure

First phase of experiment

(1) Before testees undertake the experiment, we will explain the steps and content of the overall experiment. Testees will be taught to wear the design's wearable device. Then turn on the wearable device and connect the wearable device. Before the experiment starts, they must first measure the neck blood flow with a precise blood flow meter, then follow the steps to test if the device has a problem.

(2) After the equipment is prepared and ready, testees will wear the design's wearable glove and earlobe blood flow sensor. For the tangram test, the Study provides 10 questions for the Tangram, and let testees perform the five-minute Tangram test.

(3) After the Tangram test is completed, the neck blood flow is measured by a precision instrument blood flow meter, that is, the first phase of the experiment is completed.

Second phase of the experiment

(1) In order to ensure the accuracy of the experiment, since the test of the blood flow data may cause errors and effects just after the completion of the tangram experiment, we let the testees rest until the blood flow returns to the most recent data, and then continue to the next experiment.

(2) Before proceeding to the second experiment, we still need to test the device for problems, and briefly explain how to play the Wisdom Ball. After the instrument is ready, testees will wear this study designed wearable glove and earlobe blood flow sensor for the wisdom ball experiment, and the game will take five minutes.

(3) After the wisdom ball test is completed, measure the neck blood flow with a precision instrument blood flow meter is used to complete the experiment (Fig. 7).

Fig. 7. (left) Tangram Game, (middle) Wisdom Ball game, (right) Measuring blood flow on the neck area

4 Discussion

In this section, the objective evaluation results demonstrate the performance of the new wearable device design, which are mainly divided into the performance evaluation of the earlobe-style blood flow sensor data, the data relationship between the finger activities frequency and the blood flow.

4.1 Earlobe Type Blood Flow Sensor Effectiveness Evaluation

Experiment 1: During the experiment, the subject will wear the wearable evaluation device for the tangram. The earlobe blood flow sensor detects the data in waveforms before and after the game as shown in Fig. 8 (left). The higher the wave peak represents a higher blood flow. Figure 8 (left) is the earlobe blood flow waveform presented by testees before and after the tangram game. It can be found that the earlobe blood flow after the tangram game has a slight increase in the wave peak, which indicates the tangram game positively promotes the brain's blood flow.

Experiment 2: During the experiment, the subject will wear the wearable evaluation device for the wisdom ball game. The earlobe blood flow sensor detects the data in waveforms before and after the game, as shown in Fig. 8 (middle). Before and after the wisdom ball game, the waveform of earlobe blood flow can be found to be slightly higher which indicates the testees are undertaking the wisdom ball game. After the wisdom ball game, the earlobe blood flow waveform has a slight increase, which indicates the wisdom ball game positively promotes the brain's blood flow.

This experiment also uses a sophisticated ultrasonic (Laser Doppler Spectroscopy (LDS)) blood flow instrument to measure the neck blood flow before and after the game, and evaluate the evaluation performance and accuracy of the wearable evaluation device of this Study design. Compare the subject's blood flow waveform with tangram and wisdom ball. The relative relationship is shown in Fig. 8 (right), with the data displayed by the precision Ultrasound LDS blood flow meter, indicating that the test brain blood flow increase by using the wisdom ball was higher than that of the tangram, and the same result was also found by the wearable evaluation device designed by this study. The blood flow waveform of the wisdom ball is indeed higher than the relative relationship between the blood flow waveform and the data after using the tangram.

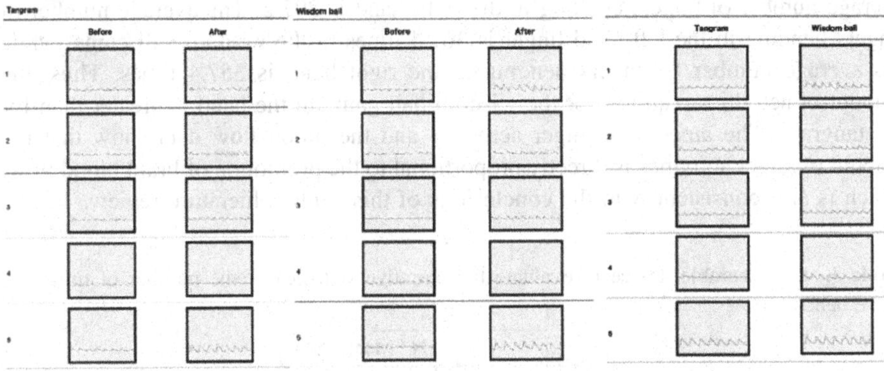

Fig. 8. The wearable device presents the blood flow waveform before and after the (left) tangram game, the (middle) wisdom ball game and the comparison between the blood flow of (right) tangram and wisdom ball

4.2   Statistical Analysis of Glove Activities' Frequency and Blood Flow

Measurement of the testees' blood flow after the tangram and wisdom ball respectively through the Doppler Ultrasound LDS blood flow meter indicates that the maximum blood flow and the average blood flow of the neck blood flow are significantly improved after the game (Tables 1 and 2).

Table 1. Descriptive statistics of three-stage experimental maximum blood flow (measured by LDS Ultrasound blood flow meter)

	Mean	Std. deviation	N
No game play	17.6	4.83	5
Tangram game	20.64	5.01	5
Wisdom ball game	30.1	6.72	5

Table 2. Descriptive statistics of three-stage experimental average blood flow (measured by LDS Ultrasound blood flow meter)

	Mean	Std. deviation	N
No game play	5.3	3.46	5
Tangram Game	7.48	3.07	5
Wisdom ball Game	12.16	3.81	5

At the same time, it can be demonstrated that using the wearable device to evaluate the degree of finger bending (Table 3), indicates the instance of finger bending during wisdom ball game is higher than that of the tangram. It can be seen that the average number of left-hand finger bending of testees in the tangram game is 204.4 times, the average number of fingers bending in the right hand is 297.2. The average number of fingers bending in the left-hand finger is 367.4 times in the wisdom ball game, while the average number of fingers bending in the right hand is 357.4 times. Thus, the amount of activities required for the wisdom ball game in the hand is greater than for the tangram. The amount of finger activities and the blood flow data show that the amount of finger activities is directly proportional to the promotion of brain blood flow, which is also consistent with the conclusions of this study's literature review.

Table 3. Use wearable device to evaluate the narrative statistics of the number of times the glove bends

		N	Mean	SD
Left	Tangram	5	204.4	67.9
	Wisdom ball	5	367.4	72.8
Right	Tangram	5	297.2	79.3
	Wisdom ball	5	357.4	60.1

According to Table 4, the results of the paired sample t-test demonstrates that the maximum blood flow before and after the tangram is in the first pair of paired samples, maximum blood flow (M = 3.48, SD = 1.31), $t(4) = -5.95$, p = 0.004, with significant difference; in the second pair of paired samples, for the wisdom ball game (M = 12.94, SD = 2.4), $t(4) = -12.07$, p = 0.0003, with significant difference, the results indicate a significant difference. In Table 5, the results of the paired sample t-test show that the average blood flow before and after the tangram is in the first pair of paired samples, average blood flow (M = 2.18, SD = 0.98), $t(4) = -4.97$, p = 0.007, with significant difference; in the second pair of paired samples, for the wisdom ball game (M = 6.86, SD = 2.22), $t(4) = -6.92$, p = 0.0002, with significant difference, the results are significant, indicating both tangram and wisdom ball games can effectively improve the brain's maximum blood flow and the brain's average blood flow.

Table 4. Conduct tangram and wisdom ball maximum blood flow paired sample t-test

	Paired differences				t	df	Sig(2-tailed)
	Mean difference	Std. error mean	95% confidence interval				
			Lower	Upper			
Tangram	3.48	1.31	1.86	5.10	5.95	4	0.004*
Wisdom ball	12.94	2.4	9.96	15.9	12.07	4	0.0003*

* The mean difference is significant at 0.05 level.

Table 5. Conduct tangram and wisdom ball average blood flow paired sample t-test

	Paired differences				t	df	Sig(2-tailed)
	Mean difference	Std. error mean	95% confidence interval				
			Lower	Upper			
Tangram	2.18	0.98	0.96	3.40	4.97	4	0.007*
Wisdom ball	6.86	2.22	4.11	9.61	6.92	4	0.0002*

* The mean difference is significant at 0.05 level.

4.3 Summary

The main purpose of the glove evaluation device is to evaluate the number of fingers used and blood flow conditions for users to facilitate future design and integration. After the above experiment, it can be demonstrated that:

- Glove performance evaluation: The evaluation device designed by this study can effectively detect the number of hand activities of users and the condition of earlobe blood flow.
- Blood sensor efficacy evaluation: The NIRS earlobe blood flow sensor can successfully display accurate blood flow waveforms, and the relative values are similar to those of precision LDS instruments.
- Finger activities and blood flow: The higher the number of finger activities, the blood flow or the precision instrument designed by this study, the corresponding blood flow shows a higher value, and the results are consistent with the conclusions of the relevant literature.

5 Conclusion

This study is mainly for the purpose of decelerating the progress of Alzheimer's disease, and the causes and possible solutions of AD are summarized by literature review and participant observation. The improvement of cerebral blood circulation and the prevention of AD are the main strategies adopted in this study. The circulation method is divided into indirect and direct stimulation, indirect is by systemic exercise or local exercise, directness is to allow the brain to directly cognize activities, this study will eventually combine hand exercise with cognition activities for game design, and currently an evaluation device that can evaluate the number of finger activities and blood flow sensing has been developed. The effectiveness and operability of the evaluation device can be seen from the above experiment. It can also be demonstrated from the experiment that the higher the number of finger activities, the better the effect of the brain stimulation, and the higher the cerebral blood flow.

The conclusions of the study are summarized as follows:
- As the global population of Dementia patients climbs yearly, there are many product designs related to AD status improvement and prevention, but less discussion of theoretical knowledge as the basis and innovation of design creation. This study informs a theoretical basis and creative ideas to elucidate breakthrough ideas for existing puzzle games.
- This study combines Arduino with human-computer interaction (HCI) to assess finger activity and blood flow status, and hopes to assess the improvement of cerebral blood circulation during human activities in an objective manner.
- The evaluation device designed by this study is a wearable device, and in addition to effectively detecting the amount of hand activities and blood status of users, does not hinder the actions of users' activities.

Follow-Up Discussion and Development
The current study points out that more brain games and exercises are important activities to slow down progression of AD. This study considers the needs of the elderly and the suitability of the device design for future lifestyles. The study aim is to develop a combination of exercise, cognition and cooperation. The wearable evaluation device game can simultaneously assess the condition of the hand activities and the adjustment of the blood flow condition, and also provide users with a large number of activities to train the fingers and cognitive thinking skills, encouraging design of a combination of games for evaluation and representing a suitable game device for improving quality of life and deceleration of neurodegeneration in patients with AD.

References

1. Cornutiu, G.: The epidemiological scale of Alzheimer's disease. J. Clin. Med. Res. 7(9), 657 (2015)
2. McKhann, G.M., et al.: The diagnosis of dementia due to Alzheimer's disease: recommendations from the National Institute on Aging-Alzheimer's Association workgroups on diagnostic guidelines for Alzheimer's disease. Alzheimer's Dement. 7(3), 263–269 (2011)
3. Albert, M.S., et al.: The diagnosis of mild cognitive impairment due to Alzheimer's disease: recommendations from the National Institute on Aging-Alzheimer's Association workgroups on diagnostic guidelines for Alzheimer's disease. Alzheimer's Dement. 7(3), 270–279 (2011)
4. Brookmeyer, R., et al.: Forecasting the global burden of Alzheimer's disease. Alzheimer's Dement. 3(3), 186–191 (2007)

5. Mendiola-Precoma, J., et al.: Therapies for prevention and treatment of Alzheimer's disease. Biomed. Res. Int. **2016**, 17 (2016)
6. Cass, S.P.: Alzheimer's disease and exercise: a literature review. Curr. Sport. Med. Rep. **16**(1), 19–22 (2017)
7. Waldemar, G., et al.: Recommendations for the diagnosis and management of Alzheimer's disease and other disorders associated with dementia: EFNS guideline. Eur. J. Neurol. **14**(1), e1–e26 (2007)
8. Thomas, T., et al.: β-Amyloid-mediated vasoactivity and vascular endothelial damage. Nature **380**(6570), 168 (1996)
9. Sevigny, J., et al.: The antibody Aducanumab reduces Aβ plaques in Alzheimer's disease. Nature **537**(7618), 50–56 (2016)
10. Wilcox, K.C., et al.: Aβ oligomer-induced synapse degeneration in Alzheimer's disease. Cell. Mol. Neurobiol. **31**(6), 939–948 (2011)
11. Park, L., et al.: Scavenger receptor CD36 is essential for the cerebrovascular oxidative stress and neurovascular dysfunction induced by amyloid-β. Proc. Natl. Acad. Sci. **108**(12), 5063–5068 (2011)
12. Morley, J.E., Farr, S.A.: The role of amyloid-beta in the regulation of memory. Biochem. Pharmacol. **88**(4), 479–485 (2014)
13. Niwa, K., et al.: Aβ-peptides enhance vasoconstriction in cerebral circulation. Am. J. Physiol. Hear. Circ. Physiol. **281**(6), H2417–H2424 (2001)
14. Tarantini, S., Tran, C.H.T., Gordon, G.R., Ungvari, Z., Csiszar, A.: Impaired neurovascular coupling in aging and Alzheimer's disease: Contribution of astrocyte dysfunction and endothelial impairment to cognitive decline. Exp. Gerontol. **94**, 52–58 (2017)
15. Sagare, A.P., Bell, R.D., Zlokovic, B.V.: Neurovascular defects and faulty amyloid-β vascular clearance in Alzheimer's disease. J. Alzheimer's Dis. **33**(s1), S87–S100 (2013)
16. Ramanathan, A., et al.: Impaired vascular-mediated clearance of brain amyloid beta in Alzheimer's disease: the role, regulation and restoration of LRP1. Front. Aging Neurosci. **7**, 136 (2015)
17. McDade, E., et al.: Cerebral perfusion alterations and cerebral amyloid in autosomal dominant Alzheimer's disease. Neurology **83**(8), 710–717 (2014)
18. Umeda, T., et al.: Intraneuronal amyloid β oligomers cause cell death via endoplasmic reticulum stress, endosomal/lysosomal leakage, and mitochondrial dysfunction in vivo. J. Neurosci. Res. **89**(7), 1031–1042 (2011)
19. Petersen, R.C., et al.: Mild cognitive impairment: clinical characterization and outcome. Arch. Neurol. **56**(3), 303–308 (1999)
20. Morris, J.C., et al.: Mild cognitive impairment represents early-stage Alzheimer's disease. Arch. Neurol. **58**(3), 397–405 (2001)
21. Sachi Ikudome, S.M., Unenaka, S., Kawanishi, M., Kitamura, T., Nakamoto, H.: Effect of long-term body-mass-based resistance exercise on cognitive function in elderly people. J. Appl. Gerontol. **36**(12), 1519–1533 (2017)
22. Yanagisawa, H., et al.: Acute moderate exercise elicits increased dorsolateral prefrontal activation and improves cognitive performance with Stroop test. Neuroimage **50**(4), 1702–1710 (2010)
23. Schelke, M.W., et al.: Mechanisms of risk reduction in the clinical practice of Alzheimer's disease prevention. Front. Aging Neurosci. **10**, 96 (2018)
24. Narme, P., et al.: Efficacy of musical interventions in dementia: evidence from a randomized controlled trial. J. Alzheimer's Dis. **38**(2), 359–369 (2014)
25. Lee, H.G., et al.: Challenging the amyloid cascade hypothesis: senile plaques and amyloid-β as protective adaptations to Alzheimer's disease. Ann. N. Y. Acad. Sci. **1019**(1), 1–4 (2004)

26. Kaplan, D.R., Martin-Zanca, D., Parada, L.F.: Tyrosine phosphorylation and tyrosine kinase activity of the trk proto-oncogene product induced by NGF. Nature **350**(6314), 158 (1991)

27. Schott, G.D.: Penfield's homunculus: a note on cerebral cartography. J. Neurol. Neurosurg. Psychiatry **56**(4), 329 (1993)

28. Murata, J., et al.: Relationship between the changes in blood flow and volume in the finger during a Braille character discrimination task. Ski. Res. Technol. **23**(4), 514–518 (2017)

29. Bronzwaer, A.S.G., et al.: Aging modifies the effect of cardiac output on middle cerebral artery blood flow velocity. Physiol. Rep. **5**(17), e13361 (2017)

30. Amaral, D.G.: The anatomical organization of the central nervous system. Princ. Neural Sci. **4**, 317–336 (2000)

Cognitive Profiles and Education of Female Cyber Defence Operators

Ricardo G. Lugo[1(✉)], Andrea Firth-Clark[2], Benjamin J. Knox[3,4],
Øyvind Jøsok[3,6], Kirsi Helkala[3], and Stefan Sütterlin[5]

[1] Faculty of Social and Health Sciences,
Inland Norway University of Applied Sciences, Lillehammer, Norway
Ricardo.Lugo@inn.no
[2] University Campus of Football Business, Burnley, UK
A.Firth-Clark@ucfb.com
[3] Norwegian Defence Cyber Academy, Norwegian Defence University College,
Lillehammer, Norway
{bknox,khelkala}@fhs.mil.no, ojosok@gmail.com
[4] Department of Information Security and Communication Technology,
Norwegian University of Science and Technology, Gjøvik, Norway
[5] Faculty for Health and Welfare Sciences, Østfold University College,
Halden, Norway
Stefan.Sutterlin@hiof.no
[6] Child and Youth Participation and Competence Development (BUK),
Inland Norway University of Applied Sciences, Lillehammer, Norway

Abstract. Rapid technological developments and definition of the cyber domain as a battlefield has challenged the cognitive attributes of its operators. In order to meet these demands, higher education programs in STEM (science, technology, engineering, and mathematics) need to recruit suited. Recruitment to STEM studies focuses on increasing the amount of females in these studies, and factors involving retention also needs to be understood. This research focused on assessing the educational setting of Norwegian Defense Cyber Academy and its factors in promoting female student retention in their computer engineering program, and profiling female officer cadets to see if any differences in personality, cognitions, and behaviours strategies exist between male and female cadets.

Keywords: Female cyber operators · Education · Cognitive profiles · Performance

1 Introduction

Rapid technological developments and definition of the cyber domain as a battlefield has challenged the cognitive attributes of its operators. In order to meet these demands, higher education programs, such as computer science, biology, and physics, are based on scientific approaches that define the science, technology, engineering, and mathematics (STEM) educations. STEM education is hypothesized to better prepare students for future work due to its inclusion of the more technical aspects that are more suited

© Springer Nature Switzerland AG 2019
D. D. Schmorrow and C. M. Fidopiastis (Eds.): HCII 2019, LNAI 11580, pp. 563–572, 2019.
https://doi.org/10.1007/978-3-030-22419-6_40

for technological advances [1]. Computer science has been included in the STEM educational umbrella due to its inclusion of scientific approaches in mathematics, engineering and maths and cyber security is a specialization within computer science that focuses on defense and protections of networks and systems. Demands for cyber security workforce is increasing and expected to rise globally to 6 million but is still 1.5 million short [2]. In 2015 the Joint Task Force on Cybersecurity Education (JTF) that was comprised of the Association for Computing Machinery (ACM), IEEE Computer Society (IEEE CS), Association for Information Systems Special Interest Group on Security (AIS SIGSEC), and International Federation for Information Processing Technical Committee on Information Security Education (IFIP WG 11.8), developed educational guidelines for cybersecurity education [3]. Mostly based on technical aspects, the aspect of human factors is mentioned in the requirements but is not expanded to include which human factor aspects need to be included.

There are recruitment differences in higher education within science, technology, engineering and mathematics (STEM). Females in universities make up only 22% of such studies, but that number drops to 12% for career choices within STEM domains of the workforce being female [4]. Within cyber security, prevalence of female professionals has risen from 11% (2013) to 20% (2019) but there is still a disparity and need for higher numbers of females. This disparity may arise from educational factors.

While recruitment to STEM studies focuses on increasing the amount of females in these studies, factors involving retention also needs to be understood. Cohoon [5] identified several factors that characterized departments that were able to retain female STEM students: (a) faculty staff included at least one female mentor and the staff shared responsibility in teaching, (b) the department had institutional support, (c) accessible job market, and (d) sufficient number of females in the study.

Females joining STEM programs may be influenced by situational factors that are not gender specific, that increase risk of drop-out. Cheryan et al. [6] showed that role models (both female and male) who project stereotypical behaviours in STEM programs may increase dissatisfaction of the program or aversion to commence studies. But with the increase of need for cyber security professionals, educational programs need to re-evaluate their approaches and retain students to fill the demands.

1.1 Understanding Cyber Security Operators Profiles

Little is known about the cognitive demands on and the profiles of cyber defence officers. Research in the area of cyber operations is scarce, and also has not reflected gender differences. Psychological determinants (i.e. decision-making, problem-solving) to understand human factors in cyber defence operations needs to be investigated to assess performance in cyber operators, especially with female officer cadets entering the domain. Female cadets may have certain psychological profiles that may be risk factors for dropping out. They also need to be examined for profiles of better performance. Lugo and Sütterlin [7] showed that cyber defence officer profiles differed from normal controls. Their emotional regulation strategies (rumination, worry) did not have the same patterns as their aged matched controls. They also found that cyber defence officers had different cognitive styles (field independent; FID) than matched controls (field dependent; FD) [8]. Knox et al. [9] and Josøk et al. [10] showed that metacognition predicted better performance in cyber domains. Cyber defence officers who also reported being more introverted were rated with higher confidence and better leadership, contrary to previous findings [11].

Psychological factors have been found to influence performance in cyber defence operations [7] but these findings did not include females. These factors may be relevant also in future selection processes. Selection processes can be time consuming, but are essential in recruiting the proper personnel for specific jobs [12]. Cognitive abilities have been found to be a strong predictor for selection and job performance.

1.2 Cognitive Factors

Perceived self-efficacy is defined as the 'beliefs in one's capabilities to organize and execute the courses of action required to produce given attainments' [13, p. 3] and is divided into a specific and a global component. General self-efficacy relates to the overall belief that one is in control over one's own life, actions, and decisions that shape one's life, while specific self-efficacy is the belief into one's performance in a certain task or described situation. Self-efficacy is also contingent on outcome expectancies, since one has to consider the desired outcome and judge if one possesses the skills necessary to reach those outcomes [13–15]. Self-efficacy can be strong and weak all within one person, as being confident in one's skills in one area of functioning does not automatically generalize to other areas. Self-efficacy is realized through four separate efficacy-activated processes [13]: (1) Cognitive processes, including goal setting, self-appraisals, anticipatory scenarios, and analytic thinking; (2) Motivational processes, which include causal attributions, self-regulatory processes, outcome expectancies, and cognized goal/reinforcements; (3) Affective processes, affected by anxiety arousal, vigilance, rumination, and situations; and (4) Selection processes, by choosing of environments. These processes work in conjunction with each other, are dynamic and can be influenced in different ways. Bandura [13] identified four influencing factors for perceived self-efficacy. The first source and most prominent affecting self-efficacy is that of mastery experience. Overcoming any demanding situation in a beneficial way increases the perception of self-efficacy, thus strengthening confidence and self-evaluations, while the opposite happens when failing.

Having an understanding of how cognitions affect behaviour requires individuals to reflect over relevant experiences and their outcomes. Reflecting can be done alone or with others (mentoring, feedback), but is an important process in consolidating experiences to long-term memory [16]. Being able to monitor and control encoding processes that arise from both negative and positive outcome but meaningful experiences leads to better long-term retention [17]. The importance of developing such metacognitive skills is essential in functioning properly within the Hybrid-Space domain [18]. Encoding experiences to long-term memory integrates both cognitive and emotional processes and strategies used. Metacognition is defined as 'awareness of one's own knowledge—what one does and does not know—and one's ability to understand, control, and manipulate one's cognitive processes' [19] and includes three components: knowledge of one's abilities, situational awareness, and behavioural regulation strategies [20]. It involves the active process of being aware of and exerting control over one's thinking to achieve present goals through planning, monitoring, and evaluating one's cognitions, emotions and behaviours, and actively adapting to the situational demands. Examples of metacognitive knowledge skills include world, technical, and experiential knowledge, and personal knowledge and awareness of one's own skills (e.g. self-efficacy), beliefs (confidence), and expected outcomes (situational knowledge). This involves the awareness of emotional and behavioural factors, and how they can be controlled and adjusted so that they can be incorporated into adaptive situational decision-making and problem solving strategies [21].

Gender differences in self-efficacy depend on the educational field [22]. Females display higher language arts self-efficacy, while males exhibited higher mathematics, computer, and social sciences self-efficacy, but these results also moderated by age, where larger effect size occurred for older respondents. But for mathematics self-efficacy, significant gender differences emerged in late adolescence. Even though these findings are significant, effect sizes were small.

1.3 Norwegian Perspective

Norway scores in the top of worldwide surveys on gender equality [23] where education, health, income, and political empowerment are measured. But Norway, and other Scandinavian countries scoring in the top of equality studies, also have a disparity and an over representation of gender inequality in traditional careers that are stereotypical, such as nursing and engineering, where females make up 89% and 17% respectively [24]. Within engineering, females make up 20% of the students, while more technological studies have a representation of 32,5%.

This research focused on assessing the educational setting of the Norwegian Defence Cyber Academy (NDCA) and its factors in promoting female student retention in their computer engineering program, and profiling female officer cadets to see if any differences in personality, cognitions, and behaviour strategies exist between male and female cadets.

2   Methods

Participants (N = 35) were recruited from the NDCA (n = 18; n_{female} = 8) and were controlled with age and gender matched non-technical students from Inland Norway University of Applied Sciences, Norway (n = 17; n_{female} = 7, as well as male (n = 9) cyber defence cadets from the NDCA. Psychological factors tested included cognitive styles, personality, emotion regulation strategies, self-efficacy, and metacognition.

2.1   Cultural Factors

Qualitative approaches were used to identify institutional factors that lead to female retention in accordance to Cohoon [5] and Cheryan [6] identified factors: (a) staff composition and behaviours, (b) institutional support, (c) accessible job market, and (d) sufficient number of females in the study. To answer a, & d, drop-out statistics, female-teacher ratio, and class composition was calculated. To answer b & c, a qualitative analysis of cultural factors relating to institutional support and need for cyber security professionals in Norway was investigated. These factors will be reported in the discussion.

2.2 Quantitative Measures

Quantitative measures were collected to identify cognitive aspects of female cyber operators and compared to males who are in the same educational route, and to other females from a non-technical but STEM education (psychological science).

Cognitive Styles. Cognitive styles were measured with the Group Embedded Figures Test (GEFT) [25]. The GEFT was developed for research into cognitive functioning, but it has become a recognized tool for exploring analytical ability, social behaviour, body concept, preferred defence mechanism and problem solving style as well as other areas. Finding common geometric shapes in a larger design is the assessment method. The results yield two cognitive styles: field dependence (FD) and field independence (FI). The GEFT is a twenty-five item assessment and scored manually. Persons with FI are considered to be detailed and analytical in their perception. They are characterized by a tendency to be able to space-orientate independently of their surroundings, and capabilities in cognitive and perceptual restructuring are considered to be the strength of FI individuals. FD individuals are described as using more "global" or overall focused perception and a lesser interest in details. They are characterized by a propensity to orient themselves in space based on their surroundings. Interpersonal abilities and emotional sensitivity are considered to characterize FD individuals [26, p. 17]. Reported reliability coefficients on GEFT test retests all fall between .78 and .92 [26].

Emotion Regulation. Emotion regulation was measured using three scales, The Rumination Styles Questionnaire, the Penn State Worry Questionnaire, and Cognitive Emotion Regulation Questionnaire.

Rumination was measured with the Response Style Questionnaire (RSQ) [27] and consists of 10 items with two subscales, brooding (five items) and reflective rumination or pondering (five items). Items are on a 4-point Likert scale from 1 – "almost never" – to 4 – "almost always". Example items for the brooding subscale include: "Why can't I handle things better" and for reflective pondering "Go away by yourself and think about why you feel this way". The RSQ shows good internal reliability (Cronbach's $\alpha = .89$).

Worry was measured with the Penn State Worry Questionnaire (PSWQ) [28]. The scale is a 5-point Likert scale ranging from 1 – "Not at all typical of me" – to 5 – "Very typical of me". The PSWQ shows good internal reliability (Cronbach's $\alpha = 0.96$).

Cognitive emotion regulation questionnaire short (CERQ-S) [29] is an 18 item 5 point Likert scale from 1 to 5 designed to evaluate nine cognitive strategies used to regulate emotions in response to negative or unpleasant events: blaming themselves, blaming others, accepting, refocusing on planning, positive refocusing, rumination, positive reappraisal, put in perspective and disaster thinking. Only the blaming themselves, blaming others, accepting, scales were used for analysis since the RSQ and PSWQ were used due to stronger links to performance. The CERQ subscales focusing on blame and acceptance were relevant for this study. Reliability analysis of internal consistency gave good Cronbach's α for the translated scale ($\alpha = .682 - .884$) except for the refocusing scale ($\alpha = .419$) which was not used in the analyses.

Metacognition. To measure metacognition, two measurements were used, The Metacognitive Awareness Inventory and the Self-regulation Questionnaire.

The Metacognitive Awareness Inventory (MCAI) [30] was used. It is a self-report scale comprising of 52 items that includes several subscales assessing knowledge of cognition (declarative knowledge, procedural knowledge, conditional knowledge) and regulation of knowledge (planning, information management strategies, monitoring, debugging strategies and evaluation). Items are assessed on bipolar responses (true/false) and then ratios are computed from the subscales. Sample items include: "I find myself using helpful learning strategies automatically" (procedural knowledge) and, "I ask myself if I have considered all options when solving a problem" (comprehension monitoring). The test shows high reliability on all subscales (Cronbach's $\alpha = .90$).

The Self-Regulation Questionnaire (SRQ-63) [31] is a 5-point Likert self-report scale, ranging from strongly disagrees to strongly agree. The scale has 7 subscales that consist of receiving, evaluating, triggering, searching, formulating, implementing, and assessing. Sample items include; "I usually keep track of my progress toward my goals" and, "I have sought out advice or information about changing". The test shows high reliability (test-retest: $r = .94$, p < .0001; $\alpha = .91$).

Trait self-efficacy was measured with the General Self-Efficacy Scale (GSES) [32]. The scale is composed of 10 Likert-scale items with scores ranging from 1 to 4, with higher scores indicating higher trait self-efficacy. The scale has shown validity in several domains and across cultures [33] and has acceptable internal validity ($\alpha = .75–.91$).

Positive affect and negative affect was measured using the Positive Affect and Negative Affect Scale (PANAS) [34]. This consists of 20 words related to positive affect (PA; 10 items) and negative affect (NA; 10 items). PANAS is a summative questionnaire with answers ranging from 1 – "not at all" – to 5 – "a lot". Positive affect items include "interested" and "excited", and negative affect items include "distressed" and "upset". Participants are asked to respond according to their usual levels of affect. Cronbach's α ranges from 0.86 to 0.90 for PA and from 0.84 to 0.87 for NA. This scale is highly correlated with depression checklists.

3 Results

3.1 Quantitative Factors

Female cyber defence officers were different on several psychological factors than non-technical control females. They showed higher degrees of maladaptive emotion regulation strategies (brooding; $t = 1.93$, $p = .040$ (1-tailed), Cohen's $d = 1.01$), less assertiveness ($t = -2.36$, $p = .038$, Cohen's $d = -1.38$), and self-efficacy ($t = 2.635$, $p = .023$, Cohen's $d = 1.44$), but had better metacognitive regulation strategies (comprehension management; $t = 2.18$, $p = .026$ (1-tailed), Cohen's $d = 1.24$).

Compared to male cyber defence officer cadets, females reported less positive affect ($t = 2.18$, $p = .044$, Cohen's $d = 0.58$), more anxiety ($t = 2.69$, $p = .016$, Cohen's

$d = 1.28$), less self-efficacy ($t = 2.71$, $p = .015$, Cohen's $d = 1.25$) and more maladaptive emotion regulation strategies (self-blame: $t = 2.10$, $p = .032$ (1-tailed), Cohen's $d = .96$; reappraisal: $t = 2.02$, $p = .032$ (1-tailed), Cohen's $d = 0.93$). Females cadets also reported lower metacognition (planning: $t = 2.246$, $p = .039$, Cohen's $d = 1.04$).

4 Discussion

This study focused on assessing the factors in promoting female student retention in their computer-engineering program, and profiling female officer cadets and the educational setting of Norwegian Defence Cyber Academy.

Results show that female cyber defence officer cadets score as other related fields (engineering) and their male counterparts in cognitive styles (field independence/dependence). The female cyber officer cadets did have some findings that could put them at risk of dropping out of schooling. They reported higher anxiety and maladaptive emotion regulation strategies than both fellow male cyber students as well as when compared to age and gender matched controls. They also reported significantly less self-efficacy than all other groups. Anxiety, low self-efficacy, and maladaptive emotion regulation styles are all risk factors in academic under-performance (see Ackerman et al., 2013 and Riegle-Crumb & King, 2010 for a review). But these factors do not seem to contribute to drop-outs, and this may be due to qualitative factors of the institution.

4.1 Cultural Factors

Some aspects of the NDCA reflects the Norwegian culture in supporting female participation in this educational field, but otherwise follow international trends of underrepresentation of females within the field. The Commandant of the institute is female and the students have access to female professors in STEM subjects. But of the 20 full time teachers at the institution, only 3 are females. The NDCA has a total of 40 students per year and since 2013 females have almost made up 20% of the cohorts but have ranged from 10% to 30% (see Table 1).

Table 1. Percentage of females at NDCA

Year	Total students	Females % (number)
2013	37	18.9% (7)
2014	38	18.4% (7)
2015	40	10% (4)
2016	35	28.6% (10)
2017	40	12.5% (5)
Total	190	17.4% (33)

However, the number of actual female dropouts at NDCA is very low. During the last 6 years, 16% of the students have been females. 8 of them, representing 21% of the total amount of females attending the NDCA in this time period, participated in this study. During this time period, there has been a 2.5% total dropout rate; 2.7% for females and 2.5% for males. The drop-out rate does not include two females that left within the first school week having been offered places at civilian engineer universities. In cases where female officer cadets decided to leave the school, poor academic performance was not the motivation.

There are several contributing factors that may explain such low female drop-out rates during the three-year bachelor degree course at the NDCA. During the first two school years, the cadets have a dedicated mentor who provides academic guidance including techniques for studying and time management. Throughout the entire three years, their closest military leader mentors the cyber officer cadets to ensure their attitude and behaviour is in accordance with expected standards. In addition, students receive mentoring and guidance from the staff during and after their attendance to military exercises. Throughout their time at the academy students attend a total of four major exercises. Class sizes at the NDCA are small due to the maximum intake each year of 40 cadets. This gives teachers the possibility to know their students and to tailor guidance to each individual needs. Attendance to all classes is also obligatory.

One additional factor is peer-support. The cohort becomes a tight-knit group over the three years. A fall-out form this is the sharing of the academic burden. Meaning individual and team workload demands can be more easily overcome. Interestingly, in 2017 a female cadet won the prize for best in military skills, and was second in the academic rankings. Then in 2018, a female cadet was awarded the first prize in academic and in military studies. This can inform that the environment is a healthy and competitive, and that females are capable of performing across domains. These two high performing females may also help motivate future female cadets to not be hindered by negative psychological factors.

The qualitative aspects of the NDCA support the model presented by Cohoon [5] and Cheryan [6]. Faculty at the NDCA has at least one female mentor and the staff shared responsibility in teaching, including the school Commandant being female. Combined with the cultural aspect that Norway is one of the most gender balanced societies in the world, institutional support is engrained in all aspects. Norway's minister of defence from 2013 to 2017 was also female and was the first woman to hold this position. The role models the female cadets are exposed too, both nationally and locally at the institution also have cultural aspects that represent equality and support. Access to job markets in Norway also follow gender equality.

The quantitative findings show that female cyber officer cadets have some risk factors that could lead to dropping out of school that reflect previous finings. Anxiety, low self-efficacy, and maladaptive emotion regulation styles are all risk factors in academic under-performance (see Ackerman et al., 2013 and Riegle-Crumb & King, 2010 for a review) and females at the NCDA displayed lower levels on similar factors, making them vulnerable to dropping out. However these results need to be seen in conjunction with the institutional factors that prevent such drop-outs occurring.

5 Conclusion

The NDCA provides novel insights on female performance in computer science domains, even when gender participation is similar to other nations. The qualitative differences of the NDCA provide support to research focusing on retention of females in computer science education. Even though female officer cadets showed worrisome scores on psychological predictors of academic outcomes, they were no different on other cognitive measurements than their male counterparts. Cyber engineering is considered a STEM degree, but due to the novelty of the domain, little is known about female operators functioning within it. Future research needs to identify how factors used in previous studies might affect female performance in cyber education and the when operating in the domain.

References

1. Breiner, J.M., Harkness, S.S., Johnson, C.C., Koehler, C.M.: What is STEM? A discussion about conceptions of STEM in education and partnerships. School Sci. Math. **112**(1), 3–11 (2012)
2. Fourie, L., Pang, S., Kingston, T., Hettema, H., Watters, P., Sarrafzadeh, H.: The global cyber security workforce: an ongoing human capital crisis (2014)
3. McGettrick, A., Cassel, L.N., Dark, M., Hawthorne, E.K., Impagliazzo, J.: Toward curricular guidelines for cybersecurity. In: Proceedings of the 45th ACM Technical Symposium on Computer Science Education, 5 March 2014, pp. 81–82. ACM (2014)
4. Sinkele, C.N., Mupinga, D.M.: The effectiveness of engineering workshops in attracting females into engineering fields: a review of the literature. Clearing House J. Educ. Strat. Issues Ideas **84**(1), 37–42 (2011)
5. Cohoon, J.M.: Toward improving female retention in the computer science major. Commun. ACM **44**(5), 108–114 (2001)
6. Cheryan, S., Siy, J.O., Vichayapai, M., Drury, B.J., Kim, S.: Do female and male role models who embody STEM stereotypes hinder women's anticipated success in STEM? Soc. Psychol. Pers. Sci. **2**(6), 656–664 (2011)
7. Lugo, R.G., Sütterlin, S.: Cyber officer profiles and performance factors. In: Harris, D. (ed.) EPCE 2018. LNCS (LNAI), vol. 10906, pp. 181–190. Springer, Cham (2018). https://doi.org/10.1007/978-3-319-91122-9_16
8. Lugo, R.G., et al.: Cognitive styles of cyber engineers - a cross cultural comparison. In: HCI International (2018)
9. Knox, B.J., Lugo, R.G., Jøsok, Ø., Helkala, K., Sütterlin, S.: Towards a cognitive agility index: the role of metacognition in human computer interaction. In: Stephanidis, C. (ed.) HCI 2017. CCIS, vol. 713, pp. 330–338. Springer, Cham (2017). https://doi.org/10.1007/978-3-319-58750-9_46
10. Jøsok, Ø., Lugo, R.G., Knox, B.J., Sütterlin, S., Helkala, K.: Self-regulation and cognitive agility in cyber operations. Front. Psychol. **10**, 875 (2019)
11. Lugo, R.G., Sütterlin, S., Helkala, K., Knox, B.J., Jøsok, Ø., Lande, N.: Emotion regulation as leadership predictors for cyber cadets. In: HCI International, 16–20 July 2018, Las Vegas, USA (2018)
12. Carretero-Dios, H., Pérez, C.: Standards for the development and review of instrumental studies: considerations about test selection in psychological research. Int. J. Clin. Health Psychol. **7**(3), 863 (2007)

13. Bandura, A.: Self-efficacy: The Exercise of Control. W.H. Freeman & Company, New York (1997)
14. Pajares, F., Schunk, D.H.: Self-beliefs and school success: self-efficacy, self-concept, and school achievement. Perception **11**, 239–266 (2001)
15. Schunk, D.H.: Self-efficacy and academic motivation. Educ. Psychol. **26**(3–4), 207–231 (1991)
16. Halpern, D.F.: Teaching critical thinking for transfer across domains: disposition, skills, structure training, and metacognitive monitoring. Am. Psychol. **53**(4), 449 (1998)
17. Bahrick, H.P., Hall, L.K.: The importance of retrieval failures to long-term retention: a metacognitive explanation of the spacing effect. J. Mem. Lang. **52**(4), 566–577 (2005)
18. Jøsok, Ø., Knox, B.J., Helkala, K., Lugo, R.G., Sütterlin, S., Ward, P.: Exploring the hybrid space. In: Schmorrow, D.D.D., Fidopiastis, C.M.M. (eds.) AC 2016. LNCS (LNAI), vol. 9744, pp. 178–188. Springer, Cham (2016). https://doi.org/10.1007/978-3-319-39952-2_18
19. Meichenbaum, D.: Metacognitive methods of instruction: current status and future prospects. Spec. Serv. Schools. **3**(1–2), 23–32 (1985)
20. Flavell, J.H.: Metacognition and cognitive monitoring: a new area of cognitive–developmental inquiry. Am. Psychol. **34**(10), 906 (1979)
21. Gross, J.J.: The emerging field of emotion regulation: an integrative review. Rev. Gen. Psychol. **2**(3), 271 (1998)
22. Huang, C.: Gender differences in academic self-efficacy: a meta-analysis. Eur. J. Psychol. Educ. **28**(1), 1–35 (2013)
23. Noland, M., Moran, T., Kotschwar, B.: Is gender diversity profitable? Evidence from a global survey (2016)
24. SSB – Statistics Norway. https://www.ssb.no/utuvh. Accessed 2 Jan 2018
25. Witkin, H.A.: Individual differences in ease of perception of embedded figures. J. Pers. **19**(1), 1–5 (1950)
26. Wapner, S., Demick, J.: Field Dependence-Independence: Bio-Psycho-Social Factors Across the Life Span, p. 2. Psychology Press, New York (2014)
27. Treynor, W., Gonzalez, R., Nolen-Hoeksema, S.: Rumination reconsidered: a psychometric analysis. Cognit. Ther. Res. **27**(3), 247–259 (2003)
28. Meyer, T.J., Miller, M.L., Metzger, R.L., Borkovec, T.D.: Development and validation of the Penn state worry questionnaire. Behav. Res. Ther. **28**(6), 487–495 (1990)
29. Garnefski, N., Kraaij, V., Spinhoven, P.: Negative life events, cognitive emotion regulation and emotional problems. Pers. Individ. Differ. **30**(8), 1311–1327 (2001)
30. Schraw, G., Dennison, R.S.: Assessing metacognitive awareness. Contemp. Educ. Psychol. **19**(4), 460–475 (1994)
31. Aubrey, L.L., Brown, J.M., Miller, W.R.: Psychometric properties of a selfregulation questionnaire (SRQ). Alcohol. Clin. Exp. Res. **18**(2), 420–525 (1994)
32. Scholz, U., Doña, B.G., Sud, S., Schwarzer, R.: Is general self-efficacy a universal construct? Psychometric findings from 25 countries. Eur. J. Psychol. Assess. **18**(3), 242 (2002)
33. Luszczynska, A., Scholz, U., Schwarzer, R.: The general self-efficacy scale: multicultural validation studies. J. Psychol. **139**(5), 439–457 (2005)
34. Watson, D., Clark, L.A., Tellegen, A.: Development and validation of brief measures of positive and negative affect: the PANAS scales. J. Pers. Soc. Psychol. **54**(6), 1063 (1988)

Self-control Strategies: Interpreting and Enhancing Augmented Cognition from a Self-regulatory Perspective

Mina Milosevic, Nicholas A. Moon, Michael W. McFerran,
Sherif al-Qallawi, Lida P. Ponce, Chris Juszczyk,
and Patrick D. Converse[✉]

Florida Institute of Technology, Melbourne, FL 32901, USA
{jmilosevic2015, pconvers}@fit.edu

Abstract. Recent work on augmented cognition has begun to highlight the importance of self-regulatory processes. In an effort to expand further in this direction, this paper outlines the nature and potential implications of an emerging view of self-regulation focusing on five major self-control strategies. In particular, we discuss how this self-control strategies perspective may help inform our understanding of current augmented cognition approaches, suggest new directions for the development of mitigation strategies, and highlight new research directions. This preliminary discussion integrating self-control strategies and augmented cognition may stimulate additional work that could benefit both research areas.

Keywords: Self-regulation · Self-control · Self-control strategies · Mitigation strategies

1 Introduction

More than three decades of augmented cognition research have undoubtedly enhanced our understanding of human-computer interactions and produced numerous advancements in technological systems capable of extending user (or operator) abilities [1–3]. Much of the research in this area has focused on identifying and overcoming cognitive bottlenecks stemming from human information-processing limitations [4, 5]. In recent years, however, there have been growing calls to expand augmented cognition research beyond the information-processing perspective and consider the role of broader mental states and processes in human-computer interactions [6]. Among those, self-regulatory processes have been identified as critical to human adaptive performance [7]. *Self-regulation* refers to an individual's ability to compare his/her current and desired states and take necessary action to resolve any discrepancy between the two [8, 9]. Some recent work has begun to answer these calls by specifically addressing the role self-regulation plays in adaptive performance [see 10, 11].

Building on this initial work exploring self-regulation in the context of augmented cognition [11], we highlight several considerations that suggest expanding further in this direction may be useful. First, existing research on self-regulation in augmented

© Springer Nature Switzerland AG 2019
D. D. Schmorrow and C. M. Fidopiastis (Eds.): HCII 2019, LNAI 11580, pp. 573–585, 2019.
https://doi.org/10.1007/978-3-030-22419-6_41

cognition has focused almost exclusively on the regulation of physiological states [12]. However, research has shown that individuals develop a variety of adaptive strategies to address discrepancies and promote behaviors leading to goal accomplishment [18]. These self-control strategies go beyond regulating one's physiological states and include regulating one's external circumstances (i.e., situational strategies) and controlling one's mental representation of the circumstances [i.e., cognitive strategies; 13]. We propose that considering the full array of self-control strategies in the context of augmented cognition can enhance human-computer interaction by (a) developing systems that more effectively support the user in initiating and exercising self-control and (b) expanding on the current mitigation strategies used by the system to address human cognitive bottlenecks. Although there is some overlap between self-control strategies and mitigation strategies currently used in HCI systems, there are self-control strategies that have not been considered in this context but could be beneficial (discussed in more detail later). Thus, expanding the view of self-regulation in this context could result in enhanced HCI systems that support and extend the individual's self-regulatory capabilities.

Second, for optimal functioning, it is necessary for the two systems–human and computer–to cooperate in a highly dynamic way [14]. This means recognizing the intrinsic need of humans to self-regulate and then designing technology to consider the self-control actions that humans inherently implement to address discrepancies between their current and desired states. In other words, successful human-computer integration requires congruent goals and alignment of control strategies initiated by either system.

Finally, adopting a broader view of self-regulation–and in particular the consideration of self-control strategies–further highlights the potential for these expanded approaches to be applied in mobile and online environments, as has recently been suggested [e.g., 15–17]. For example, mobile and online applications suggest self-regulation might be conceptualized as involving not only how individuals regulate responses in the moment, but also how they select and modify their environments to support goal pursuit.

Thus, the purpose of this paper is to begin addressing some of the suggested benefits of an expanded self-regulatory framework in human-computer interactions [7, 16]. We do so by adopting an emerging view of self-control strategies [18, 19] and exploring implications of this view for conceptualizing and developing augmented cognition efforts. More specifically, we (a) discuss how this self-regulatory perspective might inform our understanding of augmented cognition, (b) explore how this self-regulatory perspective might be used to enhance augmented cognition efforts, and (c) present ideas for future research stemming from the integration of these two areas.

1.1   Self-regulation and Self-control

Broadly, *self-regulation* is defined as "processes involved in attaining and maintaining (i.e., keeping regular) goals, where goals are internally represented (i.e., within the self) desired states" [20, p. 158]. Control theory, a prominent theoretical perspective on self-regulation, highlights three major components: standards, monitoring, and operating [21, 22]. Standards refer to an individual's ideals or goals. Monitoring refers to an individual's comparison between his/her current state and the desired state. Operating refers to an individual's response to any discrepancy between the current state and the desired state. From this control theory perspective, individuals try to reduce discrepancies between their standards and current state [9].

As a simple illustration, this control system functions similarly to a thermostat. For example, the thermostat has a goal level (e.g., 74°), measures the environmental temperature to determine if there is a discrepancy between the current environment and the goal level, and adjusts accordingly [e.g., turns on air conditioning to reduce the temperature; 23]. Control theory involves a negative feedback loop, which can cause individuals to increase effort in order to decrease the discrepancy between the current and desired state [22, 24]. Thus, self-regulation is generally seen as a controlled dynamic process in which individuals regulate their behavior in the face of external factors that may influence their current state.

Within this broad self-regulation perspective, the actions an individual takes in order to resolve any discrepancies have been conceptualized as *self-control* [13, 18, 25]. Thus, self-control takes place during the operating phase of self-regulation and can entail stopping undesirable responses or starting desirable responses related to a goal [26–28]. Researchers have found that higher self-control is related to a wide range of positive outcomes. For example, individuals with greater self-control have healthier habits, better academic success, more personal accomplishments, and fewer maladaptive behaviors [29]. On the other hand, weaker self-control has been shown to relate to negative outcomes such as criminal behaviors, obesity, drug and alcohol abuse, financial debt, impulsive spending, and procrastination [21, 30, 31]. Thus, the hallmark of a successful and healthy individual is demonstration of high self-control [32].

1.2 Self-control Strategies

Recent views of self-control have begun to take a broad perspective that considers multiple ways in which an individual can regulate his/her thoughts, feelings, and actions [18]. Building on the process model of emotion regulation [33, 34], Duckworth and colleagues [13, 18] have recently proposed that self-control strategies can be organized into a process model according to their underlying mechanisms, the stage at which they are used, and the amount of effort that is required. The resulting process model of self-control includes five families of strategies that can be divided into those that are concerned with changing external circumstances (i.e., situational control strategies) and those that are concerned with changing the mental representation of the circumstances [i.e., cognitive control strategies; 18].

Two of these self-control strategies focus on the situation: situation selection and situation modification. *Situation selection* involves intentionally selecting situations that facilitate accomplishment of valued goals. Situation selection is the most forward-thinking strategy, where individuals engaging in this strategy avoid situations that undermine goal pursuit and approach situations that facilitate goal pursuit. For example, to complete a task, an individual might choose to work in a quiet place (e.g., library) rather than in a noisy environment (e.g., bar). *Situation modification* involves changing aspects of the situation to facilitate goal pursuit. This strategy entails changing physical aspects of the environment. For instance, to complete a task, an individual might place his/her phone out of sight to avoid getting distracted by notifications.

Three of these self-control strategies focus on cognitive control: attentional deployment, cognitive change, and response modulation. *Attentional deployment* involves selectively attending to features of the situation to promote goal pursuit. In cases where individuals cannot select or modify the situation, they can attend to particular aspects of the environment to increase the chances of successful goal pursuit. In addition, individuals may purposely divert attention away from distracting stimuli. For example, students taking a test may attend to the test questions at hand, rather than the sound of the clock hands ticking as time is passing.

Cognitive change involves altering the way we think about the situation or task to facilitate goal pursuit. This can entail increasing the value of a task-related goal or decreasing the value of a potential distraction or impulse. For example, individuals completing a work task might construe the task as on opportunity to demonstrate their skills or view a potential distraction as a waste of time.

Finally, *response modulation* refers to the suppression of undesirable impulses or the amplification of desirable impulses in the moment by sheer will. This strategy thus involves just saying "no" to a disruptive impulse or forcing oneself to focus on the task or goal at hand. Response modulation can be considered a last-ditch effort to manage impulses that often fails [e.g., 36]. Based on this, Duckworth and colleagues [13, 18, 35] suggest that the other self-control strategies may be more effective than response modulation.

2   How Can a Self-regulation View Inform Our Understanding of Augmented Cognition?

The expanded view of self-regulation entailing five self-control strategies may inform our understanding of current augmented cognition approaches in three ways. First, this perspective provides an expanded framework from which to consider some of the existing approaches to augmented cognition. For example, major mitigation strategies include scheduling (including task pacing and task sequencing), modality augmentation and switching, cueing, decluttering, mixed initiative, context-sensitive help, and transposition [12]. Scheduling involves manipulating the time or the order of tasks for the user. Within this, task pacing refers to the scheduling of tasks from high-priority to low-priority, and task sequencing refers to changing simultaneous events into sequential events or dividing each task into smaller chunks and rearranging segments. Similarly, modality augmentation and switching involves two approaches: modality switching and modality redundancy. Modality switching refers to changing the sensory modality (e.g., switching from visual to auditory modality) to distribute processing load, and modality redundancy involves providing information in multiple modalities. Cueing involves capturing the user's attention by manipulating displayed information. Decluttering involves reducing the amount or complexity of information provided. Mixed-initiative systems involve the combination of user and system adjusting the level of system autonomy. Context-sensitive help involves providing necessary information to the user at the time assistance is needed. Transposition refers to changing information displayed to the user from verbal to spatial or vice-versa. Finally, some recent work in the context of augmented cognition has also focused on assisting in the regulation of physiological states [e.g., through breathing; 10, 15].

The self-control strategies framework provides a new vantage point for interpreting these mitigation approaches. For example, mitigation strategies such as scheduling (including task pacing and task sequencing), modality augmentation and switching, mixed initiative, context-sensitive help, and transposition appear to represent modifications of the situation (i.e., the context in which the user operates). The purpose of these strategies is to change various aspects of the situation in order to promote user performance. These are thus consistent with the notion of situation modification. In addition, cueing and decluttering attempt to influence the user's attention by manipulating the salience of information. Therefore, these are consistent with the notion of attentional deployment. Additionally, approaches focusing on the regulation of physiological states appear to be consistent with response modulation [e.g., deep breathing is a suggested application of response modulation; 18]. Note also that some mitigation strategies may overlap with more than one of the self-control strategies. For instance, as noted above, task sequencing appears to fit with the concept of situation modification but it may also overlap with cognitive change and attentional deployment. Similar to cognitive change, task sequencing may help with the reframing of larger tasks into smaller chunks. Similar to attentional deployment, task sequencing may also assist in directing attention by dividing each task into smaller segments. In addition, mixed initiative seems to have some overlap with situation modification but this appears to be a somewhat broader approach; thus, it may have connections with other self-control strategies depending on the particular form it takes. Overall, this analysis suggests that (a) current mitigation strategies fit largely within situation modification and attentional deployment, (b) there is limited overlap with cognitive change and response modulation, and (c) there appears to be little to no connection with situation selection (see Table 1 for a summary).

Table 1. Connections between self-control strategies and mitigation strategies.

Self-control strategy	Definition	Mitigation strategies
Situation selection	Intentionally selecting situations that facilitate accomplishment of valued goals	
Situation modification	Changing aspects of the situation to facilitate goal pursuit	Scheduling, task sequencing, task pacing, modality augmentation and switching, decluttering, context-sensitive help, transposition, mixed initiative
Attentional deployment	Selectively attending to features of the situation to promote goal pursuit	Cueing, decluttering, task sequencing, context-sensitive help, transposition
Cognitive change	Altering the way we think about the situation or task to facilitate goal pursuit	Task sequencing
Response modulation	Suppressing undesirable impulses or amplifying desirable impulses in the moment	Regulation of physiological states

Second, this expanded view of self-regulation further emphasizes the role of the human user. Although it has been proposed in the context of augmented cognition that there are two adaptive systems–the "adaptive operator" (i.e., human user) and the adaptive technical system–the primary focus has tended to be on the technical system [16, 37]. As with the information-processing approach, the main expectation is that improvements in the human-machine system will result from the technical system enhancing or controlling the user's adaptive cognitive processes [14]. In many ways, this approach overlooks the role that the human operator plays in the adaptive process. Self-regulation is so inherent in human beings that it is reasonable to expect that self-regulatory processes occur regardless of the technical system's mitigation efforts. As has been proposed, leveraging this inherent need and ability of the human operator might be useful when developing HCI systems [38]. In particular, we suggest that HCI systems could be developed to support the human user more effectively in their self-regulatory efforts. Recently, Schwartz and Fuchs (2017) argued for the use of a multidimensional approach when assessing user states [38]. They introduced the Real-time Assessment of Multidimensional User States (RASMUS) as a broad diagnostic system that is capable not only of detecting performance deteriorations but also inferring the causes of such declines by assessing six user states (workload, fatigue, motivational aspects of engagement, attention, situation awareness, and emotional states). Expanding on this multidimensional approach in assessing user states, we propose that the next logical step would be to design systems that employ mitigation strategies that are best suited to address the root causes of performance declines. For example, motivational and engagement issues are probably not effectively addressed by task sequencing or task pacing but would rather need to be addressed by increasing user engagement and motivation to allocate extra resources to the task at hand. When it comes to self-regulatory failures, we propose that self-control strategies could be useful in considering the various ways that human users can be supported in their regulatory efforts.

Finally, for human-computer adaptive functioning to be productive, it is necessary for the two systems–human and technical–to have congruent adaptive goals and strategies [14]. One way to accomplish the effective integration of the two systems is to enable the human operator to communicate his/her initial goals and preferred strategies at the onset of interaction. In addition, conceptualizing the human operator as a self-regulating individual highlights that his/her goals and strategies are likely to shift over time when facing a discrepancy between current and desired state. If the computer fails to adjust accordingly under these circumstances, mismatches in goals and strategies between the two systems may significantly undermine performance over time and even result in counterproductive interaction of the two systems [16, 37, 39]. Beyond adjusting goals and strategies as needed, keeping the human operator actively involved may require building in opportunities to initiate, modify, or override assistance from the technical system. Leveraging self-regulatory processes to promote the human user's active involvement can be especially beneficial when human-computer interaction takes place in less structured environments, such as mobile and online platforms. Thus, this more active view of the user further emphasizes the importance of developing systems that are capable of adapting over time and specifically highlights the potential need to adapt to shifting user goals and strategies.

3 How Can a Self-regulation View Enhance Augmented Cognition Efforts?

The self-control strategies perspective may also help inform the development of approaches to augmented cognition. In particular, this perspective might be leveraged to develop systems that are more capable of mimicking human self-regulatory processes, which in turn may better support users in achieving task-related goals. That is, each self-control strategy might be used to identify new or revised mitigation strategies to support user performance.

As previously discussed, situation selection refers to strategies that involve purposefully choosing to be in an environment that facilitates achievement of tasks or goals. In some cases, individuals cannot choose the situation in which they complete a task (e.g., pilots). If situation selection cannot be used, augmented cognition strategies need to focus on the remaining self-control strategies. In cases where individuals can choose the situation (e.g., educational, mobile, or online training applications), situation selection could be a highly effective strategy that might be incorporated into augmented cognition approaches. In these cases, the system could be designed to prompt or support effective selection of the situation for task completion. For instance, a variety of measurements could assist in identifying when situation selection may be particularly important. The system could use measures of the individual (e.g., performance metrics, self-ratings, physiological measures), the surrounding environment (e.g., noise, lighting), and the task (e.g., complexity, importance) to assist with situation selection. As one example, online training often involves some level of learner control (e.g., choosing when and where to initiate training). Learner control can result in increased interest and motivation for the user. However, in hypermedia environments this control can also lead to problems such as disorientation, distraction, and cognitive overload [40]. To address this, learner control could be combined with a system-assisted situation selection approach, where the system monitors the person's state, the surrounding environment, and key task characteristics, and then prompts the person to change situations if the current context is not conducive to the training at hand.

Situation modification refers to changing aspects of the situation to facilitate goal pursuit. As noted previously, several mitigation strategies (e.g., scheduling, task sequencing, task pacing, modality augmentation and switching, decluttering, context-sensitive help, transposition, mixed initiative) can be viewed as forms of situation modification. Beyond these existing mitigation strategies, the concept of situation modification suggests other approaches. For instance, although task characteristics can cause overload, so can features of the work environment (e.g., noise levels, temperature). For example, open office environments can be stressful and demotivating due to noise [41]. Additionally, past research has shown that increased office noise is associated with lower levels of job satisfaction [42]. To alleviate the effects of negative features of work environments, augmented cognition systems could monitor ambient physical conditions and the user's physiological state and then suggest environmental modifications as appropriate. For example, if the system recognized an increase in the decibel level of the room and user metrics revealed a reduction in task performance, the system could suggest the individual use noise cancelling headphones or earplugs.

Attentional deployment involves selectively attending to features of the situation to promote goal pursuit. This type of strategy is similar to situation modification but involves a different mechanism. Situation modification is aimed at facilitating overall self-control by changing aspects of the physical or social environment. On the other hand, attentional deployment does not entail changing aspects of the environment; instead, it involves focusing attention on particular features of that environment. As noted above, current mitigation strategies that map onto attentional deployment include cueing, decluttering, task sequencing, context-sensitive help, and transposition. A related approach within attentional deployment could be to have the system detect signs that the user is distracted and then manipulate the salience of displayed information to draw the user's attention back to the task. For example, if the system detects that an individual's eyes are not focusing on the task, the system could highlight particular portions of the screen. Note, however, that implementing these strategies at the wrong moment could cause a disruption in the user's work, thus preventing successful mitigation. As suggested by Afergan, Hineks, Shibata, and Jacob (2015), physiological sensors could be used to modulate these attention-related approaches in real-time as a way to combat user disruption [43].

Cognitive change refers to thinking about a situation or task differently to facilitate goal pursuit. In terms of current mitigation strategies, task sequencing has some overlap with cognitive change. Beyond this, several other strategies involving influencing the user's thinking about the situation or task might be possible. For instance, one approach to individual motivation emphasizes the concepts of valence (expected value of an outcome), instrumentality (belief regarding the link between performance and an outcome), and expectancy [belief regarding the link between effort and performance; see 44]. Based on this, the system could attempt to influence the individual's level of one or more of these variables at key points in time. For example, by monitoring the individual's engagement and performance (e.g., attention levels across task components) and task characteristics (e.g., importance, complexity), the system might be able to issue reminders of the importance of the task and how it fits in with broader objectives. Another form of cognitive change that could act to enhance performance in appropriate situations (e.g., training, education) is gamification–the application of game playing elements to encourage engagement [see 45, 46]. If the task is designed to include a gamification component (e.g., leaderboards), the system could intervene when users have low task engagement. For example, the system could use current physiological or performance indicators to determine when the user has low task engagement. Once the system has detected a need for a mitigation strategy, it could implement a gamification element by reminding the user about an underlying dynamic game (e.g., revealing the user's current ranking within a training environment). This game element could then prompt users to think about the situation differently, facilitating goal pursuit.

Finally, response modulation involves the suppression of undesirable impulses or the amplification of desirable impulses in the moment. Human control over goal-incongruent impulses is an imperfect process, but augmented cognition might assist with this. As noted, interventions focusing on regulation of physiological states [e.g., 10] appear to overlap with this self-control strategy. Building on this, if physiological indicators of disruptive impulses could be identified, they might be used in mitigation strategies, where these indicators trigger an intervention in which the user is guided through steps to reduce the strength of these impulses (e.g., mindfulness).

4 Future Directions

Integrating the self-control strategies perspective and augmented cognition work also highlights several research directions that might benefit both areas. One potential direction involves the development of measures of the self-control strategies to assess the user's initial preferences and capabilities in implementing these strategies. These measures could help provide the system with a starting point for understanding the user's potential behavior in response to the many distractions or temptations that might be experienced during task performance. This information might then be used in system customization efforts in accordance with the expected user reactions to any interruptions.

A second direction involves developing a better understanding of the dynamic physiological changes that take place as impulses are generated and different self-control strategies are implemented. By using measures included in augmented cognition-based systems, indicators related to physiological changes resulting from impulse generation and strategy implementation might be observed (e.g., cortisol level, heart rate). Understanding these underlying relationships may provide insights regarding how to objectively and dynamically assess users' reactions to various distractions, which would enable the system to implement customizations as needed. The increasing prevalence and technological advancement of wearable devices can also support the measurement of physiological changes in an efficient [e.g., 47] and accurate way [e.g., 48, 49].

A third potential direction involves exploring different ways in which mixed reality technologies [50]–such as virtual reality, augmented virtuality, and augmented reality– can directly support and enhance human self-regulatory capabilities. The virtuality continuum [50] of these technologies allows for different levels of real and virtual environments to be experienced. Future research efforts can examine how visual display devices that incorporate these technologies (e.g., virtual reality head-mounted displays, augmented reality glasses) may strengthen the implementation of self-control strategies by enhancing or limiting sensory (e.g., visual) information from the environment. For instance, virtual reality head-mounted displays can support situation selection by immersing the user in a complete virtual environment. Other self-control strategies might be enhanced by using augmented reality glasses: situation modification and attention deployment could be supported by limiting or masking disruptive elements in the environment, and cognitive change and response modulation could be supported by adding virtual layers that may act as motivational or inhibitory reminders.

Finally, a fourth direction is to capture each individual's unique pattern of self-control strategy use with the aim of integrating it later with different systems that the user may wish to use. This can be applied in two contexts: (a) relatively-similar situations and (b) relatively-novel situations. In relatively-similar situations, the user experiences a situation that is similar to previous situations in which the user's pattern of self-control strategy use was observed. In this case, standalone data about a user's preferences or capabilities in implementing self-control strategies could act as an add-on with other software programs to facilitate the customization process of systems to maximize individual performance. By capitalizing on similar aspects of these experiences, users may adapt more quickly to other systems and these systems could become more effective and versatile. In relatively-novel situations, the user experiences a situation that can be considered new in comparison to the characteristics of situations for which the user's pattern of self-control strategy use was recorded. In this case, learning about user patterns of self-control strategy use for achieving various goals in previous situations would involve building a dynamic dataset optimized by machine learning through many user-system interactions. This proposed machine learning software could then analyze novel situations or experiences and the system would be able to suggest specific self-control strategy interventions that are expected to be most effective based on each user's capacities and limitations. In this way, individual self-control strategy habits could be automated to help users adapt to and thrive in new situations where no habits have yet been formed, allowing higher levels of adaptation to be reached in an accelerated way.

5   Conclusion

Recent work has begun to highlight the potential importance of self-regulatory processes in the context of augmented cognition. The current paper attempts to expand further in this direction by discussing an emerging view of self-regulation focusing on self-control strategies. We discuss how a self-control strategies perspective may inform our understanding of current augmented cognition approaches in several ways, suggest new approaches for developing mitigation strategies, and highlight future research directions. We hope that this preliminary discussion of this integration leads to additional work that may be beneficial to both areas.

References

1. Schmorrow, D., McBride, D.: Introduction. Int. J. Hum.-Comput. Interact. **17**(2), 127–130 (2004)
2. Miller, C.A., Dorneich, M.C.: From associate systems to augmented cognition: 25 years of user adaptation in high criticality systems. In: Poster presented at the Augmented Cognition Conference (2006)
3. Raley, C., Stripling, R., Kruse, A., Schmorrow, D., Patrey, J.: Augmented cognition overview: improving information intake under stress. In: Proceedings of the Human Factors and Ergonomics Society Annual Meeting, vol. 48, no. 10, pp. 1150–1154. SAGE Publications, Los Angeles (2004)
4. Dorneich, M.C., Whitlow, S., Ververs, P.M., Rogers, W.H.: Mitigating cognitive bottlenecks via an augmented cognition adaptive system. In: IEEE International Conference on Systems, Man, and Cybernetics, vol. 1, pp. 937–944 (2003)
5. Schmorrow, D., Kruse, A., Forsythe, C.: Augmented cognition. In: Cognitive Systems, pp. 109–144. Psychology Press, New York (2006)
6. Schmorrow, D.D., Fidopiastis, C.M. (eds.): AC 2018. LNCS (LNAI), vol. 10915. Springer, Cham (2018). https://doi.org/10.1007/978-3-319-91470-1

7. Stephens, C., et al.: Biocybernetic adaptation strategies: machine awareness of human engagement for improved operational performance. In: Schmorrow, D.D., Fidopiastis, C.M. (eds.) AC 2018. LNCS (LNAI), vol. 10915, pp. 89–98. Springer, Cham (2018). https://doi.org/10.1007/978-3-319-91470-1_9

8. Powers, W.T.: Behavior: The Control of Perception. Aldine Publishing Co., Chicago (1973)

9. Vancouver, J.B.: Self-regulation in organizational settings: a tale of two paradigms. In: Boekerts, M., Pintrich, P.R., Zeidner, M. (eds.) Handbook of Self-regulation, pp. 303–341. Academic Press, San Diego (2000)

10. Brumback, H.K.: Investigation of breath counting, abdominal breathing and physiological responses in relation to cognitive load. In: Schmorrow, D.D., Fidopiastis, C.M. (eds.) AC 2017. LNCS (LNAI), vol. 10284, pp. 275–286. Springer, Cham (2017). https://doi.org/10.1007/978-3-319-58628-1_22

11. Schwarz, J., Fuchs, S., Flemisch, F.: Towards a more holistic view on user state assessment in adaptive human-computer interaction. In: 2014 IEEE International Conference on Systems, Man and Cybernetics (SMC), pp. 1228–1234 (2014)

12. Fuchs, S., Hale, K.S., Stanney, K.M., Juhnke, J., Schmorrow, D.D.: Enhancing mitigation in augmented cognition. J. Cognit. Eng. Decis. Making 1(3), 309–326 (2007)

13. Duckworth, A.L., Gendler, T.S., Gross, J.J.: Situational strategies for self-control. Perspect. Psychol. Sci. 11(1), 35–55 (2016)

14. Fuchs, S., Schwarz, J., Flemisch, F.O.: Two steps back for one step forward: revisiting augmented cognition principles from a perspective of (social) system theory. In: Schmorrow, D.D., Fidopiastis, C.M. (eds.) AC 2014. LNCS, vol. 8534, pp. 114–124. Springer, Cham (2014). https://doi.org/10.1007/978-3-319-07527-3_11

15. Elkin-Frankston, S., Wollocko, A., Niehaus, J.: Strengthening health and improving emotional defenses (SHIELD). In: Schmorrow, D.D., Fidopiastis, C.M. (eds.) AC 2018. LNCS (LNAI), vol. 10916, pp. 58–66. Springer, Cham (2018). https://doi.org/10.1007/978-3-319-91467-1_5

16. Fuchs, S.: Session overview: adaptation strategies and adaptation management. In: Schmorrow, D., Fidopiastis, C. (eds.) AC 2018. LNCS, vol. 10915, pp. 3–8. Springer, Cham (2018). https://doi.org/10.1007/978-3-319-91470-1_1

17. Jøsok, Ø., Hedberg, M., Knox, B.J., Helkala, K., Sütterlin, S., Lugo, R.G.: Development and application of the hybrid space app for measuring cognitive focus in hybrid contexts. In: Schmorrow, D.D., Fidopiastis, C.M. (eds.) AC 2018. LNCS (LNAI), vol. 10915, pp. 369–382. Springer, Cham (2018). https://doi.org/10.1007/978-3-319-91470-1_30

18. Duckworth, A.L., Gendler, T.S., Gross, J.J.: Self-control in school-age children. Educ. Psychol. 49(3), 199–217 (2014)

19. Duckworth, A.L., White, R.E., Matteucci, A.J., Shearer, A., Gross, J.J.: A stitch in time: strategic self-control in high school and college students. J. Educ. Psychol. 108(3), 329–341 (2016)

20. Vancouver, J.B., Day, D.V.: Industrial and organisation research on self-regulation: from constructs to applications. Appl. Psychol. 54(2), 155–185 (2005)

21. Baumeister, R.F., Heatherton, T.F.: Self-regulation failure: an overview. Psychol. Inq. 7(1), 1–15 (1996)

22. Carver, C.S., Scheier, M.F.: On the Self-regulation of Behavior. Cambridge University Press, New York (1998)

23. Vancouver, J.B., Thompson, C.M., Williams, A.A.: The changing signs in the relationships among self-efficacy, personal goals, and performance. J. Appl. Psychol. 86(4), 605–620 (2001)

24. Klein, H.J.: An integrated control theory model of work motivation. Acad. Manage. Rev. 14(2), 150–172 (1989)

25. Gillebaart, M.: The "operational" definition of self-control. Front. Psychol. **9**, 1231 (2018)
26. de Boer, B.J., van Hooft, E.A., Bakker, A.B.: Stop and start control: a distinction within self-control. Eur. J. Pers. **25**(5), 349–362 (2011)
27. de Boer, B.J., Van Hooft, E.A., Bakker, A.B.: Self-control at work: its relationship with contextual performance. J. Managerial Psychol. **30**(4), 406–421 (2015)
28. de Ridder, D.T., de Boer, B.J., Lugtig, P., Bakker, A.B., van Hooft, E.A.: Not doing bad things is not equivalent to doing the right thing: distinguishing between inhibitory and initiatory self-control. Pers. Individ. Differ. **50**(7), 1006–1011 (2011)
29. Tangney, J.P., Baumeister, R.F., Boone, A.L.: High self-control predicts good adjustment, less pathology, better grades, and interpersonal success. J. Pers. **72**(2), 271–322 (2004)
30. Gottfredson, M.R., Hirschi, T.: A General Theory of Crime. Stanford University Press, Stanford (1990)
31. Vohs, K.D., Faber, R.J.: Spent resources: self-regulatory resource availability affects impulse buying. J. Consum. Res. **33**(4), 537–547 (2007)
32. Vohs, K.D., Baumeister, R.F.: Understanding self-regulation: an introduction. In: Handbook of Self-regulation: Research, Theory, and Applications, pp. 1–9 (2004)
33. Gross, J.J.: The emerging field of emotion regulation: an integrative review. Rev. Gen. Psychol. **2**(3), 271–299 (1998)
34. Gross, J.J., Thompson, R.A.: Emotion regulation: conceptual foundations. In: Gross, J. J. (ed.) Handbook of Emotion Regulation, pp. 3–24. Guilford Press, New York (2007)
35. Duckworth, A.L., Taxer, J.L., Eskreis-Winkler, L., Galla, B.M., Gross, J.J.: Self-control and academic achievement. Ann. Rev. Psychol. **70**, 373–399 (2019)
36. Hofmann, W., Schmeichel, B.J., Baddeley, A.D.: Executive functions and self-regulation. Trends Cognit. Sci. **16**(3), 174–180 (2012)
37. Veltman, H.J., Jansen, C.: The adaptive operator. Hum. Perform. Situation Awareness Automat. Current Res. Trends **2**, 7–10 (2004)
38. Schwarz, J., Fuchs, S.: Multidimensional real-time assessment of user state and performance to trigger dynamic system adaptation. In: Schmorrow, D.D., Fidopiastis, C.M. (eds.) AC 2017. LNCS (LNAI), vol. 10284, pp. 383–398. Springer, Cham (2017). https://doi.org/10.1007/978-3-319-58628-1_30
39. Galster, S.M., Parasuraman, R.: Task dependencies in stage-based examinations of the effects of unreliable automation. In: Proceeding of the Second Human Performance, Situation Awareness and Automation Conference (HPSAA II), vol. 2, pp. 23–27 (2004)
40. Scheiter, K., Gerjets, P.: Learner control in hypermedia environments. Educ. Psychol. Rev. **19**(3), 285–307 (2007)
41. Evans, G.W., Johnson, D.: Stress and open-office noise. J. Appl. Psychol. **85**(5), 779–783 (2000)
42. Sundstrom, E., Town, J.P., Rice, R.W., Osborn, D.P., Brill, M.: Office noise, satisfaction, and performance. Environ. Behav. **26**(2), 195–222 (1994)
43. Afergan, D., Hincks, S.W., Shibata, T., Jacob, R.J.K.: Phylter: a system for modulating notifications in wearables using physiological sensing. In: Schmorrow, D.D., Fidopiastis, C. M. (eds.) AC 2015. LNCS (LNAI), vol. 9183, pp. 167–177. Springer, Cham (2015). https://doi.org/10.1007/978-3-319-20816-9_17
44. Pinder, C.C.: Work Motivation in Organizational Behavior. Psychology Press, New York (2008)
45. Landers, R.N., Auer, E.M., Collmus, A.B., Armstrong, M.B.: Gamification science, its history and future: definitions and a research agenda. Simul. Gaming **49**(3), 315–337 (2018)
46. Landers, R.N., Bauer, K.N., Callan, R.C.: Gamification of task performance with leaderboards: a goal setting experiment. Comput. Hum. Behav. **71**, 508–515 (2017)

47. Vaughn, J., Gollarahalli, S., Summers-Goeckerman, E., Jonassaint, J., Shah, N.: Integrating mobile health technology for symptom management in acute pediatric blood and marrow transplant patients. Blood **132**(Suppl 1), 4726 (2018)
48. Hernando, D., Roca, S., Sancho, J., Alesanco, Á., Bailón, R.: Validation of the apple watch for heart rate variability measurements during relax and mental stress in healthy subjects. Sensors (Basel, Switzerland) **18**(8), 2619 (2018)
49. Falter, M., Budts, W., Goetschalckx, K., Cornelissen, V., Buys, R.: Accuracy of apple watch measurements of heart rate and energy expenditure in patients with cardiovascular disease. JMIR mHealth uHealth **7**, e11889 (2018)
50. Milgram, P., Kishino, F.: A taxonomy of mixed reality visual displays. IEICE Trans. Inf. Syst. **77**(12), 1321–1329 (1994)

Guided Mindfulness: New Frontier to Augmented Learning

Nisha Quraishi, Nicholas A. Moon, Katherine Rau, Lida P. Ponce, Mina Milosevic, Katrina Merlini, and Richard L. Griffith$^{(\boxtimes)}$

Florida Institute of Technology, Melbourne, FL 32901, USA
griffith@fit.edu

Abstract. This paper presents Guided Mindfulness (GM), a technology assisted platform that optimizes experiential learning and guides its users to recollect, recognize, decide, and be aware of new learning experiences through specific prepare and reflect experiences. The purpose of this paper is to introduce GM as a tool that can be utilized for skill acquisition through experiential learning in complex environments. Theories of skills complex and technical skill acquisition are linked to learning in complex dynamic environments to best explain the vision of GM.

Keywords: Guided Mindfulness · Experiential learning · Skill acquisition · Complex skills · Technical skills

> *The side that learns faster and adapts more rapidly—the better learning organization—usually wins.*
>
> *- General David Petraeus*

1 Introduction

The environment in which the U.S. military now operates in is characterized by increased complexity. In order to address this changing environment, U.S. military training doctrine now emphasizes the development of agile, adaptable leaders with broad critical thinking skills. Conventional wisdom suggests that the development of these skills is not suited to classroom activity and is best achieved through experiential learning. While experiential learning has major advantages over formal training methods, in practice the benefits can be difficult to achieve. Even with exposure to the appropriate experiences, without proper support skill acquisition can be hampered.

Skill acquisition refers to the development of technical and complex skills through four phases; recollection, recognition, decision, and awareness [1]. Skill acquisition is done consistently throughout the course of a life cycle, and the complexity of the skill acquired depends on the task at hand. Skill acquisition begins with the recollection phase, where a learner is able to understand parts of a situation. Leading to the recognition phase, where a learner holistically links parts of a situation together to understand the situation as a whole. Followed by the decision phase, where the learner makes analytical or intuitive decisions about the learning experience. Finally leading to

D. D. Schmorrow and C. M. Fidopiastis (Eds.): HCII 2019, LNAI 11580, pp. 586–596, 2019.
https://doi.org/10.1007/978-3-030-22419-6_42

the awareness phase, where the learner is able to be consumed in his or her performance [1].

Skill acquisition consists of the development of all kinds of skills. Simple skills, technical skills, hard skills, soft skills, complex skills, etc. are all skills developed in skill acquisition. In this paper, we will be focusing on two specific sets of skills; technical skills and complex interpersonal skills. For the rest of this paper, we will be referring to complex interpersonal skills as complex skills. Technical skills refer to skills that are needed in specific areas of expertise [2]. Complex skills, on the other hand, are the skills that are needed when interpreting highly volatile, uncertain, complex, and ambiguous (VUCA) environments [3]. See Table 1 for further distinction between technical and complex skills.

Table 1. List of Technical vs Complex skills

Technical skills	Complex skills
Tool usage	Adaptive thinking
Application	Adaptive performance
Maintenance	Resilience

The purpose of this paper is to introduce Guided Mindfulness (GM) as a tool that can be utilized for skill acquisition through experiential learning in complex environments. GM is a technology assisted platform that optimizes experiential learning and guides its users to recollect, recognize, decide, and be aware of new learning experiences through specific prepare and reflect experiences. These experiences are facilitated through an artificial intelligence feature in the app that prompts its users to answer a series of questions to prepare them for a new learning experience followed by a series of reflection questions after the completion of the learning experience. Through this process, GM can more effectively develop technical and complex skills and more systematically facilitate the skill acquisition process. GM can aid in developing these skills in complex dynamic and organizational environments, where learning opportunities can often be lost in rapid operational tempo. Overall, GM will aid its users by reinforcing effective memory and event preparation strategies across a potentially wide range of competencies.

1.1 Overview of Guided Mindfulness (GM)

Although approximately 70% of learning is expected to stem from experiential processes [4], whether individuals actually learn from their experiences is not guaranteed [5]. For instance, various individual differences (e.g., conscientiousness) may play a role in whether individuals have a propensity to reflect meaningfully upon their experiences and effectively learn from them [6]. Therefore, a more systematic, or guided, approach to experiential learning may be required to facilitate learning across individuals with varying degrees of learning dispositions and abilities. GM is proposed as a technology assisted platform intended to facilitate its users' abilities to effectively and strategically learn from experiences.

GM capitalizes on complex learning from experiences by engaging the users in event-based preparation and reflection activities grounded in self-regulatory and mindfulness principles [3]. Users are first assessed on various competencies deemed relevant (based on the Service's mission, job requirements, the user's career trajectory, etc.). The users' current standings on these competencies aid in their self-awareness and become the baseline for learning improvements. GM then initializes the prepare phase upon the users' identification of an important event where learning opportunities may take place. In the prepare phase, GM may prompt the user to consider which of the assessed competencies are necessary for successful performance, their level of proficiency on those competencies, and possible barriers or roadblocks that may interfere with successful performance. Through this preparation phase, GM helps the users become more self-aware by understanding their strengths and weaknesses and by identifying discrepancies between their current standings and their competency goals, which is an important aspect of self-regulation. This phase also draws attention to situational and social cues that are critical to learning. Directing the users' attention to these cues can heighten their salience during the event, increasing the users' ability to be mindful of them while in action. Being mindful of these cues while engaging in the event can help users recognize when adjustments to their actions are required (e.g., based on non-verbal feedback from the social environment) and will help them more accurately assess the event during the reflection phase.

After the event, the reflection phase is initiated. Specifically, GM prompts the users to reflect by providing a structured sensemaking process that allows the users to think through and analyze important aspects of the event. Through this process, users make meaning of what happened during event, including what they did and did not do well, resulting in new or revised mental models [3]. Upon this sensemaking, GM prompts the users to simulate alternative actions that could prove useful in future, similar events. This simulation is expected to result in greater internalization of what was learned as well as a greater number of strategies the users have in their repertoires for application in future events [3]. All data gathered from the prepare and reflect phases is stored in a GM database and can be sorted and reviewed to aid in future meta-learning (e.g., identification of problem areas across various events or certain types of events). Further, through ongoing engagement with GM, users are expected to become more efficient and effective in learning from their experiences overtime; ultimately enhancing their self-regulatory and mindfulness capabilities.

1.2   GM and Technical Skill Acquisition

Technical skills, also known as "hard skills", generally refer to skills or competencies related to an individual's particular area of expertise. Although they are often associated with the use of machinery, tools, or equipment, these competencies go beyond just engineering or mechanical tasks [2]. Technical skills are considered critical aspects of performance among surgeons, pilots, graphic designers, etc. [7, 8]. Technical skills can also range from simple to complex. Simple technical skills are fairly straightforward, can be learned quickly, and usually involve one or very few motor activities. Complex technical skills, however, require a combination of specific knowledge and psychomotor performance; these skills take longer than one session to learn and can be a clear differentiator between novices and experts [9]. In challenging modern organizations, technological advancements and dynamic work environments have made almost every technical skill a complex one.

Acquiring technical skills usually proceeds in three phases [10]. In the first phase, trainee develops basic cognitive understanding of the task. This cognitive phase is characterized by erratic performance that proceeds in distinctive steps. Through practice and feedback, the learner moves from cognitive knowledge to motor performance, also known as integrative phase. This phase is characterized by more stable performance and blurring of the distinctive steps into one fluid action; however, the learner still has to pay attention to the task. In the final, autonomous phase, the learner is able to perform smoothly and with minimal, if any, attention to the activity [11].

More recently, it has been proposed that acquisition of technical skills is followed by a fourth phase: the application of the knowledge in new and complex situations. In this phase, experiential learning has been identified as a critical training mechanism that enables knowledge application [12]. In order for experiential learning to be effective, the learner must have an opportunity to reflect on his own experience, and through reflection, broaden the understanding of self, her/his action, and the context in which the action occurs [13]. This translates into an ability to apply knowledge beyond the training context and effectively adapt that application of knowledge when necessary, at which point the trainee becomes an expert.

As critical as experiential learning is to the development of expertise, this part of technical skills training is often neglected. We propose that GM can be a powerful tool that can assist learner capitalize on experiential learning by leveraging preparation and reflection during skill acquisition as well as during application of skills in real-life settings.

1.3 GM and Complex Skill Acquisition

As task complexity increases, so does the gap in performance between practiced professionals and beginners [14]. As such, is it important that the learning process of complex skills is fostered within novices, especially when such acquisition is necessary for success or must occur within a short period of time. It is a commonly held belief that the majority of complex skills are most successfully acquired through experiential learning methods [15]. GM aids in complex skill acquisition through experiential methods, namely by improving learners' self-regulatory mechanisms, such as awareness (self, situational, social), sensemaking, and simulation.

The importance of self-regulation to learning outcomes is rooted in control theory [16, 17]. Control theory describes a negative feedback loop where an individual deliberately compares his or her current state to a desired state [18]. As such, a plan is set into motion to reach the desired state if there is a discrepancy. For complex skills, individuals must take stock of skill and competency states in addition to social and situational factors external to the individual to reach their desired end state. The individual then uses the feedback gathered from the assessment of these states and factors to reflect and integrate their learning experiences and further improve [3]. However, as task and situation complexity increases, individuals may be unable to effectively synthesize learning in real-time due to cognitive overload and new task demands. GM thus serves as a facilitating mechanism by which individuals can effectively manage their learning time to optimize experiential learning. The GM platform is designed to prompt users with insightful preparation and reflection questions that prompt users to formulate internal feedback, providing insight into their performance and learning.

In order for self-regulation to be successful in terms of behavior, cognition, and affect, individuals must first be self-aware [19]. Self-awareness is defined as an individual's ability to understand his or her own skills, affect, values, strengths, weaknesses, assumptions of the world and others, and biases [20]. As control theory is predicated on the ability to compare desired states to current states [18], self-awareness is the next step in directing an individual to achieve their goals. Through pre-event preparation, GM guides its users to assess their own standing on skills and competencies, allowing users to become more aware of the automatic ways in which they interact with the environment and pinpoint strengths and ways to improve [3]. Post-event reflection allows for a comparison to pre-event expectations of the event itself and anticipated vs. actual behavior in the event. Such reflection allows for learners to seek and receive feedback from themselves and the environment, improving overall self-awareness. When tasked with acquiring complex skills, individuals face higher levels of cognitive load and may be required to access various bodies of knowledge, skills, and competencies at once to effectively perform [14, 21]. In this case, pre-event preparation allows learners to assess their abilities while refraining from judgment [22], making them better prepared for the reality of the upcoming learning event. Additionally, making feedback self-guided was found to be beneficial for performance since learners are forced to learn how to identify and correct their own errors [21, 23]. Thus, providing support that the self-guided nature of post event reflection is advantageous to users who wish to learn complex skills.

Situational awareness refers to an individual's understanding of a situation and the ability to rapidly adjust their understanding as the environment changes [24]. Situational awareness is paramount in VUCA environments. In today's high-speed business world, acquiring complex skills oftentimes happens in VUCA environments, and the process of acquiring complex skills is often filled with uncertainty, complexity, and ambiguity. Situational awareness requires individuals to have accurate mental models of the given circumstance and all related components [25], allowing for thorough understanding of the environment where possible. Situational awareness, similar to self-awareness, is critical to self-regulation because it allows individuals to process in-time environmental feedback and direct cognitive resources to the learning event [3]. Prior to an event, GM directly prompts individuals to identify contextual factors relevant to his or her ability to perform in the event and asks, more specifically, how each factor may influence the event. Following the event, the user is then asked to confirm or disconfirm how each situational factor mentioned previously affected the event. In doing so, the learner's attention is brought to situational cues critical to performance in the future and allows for the restructuring and improvement of associated mental models [3].

Social awareness focuses on the understanding of appropriate behaviors between individuals in addition to group-level dynamics [26]. Learning, especially complex skills such as global leadership or cross-cultural competence, rarely, if ever, occurs within a vacuum. Navigating social interaction and obtaining feedback from such environments is a crucial aspect of the learning process. Complex social environments require more than automatic processing; individuals must dedicate mindful attention to social interactions to notice and successfully perceive social cues, recognize patterns of relationships between others, and understand networks of power and social influence [27]. A better understanding of the social climate is likely to lead to better outcomes for individuals engaging in social interaction [28]. GM implores users to investigate the social environment by asking users to determine relevant stakeholders and other players in given contexts relevant to their goals and learning. Following events, GM helps guide mindful reflection of social interaction to provide feedback on an individual's behavior and overall performance during the event.

Sensemaking and simulation are two of the more forward-focused components to the GM platform and as such, have especially unique implications for complex skill acquisition. Sensemaking is accomplished through reflection and is defined by Griffith and colleagues (2017) as "the process by which people infer meaning from an event and use that derived meaning to decide on a future course of action," (p. 8) [3]. Sensemaking necessitates thorough and effective self, situational, and social awareness [29]. Sensemaking involves the complex process of inferring meaning from relevant cues (i.e., feedback from various sources) to adjust mental models and inform future decisions [3]. Sensemaking aids individuals in meaningfully categorizing and clarifying inputs [30]. This process allows tacit knowledge to develop following experiences individuals have, making learned knowledge explicit and viable. As such, sensemaking is a vastly important process for those learning complex skills [3]. GM aims to improve individual sensemaking by utilizing structured and unstructured reflection strategies that require critical thinking and recognition of interrelated concepts and mental models. Pre-event prompts call attention to salient contextual cues that users can attune to in real-time, consistently updating their cognitive frameworks and adapting as needed. Mental models that are updated in real-time may expedite the learning process of complex skills by allowing users to know which elements need to be practiced or more fully understood and where to ask questions or seek direct feedback from others.

GM considers simulation as a sort of "mental trial" of relevant events. Successful simulation occurs following a thorough sensemaking process, where individuals can more effectively answer the various, "what if's" of any given situation [3]. This forward-thinking exercise allows individual to prepare for any upcoming complex situation. For complex skill learners, simulation is especially important as such learners are often forced to operate "on all cylinders" in terms of the variety of knowledge, skills, and competencies. Simulating an upcoming event, GM prompts its users to consider which knowledge, skills, and competencies they will have to incorporate and account for in the upcoming event, making them better prepared and better positioned to obtain favorable outcomes. Simulation is also prompted following the event, where users are tasked with considering alternative courses of action and their possible ramifications. This sort of debrief aims to more deeply embed the learning associated with the event, further clarifying and expanding upon cognitive frameworks across environments.

1.4   Adapting to Complex/Dynamic Environments

Complex dynamic environments are environments where tasks are more intricate, leading to tasks that take longer to complete, increasing the variety of possible outcome decisions [31]. There are a myriad of environments that can be classified as complex dynamic environments that include the following: uncertainty, complexity, ill structured problems, time pressures, shifting opposing goals, poorly defined goals, high stakes, feedback loops, multiple players, multiple elements, under vigilance or pressure to perform to organizational goals and norms [32]. Typically individuals in complex dynamic environments engage in complex dynamic control tasks (CDC tasks) that use some form of cognitive activity such as decision making, critical thinking, problem solving, or cognitive flexibility [33].

Complex dynamic environments are the "normal" milieu of the military. Warfighters operate in complex situations riddled with ambiguity and volatility. When warfighters are sent into a VUCA environment, in many instances their mission is not easily accomplished due to a variety of contextual factors (the elements, lack of information and equipment, etc.). Each of these factors can change at any given time leading to different consequences for each chosen action. Another aspect to the complexity in the environment of the warfighter are the series of mission changes and new environments that consist of novel elements and variables that require immediate transfer of skills and knowledge.

Failure to acquire the skills needed to adapt to complex dynamic environments is due to a multitude of factors. Spiro, Feltovich, Coulson [33] explain a factor as individuals possessing a reductive world view known as the "inappropriate lessening or oversimplification of complexity" in which "parts of complex systems are assumed to 'add up' to wholes" (p. S52). Another factor is the lack of experts sent into these complex dynamic environments. Many studies support the notion that experts do better than novices in complex and dynamic environments and require less effort to be proficient [34–40]. An expert is defined as someone with more experience and knowledge whereas, a novice is defined as someone less skilled with less experience and knowledge [34]. This factor may be influenced by the lack of training time needed to effectively adapt to a complex dynamic environment. As such, even if there is time and training for novices to become experts, a number of studies demonstrate that novices do just as well as experts in decision-making tasks [34]. This is only the case if novices are capable of having a more expansive and flexible cognitive view on processing complexity [33]. GM can aid in creating expertise and cognitive flexibility which can support performance in uncertain dynamic environments [3, 41].

1.5 Integrating GM Over the Learning Cycle-Adapting to the Environment, not so Much Learning and Improving Complex Skills

Based on the previously mentioned skill developments, GM is expected to be best utilized by integrating over a full learning or training cycle of these complex and technical skills. Therefore, organizations can capitalize on learning, training and development through implementing GM as an ongoing tool to manage learning and acquiring skills. GM can help with both the individual's learning and the organization's learning by capitalizing on the experiential aspect of learning for the former and the organizational learning cycle for the latter.

Individual's learn in several ways: direct experience, verbal transmitting of information (e.g., reading and writing), and reorganizing previously learned information

[42, 43]. Experiential learning consists of four stages: concrete experience, reflective observation, abstract conceptualization, and active experimentation [13]. Unfortunately, information is lost over the experiential learning cycle due to the lack of time to properly engage in reflective observation and abstract conceptualization; therefore, GM is a possible solution to this loss of information by prompting individuals to reflect and make sense of experiences in real-time after a learning event occurs [13].

Organizational learning typically occurs over four stages: generating, integrating, interpreting, and acting [42]. Generating refers to the collection of external data and the generation of new ideas within the organization. Integrating refers to the combination of the information into an organization's context so that departments and individuals within an organization share this information and can act in concert with each other. Interpreting refers to when organizational members digest and clarify the information generated collectively. Acting refers to when organizational members have access to the previously mentioned information and make decisions based on this information. Often organizations fail at this stage by placing decision making in the hands of leaders therefore undermining the learning process [42]. Integrating GM across the latter stages of organizational learning will serve organizational learning well since it will allow for organizational members to take the information generated during the generation phase and allow for the integration and interpretation of that information. However, organizations will still need to empower employees to make decisions at the final stages to allow for the full organizational learning cycle to be in effect (i.e., allow employees to act on information learned during the former stages).

When GM is integrated over a full experiential learning cycle, it can allow for complex interpersonal skills to be developed [44]. The GM approach is designed to be competency neutral, so competencies can be either complex or technical; however, it is expected that learning complex skills will benefit from GM more than technical skills, due to the capitalization on the experiential learning of the individual.

Since GM is best used to capitalize on experiential learning, it is expected that GM will be best served over a period of time throughout the learning cycle rather than just one point of intervention. Experiential learning should make up 70% of learning in an organization through the on-the-job experiences [4]. Since GM capitalizes on self-regulated learning, GM would best be integrated over a full cycle of learning where time is required for feedback loops. Therefore, GM is not expected to be a one-time intervention during skill acquisition, rather GM should be implemented into daily experiences and learning on-the-job as it enhances the latter stages of organizational learning (i.e., integrating and interpreting).

1.6 Future Directions

This paper addressed the use of GM for skill acquisition. Future direction for GM can be geared towards developing the platform for specific skills. GM can be used to examine what types of skills are best learned through the platform; complex skills or technical skills. In a time where cross-cultural research and practice is at a peak, GM can assess skill acquisition among different cultures. GM can assess if there are cultural differences in skill acquisition. This may prove to be beneficial for leaders that lead cross cultural teams – GM may help certain team members to develop skills needed for team improvement. GM may also be used to examine age group differences in skill acquisition. It is clear that adults learn differently from children and adolescents. What can be further examined is how well adults in different age groups learn comparatively.

References

1. Dreyfus, S.E., Dreyfus, H.L.: A five-stage model of the mental activities involved in directed skill acquisition (No. ORC-80-2). California University Berkeley Operations Research Center (1980)
2. Medina, R.: Upgrading yourself-technical and nontechnical competencies. IEEE Potentials **29**(1), 10–13 (2010)
3. Griffith, R.L., Steelman, L.A., Wildman, J.L., LeNoble, C.A., Zhou, Z.E.: Guided mindfulness: a self-regulatory approach to experiential learning of complex skills. Theor. Issues Ergon. Sci. **18**(2), 147–166 (2017)
4. Lindsey, E.H., Homes, V., McCall Jr., M.W.: Key Events in Executives' Lives. Center for Creative Leadership, Greensboro, NC (1987)
5. Heslin, P.A., Keating, L.A.: In learning mode? The role of mindsets in derailing and enabling experiential leadership development. Leadersh. Q. **28**, 367–384 (2017)
6. DeRue, D.S., Nahrgang, J.D., Hollenbeck, J.R., Workman, K.: A quasi-experimental study of after-event reviews and leadership development. J. Appl. Psychol. **97**, 997–1015 (2012)
7. Nasir, A.N.M., Ali, D.F., Noordin, M.K.B., Nordin, M.S.B.: Technical skills and non-technical skills: predefinition concept. In: Proceedings of the IETEC 2011 Conference, January 2011
8. Katz, L.F., Margo, R.A.: Technical change and the relative demand for skilled labor: the united states in historical perspective. In: Human Capital in History: The American Record, pp. 15–57. University of Chicago Press, Chicago (2014)
9. Damooei, J., Maxey, C., Watkins, W.: A Survey of Skill Gaps and Related Workforce Issues in Selected Manufacturing Sectors: Report and Recommendations. Workforce Investment Board of Ventura County, USA (2008)
10. Fitts, P.M., Posner, M.I.: Human performance (1967)
11. Reznick, R.K., MacRae, H.: Teaching surgical skills—changes in the wind. N. Engl. J. Med. **355**(25), 2664–2669 (2006)
12. Aggarwal, R., Grantcharov, T.P., Darzi, A.: Framework for systematic training and assessment of technical skills. J. Am. Coll. Surg. **204**(4), 697–705 (2007)
13. Kolb, D.A.: Experiential Learning: Experience as the Resource of Learning and Development. Prentice Hall, Englewood Cliffs (1984)
14. Gopher, D., Weil, M., Siegel, D.: Practice under changing priorities: an approach to the training of complex skills. Acta Psychol. **71**, 147–177 (1989)
15. McCall, M.W.: Recasting leadership development. Ind. Organ. Psychol. **3**, 3–19 (2010)
16. Powers, W.T.: Behavior: The Control of Perception. Aldine, Chicago (1973)
17. Carver, C.S., Scheier, M.F.: Control theory: a useful conceptual framework for personality social, clinical, and health psychology. Psychol. Bull. **92**(1), 111 (1982)
18. Vancouver, J.B.: Self-regulation in organizational settings: a tale of two paradigms. In: Boekaerts, M., Pintrich, P., Zeidner, M. (eds.) Handbook of Self-Regulation, pp. 303–341. Academic Press, San Diego (2000)
19. Wagner, D.D., Heatherton, T.F.: Self-regulation and its failure: seven deadly threats to self-regulation. In: Borgida, E., Bargh, J. (eds.) APA Handbook of Personality and Social

Psychology, Attitudes and Social Cognition, vol. 1, pp. 805–842. American Psychological Association, Washington, DC (2015)

20. Goleman, D.: Working with Emotional Intelligence. Bantam Books, New York (1998)
21. Wulf, G., Shea, C.H.: Principles derived from the study of simple skills do not generalize to complex skill learning. Psychon. Bull. Rev. 9(2), 185–211 (2002)
22. Glomb, T.M., Duffy, M.K., Bono, J.E., Yang, T.: Mindfulness at work. Res. Pers. Hum. Resour. Manag. 30, 115–157 (2011)
23. Schmidt, R.A.: Frequent augmented feedback can degrade learning: evidence and interpretations. In: Requin, J., Stelmach, G.E. (eds.) Tutorials in Motor Neuroscience, pp. 59–75 (1991)
24. Rico, R., Sanchez-Manzanares, M., Gil, F., Gibson, C.: Team implicit coordination processes: a team knowledge-based approach. Acad. Manag. Rev. 33(1), 163–184 (2008)
25. Cooke, N.J., Stout, R.J., Salas, E.: A knowledge elicitation approach to the measurement of team situation awareness. In: McNeese, M., Endsley, M., Salas, E. (eds.) New Trends in Cooperative Activities: System Dynamics in Complex Settings, pp. 114–139. Human Factors, Santa Monica, CA (2001)
26. Hilton, R.M., Shuffler, M., Zaccaro, S.J., Salas, E., Chiara, J., Ruark, G.: Critical Social Thinking and Response Training: A Conceptual Framework for a Critical Social Thinking Training Program. (ARI Research Report). Army Research Institute for the Behavioral and Social Sciences, Arlington, VA (2009)
27. Gilbert, J.A., Kottke, J.L.: Developing a measure of social perceptiveness. Paper Presented at the Annual Conference of the Association for Psychological Science, San Francisco, CA (2009)
28. Grossman, R., Thayer, A.L., Shuffler, M.L., Burke, C.S., Salas, E.: Critical social thinking a conceptual model and insights for training. Organ. Psychol. Rev. 5(2), 99–125 (2015)
29. Sandberg, J., Tsoukas, H.: Makin sense of the sensemaking perspective: it's constituents, limitations, and opportunities for further development. J. Organ. Behav. 36(1), 6–32 (2015)
30. Weick, K.E., Sutcliffe, K.M., Obstfeld, D.: Organizing and the process of sensemaking and organizing. Organ. Sci. 16(4), 409–421 (2005)
31. Freed, M.: Managing multiple tasks in complex, dynamic environments. In: AAAI/IAAI, pp. 921–927, July 1998
32. Osman, M.: Controlling uncertainty: a review of human behavior in complex dynamic environments. Psychol. Bull. 136(1), 65 (2010)
33. Spiro, R.J., Feltovich, P.J., Coulson, R.L.: Two epistemic world-views: prefigurative schemas and learning in complex domains. Appl. Cogn. Psychol. 10(7), 51–61 (1996)
34. Cellier, J.M., Eyrolle, H., Mariné, C.: Expertise in dynamic environments. Ergonomics 40(1), 28–50 (1997)
35. Schvaneveldt, R.W., et al.: Measuring the structure of expertise. Int. J. Man Mach. Stud. 23(6), 699–728 (1985)
36. Leplat, J.: The elicitation of expert knowledge. In: Hollnagel, E., Mancini, G., Woods, D.D. (eds.) Intelligent Decision Support in Process Environments. NATO ASI Series (Series F: Computer and Systems Sciences), vol. 21, pp. 107–122. Springer, Heidelberg (1986). https://doi.org/10.1007/978-3-642-50329-0_7
37. Glaser, R., Chi, M.: Overview. In: Chi, M., Glaser, R., Farr, M. (eds.) The Nature of Expertise (1988)
38. Ericsson, K.A., Charness, N.: Expert performance: its structure and acquisition. Am. Psychol. 49(8), 725 (1994)
39. Sweller, J.: Cognitive load during problem solving: effects on learning. Cogn. Sci. 12(2), 257–285 (1988)

40. Sternberg, R.J., Frensch, P.A.: On being an expert: a cost-benefit analysis. In: Hoffman, R.R. (ed.) The Psychology of Expertise, pp. 191–203. Springer, New York (1992). https://doi.org/10.1007/978-1-4613-9733-5_11

41. Griffith, R.L., Steelman, L.A., Moon, N., Al-Qallawi, S., Quraishi, N.: Guided mindfulness: optimizing experiential learning of complex interpersonal competencies. In: Schmorrow, D.D., Fidopiastis, C.M. (eds.) AC 2018, Part II. LNCS (LNAI), vol. 10916, pp. 205–213. Springer, Cham (2018). https://doi.org/10.1007/978-3-319-91467-1_17

42. Dixon, N.M.: The Organizational Learning Cycle: How We Can Learn Collectively. Routledge, London (2017)

43. Noe, R.A., Peacock, M.: Employee training and development (2002)

44. McCall, M.W.: Leadership development through experience. Acad. Manag. Exec. **18**, 127–130 (2004)

Reading Behavior and Comprehension of C++ Source Code - A Classroom Study

Jonathan A. Saddler[1], Cole S. Peterson[1], Patrick Peachock[2], and Bonita Sharif[1(✉)]

[1] University of Nebraska - Lincoln, Lincoln, NE 68588, USA
{saddler,cole.scott.peterson}@huskers.unl.edu, bsharif@unl.edu
[2] Youngstown State University, Youngstown, OH 44555, USA
prpeachock@outlook.com

Abstract. This paper presents an eye-tracking study conducted in a classroom setting with seventeen students enrolled in a Computer Science program. The students were a mix of twelve first-year undergraduates (novices) and five masters students (non-novices). Students were asked to answer a comprehension question for each of thirteen C++ programs after reading them. Each program is split into a series of chunks which logically break down meaningful parts where eye gazes hint at cognition about parts of programs useful to solving problems. We analyze these patterns across chunks for the stories they tell about how participants went about searching for cues, and to learn whether their gaze patterns predicted accurate answers to three types of questions The results show that novices tend to visit print output statements and declaration statements the same amount as they do other statements in code with the exception of control block headers, which both groups tend to focus on the most across all categories. We also find that non-novices spend longer fixating inside of chunks of code before transitioning to other chunks, and tend to transition to chunks that are further away from their original position than novices.

Keywords: Eye tracking · Program comprehension · Expertise · Gaze transitions

1 Introduction

Program Comprehension is an essential activity [1] for students while they try to read and understand source code. It is important for students to develop code reading skills [2] early in the program because it forms the basis of many other activities. To better develop teaching tools for programming, it is critical to understand how students learn programming concepts. Source code is a rich combination of syntax and semantics. Determining either the importance of the syntax or semantics for a programmer (especially a student learning programming) requires a better understanding of how programmers read and understand

© Springer Nature Switzerland AG 2019
D. D. Schmorrow and C. M. Fidopiastis (Eds.): HCII 2019, LNAI 11580, pp. 597–616, 2019.
https://doi.org/10.1007/978-3-030-22419-6_43

code. From a programmer's own perspective, the question of "Where can I go to find what is important?" is an important research problem that is heavily task dependent. As researchers help develop better teaching and learning tools, we propose that the answers to these questions are perhaps stronger when quoted from the experiences of students who are learning in their field. To add to the evidence of how students learn, we present an eye tracking study conducted with students in a classroom setting using thirteen C++ short code snippets that were chosen based on concepts students learnt in the class. There has been an increase in the number of studies being conducted using an eye tracker in recent years [3]. However, there is still much work to be done to understand what students actually read while comprehending code. We see this paper among many others that will eventually be used in a meta-analysis several years from now. In this paper, we focus on C++ as most previous studies were done mostly on Java. Another unique aspect of this paper is the method used to analyze the data. Instead of simply looking at line level analysis of what students look at, we study how they read chunks of code and how they transition between them to answer comprehension questions. The research questions we seek to analyze are:

- RQ 1: How do students perform on comprehension questions related to short C++ code snippets?
- RQ 2: What sections of code (chunks) do students fixate on and if this changes with program size?
- RQ 3: What chunks do students transition between during reading?

Our first research question seeks to determine how accurately students perform on the comprehension tasks. In the second and third research questions, we analyze the eye tracking data collected on the C++ programs by segmenting the programs into chunks we are interested in analyzing and link them to the students' performance from our first research question.

2   Related Work

In prior work, we discuss the role eye tracking can have in computing education [4]. In this section, we present selected work on program comprehension done using an eye tracker. For a more exhaustive list, we direct the reader to Obaidellah et al. [3].

An eye tracker is a device that is used to monitor where a user is looking at on a screen. Eye trackers record raw gazes as they occur at various speeds. Later, via event detection algorithms, fixations and saccades are identified. A fixation is a point on the screen where the eyes are relatively stable while a saccade is the movement from one fixation to another fixation indicating navigation. Most saccades frequently last between 200 and 300 ms, but the time may vary. A group of saccades makes up a scan path [5]. A computer program is a set of instructions written to perform a specified task. Comprehension of a program is defined as understanding lines of code. This programming code can be in any language, C++, Java, or C# for example. To investigate one way programmers focus on code, studies have been done that look into different fragments of code, also known as *beacons*. Beacons can differ from user to user, thus giving us the knowledge that not all programmers look at the same code the same way [1].

In order to find a connection between the way programmers read code, two different studies were performed where the programs had specific syntax highlighting. In the first study [6] by Bleeders et al., 31 participants reading C# programming code were made part of either a black and white team or syntax highlighting team. They were given programs with no errors and unlimited time to read the program. Using an eye tracker, in the data recorded they reported on regressions, fixations, and scan percentage. While this study showed minimal difference, another study by Sarkar [7] found that highlighting was more effective with novice programmers but became less effective as a programmer accrued more experience. A continued effort into how a programmer explores code was performed by Raina et al. [8]. The study was focused on finding how students can retain information by reading in a less linear pattern. Instead of having students read code left to right, top to bottom, they gave students code in a segmented patterns. With an eye tracker they took a look at two metrics, reading depth and reading scores. The 19 students were split into a control group and a treatment group, both given the same C++ module. The treatment group was given segmented code while the control group was given linear code. Results of the study showed that subjects given the segmented code had higher scores in both reading and depth. They were able to focus and understand code better than those who read it linearly.

Sharif et al. [9] performed a study that focused on the comparison of Python and C++. Participants were split into groups based on their knowledge of each given language. Students were given tasks that consisted of finding bugs. Metrics used included fixation duration, fixation counts, time, and accuracy. The study showed that although C++ debugging took longer, that there was higher accuracy in the output matching specifications. Even though the study did show these differences, the overall analytical results came to the conclusion that there was no significant difference found between the programming languages. Note that this does not mean that there is no difference.

Using an eye tracker can help to better understand how code is reviewed. In 2002, a study was performed that looked at code reviewing [10]. Using an in-house developed tool, the authors looked at fixations on lines. The six different programs were reviewed by five different programmers. The reviewers all had a similar reading patterns in that they read the entire code once called a "Scan". After scanning the code, each would then go back and focus on certain parts of the code that they considered important. While this was a recurring instance for all reviewers, the results show that the reviewers had different patterns that involved recursive styles and that each focused on different variables.

Studies have been done that gather data on the effect of identifier styles such as underscores and camel case on comprehension [11,12]. Results showed that naming styles had an effect on the time and effort it took to find different identifiers. The overall conclusion found that when using underscores, speed to find identifiers was improved however even though camel case took longer it was more accurate.

3 Experimental Design

This study seeks to investigate what students read while they try to understand C++ code snippets. We study reading by analyzing the eye movements of students using an eye tracker. Each student was first asked to take as much time as needed to read a snippet of C++ code presented to them in random order. After each code snippet a random comprehension question was given (related to the corresponding C++ code fragment). We randomized the order of tasks to avoid any order biases. Interested readers can find our complete replication package at http://seresl.unl.edu/projects/hcii2019.

3.1 Tasks

The C++ tasks given to participants had varying degrees of constructs used with varied levels of difficulty. The 13 C++ programs used are shown in Table 1 with their corresponding difficulty level. Participants were given as much time as they needed to complete their task. After each task they were asked to answer one of three randomly assigned comprehension questions. Each comprehension question was followed by a question asking about confidence in their answer and their perceived difficulty of each task. At the end of the session, participants were also asked if they had any problems during the test, if they were given enough time, and the overall difficulty of all tasks. The comprehension question was one of the following: a question about what the program outputs, a short answer question, or a multiple choice question. The main purpose of randomly assigning questions was to deter students from sharing answers with each other, although our methods also ensured a well-distributed random assignment.

Table 1. C++ programs with constructs used, number of lines of code, and a difficulty rating based on how easy the concepts are for students to grasp.

Program Name	Constructs Used	LOC	Difficulty
StreetH.cpp	Classes, Get and set, parameter passing, this pointer	25	Medium
Student.cpp	Classes, Get and method, this pointer, constructor	25	Medium
Rectangle.cpp	Constructor, Inline methods, this pointer, parameter passing	24	Difficult
Vehicle.cpp	Class, constructor, parameter passing, if statement	34	Medium
StringDemo.cpp	Std String class, Replace, Find, Length, for loop,	17	Medium
TextClass.cpp	Std string class, string find, string length, string substr, string replace	12	Medium
WhileClass.cpp	String class, while loop, if statement, && operator	21	Difficult
Between.cpp	&& operator, functions, parameter passing, if statement	15	Medium
Calculation.cpp	Parameter passing, for loop, running total	16	Medium
SignCheckerClassMR.cpp	Constructor, nested ifs	33	Difficult
PrintPatternR.cpp	Nested for loops	13	Difficult
ReversePtrH.cpp	One dimensional arrays, for loop, swap, functions, parameter passing	23	Difficult
CalculatorRefH.cpp	Function prototypes, switch statement, parameter passing, pass by reference	23	Difficult

3.2 Areas of Interest

In order to analyze the students' eye movements in a more structured way, we broke down the program into different AOIs (areas of interest). AOIs were created for each line we found in every stimulus, and the fixations were mapped to the appropriate AOI. Next, we grouped these AOIs together to form *"chunks"* whose contents logically fit together into a unit that may be of interest to a programmer. We customized the selection of each of these chunks down to both the stimulus and task given to the participant. We further grouped these chunks into cross-stimulus "code categories", which we then used to discover constructs that groups of participants looked at with the highest frequency across all stimuli. This process is detailed in Sect. 4.4. In our selective mapping, the contents of each chunk are groups of contiguous lines logically suited to, as a unit, be a cue of interest to a programmer.

3.3 Study Variables

The independent variable is the expertise of the test participants with two treatments - novice and non-novice. The type of question asked per participant was a mixed group factor randomly assigned to each individual and recorded, and we use it to report on the "traits" of accuracy for each participant. The nine dependent variables measured in this study are given below. Transition refers to gaze transitions made between two chunks. Fixations were detected at 60ms using the Olsson fixation filter [13].

- Accuracy: Number of questions answered out of those presented. (Each participant was presented a total of 13 questions, and answered between 2 and 8 questions each from the three categories)
- Fixation Counts: The number of fixations made in a given chunk.
- Fixation Duration: The total sum of all fixations made in a given chunk.
- Chunk Fixation Duration Prior Exit: The average time spent fixating on a chunk before a transition is made.
- Transition Count: The sum of all transitions between chunks for a given code snippet.
- Vertical Later Chunk: The percentage of transitions between chunks that transition to a chunk lower in the code.
- Vertical Early Chunk: The percentage of transitions between chunks that transition to a chunk higher up in the code.
- Average Chunk Distance: The number of chunks in between a transition (including the final chunk).
- Mean Fixation Frequency per category: in terms of duration and visit count, where did a participant's eyes rest most often.

3.4 Participants

Students were given a questionnaire prior to the study to gather information on their skill level. A total of seventeen students volunteered for the study. Of the students involved there was a single student that did not speak English as their primary language. All students other than one had high proficiency in English. Only two of the students tested were over the age of 27, with 6 being between the ages of 23 and 27 and 9 being between the ages of 18 and 22. 10 of the students were female, and 7 were male. We split the participants into two groups, *novices and non-novices*, based on their years in the program. Individuals who had completed at least the first semester of their program up to their junior year were placed in the novice group. Those who had completed at least 3 out of the 4 years of their undergraduate program, in addition to participants enrolled in the graduate program, were considered beyond novice level, and were placed in the non-novice group.

3.5 Eye Tracking Apparatus

We used a Tobii X60 eye tracker. It is a binocular, non intrusive, remote eye-tracker that records 60 frames per second. We used it to record several pieces of information including gaze positions, fixations, timestamps, duration, validity codes, pupil size, start and end times, and areas of interest for each trial. The eye tracker was positioned on a desk in front of the monitors where students read the programming code. With an accuracy of roughly 15 pixels as well as being able to gather 60 samples of eye data per second, the Tobii X60 was a design choice that mapped well to what we needed to measure our study variables accurately. The monitors were 24" displays and set at a 1920 × 1080 resolution.

3.6 Study Instrumentation

To assist in the study, we used a Logitech webcam to record both video and audio. Each student was positioned in front of a dual-monitor configuration with the code snippets displayed on one, while the other was used by the researcher to control and monitor the study. After the tasks were performed, students were given a post questionnaire. In it we asked all students if they had enough time to complete the study, and all participants replied that they did. 11 students stated that the overall difficulty of the tasks was average, 6 stated that they were difficult and 1 stated that the tasks were very difficult. We also asked the students to describe any difficulties that they had: several stated that the code was difficult to remember to determine outputs. Additional comments included that the study was interesting and enjoyable while others thought that it was intense.

4   Post Processing

After the data was collected, we conducted three post processing steps. The first step involved correcting the eye tracking data for any drift that might have occurred with the tracker. The second step involved mapping gaze to lines of code and finally the identification of chunks. The third step involves identifying and regrouping lines into chunks with similar code structures across all stimuli, into "coded categories" that would enable us to analyze gaze patterns across multiple stimuli.

4.1 Data Correction

We used the open source tool *Vizmanip* to visually locate strands of fixations that were made on code snippet images. Vizmanip is a tool that allows the user to adjust and manipulate strands of contiguously recorded fixations available at https://github.com/SERESLab/fixation-correction-vizmanip. Identified fixations directly recorded from the eyetracker can sometimes drift [14] some distance away from the defined areas of interest. Given that we had a standard for the definition of AOI's this problem was mitigated by selecting sets of 10 or more contiguous fixations from the dataset and shifting them all a set number of pixels to better align with the identified areas.

4.2 Mapping of Eye Gaze to Lines

After all the corrections were done, we used *eyecode* [15], a Python library with a key focus on parsing and manipulating areas of interest in images. In addition, it also maps eye gaze fixations to areas of interest (in our case these were lines). After demonstrating that eyecode could appropriately handle the creation of AOI's from various formats of images, we ran its automated image parser to generate areas of interest of line level granularity, which emits a tabular list of rectangles that map to lines. Two graduate students manually inspected each generated AOI file to make sure the AOI's were correctly generated as *eyecode* does not always work as intended. Six stimuli needed manual readjustment where the generated AOI's did not represent a line. Other parameters in *eyecode*, such as vertical padding, needed to be set via trial and error. Such tedious post processing is unfortunately required when using code as images. A better approach would have been to use an IDE that enables eye tracking natively. In the future, we plan to use iTrace [16] for this purpose as it enables implicit eye tracking within the IDE and automatically maps gaze to code elements eliminating such manual mapping.

4.3 Motivation for Chunks

These line-level AOI's alone can provide interesting results however, we wanted to explore how the participants read groups of related lines or "chunks". Code snippets are not read the same way as natural language [2]. Thus each chunk was selectively customized down to the code snippet and task given to the participant. Because of these differences, analyzing the chunks of related code could provide more insight into the behavior of the participants than restricting the analysis to line-level AOIs. This practice of chunking has some credence in the study of program comprehension. The bottom-up comprehension model [17] sees participants read code and mentally group them together into an abstract representation of multiple lines. In the top-down model [17], participants use their knowledge of the program domain to understand its function. One of the ways that they can do this is through beacons, recognizable features in the code such as a series of lines to swap two variables [18]. Both models rely on participants to process the code not as a series of lines, but as sets of related lines and functionality.

4.4 Identifying Chunks and Categories

Since our focus was to study method level program comprehension, we had to make the chosen chunk areas granular enough to decisively determine whether cognition was leading to comprehension. Three of the authors rated two independently formed chunk mappings. Any disagreements were discussed by at least two authors to come to a 100% agreement. After 90 min of discussing 13 programs, the amount of lines to be grouped per chunk was decided. It took another 60 min of conferring between a different group of authors to decide what names and categorizations to grant to each region in each stimulus. For an example of these chunks, see Fig. 3.

The boilerplate left at the top of files (#INCLUDE and namespace statements fall in this category) were all grouped into one chunk in every file. Other notable C++ code elements that fall in this category are *method signature lines, method prototypes,* public and private *variable declarations,* (without their access modifiers), and return statements found inside control blocks. We felt that transitions among lines within these would be too minor to study for our tasks. Fixations on prototype access modifiers and the main method signature, if they exist in a stimulus, are completely omitted from our analysis as they were not the focus of our study. More interesting are tokens that play a key role in understanding assignment and data flow. Data flow tokens such as control loop parameters and branch statement parameters, are all included in our block categorizations for every stimulus. Data flow patterns played a role in our choice for grouping areas of interest. If a stimulus contains two related method-calls or def-use flows rooted in the main method, we try to separate into chunks two or more method calls that appear to have disjoint data flow chains, especially if the file is complex enough. This dataflow analysis was conducted and agreed upon via manual inspection by two authors.

We further categorize each chunk pattern into code feature categories. These categories represent groupings of certain code features that exist across many types of stimuli. In theory, these would be important places where participants would look in code for important information about how the code works. We put effort to reduce this set to 5 groups that would be common enough to be tracked across many stimuli. The code features we selected include the following:

- **control blocks** include if statements, switch statements and loop statements (typically their predicates only),
- **signatures** include method signatures and constructor signatures.

- **initializers** include constructor and method declarations, and statements or statement groups that initialize variables.
- **calls** include method calls and constructor calls
- **output** include statements that generate output printed to the console

Boilerplate lines, return statements, and inline methods were not grouped into these categories. Though they might provide value, we had to keep the groups under study to a minimum to be able to compare all the means.

5   Experimental Results

5.1   Results for RQ1: Accuracy

The number of questions participants answered correctly is shown in Fig. 1. On average, it took a participant 61.20 s to finish reading the code snippet before moving on to the comprehension question.

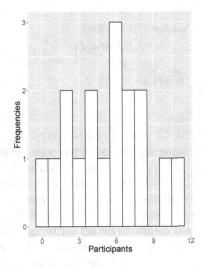

Fig. 1. Number of questions answered correctly by each participant

We provide the data in Table 2 to compare the results in different groups of our sample. We use the ANOVA test as it is quite robust and reliable way to compare means of two or more samples. We discuss the results of comparing the means of three sets of responses across the two groups (novices and non-novices). Each mean represents the responses gathered from the three types of questions, "Program Overview (Overview)", "What is the Output? (Output)", and "Give a Summary (Summary)". First, post-hoc analysis was able to confirm that, all participants considered, a fairly equivalent amount of questions got answered among

Table 2. Question accuracy non-novice/novice breakdown: inner cells show means by category and their comparisons. The estimated marginal mean (EMMean) shown for each category gives a fairer value to compare groups than the unweighted means of the inner cells by applying a few statistical corrections, including weighting the means according to how many questions were answered in a category. They are shown for replication purposes, though we do not use them to draw conclusions at this time.

Non-novice/novice means breakdown (ANOVA) (Standard deviation in [brackets], N in (parentheses))			
		Est. marginal mean	
	Expertise	Non-novice 57.4%	Novice 35.6%
Question type		MEANS	
EMMean 64.4%	Overview	76.0% [27.7%] (21)	= 52.8% (53) [32.0%]
		=	>
EMMean 36.6%	Output	55.3% [34.4%] (30)	> 17.9% (49) [30.6%]
		=	=
EMMean 33.4%	Summary	40.0% [37.9%] (14)	= 26.7 (50)% [22.6%]
		<	<
	(Overview)	76.0%	= 53.5%

all three question types (70, 74, and 64 respectively). The ANOVA Omnibus F-test indicates there exist some significant differences between the means of the novices and non-novices, taking into account weighted means existing across all three categories. ($F(1, 15) = 4.618$, $p = .048$, with effect size or $r = .485$). As expected, non-novices scored significantly higher than novices across all three questions (mean difference $= 24.7\%$, $p = .048$). Upon learning this, we took a closer look at the individual means to detect patterns, whether this trend holds across all question types. In particular, we found that novices did better on program overview questions than on output questions by 34.9% ($p = .002$). This pattern does not carry across the same to non-novices, where they performed statistically the same on overview questions as they did output questions ($p = .165$). However, we found a statistically significant difference in the amount of questions that non-novices answered correctly compared to the novice participants in terms of output questions ($p = .042$).

5.2 Results for RQ2: Fixations in Chunks

Table 3 gives the results of the Mann-Whitney test on each of the dependent variables. Simple mean comparisons revealed that novices looked at method signatures significantly longer than non-novices ($p = .036$). Non-novices however, looked at output statements significantly longer than novices by 22.8% ($p = .031$). The first two metrics, fixation duration and fixation counts are relevant to RQ2.

Table 3. Eye movement metrics calculated over all participants, non-novices, and novices. The p-values for the differences between the non-novices and novices mean (using Mann Whitney test) are shown along with effect size

Dependent variable	Mean	Non-novice	Novice	p-value	Cliff's delta
Fixation duration	45.45 s	46.33 s	45.05 s	0.7647	
Fixation counts	195.7	196.5	195.4	0.8224	
Transition count	48.63	50.84	47.64	0.5091	
Chunk fix. dur. prior exit	0.82 s	0.69 s	0.88 s	<0.001	0.1952
Vertical later chunk	45.00%	44.51%	45.22%	0.7945	
Vertical earlier chunk	38.79%	41.20%	37.71%	0.0151	0.2245
Avg chunk distance	1.49	1.57	1.46	0.0080	0.2448

We found the average total fixation duration across all snippets to be 45.445 s. We observe that non-novices on average had a longer fixation duration with an average code snippet fixation duration of 46.325 s while novices had an chunk fixation duration of 45.049 s. After running a Mann Whitney test, we did not find this grand mean difference to be statistically significant between novice's and non-novice's fixation durations ($p = 0.7647$).

Table 4. WhileClass chunks ranked by count of participants with highest and second highest total fixation visits and total fixation duration

Top visited	5's - "letter" if block	12	92%	Longest duration	5's - letter if block	12	92%
	8's - main method	1	8%		8's - main method	1	8%
2nd top visited	1's - boilerplate	1	8%	2nd longest duration	2's - method sig	1	8%
	3's - variable init	1	8%		3's - variable init	1	8%
	4's - while block	2	15%		4's - while block	3	23%
	5's - letter if block	1	8%		5's - letter if block	1	8%
	8's - main method	8	62%		8's - main method	7	54%

We now move to a discussion of the results we found while observing fixation patterns among named chunks. We chose four stimuli to break down fixation patterns - two with fewer lines of code WhileClass and PrintPatternR (Tables 4 and 5), and two with greater amounts of code Rectangle and SignCheckerClassMR (Tables 7 and 6). We chose to discuss programs with significant complexity with the potential to facilitate deeper discussion: both small programs have at least one loop construct, and the larger ones employ def-use flows that flow through multiple methods. See Figs. 2 and 3 for snapshots of selections from both groups.

Table 5. `PrintPatternR` chunks ranked by count of participants with highest and second highest total fixation visits and total fixation duration

Category	Chunk	Count	Pct.	Category	Chunk	Count	Pct.
Most visited	Boilerplate	1	7%	Longest total duration	Outer for	1	7%
	Inner for	14	93%		Inner for	14	93%
2nd most visited	Boilerplate	1	7%	2nd longest total duration	Boilerplate	1	7%
	Outer for	10	67%		Outer for	8	53%
	Output	5	33%		Inner for	1	7%
					Output	5	33%

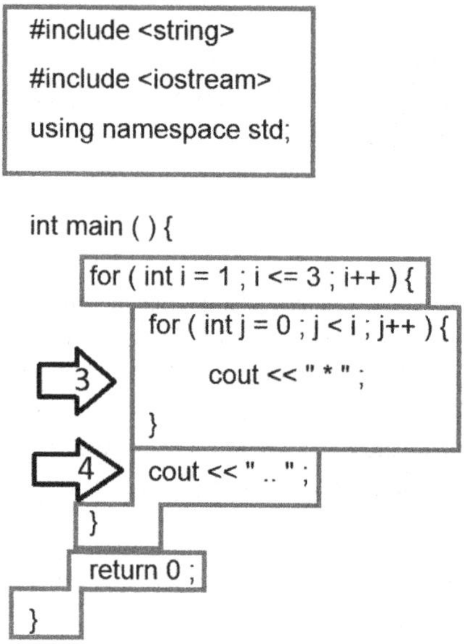

Fig. 2. Chunks of `PrintPatternR` with chunk 3 and 4 highlighted

After studying the fixation durations of participants, we noticed in small programs like `PrintPatternR` and `WhileClass` that regions of fixations tended to converge to the exact same point, regardless of whether the participant scored correct or incorrect, and regardless of expertise. See Table 5 for our records of chunks on which participants gazed at the longest. 93% of participants all fixated *the most* **and** *the longest* on chunk 3, the inner for loop with the print statement, responsible for printing the asterisk pattern. Notably this chunk was designed to contain not one but two important code categories, namely loops and print statements, but participants potentially look here due to its relevance to the overall function of the program. Chunks 2, 3 and 4 from this program stand out as retaining the longest fixation durations and highest visit count for most participants, boilerplate only scoring at the top of one participant's focal point of

Table 6. `SignCheckerClassMR` chunks ranked by count of participants with highest and second highest total fixation visits and total fixation duration

Most visited	If block 1	7	47%	Longest duration	If block 1	7	47%
	If block 2	1	7%		If block 2	2	13%
	Construct call 1	8	53%		Construct call 1	6	40%
2nd most visited	Method declare	2	13%	2nd longest duration	Method declare	3	20%
	Constructor	2	13%		Constructor	2	13%
	If block 1	7	47%		If block 1	7	47%
	Construct call	1	427%		Construct call 1	3	20%
	Construct call	2	17%				

Table 7. `Rectangle` chunks ranked by count of participants with highest and second highest total fixation visits and total fixation duration

Category	Chunk	Count	Pct.	Category	Chunk	Count	Pct.
Most visited	Dim methods	5	36%	Longest total duration	Dim methods	5	36%
	Area method	4	29%		Area method	3	29%
	Constr. sig	2	14%		Constr. sig	3	14%
	Constr. body	2	14%		Constr. body	2	14%
	Constr. call 2	1	7%		mm. consr. call 2	1	7%
2nd most visited	Dim methods	4	29%	2nd longest total duration	Constr. call 1	3	21%
	Area methods	3	21%		Output 1	1	7%
	Constr. sig	1	7%		Area method	3	21%
	Constr. body	2	14%		Constr. body	2	14%
	Constr. call 1	3	21%		Dim method	4	29%
	Output 1	1	7%		Constr. sig	1	7%

attention. A few chunks were tied for second for certain chunks in the second-top visited category.

We find a few contrasts to small programs like `PrintPatternR` when we look at large programs such as `Rectangle` (Table 7) and `SignCheckerClassMR` (Table 6). We see trends that occurs in programs with more information that do not occur in these small programs. As for `Rectangle`, we saw most participants focus on bodies of inline methods and constructors. See Table 7. The dimension methods received the most fixations and the longest duration times for most participants, followed closely by either the area calculation method, or constructor method. What this seems to show is a concern by most participants for the information that the *statement code* and not the *declarations and prototypes* offer. In Fig. 3, we see the program numbered by chunk with shaded regions. The darker hues represent regions that more participants visited the most times throughout their session. We note that variable or method declarations (outside signatures) did not get the highest attention of any of our participants. The results shown here for these programs do not show the main method as gaining much attention either. These are promising results that our analysis was able to capture.

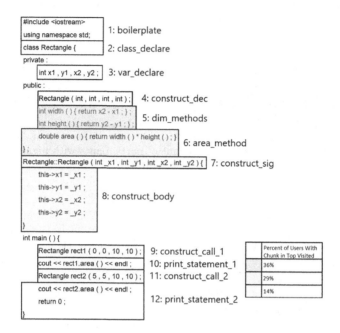

Fig. 3. Chunks of related code for `Rectangle.cpp` with top visited chunks highlighted

Looking closely, the top most looked at chunks cover the constructor, its method signature, and its helper method definitions. Our results do not greatly concern the main method. At least one element of the boilerplate code and one from the main method chunks scored in the bottom-three most-fixated chunks for eight of our participants. This last trend seemed to cover both high and low scorers. 4 out of 6 top scorers for `Rectangle` all had chunk 5 as their most fixated chunk.

5.3   Results for RQ3: Chunk Transitions

We address RQ3 by observing up close the transitions made between various stimuli, by looking at other dependent variables such as fixation counts more closely, and by looking for the trends that exist across gaze data for multiple stimuli. The first metric we investigate is number of transitions between chunks made by a participant during a single task. We found that on average 48.63 of these transitions between chunks were made by a participant during a single task. We observe that non-novices made more transitions on average (50.84 transitions) than novices (47.64). After running a Mann Whitney test, we did not find this difference to be statistically significant ($p = 0.5091$).

Next we analyzed Chunk Fixation Duration Prior Exits. We found that on average participants spent 0.82 s fixating on a chunk before transitioning to another chunk. Non-novices had a shorter Chunk Fixation Duration Prior Exit with an average of 0.69 s before a transition was made, and novices looked at the chunks for a longer Chunk Fixation Duration Prior Exit of 0.88 s. After running a Mann Whitney test, we found this difference to be statistically significant ($p < 0.001$). The effect size was found to be small according to Cliff's delta ($d = 0.1952$).

For the Vertical Later Chunk, we found that on average 45.00% of transitions were made to a vertically lower chunk. For non-novices, we found that they made less transitions to vertically lower chunks with an average of 44.51% of transitions. For novices, we found that transitions to a vertical later chunk accounted for on average 45.22% of transitions. After running a Mann Whitney test, we find that these differences are not statistically significant ($p = 0.7945$). Next we analyzed a related metric, Vertical Earlier Chunk, for the transitions. We found that on average 38.79% of transitions were made to a vertically earlier chunk. The reason that the Vertical Later Chunk and Vertical Earlier Chunk percentages do not add to 100% is because some transitions are made to lines that are not included in a chunk or to points that are not mapped to lines. For non-novices, we found that they made more transitions to vertically earlier chunks with an average 41.20% of transitions. For novices, we found the Vertical Later Chunk was on average 37.71% of transitions. After running a Mann Whitney test, we find that these differences are statistically significant ($p = 0.0151$). The effect size was found to be small according to Cliff's delta ($d = 0.2245$). The two previous metrics show us that non-novices are less likely to read code from the top chunk to the bottom chunk, and non-novices are more flexible in the direction they transition to. In addition, we can also see that non-novices transitions from chunks to chunks instead of between lines not included in a chunk more than novices.

We found that the average chunk distance of a transition made between chunks was 1.49. Non-novices transitioned to chunks that were on average farther away with an average chunk distance of 1.57, and novices transitioned to chunks that had on average a chunk distance of 1.46. After running a Mann Whitney test, we find that these difference are statistically significant ($p = 0.0080$). The effect size was found to be small according to Cliff's delta ($d = 0.2448$). The most common chunk distance for a transition between chunks was 1 which shows that participants most commonly transitioned to chunks that are close to the current chunk being fixated on.

We now combine the results obtained from the eye tracker, namely the fixation regions of each participant and the length of each fixation duration, with the data that we have on the locations of chunks in files. We use a tool, named the Radial Transition Graph Comparison Tool (RTGCT), that was provided by researchers at the University of Stuttgart Institute of Visualization and Interactive Systems. This tool is used to display data from fixation files and materialize visual data on a computer screen in a tree-annulus style fashion, in a way shows how long participants gaze was on a certain part of the code and that allows users to view activity from a whole task at once in a single image. Each stimulus is colored differently and positioned adjacent to other stimuli along an annulus, the arc length of its color showing the percentage of the total duration of the participant's task taken up by his accumulated fixations on that stimulus. See Fig. 4.

We observe the output of the tool for two of our largest programs, where we can find some interesting transitions. We first discuss the `Rectangle` example.

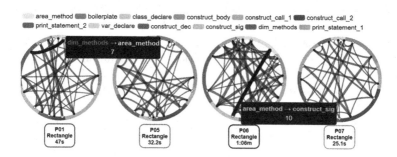

Fig. 4. Output of RTGCT for `Rectangle`, highlighting inter-chunk transitions between constructor, dimension methods, and the area method.

The top scorers in the non-novice category were P01 and P06, and a few notable trends appear in their results. See Fig. 4 for the transition rate between P01 and P06 between the *constructor signature* and both the area function and the chunk named *dimension methods*" (containing width and height functions for the rectangle), are greater in comparison to transitions between main method, boilerplate and other regions of the program. P01 a high scorer, made 7 transitions between the dimension methods and the area method. P06, the other high scorer, made a fascinating 10 transitions between the constructor signature and the area method. These patterns do not appear in other non-novice eye gaze patterns. These transitions are either non-existent or diminished in comparison to other non-novice participants indicating to us that these two points of the program might have been important for these participants.

The `SignCheckerClassMR` code snippet transitions are visualized in Fig. 5. In order to properly depict transitions and not hide any, we chose to use the RTGCT's "Equal Sectors" mode to show all chunks as equivalent segments along the outer ring. In this example P01 and P07 performed poorer than other participants. We can see a trend that transitioning between methods and the constructor may have led to this.

5.4   Threats to Validity

We describe the main threats to validity to our study and measures taken to mitigate them.

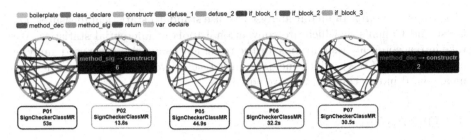

Fig. 5. Output of RTGCT for `SignCheckerClassMR` which indicate trends in method declaration lookups with ring sectors sized equally regardless of duration percentages

Internal Validity: The 13 C++ programs used in this study are code snippets and might not be representative of real-world programs. To mitigate this, we had code snippets vary in length, difficulty, and constructs used to add variety to our independent variables. Correcting the eye tracking data to account for drift can introduce bias to the data. To mitigate this, only groups of ten fixations were moved at a time and the new location had to be agreed on by two of the authors.

External Validity: A threat to the generalization of our results is that all our participants were students. This was mitigated by the inclusion of students with widely varying degrees of expertise, ranging from 1 year of study to 5+ years (4 years of baccalaureate plus some years in a graduate program). Another threat is our sample size. We ended our study with comprehension data from 17 participants, and with viable eye tracking data from 15 participants. However, the fact that results we analyzed for non-novices came from only 5 participants may raise questions. In response, the fact we successfully gathered from all participants repeated measures on at least 10 stimuli per participant, and that we collect a total of 57 eye-gaze patterns and 65 question responses from these participants alone is suggestive of the rigor that went into our assessments of how each participant did.

Construct Validity: A threat to the validity of this study is that the method we chose to use to break lines into chunks was done using standards agreed upon by the authors of whether certain chunks would remain relevant by the end of our study. However, these decisions may not generalize to all potential code comprehension analyses, as these choices were made subjective to the data authors had at their disposal at different points of the study. To mitigate this threat, we carefully synchronized each decision on how to divide lines into chunks for each of our 13 stimuli, and two of the authors met for 90 min before the final decision was made on which chunks would remain. Since we only are only measuring our participants on program comprehension, a mono-operation bias can occur. In order to mitigate this, we used three different types of program comprehension questions, summarization, output, and overview, in order to vary the exact task being performed.

Conclusion Validity: In all our analyses we use standard statistical measures, i.e. t-test and Cohen's d, which are conventional tools in inferential statistics. We take into account all assumptions for the tests. For comparisons we used analysis of variance (ANOVA), which includes an F test in order to decide whether the means used in our comparisons are equal.

6   Discussion

We found differences between the two levels of expertise in frequency of eye movements among the chunks we coded. Non-novices looked at chunks longer before transitioning to others, tended to transition to chunks that were further away from their original position, and had more transitions to earlier chunks than novices. Looking closer at the data for what participants took most interest in, we found that for smaller programs (`PrintPatternR` and `WhileClass`) over 90% of all participants from both groups fixated on a single segment of code. Larger programs like `Rectangle` brought up situations where there was little agreement, especially among non-novices, about which chunk got either the most fixations, the longest fixation durations, or both. These results were not necessarily isolated to `Rectangle`.

When looking at fixation data (without considering question responses), non-novices tended to shun other elements (other than control blocks) in stronger favor of output statements most of the time. However, interestingly in our data, novices tended to allocate equal amounts of their attention to visiting areas other than control blocks. They tended to hold their fixations on declarations more than signatures, but this is the only deviation from that pattern we could find. Output statements was the 2nd-least visited category among all the coded categories for novices and method signatures were the least visited category for both novices and non-novices. For over 50% of the questions for non-novice participants we saw non-novices focus on output statements in their top two most visited categories.

When looking at responses to questions, we realized that we cannot say much to what fixation categories generally lead to better answers on questions. This is because the better areas to fixate upon *depend heavily on the content of the stimulus.* We were able to show in our data that for some stimuli – those which had more complex-structured helper methods – participants focusing on method calls longest received better scores, but that focusing on method calls predicted worse scores for a stimulus with more complex control blocks. Future work will need to be done that controls *across multiple stimuli* for the complexity of code within a stimulus, perhaps evening out complexities of control blocks and of the def-use method call chains within stimuli, in order to ensure that comparisons can be drawn fairly when gathering what fixation patterns might lead to better performance. We did not have enough stimuli to make this kind of comparison, even though we noticed differences in performance when two stimuli had these differences in their structure.

7 Conclusions and Future Work

The paper presents an eye tracking study on thirteen C++ programs done in a classroom setting with students during the last week of a semester. We find that the link between the expertise of a student and how accurately they answer questions, is made much clearer when paired with the insight of what visual cues were used by students the most. The visual cues led us to discover that students agree less on which areas to focus on the most when the program size grows to be large. These insights also showed us that the frequency of incorrectly answered questions is only significantly affected in certain stimuli by the areas participants looked at – or perhaps what they did not look at. Finally, we saw that performance of non-novice students can be intrinsically linked to both the number of fixations and the transitions made between important segments of the code. More research will be required to determine whether it is the data flow through the constructs or simply the types of constructs available that drive where participants look. We were able to uncover and visualize patterns among top performers that showed what transitions may have mattered the most as cues perhaps leading to better understanding. In addition, more research will be required to learn whether more frequent transitions amongst coded categories within stimuli are truly linked to better performance, or whether other factors we did not observe more closely contributed more to success. As part of future work, we would like to use the iTrace infrastructure [16] to conduct experiments with industry professionals on real large-scale systems.

References

1. Brooks, R.E.: Towards a theory of the comprehension of computer programs. Int. J. Man-Mach. Stud. **18**(6), 543–554 (1983)
2. Busjahn, T., et al.: Eye movements in code reading: relaxing the linear order. In: Proceedings of the 2015 IEEE 23rd International Conference on Program Comprehension, ICPC 2015, Piscataway, NJ, USA, pp. 255–265. IEEE Press (2015)
3. Obaidellah, U., Al Haek, M., Cheng, P.C.H.: A survey on the usage of eye-tracking in computer programming. ACM Comput. Surv. **51**(1), 5:1–5:58 (2018)
4. Busjahn, T., et al.: Eye tracking in computing education. In: International Computing Education Research Conference, ICER 2014, Glasgow, United Kingdom, 11–13 August 2014, pp. 3–10 (2014)
5. Rayner, K., Chace, K., Slattery, T., Ashby, J.: Eye movements as reflections of comprehension processes in reading. In: Proceedings of the 2018 ACM Symposium on Eye Tracking Research & Applications, pp. 543–554 (2006)
6. Beelders, T., du Plessis, J.P.: The influence of syntax highlighting on scanning and reading behaviour for source code. In: Proceedings of the Annual Conference of the South African Institute of Computer Scientists and Information Technologists, SAICSIT 2016, pp. 5:1–5:10. ACM, New York (2016)
7. Sarkar, A.: The impact of syntax colouring on program comprehension. In: PPIG, July 2015
8. Raina, S., Bernard, L., Taylor, B., Kaza, S.: Using eye-tracking to investigate content skipping: a study on learning modules in cybersecurity. In: 2016 IEEE Conference on Intelligence and Security Informatics (ISI) (2016)

9. Turner, R., Falcone, M., Sharif, B., Lazar, A.: An eye-tracking study assessing the comprehension of C++ and python source code. In: Proceedings of the Symposium on Eye Tracking Research and Applications, ETRA 2014, pp. 231–234. ACM, New York (2014)

10. Uwano, H., Nakamura, M., Monden, A., Matsumoto, K.I.: Analyzing individual performance of source code review using reviewers' eye movement. In: Proceedings of the 2006 Symposium on Eye Tracking Research Applications, ETRA 2006. ACM, New York (2006). http://doi.acm.org/10.1145/1117309.1117357

11. Sharif, B., Maletic, J.: An eye tracking study on camelcase and under score identifier styles. In: Proceedings of the 2010 IEEE 18th International Conference on Program Comprehension (2010)

12. Binkley, D., Davis, M., Lawrie, D., Maletic, J., Morrell, C., Sharif, B.: The impact of identifier style on effort and comprehension. Empirical Softw. Eng. 18(2), 219–276 (2013)

13. Olsson, P.: Real-time and offline filters for eye tracking (2007)

14. Palmer, C., Sharif, B.: Towards automating fixation correction for source code. In: Proceedings of the Ninth Biennial ACM Symposium on Eye Tracking Research & Applications, ETRA 2016, Charleston, SC, USA, 14–17 March 2016, pp. 65–68 (2016)

15. Hansen, M.: Github - synesthesiam/eyecode-tools: a collection of tools for analyzing data from my eyecode experiment. https://github.com/synesthesiam/eyecode-tools

16. Guarnera, D.T., Bryant, C.A., Mishra, A., Maletic, J.I., Sharif, B.: iTrace: eye tracking infrastructure for development environments. In: Proceedings of the 2018 ACM Symposium on Eye Tracking Research & Applications, ETRA 2018, pp. 105:1–105:3. ACM, New York (2018). http://doi.acm.org/10.1145/3204493.3208343

17. Storey, M.A.: Theories, methods and tools in program comprehension: past, present and future. In: Proceedings of the 13th International Workshop on Program Comprehension, IWPC 2005, pp. 181–191. IEEE (2005)

18. Von Mayrhauser, A., Vans, A.M.: Program comprehension during software maintenance and evolution. Computer 28(8), 44–55 (1995)

Self-regulated Learning and Expertise: Dual Cognitive Processes

Webb Stacy[1,2(✉)], Jeffrey M. Beaubien[1,2], and Tara Brown[1,2]

[1] Aptima, Inc., Woburn, MA, USA
{wstacy, jbeaubien, tbrown}@aptima.com
[2] Aptima, Inc., Indianapolis, IN, USA

Abstract. The human brain processes cognitive information in two ways [1]: Type 1 processes—fast, effortless, and non-conscious—and Type 2 processes—slower, requiring cognitive effort, and available for conscious introspection. These two processes combine to produce four kinds of metacognition. When considered in the context of learner motivation, they also produce new ways of thinking about self-regulated learning. In this paper, we apply these notions to three domains: the classroom, the workplace, and a use case that provides an example of learning an advanced perceptual-motor-cognitive skill, namely landing a jet on an aircraft carrier. We conclude by discussing opportunities for further research.

Keywords: Self-regulated learning · Metacognition · Dual-systems theory

1 Introduction

This paper urges an integration of findings from the field of Cognitive Science with the rich literature on self-regulated learning from the fields of Educational and Industrial/ Organizational Psychology, respectively. Specifically, we will examine the implications of dual modes of cognitive processing for two important components of self-regulated learning: metacognition and motivation. We believe that these three domains of study will mutually benefit from cross-fertilization of this sort, and hope that other scholars will join us in this endeavor. The authors are two Industrial/Organizational Psychologists (JB and TB), a field from which much research on self-regulated learning has emerged, and a Cognitive Scientist (WS), the field which spawned the dual-system theory of cognition. As we hope the reader of this article will discover, we have identified several interesting and important points of intersection among these disciplines.

2 Four Kinds of Metacognition

A key finding from Cognitive Science is that the human brain processes information in two ways. In [1] they are called Type 1 processes—fast, effortless, non-conscious, and often automatic—and Type 2 processes—slower, requiring cognitive effort, and

© Springer Nature Switzerland AG 2019
D. D. Schmorrow and C. M. Fidopiastis (Eds.): HCII 2019, LNAI 11580, pp. 617–630, 2019.
https://doi.org/10.1007/978-3-030-22419-6_44

available for conscious introspection.[1] Type 1 processes are sometimes described as pattern recognition, intuition, or "gut feel", and Type 2 processes are sometimes known as analytic or critical thinking. [2] provides an explanation of a variety of cognitive heuristics and biases in terms of dual processes. [3] shows that dual processes can help reconcile the apparent conflict between naturalistic decision making (where Type 1 processes are strongly valued because of their role in expert performance) and probabilistic thinking (where Type 1 processes are viewed with skepticism because they sometimes result in logical errors.)

A second finding from Cognitive Science is that expertise develops as a function of deliberate practice [4, 5]. This is true for learners on their way to being experts, as well. The "deliberate" part of deliberate practice, of course implies that learners – often working with a coach or mentor – choose the nature of the practice they engage in, and therefore it implies a degree of self-regulation. This kind of self-regulated practice is key to the development of expertise across a range of disciplines including music, dance, sports, martial arts, and chess among others.

Deliberate practice has interesting and differential effects on these two cognitive processes. Knowledge, skills, and aptitudes (KSAs) start out as declarative information, typically learned in didactic environments such as a classroom, and are learned via Type 2 processes. As learners gain domain-specific experience, though, the KSAs become "chunked" and "compiled" into more efficient Type 1 processes [6–8].

Expert-like performance requires several years of sustained deliberate practice, which involves: skills practice that is specifically targeted toward one's greatest areas of weakness, diagnostic performance feedback, deliberate self-reflection on that feedback, and strategically-timed periods of recuperation [5]. During this extended period of training, learners set specific, difficult goals for improving their future performance, such as running a mile in under 8 min, or completing a half-marathon within 2 h. Goals are critical to any theory of self-regulated learning because they direct the learner's attention, increase task-related effort, and help the learner to persist in the face of adversity [10]. Along with increased skills, the learner's self-confidence invariably increases with expertise. This further helps to motivate the learner during the long, difficult, and sometimes painful process of deliberate practice [11]. Therefore, any discussion of self-regulated learning must also include elements of motivation as well.

Type 1 and Type 2 cognitive processes are not mutually exclusive; they often work together. For example, Type 2 processes often prime the Type 1 processes by providing critical contextual information that is needed for successful task performance. For example, baseball players have very limited time to decide where and how to swing at a pitched ball [12], so they rely on Type 1 processing to make that decision. But that decision is heavily influenced by game context—the score, the stage of the game, the known style of the pitcher—which is provided by Type 2 processes.

Figure 1 shows a notional overview of the development of Type 1 and Type 2 cognitive processing with increased domain experience.

---

[1] Type 1 and Type 2 processes are sometimes referred to as "System 1" and "System 2", as in [2]. In this paper we prefer to call these modes of cognition processes rather than systems because we follow [1] in acknowledging the existence of the two modes without requiring that they be accomplished with separate brain systems.

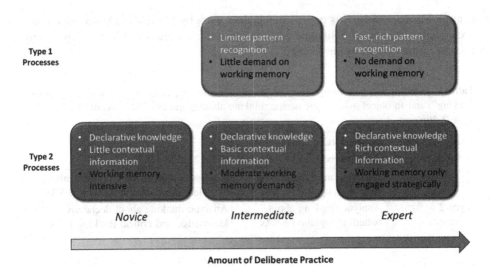

Fig. 1. Notional overview of the development of Type 1 and Type 2 processing as a function of the accumulated amount of deliberate practice.

The process of thinking about one's thinking is generally referred to as metacognition [9], and is a critical hallmark of domain expertise. The challenge is that Type 1 processes are generally not available for introspection. They occur without the learner's conscious knowledge and often involve automatic pattern recognition. This gives rise to four distinct kinds of metacognition, as shown in Table 1. In the upper left hand quadrant (Type 1/Type 1) might be a fire ground commander who immediately recognizes that the fire has spread into the structure's walls and knows instinctively that he can recognize this category of patterns, and orders all personnel out of the building. In the lower left quadrant (Type 2/Type 1), the fire ground commander realizes that he has never seen this type of fire behavior before, and begins sequentially reviewing alternatives to find one that fits the current circumstances. In the upper right hand quadrant (Type 1/Type 2), a student realizes that he has been reading the same sentence repeatedly, but has not been able to comprehend the content. He therefore decides to put the book down and return to it after a break. In the lower right hand quadrant (Type 2/Type 2), a student identifies a gap in his understanding and deliberately seeks out the instructor's assistance to help resolve the gap.

Metacognition using a Type 1 process, that is, automatically recognizing patterns in either other pattern recognition domains or in declarative knowledge and critical thinking domains, is currently a kind of black box, not only because Type 1 processes are not available for introspection, but also because they have not been widely discussed. [13] discuss metacognition in the context of dual-process cognition, but they focused only on the second column in Table 1 (that is, they considered Type 1 and Type 2 thinking about Type 2 thinking only). They made the astute observation (in terms of Table 1) that only Type 2 thoughts can be communicated to others, and therefore they viewed Type 1 metacognition as useful for intra-personal cognitive control, and Type 2 metacognition as useful for shared supra-personal cognitive control. The analysis in this paper extends their work by explicitly acknowledging that accumulated amount of deliberate practice leads to an increased role for Type 1 processing in the domain being learned.

Table 1 Four kinds of metacognition. Metacognition has a process type *(how* is the person thinking?) and an object *(what* is the person thinking about?), either of which could be Type 1 or Type 2 thinking.

Metacognition using....	Metacognition about....	
	Type 1 processes	Type 2 processes
Type 1 processes	Recognition of patterns in pattern recognition KSAs	Recognition of patterns in declarative knowledge and critical thinking KSAs
Type 2 processes	Analytic thinking about pattern recognition KSAs	Analytic thinking about declarative knowledge and critical thinking KSAs

In any case, both cells in Column 1 provide good fodder for future research, because there are well-developed methodologies for investigating rapid pattern recognition (for example, temporal occlusion [14–16] and drift diffusion modeling [17–19] to name two of many.)

Metacognition involving Type 1 processes is thus necessarily indirect, more so as the learner gains proficiency and more Type 1 processes come into play. Since metacognition is one component of self-regulated learning (along with motivation, as in [20, 21]), the nature of the metacognitive component of self-regulated learning must therefore evolve with increasing amounts of domain-specific deliberate practice.

2.1   Guide to the Rest of the Paper

In the remainder of this paper, we will discuss self-regulated learning from a dual-process perspective in three contexts: in the classroom, in the workplace, and in learning an advanced perceptual-motor-cognitive skill, namely landing a jet on an aircraft carrier. We believe those environments provide a good sample of real-world human learning situations in which self-regulated learning comes into play. After those discussions, we conclude with recommendations for future research.

3   Self-regulated Learning in the Classroom from a Dual-Process Perspective

In the field of Educational Psychology, much of the research on self-regulated learning emanates from two distinct schools of thought. The first focuses on the learning of specific domain knowledge, such as learning specific math or science concepts. The second focuses on generalized principles of "learning to learn". The discussion that appears below draws heavily from this second school of thought.

3.1 Metacognition in the Classroom

Because learning new declarative knowledge is a Type 2 cognitive process, metacognitive research in the classroom has focused entirely on the right-hand column of Table 1. The upper right-hand quadrant depicts what happens when a learner uses pattern-recognition (Type 1) processes to dynamically monitor and regulate the learning process in real time. For example, while studying for a test, the learner may recognize that she has just re-read the same sentence 3 times, but still does not understand the concept. As a result, the learner puts the book down, takes a break, and returns to the material with a fresh perspective. Similarly, while taking an exam, the learner may come to a test question for which they do not know the answer. Recognizing that the exam is timed, the learner skips that question in the hopes that answering the subsequent test questions will cue recall of the forgotten information.

The lower right-hand quadrant depicts what happens when a learner uses deliberate or analytic (Type 2) processes to engage in metacognitive planning. For example, before reading a textbook chapter, the learner may: actively seek out a quiet place to study; skim the chapter headings beforehand to identify key concepts, and; review the end-of-chapter self-assessment questions before reading the chapter text. Similarly, while reading the chapter text, the learner may identify a critical gap in her understanding of the material and then actively seek out the instructor's assistance to help resolve that gap.

Research on metacognition has identified three inter-related components: metacognitive planning, self-monitoring, and self-regulation [22, 23]. The first component, metacognitive planning, appears to be a Type 2 process. It includes behaviors such as setting goals for studying, skimming the chapter headings to identify key concepts, and reviewing the end-of-chapter assessment questions before reading the chapter text. It is proactive and future-oriented, rather than reactive and present-oriented. The last two components, self-monitoring and self-regulation, appear to be Type 1 processes. Examples of self-monitoring include: monitoring one's attention in real time, self-testing to assess one's comprehension of the material, and using "smart" test-taking strategies such as process-of-elimination to rule out potential incorrect responses. Examples of self-regulation include: modulating one's reading speed as a function of the material difficulty, re-reading key text passages to improve comprehension, and resolving perceived inconstancies by systematically comparing one's class notes to the chapter text [22, 23].

While they are theoretically distinct, in practice these three metacognitive components are highly inter-correlated [24, 25]. As a result, they are often measured together.

3.2 Motivation in the Classroom

The field of Educational Psychology has also made a number of important contributions to the study of motivation. Perhaps one of the most critical is recognizing the dynamic interplay between "cold" cognition and "hot" motivation [24]. Simply put, learners are not machines. Their feelings and emotions influence their ability to self-regulate the learning process. Another is the field's conceptualization of the learner as an active information processor whose feelings and beliefs can be contextually-activated [24]. While the student may have learned though extended experience certain automatic dispositions that cause them to respond in a particular manner in ambiguous situations (Type 1 processes), those characteristics can be over-ruled by the classroom "culture" (Type 2 processes). As a result, highly-skilled instructors systematically modify the classroom environment to their advantage [26], especially when teaching difficult courses like mathematics. In the following paragraphs, we describe two motivational constructs—goal orientation and self-efficacy— both of which appear to have both automatic (Type 1) and situationally malleable (Type 2) components.

The term "goal orientation" refers to one's goal-related preferences in achievement-related situations. Research suggests that there are three different, although non-exclusive, types of goal orientation [27]. Individuals with a strong "Learning Goal Orientation" (LGO) approach new situations as an opportunity to learn, grow, or develop themselves. As a result, they interpret failure as a learning opportunity, and welcome it. Individuals with a strong "Performance Goal Orientation" (PGO) approach new situations as an opportunity to demonstrate their skills and abilities to others. In essence, it is an opportunity to "show off" in front of one's peers. Such individuals tend to interpret the course grade or exam score as an end in itself; they tend to be ambivalent to performance feedback. Finally, individuals with a strong "Fear of Failure" (FOF) approach new situations as potential opportunities to demonstrate their lack of knowledge or skills to their peers. As a result, they try to avoid receiving negative performance feedback. They often do this by choosing easy tasks (because doing so guarantees success) and choosing tried-and-true methods (which have worked well in the past) rather than trying new methods. Previous research suggests that goal orientation has some automatic (Type 1) characteristics; however, it can be situationally induced (Type 2). For example, an instructor can deliberately induce a LGO classroom culture by: teaching the learners effective goal-setting strategies; involving students in classroom decision making process; making grades contingent on improvements vis-à-vis one's prior performance, or; rewarding learners when they try new approaches regardless of their performance outcomes [26, 28].

The term "self-efficacy" refers to a learner's self-confidence in their ability to successfully perform a specific learning-related task [11]. Previous research shows that self-efficacy influences the goals that learners set for themselves, how much effort they mobilize toward their goals, how long they persevere in the face of difficulty, and whether their thought processes are self-aiding or self-hindering [11]. Although self-efficacy was originally conceptualized as a situationally-specific belief [11, 29], there is some evidence to suggest that it may also have stable components. In essence, some learners have a generalized belief in their ability to be successful across a wide range of learning-related tasks and situations [30]. Also, like goal orientation, instructors can deliberately induce high levels of self-efficacy among their students, for example by systematically engineering the learning experience by starting with relatively simple tasks (to ensure initial success) and then slowly building up to more complex ones, by actively modeling the correct way to perform the task, by verbally persuading learners, as needed, when they are struggling to perform the task, and by helping the learner to correctly attribute their physiological states (e.g., anxiety, stress, fatigue) as being both normal and common even among high-performers [29].

Previous research suggests that regardless of how learning is measured (e.g., tests of basic skills, classroom performance, scores on standardized tests), self-efficacy exhibits strong positive correlations with learning-related outcomes. Moreover, the results hold for learners in elementary school, high school, and college [31].

4 Self-regulated Learning in the Workplace from a Dual-Process Perspective

Historically, researchers in the field of Industrial/Organizational Psychology have focused on how to maximize the effectiveness of formal, employer-provided training courses. Over the years, the field has amassed an impressive body of literature on how to accomplish this [32]. However, there is an emerging body of research which suggests that for many employees, a substantial majority (70–90%) of their on-the-job learning experiences occur via informal, semi-structured, and self-directed means [33].

While this shift in emphasis from organizationally-provided training to self-directed learning places greater control in the hands of the learner, it requires a significant amount of self-regulation by the learner, both in terms of metacognition and motivation. However, there is a wealth of evidence that learners have difficulty when it comes to self-regulated learning. For example, they often make poor decisions about what to focus on and how much effort to invest in the learning process [34–36]. Therefore, it is critical to understand the factors that might influence employees' ability to effectively self-regulate, as well as what interventions can support them.

As noted previously, motivation and metacognition are central to self-regulated learning in general, and to the deliberate practice approach, in particular. To highlight the key points of intersection, we first need to consider the conditions under which effective self-regulation takes place, including goal setting, task selection, feedback, concentrated periods of effort, and motivation [37]. Inherently, the first three conditions align more with metacognition. That is, to set effective goals and identify appropriate tasks to focus on, learners need sufficient domain expertise to know what they know, to know what they don't know, and to prioritize what they should focus on next [38]. Without some baseline level of expertise, learners can easily set goals and identify tasks that lead to shallower learning curves or that result in bad habits.

The latter two conditions focus more on the motivation, or effort regulation, aspect of self-regulated learning. That is, learners need to have the discipline and motivation to engage in concentrated, effortful practice and to choose to invest effort into developing new skills, rather than practicing what they already know or enjoy doing. In general, this suggests that employees who are at an expert-level of learning might be better equipped and more capable of engaging in self-directed learning, while novice employees might need more support and guidance to engage in productive self-regulation. However, regardless of where a learner falls on the expertise continuum, there are challenges to engaging in effective self-regulatory learning as seen in Table 2.

As depicted in the table, understanding the intersection of self-regulatory learning and Type 1/Type 2 processes provides more focused insight as to the likely strengths and limitations of different types of learners for engaging in self-regulation, as well as what types of strategies or interventions might be needed to support their ability to

Table 2. Implications for self-regulatory learning based on the current state of the learner.

Current state of the learner	Implications for self-regulatory learning	
	Metacognition	Motivation
Type 1 (Expert)	**Capacity:** Possess the cognitive capacity and expertise to set goals and identify tasks to extend understanding (i.e., identify Type 2 KSAs that should be practiced to supplement current expertise)	**Capacity:** Possess the attentional resources to put effort into the deliberate practice needed, as they are not bogged down by learning or performing by Type 2 processing in their task
	Potential challenge: May not stop and reflect on what improvements they can make to an already automatic set of skills	**Potential challenge:** May be comfortable in their current level of expertise or prefer to engage in tasks that they are already good at or enjoy, rather than putting effort into learning new things
	Recommendation: Provide explicit prompts that pull experts briefly out of their Type 1 processing to briefly engage in Type 2 reflecting on whether their automated strategy is still appropriate and/or where they could grow	**Recommendation:** Incentivize effort instead of performance or outcomes. Engage experts in identifying experiences or learning opportunities that are engaging to them, while still focusing on new skill development
Type 2 (Novice/Intermediate)	**Capacity:** Limited cognitive capacity and expertise to set goals and identify appropriate tasks	**Capacity:** Limited attentional resources to put into effortful deliberate practice activities
	Potential challenge: Metacognition may interfere with task learning/execution or be overwhelming given the already high demand placed on learners engaged in Type 2 processing. May also be unaware of what they should be thinking about with respect to their cognition	**Potential challenge:** May lack the time and energy to engage in intentional, effortful practice, or may not know where to direct effort or how much effort to direct (inefficient effort allocation). May shut down or withdraw from learning if they feel overwhelmed or burned out
	Recommendation: Provide support from an experienced mentor/supervisor to set appropriate goals and identify strategic learning opportunities in more structured environments. Provide more immediate feedback with clear guidance	**Recommendation:** Provide structure for the learner in the form of guidelines for how much effort to engage in, removing unnecessary distractions, and ensuring goals are specific and difficult, yet doable, so learners are not setting themselves up for failure

engage in self-regulated learning. However, while [39] provides a rich framework of strategies for promoting more effective self-directed learning, noticeably missing from the discussion is how these strategies may be more or less effective for individuals at different levels of expertise. [39] explicitly notes that research is lacking on how the effectiveness of different self-regulatory strategies (or the need for them) varies based on individual differences, with the seeming assumption that the strategies hold across all individuals. However, the limited research on other individual differences, including self-efficacy [35], goal orientation [40], and personality [41] suggests that this assumption is likely not true. As such, the intersection highlighted in Table 2 above is believed to be an important extension of this work, as it can help inform how organizations utilize and support self-directed learning of their employees, when considering their level of expertise, to ensure maximum benefits, while minimizing the potential negative side-effects (e.g., withdrawal, attrition).

5 Self-regulated Learning of an Advanced Perceptual-Motor-Cognitive Skill from a Dual Process Perspective

An extension to self-regulated learning in the workplace is self-regulated learning that involves difficult perceptual-motor-cognitive skills. A good example of such learning is learning to land a jet on an aircraft carrier. In addition to the normal challenges associated with landing an aircraft, this task requires landing a fast-moving aircraft on a very short runway that moves, both because the carrier is in motion and because the carrier is affected by the motion of the ocean. Pilots have considerable training in the classroom and landing on a fixed airfield before they attempt a carrier landing, and the initial attempts at landings of the typical pilot learning to land on the carrier can often be described as "colorful".[2]

Because of the extensive amount of training involved, many of the skills are Type 1 skills: they are practiced enough that they do not require cognitive resources and they are more-or-less automatic. Yet there are substantial Type 2 components, as well. For example, expert pilots cue on the carrier's wake to infer the wind conditions over the deck: if the wake is frothy, the carrier must be moving fast enough to generate its own wind, meaning that the wind will be in a known direction at a known range of speeds. Normally this tells the pilot that they will have to keep the left wing down on landing. On the other hand, if the wake is minimal, then environmental wind will prevail, which could mean that the carrier has turned into the wind to aid with the recovery, or could mean something else—the pilot will need to stay alert to the possibilities during landing.

Pilot motivation is strong and straightforward. They need to land safely on the carrier to ensure their own safety, the safety of other personnel on the carrier, and the aircraft itself. Further, there is a spirit of sporting competition among pilots. The quality of every landing is evaluated by a Landing Signals Officer (LSO), and average scores are posted in the ward room on a "greenie board", providing a way for pilots to

---

[2] CDR Beth "Gabby" Creighton (Ret.), personal communication to WS, 2014.

compare their performance with one another, and offering motivation to achieve high scores. This incentivizes expert pilot effort (as discussed in the Type 1/motivation cell in Table 2) and suggests that expert pilots have a Performance Goal Orientation [27], as described in Sect. 3.2. On the other hand, LSOs provide explicit structure and guidance to less experienced pilots (as discussed in the Type 2/motivation cell in Table 2), suggesting that novice pilots have a Learning Goal Orientation.

[14] conducted an experiment in a simulator-based training study using a temporal occlusion paradigm [42] looking at several issues surrounding optimal training environments for carrier landing. One such issue was potential differences between the cognitive processes used by expert and by novice pilots. Pilots were shown an 8-second video of a portion of a carrier landing and were asked effectively if they were where they should be or if they anticipated needing to make an aggressive correction to ensure a safe landing. Correct responses were identified by a panel of subject matter experts, and the speed and accuracy of the pilot participants was assessed via computer. This temporal occlusion measurement was made both before and after a simulation-based carrier landing training session in order to assess the amount of learning.

One of the hypotheses under test was that experts use better-developed Type 1 responses because of their greater experience and pattern recognition capabilities, and that this should show up as faster and more correct responses compared to the novice pilots. Figure 2 shows the results. As predicted, experts were faster and more accurate[3], more so after the training session.[4] This is consistent with the idea that experts engage in Type 1 pattern recognition more than novices, who must primarily reason using slower Type 2 cognitive processes.

What does this say about the kinds of metacognition that pilots engage in? Definitive answers await future research, be we can see evidence of all three of the kinds of metacognition described in [21] and Sect. 3.1 at different levels of experience.

Metacognitive planning is likely a Type 2 endeavor, and likely of central importance for novices. For example, one pilot described a time when he was a novice and had just completed three landings in a row where he was much higher than the normal glideslope during landing. That pilot recalled deliberately thinking through what he needed to do in order to correct the problem before his next attempted carrier landing. On the other hand, experts are more likely to engage in self-monitoring and, should they detect unwelcome patterns, dynamically self-regulate their behavior. For one thing, Type 1 processing is prominent during their landings, so automatically recognizing and correcting problematic patterns in their own performance is also a task at which they have had considerable practice. For another, experts—unlike novices—do not generally get extensive guidance from the LSOs about ways they could improve their landing skills. For experienced pilots, LSOs generally limit their debriefs to a simple description of what happened during the landing. It is expected that the expert pilots will figure out for themselves how they need to improve.

---

[3] An Analysis of Variance revealed that expert-novice differences were significant for reaction time ($F_{(1,139)} = 4.81$, $p < .05$) and accuracy ($F_{(1,175)} = 7.01$, $p < .01$). Pre-post differences showed a learning trend for both measures, though the difference did not approach significance.

[4] Since error reaction times were longer than correct reaction times, we can be confident that none of the results represented a speed-for-accuracy tradeoff.

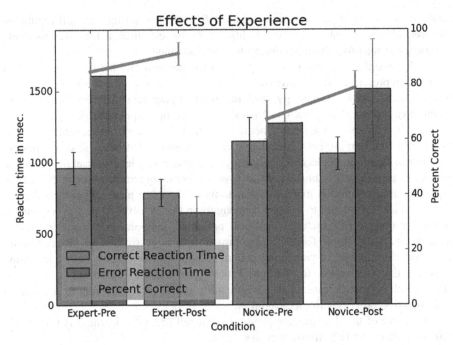

Fig. 2. Results of temporal occlusion experiment for expert and novice pilots, before and after a simulation-based training session.

6 Conclusions and Future Research

Dual-processing theory makes interesting contributions to the understanding of metacognition. There are certain predictable changes—in cognitive processing, metacognition, and motivation—that occur as learners accumulate increasingly greater amounts of domain expertise. Having little or no direct hands-on experience, novices generally understand the domain based on what they have read, heard in a lecture, or learned vicariously. Their knowledge is abstract, poorly organized, and decays quickly. During skills practice, their performance is slow and effortful. They perform the task in discrete steps following the textbook description, their movements are "jerky" rather than "smooth", and they make common errors such as omitting critical steps or performing steps in the wrong order. Because they must consciously monitor their performance to avoid making errors, novices experience high levels of workload thereby making it difficult to self-monitor or self-self-regulate. Finally, given that they are also prone to failure, their self-efficacy may be fragile, thereby leading to intrusive thoughts during task performance.

With increased domain-specific deliberate practice, the learners' performance becomes more expert-like, their behavior becomes goal-directed and contextualized. Depending on their particular goal (or mission), certain situational cues now take on greater relevance, thereby priming their Type 1 pattern recognition processes. Moreover, as the task becomes increasingly "compiled", additional cognitive resources are freed up—these resources can now be used for self-monitoring and self-regulation purposes. Finally, given their relatively high levels of self-efficacy, they are less likely to experience intrusive thoughts during task performance.

We believe that the four distinct "types" of metacognition that have described in this paper represent a critical extension of prior research from the fields of Cognitive Science, Educational Psychology, and Industrial/Organizational Psychology—all of which have looked at very narrow "slices" of this complex phenomenon.

We fully recognize that, despite being consistent with the literature, some our ideas about metacognition are both speculative and anecdotal. That being said, we believe this kind of theoretic exploration is healthy, and that it can lead to useful hypothesis testing and modeling. Fortunately, since Type 1 processing requires little in the way of cognitive resources and is more-or-less automatic, there are many experimental paradigms that can be brought to bear on metacognition that involves Type 1 processing as method or object. This includes dual-task protocols, temporal and spatial occlusion protocols, two-alternative forced choice protocols, a large fraction of the techniques that have been applied to perceptual pattern recognition, and many others, including neuroscientific techniques. In addition, Table 2 provides the beginnings of the guidance for self-regulated learning that will result a deeper understanding of the role of dual cognitive processes in metacognition.

The future of the dual-process perspective in self-regulated learning is bright. We look forward to the rich discoveries that await.

References

1. Evans, J., Stanovich, K.: Dual-process theories of higher cognition: advancing the debate. Perspect. Psychol. Sci. **8**, 223–241 (2013)
2. Kahneman, D.: Thinking, Fast and Slow. Farrar, Straus and Giroux, New York (2011)
3. Kahneman, D., Klein, G.: Conditions for intuitive expertise: a failure to disagree. Am. Psychol. **64**, 515–526 (2009)
4. Ericsson, A., Pool, R.: Peak: Secrets from the New Science of Expertise. Houghton Mifflin Harcourt, Boston (2016)
5. Ericsson, K.A., Krampe, R.T., Tesch-Romer, C.: The role of deliberate practice in the acquisition of expert performance. Psychol. Rev. **100**, 363–406 (1993)
6. Ericsson, K.A., Charness, N.: Expert performance: its structure and acquisition. Am. Psychol. **49**, 725–747 (1994)
7. Klein, G.A.: Sources of Power: How People Make Decisions. MIT Press, Cambridge (2017)
8. Gobet, F., Chassy, P.: Expertise and intuition: a tale of three theories. Mind. Mach. **19**, 151–180 (2009)
9. Clark, R.: Metacognition and human performance improvement. Perform. Improv. Q. **1**, 33–45 (1988)
10. Locke, E.A., Latham, G.P.: A Theory of Goal Setting and Task Performance. Prentice-Hall Inc., Englewood Cliffs (1990)
11. Bandura, A.: Social cognitive theory of self-regulation. Organ. Behav. Hum. Decis. Process. **50**, 248–287 (1991)
12. Libet, B.: Mind Time: The Temporal Factor in Consciousness. Harvard University Press, Cambridge (2009)

13. Shea, N., Boldt, A., Bang, D., Yeung, N., Heyes, C., Frith, C.D.: Supra-personal cognitive control and metacognition. Trends Cogn. Sci. **18**, 186–193 (2014)
14. Stacy, E.W., Beaubien, J.M., Wiggins, S.M., Walwanis, M.M., Bolton, A.E.: Using temporal occlusion to assess carrier landing skills. In: Proceedings of the 2014 Annual Meeting of the Interservice/Industry Training, Simulation, and Education Conference (I/ITSEC), p. 14171. National Training and Simulation Association, Alrington (2014)
15. Abernethy, B., Gill, D.P., Parks, S.L., Packer, S.T.: Expertise and the perception of kinematic and situational probability information. Perception **30**, 233–252 (2001)
16. Ward, P., Ferrow, D., Harris, K., Williams, A., Eccles, D., Ericcson, K.A.: Training perceptual-cognitive skills: can sport psychology research inform military decision making? Mil. Psychol. **20**, S71–S102 (2008)
17. Stacy, E.W., Beaubien, J.M., Wiggins, S.M.: Assessing military perceptual expertise with drift diffusion modeling. In: Proceedings of the 2017 Annual Meeting of the Interservice/Industry Training, Simulation, and Education Conference (I/ITSEC), p. 17202. National Training and Simulation Association, Alrington (2017)
18. Ratcliff, R., Rouder, J.N.: Modeling response times for two-choice decisions. Psychol. Sci. **9**, 347–356 (1998)
19. Ratcliff, R., Tuerlinckx, F.: Estimating parameters of the diffusion model: approaches to dealing with contaminant reaction times and parameter variability. Psychon. Bull. Rev. **9**, 438–481 (2002)
20. Zimmerman, B.J., Campillo, M.: Motivating self-regulated problem solvers. In: The Psychology of Problem Solving, pp. 233–262. Cambridge University Press, Cambridge (2003)
21. Pintrich, P.R.: An achievement goal theory perspective on issues in motivation terminology, theory, and research. Contemp. Educ. Psychol. **25**, 92–104 (2000)
22. Pintrich, P.R.: The dynamic interplay of student motivation and cognition in the college classroom. In: Advances in Motivation and Achievement, pp. 117–160. JAI Press, New York (1989)
23. Pintrich, P.R., De Groot, E.V.: Motivational and self-regulated learning components of classroom academic performance. J. Educ. Psychol. **82**, 33–40 (1990)
24. Duncan, T.G., McKeachie, W.J.: The making of the motivated strategies for learning questionnaire. Educ. Psychol. **40**, 117–128 (2005)
25. Pintrich, P.R., Smith, D.A., Garcia, T., McKeachie, W.J.: Reliability and predictive validity of the motivated strategies for learning questionnaire. Educ. Psychol. Measur. **53**, 801–813 (1993)
26. Pintrich, P.R.: The role of goal orientation in self-regulated learning. In: Handbook of Self-Regulation, pp. 451–529. Academic Press, Cambridge (2000)
27. Elliot, A.J., Church, M.A.: A hierarchical model of approach and avoidance achievement motivation. J. Pers. Soc. Psychol. **72**, 218 (1997)
28. Heyman, G.D., Dweck, C.S.: Achievement goals and intrinsic motivation: their relation and their role in adaptive motivation. Motiv. Emot. **16**, 1992 (1992)
29. Bandura, A.: Self-efficacy mechanism in human agency. Am. Psychol. **37**, 122–147 (1982)
30. Chen, G., Gully, S.M., Eden, D.: Validation of a new general self-efficacy scale. Organ. Res. Methods **4**, 62–83 (2001)
31. Multon, K.D., Brown, S.D., Lent, R.W.: Relation of self-efficacy beliefs to academic outcomes: a meta-analytic investigation. J. Couns. Psychol. **38**, 30–38 (1991)
32. Arthur, W., Bennett, W., Edens, P., Bell, S.: Effectiveness of training in organizations: a meta-analysis of design and evaluation features. J. Appl. Psychol. **88**, 234–245 (2003)

33. Cerasoli, C.P., Alliger, G.M., Donsbach, J.S., Mathieu, J.E., Tannenbaum, S.I., Orvis, K.A.: Antecedents and outcomes of informal learning behaviors: a meta-analysis. J. Bus. Psychol. **33**, 203–230 (2018)

34. Brown, K.G.: Using computers to deliver training: which employees learn and why? Pers. Psychol. **54**, 271–296 (2001)

35. Sitzmann, T., Ely, K.: Sometimes you need a reminder: the effects of prompting self-regulation on regulatory processes, learning, and attrition. J. Appl. Psychol. **95**, 132–144 (2010)

36. Welsh, E.T., Wanberg, C.R., Brown, K.G., Simmering, M.J.: E-learning: emerging uses, empirical results and future directions. Int. J. Training Dev. **7**, 245–258 (2003)

37. Schwartz, D.L., Tsang, J.M., Blair, K.P.: The ABCs of How We Learn: 26 Scientifically Proven Approaches, How They Work, and When to Use Them. Norton, New York (2016)

38. Bransford, J.D., Schwartz, D.L.: It takes expertise to make expertise: Some thoughts about why and how and reflections on the themes in chapters 15–18. In: Development of Professional Expertise: Toward Measurement of Expert Performance and Design of Optimal Learning Environments, pp. 432–449. Cambridge University Press, Cambridge (2009)

39. Bell, B.S.: Strategies for supporting self-regulation during self-directed learning in the workplace. In: Autonomous Learning in the Workplace, pp. 117–134. Routledge, New York (2017)

40. Schmidt, A.M., Ford, J.K.: Learning within a learner control training environment: the interactive effects of goal orientation and metacognitive instruction on learning outcomes. Pers. Psychol. **56**, 405–429 (2003)

41. DeRue, D.S., Ashford, S.J., Myers, C.G.: Learning agility: in search of conceptual clarity and theoretical grounding. Ind. Organ. Psychol. **5**, 258–279 (2012)

42. Farrow, D., Abernethy, B., Jackson, R.C.: Probing expert anticipation with the temporal occlusion paradigm: experimental investigations of some methodological issues. Mot. Control **9**, 330–349 (2005)

Clarifying Cognitive Flexibility from a Self-regulatory Perspective

Melissa M. Walwanis[1(✉)] and Shelby-Jo Ponto[2]

[1] Florida Institute of Technology, Melbourne, USA
Melissa.Walwanis@navy.mil
[2] StraCon Services Group, Fort Worth, USA
shelbyjo.ponto@stracongroup.com

Abstract. Cognitive flexibility is a definitionally varied construct utilized across disciplines that revolves around an individual's ability to suitably adjust to changes in a task-based context [5]. The diversity of disciplines exploring cognitive flexibility have resulted in a number of operationalizations that function more or less optimally depending upon the context in which the construct is measured. Each measurement technique holds both promise toward explicating the nature of this construct while simultaneously stalling progress, as crosstalk between disciplines is stilted. To support this effort in understanding how cognitive flexibility arises, the link between it and self-regulation is structurally explored.

This paper reviews findings throughout the literature (1) for different cognitive flexibility definitions and measures (e.g., the Wisconsin Card Sort Task [8], Trail Making Test [19], Cognitive Flexibility Scale [14]), (2) across multiple disciplines (e.g., psychology, management, communications), and (3) within various populations (e.g., clinical, adolescents, young adults, older adults) in an effort to initiate conversation about increasing sensitivity of utilized measures to better analyze and understand results between cognitive flexibility and other key constructs relevant to a work environment (e.g., learning [23]). Lastly, this paper proposes an explanatory link between cognitive flexibility and self-regulation, utilizing Sitzman and Ely's [22] parsimonious framework of self-regulatory learning constructs and Pintrich's [17] four phases of self-regulation as applicable to each measure. This work lays the groundwork for advancing the science underpinning this construct.

Keywords: Cognitive flexibility · Self-regulation · Construct clarity · Construct measurement

1 Introduction

Given inspection from multiple fields of science, it is common for constructs to take on discipline-specific operationalizations. As each scientific lens has unique aspects and can contribute discrete information, it is meaningful to evaluate all perspectives and consider how the resulting measures were developed. Given its importance to the context of learning [23], cognitive flexibility is deserving of further investigation in this holistically evaluative manner. That is, each measurement technique within the realm

This is a U.S. government work and not under copyright protection in the United States; foreign copyright protection may apply 2019
D. D. Schmorrow and C. M. Fidopiastis (Eds.): HCII 2019, LNAI 11580, pp. 631–643, 2019.
https://doi.org/10.1007/978-3-030-22419-6_45

of cognitive flexibility holds both promise toward further explicating the nature of the concept while simultaneously stalling progress, as crosstalk is stilted, as exemplified below.

Findings regarding the relationship between learning and cognitive flexibility are mixed. Existing studies use a range of research designs, measures, and populations. Unfortunately, the diversity in populations further confounds the ability to come to consensus regarding the relationship as the nature of cognitive flexibility changes over the course of human life peaking between the ages of 21–30 (e.g., [5, 18]). The studies explore different parts of the nomological network (e.g., learning agility, cognitive reasoning, critical thinking, intelligence quotient). Finally, the studies utilize a mix of state, trait, and state and trait measures. Behavioral measures used followed set-shifting tasks, which are a lower form of cognitive flexibility and easier to perform [5]. Therefore, the lack of findings may be an artifact of measurement chosen. With the literature base rife with examples such as this, the sections to follow highlight the various delicacies of this construct and self-regulatory constructs to further explain the nature of how research can be better structured in future efforts.

2 Cognitive Flexibility

Cognitive flexibility has been described as "an emergent property of efficient executive function" by Dajani and Uddin ([5], p. 579). Cognitive flexibility is of interest to many academic disciplines such as psychology (e.g., clinical, education, experimental, industrial/organizational, neurocognitive), management, and communications. This is not surprising as flexibility of thought holds promise to arrive at better decisions, which is important in all of the practical domains served by these disciplines. The variety of academic disciplines has driven diversity in treatment of the construct and follow-on models. While these departures do make it difficult to come to consensus on many aspects of the construct, they are useful in painting a broad picture of the mechanisms that underpin cognitive flexibility. The following sections, ordered from executive function's role in cognitive flexibility to the measures themselves, further explain the general details of each of these conceptualizations and operational aspects of cognitive flexibility.

2.1   Executive Function

Dajani and Uddin [5] break down the construct in the ways that the executive function of the brain works with cognitive flexibility, including the interactivity of (1) cue-salience detection, (2) goal-directedness of attention, (3) inhibition to readjust behavior, and (4) working memory representations of task-behavior. First, in salience detection, the salience of a cue determines if attention is captured. Second, attention is either goal-directed, involving top-down processing, or bottom-up directed in response to cues in the environment. Third, in response to an update in the necessary actions or goals to achieve success, inhibition of a learned response occurs. Fourth, cognitive flexibility necessarily requires two or greater representations in working memory for successful task completion. Further, certain operationalizations of cognitive flexibility are more difficult and place greater demand on working memory (e.g., task-switching in state behavioral measures). The intricacies of state and trait conceptualizations and their prevalence in attitudinal and behavioral measures is further explained below.

2.2   State and Trait

The construct has been treated as both a trait and a state construct absent explicit acknowledgement supporting that fact. Spiro and colleagues [23] proposed the need to determine if training could facilitate gaining skill in cognitive flexibility as both a state and a trait construct. Additionally, the authors proposed that the effects of shifts in epistemic views following a training intervention is a key, unanswered question [23]. Similarly, Dajani and Uddin [5] propose that scientific inquiry should address if training inventions can alleviate cognitive inflexibility and ruminative thought patterns in clinical populations; indeed, this is a question just as applicable to the general population.

Finally, studies that have attempted to administer a comprehensive, cognitive flexibility test-battery have failed to find significant correlations between state and trait operationalizations of cognitive flexibility (e.g., [10, 26]). We hold the position that cognitive flexibility can be both a trait and a state construct. Measurement of trait cognitive flexibility needs to be carefully considered since there is overlap with intelligence (i.e., considered as a control variable). Cognitive flexibility has been shown to be influenced by practices such as mindfulness [15], lending support for it also serving as a state construct, malleable when executive function is manipulated. Another dichotomization in the construct's literature includes attitudinal and behavioral measures.

2.3   Attitudinal and Behavioral Measurements

Existing operationalizations of cognitive flexibility can be crudely grouped as attitudinal or behavioral in nature. First, attitudinal self-report instruments are expressed at a trait-level wherein lies a lack of consistency in self-report measures with variations existing due to practical applications (e.g., decision-making, communications). Next, behavioral operationalizations are expressed as a state. Dajani and Uddin [5] identify two operationalizations: set-shifting and task-switching. Set-shifting is an attentional shift in the schema to different features or cues to complete the same instruction successfully. Task-switching is a switch between tasks with different instructions for successful completion. There is a differential effect on switch-costs, or the slowing in response time and decrease in accuracy, between task- and set-shift operationalizations, with task-shift representing the most difficult form of cognitive flexibility as it is more cognitively demanding to jump between a set-shift and a task-shift. Additionally, there are developmental transitions across human development that necessitate understanding how attitudinal and behavioral measurements operate.

2.4   Developmental Transitions

Munakata, Snyder, and Chatham [16] posit three developmental transitions of cognitive flexibility that adolescents progress through as they grow. First, adolescents tend to control required behavioral changes via environmental cues (a reactive response). For example, a child looking to grab a toy will continue to attempt to do so even when the toy may be hidden under a blanket. To maintain continuity, the child has to attend to the information of where the toy was last placed while also avoiding attending to competing thoughts or interruptions (e.g., satiating hunger by eating the banana on the plate). Second, they will tend to adjust from a reactive response to a proactive response where they are ready to change tactics. The transition from a reactionary perspective to a proactive one is due to the development of the ability for active maintenance of information (e.g., recalling that the toy is still hidden under the blanket after several days).

Third, children will move away from requiring external cues and begin utilizing more self-directed control. This is exemplified when internally-generated goals begin to come into effect. Instead of being told to do a task, an individual will do the task without prompting (e.g., brushing teeth). Munakata and colleagues [16] further posit that these transitions are driven by the progressive development of goal representations over time, which is covered further in later sections of this paper (see Sect. 3.1). While these transitions are typical transitions across adolescents, the authors also acknowledge that adults may still utilize the second phase and rely on external cues. This information is useful in further understanding the application of cognitive flexibility measures across populations and gives a perspective of goal-relevance across the phases.

2.5   Goal Achievement

Maddox, Baldwin, and Markman [13] take a similar simplified approach to defining cognitive flexibility. In their view, cognitive flexibility is an individual's skill or willingness to utilize a variety of strategies to achieve an objective or goal. They subdivide the construct into three factors in their model. First, cognitive flexibility is characterized by the ability to adapt to general change. Second, cognitive flexibility is the ability to think of a variety of categories or concepts (e.g., cognitive sets). Third, they depart from the original theory in adding that cognitive flexibility is enabled by individual self-efficacy in being flexible and adaptive [14]. Self-efficacy has a long history of being linked to individual's learning, making it a positive addition to understanding learner engagement and willingness to be flexible. This research with goal-influenced developmental transitions and goal-achievement has begun to explicate the nature of the effect of self-regulation with cognitive flexibility, but more research is called for [13].

2.6   Measures

Cognitive flexibility measures observed in the literature include assessment through set-switching tasks, task-switching tasks, psychophysiological measures of executive brain function, and attitudes. The following Tables 1 and 2 show, respectively, a variety of attitudinal and behavioral measures. These measures are shown here as a starting point for others to consider which versions may work best for research. For example, psychophysiological measurement techniques (e.g., fMRIs) offer promise to test Dajani and Uddin's [5] proposition that cognitive flexibility is an executive function, which would be demonstrable through a task or set shift task hypothesized to result in a particular pattern of action in the brain. Furthermore, these measures will be utilized in later sections involving their particular connection with self-regulatory constructs and phases.

Table 1. Attitudinal (trait) measures of cognitive flexibility

Citation	Description	Targeted population
[1]	• Cognitive flexibility scale • Reliability = .78 • Survey: 14 items, five-point scale of situational-relatability	Work population
[4]	• Alternative Uses Task (AUT) • Participants list uses of six items • Flexibility is the number of different categories used	Meditative population
[6]	• Cognitive flexibility inventory • 20 items, 7-point agreement scale • Two-factor structure: alternatives & controls • Focuses on how individuals adapt (positively or maladaptively) to situations	Clinical, generalizable to work population
[7]	• Chinese Remote Association Test (CRAT) • Create compound words using the CRAT while in an fMRI	Clinical neurology
[14]	• Cognitive flexibility scale • Survey: 12 statements, six-point agreement scale on personal behaviors	Undergraduates, general population
[24]	• Cognitive flexibility inventory • Items are ambiguously abstracted, poetic verses • 19 paired statement-items; 7-point agreement scale • Learning in ill-structured instructional environments	Medical students, general population
[29]	• Cognitive flexibility scale revised • Epistemological belief measure • 14 pairs of statements, six-point scale of agreement • Higher score means more complex epistemic beliefs	General population

Table 2. Behavioral (state) measures of cognitive flexibility

Citation	Description	Targeted population
[8]	• Wisconsin card sorting task • Reliability = .60 • Strategy shift task – measures shifts of attention as cards are sorted based upon color, shape, and number	Undergraduates, average population
[19]	• Trail making task • Motor control task & ascending alphabet	Clinical neurology
[25]	• Stroop task • Candidates are instructed over three phases to (1) read color words, (2) name the color of a word, and then (3) read the color of color words (e.g., the word "yellow" in red ink)	Undergraduates, average population
[2]	• Brixton test • Predict movement of a blue circle (strategy shift necessary for prediction as movement algorithm changes)	Clinical neurology
[26]	• Picture set test • Set flexibility tested through picture groupings	Clinical neurology
[11]	• Verbal fluency task • Participant vocalizes words starting with f, a, or s and excludes nouns, numbers, and sequences over a minute	Clinical neurology
[27]	• Cat bat task • Fill missing letters in a written short story as fast as possible	Clinical neurology
[28]	• Haptic illusion task • Perceptual set-shifting task involving holding balls of different shapes while eyes are closed	Clinical neurology

3 Self-regulation

The following sections concerning self-regulation, which is loosely defined as the control of one's behavior as relevant to one's goals. Constructs underpinning self-regulation are considered as antecedents to cognitive flexibility. For example, self-efficacy or the belief that one can succeed at performing a task could reasonably be hypothesized to result in greater cognitive flexibility. A better understanding of the nomological network surrounding cognitive flexibility as it relates to self-regulation could facilitate understanding how cognitive flexibility operates.

3.1 A Process Model

Self-regulation is proposed as a process model involving the reciprocal, interdependent interaction of three open feedback loops resident in the person, environment, and behavior. These interactions result in self-management of environmental contingencies and the application of knowledge and skills during periods of action [21, 30].

Monitoring of these feedback loops allows an individual to better triangulate their current state to their desired state (i.e., goal), thus performance is contingent upon consistent monitoring and accurate interpretation.

However, development of self-regulatory skills can be challenging, taxing both physical and mental resources [30]. Four levels of self-regulatory skills include (1) observation of models, (2) emulation of a model's skill with feedback to facilitate refinement, (3) self-control in structured situations, and (4) independent self-regulation of personal and environmental conditions to meet performance goals. These skills interact with Munakata and colleagues' [16] posited transitions for cognitive flexibility development. The observation of models and subsequent emulation are tied into the developmental stage of reactive responses. As a child develops and learns these behaviors, they are able to incorporate them into their reactionary responses and grow. As one considers the third skill, it is reflected in the second phase of development. That is, one would be better able to control oneself with being able to actively manage information in relation to goals. The fourth skill lends itself to both the second and third developmental stages. Just as self-regulation is important to externally generated goals, so it is to internally generated ones. Beyond a development of a regulatory skillset, research has also produced two important aspects as reflected in self-regulated learning, explained below: (1) Pintrich's [17] four phases and (2) Sitzman and Ely's [22] meta-analytic, heuristic framework of self-regulation.

3.2   Self-regulated Learning

Self-regulated learning theories all share four assumptions [17]. First, learners are assumed to be active in the process of learning, constructing meaning and goals. Second, learners possess the potential for control. Through monitoring, learners control, or regulate, their cognition, motivation, behavior, and environment. Third, learners set goals, criterion, or standards against which cognition, motivation, behaviors, and environments are regulated. Fourth, self-regulatory activities are mediators between personal and contextual characteristics and performance outcomes.

Self-regulation is posited to follow four phases of (1) planning, (2) monitoring, (3) regulating self and context, and (4) reflecting. These phases are not linear in nature and any phase can be active at any time. Although not restricted linearly, each phase still usually happens in one's attempt to perform a task. In light of all of these characteristics, Pintrich [17] defined self-regulated learning as: "an active, constructive process whereby learners set goals for their learning and then attempt to monitor, regulate, and control their cognition, motivation, and behavior, guided and constrained by their goals and the contextual features of the environment" (p. 453). Conceptually, this definition and framework has presented abundant opportunities to test many constructs attached to the assumptions and learning strategies across the phases.

The first phase is the preparatory development of perceptions or knowledge and goal setting. Here, an individual gathers information about both task and context as it relates to him or her. In the second phase, monitoring, individuals process this information metacognitively, highlighting self-awareness. Within the third phase, an individual exerts a change or effort to control. These exertions will center around both the self and context. The fourth phase is primarily concerned with the reactions of both the aforementioned aspects (self and context).

Sitzman and Ely [22] undertook the effort of reviewing the self-regulated learning literature to synthesize findings into a broad heuristics framework and conduct a meta-analysis to understand the effectiveness of constructs addressed as supporting self-regulated learning. They identified 16 constructs covered both theoretically and empirically in the literature. Meta-analysis revealed intercorrelations between the constructs. Based upon their findings, they proposed the parsimonious framework for self-regulated learning to provide a manageable framework upon which to base future research. This paper focuses on the nine constructs that were found to be most related to learning outcomes.

The original framework contains sixteen constructs spanning across three classifications (i.e., regulatory agents, appraisals, and mechanisms). The first classification, regulatory agents, are primarily goal levels. That is, they allow learners to initiate self-regulation in reference to an objective. Appraisals are evaluations of progress towards goals. Mechanisms are actual processes. These processes allow learners to efficiently progress towards goals. The outer band of the model contains (1) goal-level (agent) as well as (2) attributions and (3) self-efficacy (appraisals) to predict learning at a moderate to strong level. The inner band of the model of mechanisms contains weak to moderate predictors of learning, which are (4) metacognition, (5) attention, (6) time management, (7) environmental structuring, (8) motivation, and (9) effort.

First, and as previously explained, the agent classification of goal level simply refers to a learner's intended objective. Second, attributions refer to a learner's opinions about causality of achievement. Commonly, unless a learner considers their own work on the task subpar (i.e., an internal attribution) with regards to malleable aspects (e.g., effort instead of a stable construct like low ability), it will not be readdressed via self-regulation (e.g., an increase in effort to retry the task [17]). Third, self-efficacy takes this a step further, referring to a learner's opinions about personal training and performance capabilities. Self-efficacy is a predictor for how challenging of a goal a learner will set. Those with high self-efficacy often set more challenging goals [17].

A wide variety of regulatory mechanisms are also discussed. The fourth construct, metacognition itself, is a difficult, semi-umbrella term in self-regulation as it spans aspects of personality, self-regulation, and self-awareness. Sitzman and Ely [22] define the construct as "planning and monitoring goal-directed behavior and devoting attention toward the course material" (p. 429). Fifth, attention is the learner's concentration during training. Resource allocation theory primarily focuses on how learners dedicate their attention in terms of on- or off-task allocation. A portion of the resources are also allocated towards self-regulatory processes [12]. Sixth, time management revolves around a learner's scheduling capabilities concerning studying. Issues resulting from poor time management (e.g., procrastination) include limited personal growth [30].

Seventh, environmental structuring reflects what kind of location in which a learner studies. This may include a quiet venue to concentrate while reviewing. Additionally, this construct is often one of the fewer studied constructs in motivation [22]. Eighth, motivation is a learner's "willingness to engage in learning and desire to learn" as it relates to the content ([22], p. 430). If a learner does not find inherent value in the training, he or she will not retain the information [20]. Finally, ninth, effort is purely time used learning. An example is a blocked-out hour a day for studying. Effort is often increased when a learner notices a discrepancy in their feedback-loop from the intended goal-level [3]. These structuralizations of self-regulation (phases and constructs) are utilized below to explain the connection between cognitive flexibility measures and where these areas most likely overlap or correlate with self-regulatory concepts.

4 Cognitive Flexibility and Self-regulation

The following tables (Tables 3, 4, and 5) highlight where the authors believe there is the most representative (phasic) placement or conceptual correlation (self-regulatory constructs) for each cognitive flexibility measure. Within Table 3, both attitudinal and behavioral measures were divided and plotted along the four phases of self-regulation as explained by Pintrich [17]. Each measure was evaluated by the authors and placed in a phase if it was agreed to that it was representative of the characteristics of each phase. Disagreements on placement were addressed for 100% agreement. The most outstanding difference in this table is where none of the current measures of one kind (i.e., either attitudinal or behavioral) fill a phase. While both kinds of measures fall into the monitoring phase, the other three phases only contain one or the other: (1) planning and (3) regulating self and context contained only behavioral and (4) reflecting only one attitudinal measure.

Table 3. Cognitive flexibility measures plotted along four phases of self-regulation

Measure type	1. Planning	2. Monitoring	3. Regulating self & context	4. Reflecting
Attitudinal measures		CF scale [1]		CF inventory [6]
		AUT [4]		
		CRAT [7]		
		CF scale [14]		
		CF inventory [24]		
		CF scale revised [29]		
Behavioral measures	Stroop task [25]	Haptic illusion task [28]	TMT [19]	
	WCST [8]	Picture set test [26]	Cat bat task [27]	
			Brixton test [2]	
			Verbal fluency task [11]	

Tables 4 and 5 are divided by attitudinal and behavioral measures. Each measure was evaluated by the authors and an x was placed under a self-regulation concept if it was agreed to that it was represented in the measure. Disagreements on placement were addressed for 100% agreement in the following tables. Within Table 4, many of the attitudinal measures were representative of various self-regulation constructs. However, none of the attitudinal measures discussed in this paper were found to conceptually overlap with the construct of time management. Within Table 5, a similar pattern emerged where none of the discussed behavioral cognitive flexibility measures were representative of attributions, environmental structuring, or motivation.

Table 4. Attitudinal measures are plotted along nine self-regulation constructs. An x indicates theoretical correlation between a measure and self-regulatory construct.

Self-regulation construct	CF scale [1]	AUT [4]	CRAT [7]	CF scale [14]	CF inventory [24]	CF scale revised [29]	CF inventory [6]
Goal level		x	x		x	x	x
Attributions	x				x	x	x
Self-efficacy	x			x	x	x	x
Metacognition	x	x	x	x	x	x	x
Attention	x	x	x	x	x	x	
Time management							
Environmental structuring	x			x	x	x	
Motivation	x	x	x	x	x	x	x
Effort		x	x				

Table 5. Behavioral measures are plotted along nine self-regulation constructs. An x indicates theoretical correlation between a measure and self-regulatory construct.

Self-regulation construct	Stroop task [25]	WCST [8]	Haptic illusion task [28]	Picture set test [26]	TMT [19]	Cat bat task [27]	Brixton test [2]	Verbal fluency task [11]
Goal level	x	x		x	x	x	x	x
Attributions								
Self-efficacy	x	x		x				x
Metacognition	x	x	x	x	x	x	x	x
Attention	x	x	x	x	x	x	x	x
Time Management			x		x	x	x	x
Environmental structuring								
Motivation								
Effort	x		x	x	x	x	x	x

When comparing both of the tables, three patterns stand out. First, it is of interest to note that no single measure was thought to overlap with every self-regulation construct. Next, only one measure out of those evaluated was rated as *not* representing metacognition and attention. Lastly, only two of the attitudinal measures were considered to contain effort constructs while nearly all of the behavioral measures measured effort. These general results are briefly considered in the next sections.

5 Conclusion

This paper identified existing measures in the literature for cognitive flexibility that may facilitate researching the relationships between self-regulation phases and cognitive flexibility across the stages of human development. Gaps in measures available to support research were identified that point to the need for development of psychometrically sound instruments. Further, theoretical relationships between self-regulation constructs and existing attitudinal and behavioral measures were identified. These activities lay the ground work for comprehensively researching the relationship between cognitive flexibility and the constructs underpinning self-regulation regardless of discipline.

As is characteristic of any study attempting to strike a balance between brevity and richness of detail, this paper does not stand to be the last voice on how these correlations may come about or that every cognitive flexibility measure was discovered. This paper does attempt to create a structure to build from. Additionally, the authors evaluating these theoretical overlaps may be susceptible to human error or biases. Possible areas for future research are expanded below.

5.1 Future Research

We suggest that developing psychometrically sound measures to assess both attitudinal and behavioral dimensions of cognitive flexibility for use in a neurotypical population is necessary. Performance of some tests has been questionable in existing studies because practicality often overrides the power necessary to detect a statistically significant result due to test construction for populations or purposes other than that intended. Behavioral tests likely need added complexity resulting from both set- and task-shifting [5]. Attitudinal tests need to be developed with items that are easily understood by the general population (e.g., void of clinical or academically inclined jargon or at a more common reading level).

Furthermore, it would be fruitful to investigate the use of multiple, various cognitive flexibility measures. For example, non-invasive, psychophysiological measurement techniques (e.g., heart rate monitoring, breathing patterns) could be used to assess training effectiveness in combination with attitudinal and behaviors of cognitive flexibility to detect the effects of self-regulation on short- and long-term training outcomes [9]. Along these lines, qualitative measures obtained from various sources such as coworkers, supervisors, or customers could support understanding cognitive flexibility. Overall, a more comprehensive or holistic perspective may help researchers better understand the complex interactions of these constructs.

References

1. Allen, J.: Conceptualizing learning agility and investigating its nomological network. Doctoral dissertation. Retrieved from ProQuest (10259350) (2016)
2. Burgess, P.W., Shallice, T.: The Hayling and Brixton Tests. Thames Valley Test Company Ltd., Bury St. Edmonds (1997)
3. Carver, C.S., Scheier, M.F.: On the structure of behavioral self-regulation. In: Boekaerts, M., Pintrich, P.R., Zeidner, M. (eds.) Handbook of Self-regulation, pp. 41–84. Academic Press, San Diego (2000)
4. Colzato, L.S., Ozturk, A., Hommel, B.: Meditate to create: the impact of focused-attention and open-monitoring training on convergent and divergent thinking. Front. Psychol. **3**, 1–5 (2012). https://doi.org/10.3389/fpsyg.2012.00116
5. Dajani, D.R., Uddin, L.Q.: Demystifying cognitive flexibility: implications for clinical and developmental neuroscience. Trends Neurosci. **38**(9), 571–578 (2017). https://doi.org/10.1016/j.tins.2015.07.003
6. Dennis, J.P., Vander Wal, J.S.: The cognitive flexibility inventory: instrument development and estimates of reliability and validity. Cogn. Ther. Res. **34**(3), 241–253 (2010). https://doi.org/10.1007/s10608-009-9276-4
7. Ding, X., et al.: Short-term meditation modulates brain activity of insight evoked with solution cue. Soc. Cogn. Affect. Neurosci. **10**(1), 43–49 (2014). https://doi.org/10.1093/scan/nsu032
8. Grant, D.A., Berg, E.: A behavioral analysis of degree of reinforcement and ease of shifting to new responses in Weigl-type card-sorting problem. J. Exp. Psychol. **38**(4), 404–411 (1948). https://doi.org/10.1037/h0059831
9. Grubb, J.D., Walwanis Nelson, M.M., Fatolitis, P., Shutty, M., Knicely, J., Alicia, T.: Evaluation of psychophysiological sensor data's contribution to enhancing training effectiveness: what do we know? Poster Session Presented at the 4th International Augmented Cognition Conference, Baltimore, MD, October 2007
10. Johnco, C., Wuthrich, V.M., Rapee, R.M.: The influence of cognitive flexibility on treatment outcome and cognitive restructuring skill acquisition during cognitive behavioural treatment for anxiety and depression in older adults: results of a pilot study. Behav. Res. Ther. **57**(1), 55–64 (2014). https://doi.org/10.1016/j.brat.2014.04.005
11. Lezak, M.D.: Neuropsychological Assessment. Oxford University Press, Oxford (1983)
12. Locke, E.A., Latham, G.P.: Building a practically useful theory of goal setting and task motivation: a 35-year odyssey. Am. Psychol. **57**(9), 705–717 (2002). https://doi.org/10.1037/0003-066X.57.9.705
13. Maddox, W.T., Baldwin, G.C., Markman, A.B.: A test of the regulatory fit hypothesis in perceptual classification learning. Mem. Cogn. **34**(7), 1377–1397 (2006). https://doi.org/10.3758/BF03195904
14. Martin, M.M., Rubin, R.B.: A new measure of cognitive flexibility. Psychol. Rep. **76**(2), 623–626 (1995). https://doi.org/10.2466/pr0.1995.76.2.623
15. Moore, A., Malinowski, P.: Meditation, mindfulness and cognitive flexibility. Conscious. Cogn. **18**(1), 176–186 (2009). https://doi.org/10.1016/j.concog.2008.12.008
16. Munakata, Y., Snyder, H.R., Chatham, C.H.: Developing cognitive control: three key transitions. Curr. Dir. Psychol. Sci. **21**(2), 71–77 (2015). https://doi.org/10.1177/0963721412436807
17. Pintrich, P.R.: The role of goal orientation in self-regulated learning. In: Zeidner, M., Pintrich, P.R., Boekaerts, M. (eds.) Handbook of Self-Regulation, pp. 451–529. Elsevier Academic Press, Burlington (2000)

18. Purichia, H.R.: An investigation of the theory of cognitive flexibility and an estimated measure of the construct. Doctoral dissertation. Retrieved from ProQuest (621047572) (2004)

19. Reitan, R.M.: Validity of the trail making test as an indicator of organic brain damage. Percept. Mot. Skills **8**(3), 271–276 (1958). https://doi.org/10.2466/pms.1958.8.3.271

20. Schunk, D.H., Ertmer, P.A.: Self-regulation and academic learning: self-efficacy enhancing interventions. In: Boekaerts, M., Pintrich, P.R., Zeidner, M. (eds.) Handbook of Self-Regulation, pp. 631–649. Academic Press, San Diego (2000). https://doi.org/10.1016/b978-012109890-2/50048-2

21. Schunk, D.H., Zimmerman, B.J.: Self-regulation and learning. In: Reynolds, W.M., Miller, G.E., Weiner, I.B. (eds.) Handbook of Psychology, pp. 59–78. Wiley, Hoboken (2005)

22. Sitzman, T., Ely, K.: A meta-analysis of self-regulated learning in work-related training and educational attainment: what we know and where we need to go. Psychol. Bull. **137**(3), 421–442 (2011). https://doi.org/10.1037/a0022777

23. Spiro, R.J., Collins, B.P., Thota, J.J., Feltovich, P.J.: Hypermedia for complex learning, adaptive knowledge application, and experience acceleration. Educ. Technol. **43**(5), 5–10 (2003). http://www.jstor.org/stable/44429454

24. Spiro, R.J., Feltovich, P.J., Coulson, R.L.: Two epistemic world-views: prefigurative schemas and learning in complex domains. Appl. Cogn. Psychol. **10**(7), 51–61 (1996). https://doi.org/10.1002/(SICI)1099-0720(199611)10:7<51::AID-ACP437>3.0.CO;2-F

25. Stroop, J.R.: Studies of interference in serial verbal reactions. J. Exp. Psychol. **18**(6), 643–662 (1935). https://doi.org/10.1037/h0054651

26. Tchanturia, K., et al.: Cognitive flexibility in anorexia nervosa and bulimia nervosa. J. Int. Neuropsychol. Soc. **10**(4), 513–520 (2004). https://doi.org/10.1017/S1355617704104086

27. Tchanturia, K., Morris, R., Surguladze, S., Treasure, J.: An examination of perceptual and cognitive set shifting tasks in acute anorexia nervosa and following recovery. Eating Weight Disord. **7**, 312–316 (2002). https://doi.org/10.1007/BF03324978

28. Tchanturia, K., Serpell, L., Troop, N., Treasure, J.: Perceptual illusions in eating disorders: Rigid and fluctuating styles. J. Behav. Ther. Exp. Psychiatry **32**(3), 107–115 (2001). https://doi.org/10.1016/S0005-7916(01)00025-8

29. Zhang Ulyshen, T., Koehler, M.J., Gao, F.: Cognitive Flexibility Inventory–Revised [Database record] (2015). PsycTESTS. https://doi.org/10.1037/t54991-000

30. Zimmerman, B.J.: Attaining self-regulation: a social cognitive perspective. In: Zeidner, M., Pintrich, P.R., Boekaerts, M. (eds.) Handbook of Self-Regulation, 2nd edn, pp. 13–39. Elsevier Academic Press, Burlington (2005)

Enhancing Simulated Students with Models of Self-regulated Learning

Robert E. Wray[(✉)]

Soar Technology, Inc., 3600 Green Court Suite 600, Ann Arbor, MI 48105, USA
wray@soartech.com

Abstract. A practical constraint in the design and development of algorithms and tools for personalized learning is the need to design, implement and integrate adaptive algorithms without the benefit of a priori large-scale user testing. The complexity of software integration and limited access to physical devices can result in commitment to designs that turn out in practice to fall short of envisioned training benefits. We are developing methodology and tools that employ simulated students and software verification methods to attempt to understand, prior to full-scale development, the potential benefits of adaptive algorithms and the requirements they impose on students and instructors. Simulated students are used to evaluate design choices and estimate impacts on learning. These tools typically model low-level interaction with a learning environment. Learning science research shows, however, that the overall learning context, rather than just the learning content itself, has significant impact on what is learned and how readily it is learned. We overview the use of simulated students in learning research and evaluation and then review constructs identified as relevant to self-regulated learning. We discuss which constructs are likely to have the greatest impact on simulated students and how these constructs can be incorporated into future simulated students to improve the utility and comprehensiveness of these tools.

Keywords: Simulated students · Learner models · Self-regulated learning

1 Introduction

The design and development of adaptive learning systems for complex cognitive skills is challenging. For instance, empirical consideration and evaluation of design tradeoffs in complex learning environments in relatively uncommon. Multiple factors contribute to the relative immaturity of evidence-based designs, especially cost and time. The costs of strongly controlled and well-designed studies to evaluate design alternatives can significantly increase overall development costs, in part because functional training prototypes must be developed to support empirical evaluation. The time needed to plan, to conduct, and to evaluate results from human-subjects studies also slows the speed of deploying new training solutions.

As we discuss further in this paper, the use of simulated students has potential to mitigate the challenge of building timely and cost-effective while also evidence-based adaptive training. Simulated students are computational models of learners [1–4]. They

© Springer Nature Switzerland AG 2019
D. D. Schmorrow and C. M. Fidopiastis (Eds.): HCII 2019, LNAI 11580, pp. 644–654, 2019.
https://doi.org/10.1007/978-3-030-22419-6_46

can be used to predict learning effects [1, 5, 6], to simulate learner errors to test adaptive responsiveness [7], and to test system function generally when human subjects are not available to test those functions [8, 9]. Although these and other potential uses abound, we are particularly focused in our work on the use of simulated students to support early-stage verification and assessment of adaptive system functions. That is, we are applying simulated students exactly where the resource and time cost of human-subjects evaluation is a barrier to evidence-based design choices. The goal is to make design choices informed by empirical evidence, but using evidence encoded in the simulated student rather than via direct evidence from human subjects.

Simulated students to-date, have largely focused on modeling a learner's direct interaction with the learning content. Thus, the typical variables that are manipulated in studies with simulated students are various instructional design and delivery choices. However, individual learners bring different learning strategies to a learning environment [10]; attitudes and social context modulate learning as well [11–13]. The overall learning context, rather than just the learning content itself, has significant impact on what is learned and how readily it is learned.

As learning is increasingly occurring outside of traditional, highly structured learning environments of the past (e.g., classrooms), these factors have larger impact on learning outcomes [14]. In less highly structured settings, the results of learning may be shaped as much by the learner's self-regulation of the learning experience than the content itself. Such factors are largely absent in today's models of simulated students. Thus, the lack of models of self-regulation in simulated students makes them less

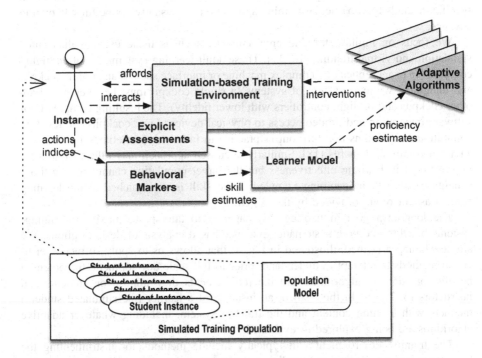

Fig. 1. Testbed for evaluation and verification of adaptive training design alternatives.

useful for estimating the potential learning that will occur when self-regulation may play a significant or even dominate role in outcome. It also inhibits their utility in evaluating the relative impact of tools and methods that could scaffold self-regulatory behavior in less structured learning environments.

In this paper, we overview the use of simulated students in learning research and evaluation. We review the constructs identified in Sitzmann's and Ely's [15] meta-analysis of self-regulated learning in the context of simulated students. We map those constructs to specific modeling opportunities and research needs based on the potential effectiveness for verification testing. This analysis provides direction for extending simulated student implementations in order that they can demonstrate self-regulation strategies to support future learning research and systems design.

2 Context: Software Verification of Adaptive Training Systems

As above, a practical constraint in the design and development of algorithms and tools for personalized learning is the need to design, implement and integrate adaptive algorithms, oftentimes within complex software environments, without the benefit of a priori large-scale user testing. User testing (which can include piloting and A/B testing, as well as formal experimentation), can directly inform what adaptive methods are more (and less) beneficial within a particular training setting. Because the space of adaptive interventions is large and the field is relatively new, the most beneficial, specific methods for a particular training application will usually not be fully known in advance.

We focus on military training environments, such as those used in distributed simulation and virtual training [16, 17]. These adult-learning systems are often highly complex (an exact replica of a complex machine or interface such as an aircraft cockpit) and integrated within a system of systems (a single cockpit must be integrated with other cockpits and/or flight controllers with lower fidelity). The resulting complexity of software integration and limited access to physical devices (e.g., cockpit) can result in commitment to designs that turn out in practice to fall short (sometimes far short) of envisioned training benefits [18]. Similarly, a chosen approach may offer a significant improvement in learning effectiveness but the target population cannot realize those benefits because their incoming knowledge and skill is not matched to the learning context assumed and provided by the system.

The long-term goal motivating this paper is to attempt to maximize training-systems benefits via more systematic, evidence-based analysis of design options. We are developing a testbed, illustrated in Fig. 1, that allows us to apply software verification methods to attempt to understand, prior to full-scale development, the potential benefits of adaptive algorithms and the requirements they impose on students and instructors [6, 7, 19]. In the testbed, an instance of a learner (the simulated student) interacts with training content and the training system, including whatever adaptive algorithms are being explored or evaluated.

The training environment will typically include methods for instrumenting the learner via various explicit and implicit assessments and creating a learner model that informs the choices of the system's adaptive algorithms. As the diagram suggests, one of the advantages of the testbed is that various adaptive algorithms can be tested against the identical learner instance in order to evaluate the differential effects of various adaptive choices. This differential testing can only be reliably conducted with human subjects when a large pool of test subjects is available. Exposure to the learning content changes subject knowledge. Such large subject pools are often available for validation tests, but are not generally feasible for evaluating individual design choices.

3   The Anatomy of a Simulated Student

Ideally, a simulated student would offer fully representative computational models of learners. In work to-date, the simulated student is much more limited than this vision. Our simulated students encapsulate established patterns of learning as derived from the literature (e.g., incremental models of the power law of learning [20]) and encode patterns of actual learner behavior as described by instructors and subject-matter experts [7].

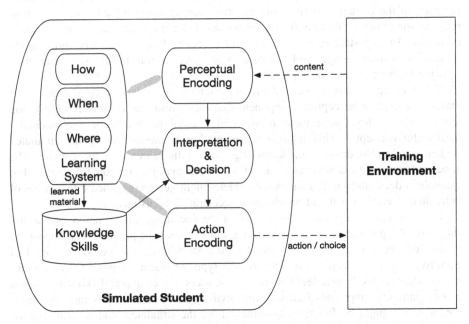

Fig. 2. Conceptual components within a simulated student.

There are many opportunities to develop simulated students further. For example, one specific technical challenge for adaptive-systems verification is making it easier and faster to build and to customize them for a particular domain application [3]. In this analysis, however, we focus on the breadth of the population of learners that may participate in a learning environment. In Fig. 1, a pool of simulated student "test subjects" are generated from a population model. Similar to the simulated students themselves, the current population model is quite limited. It assumes normal distributions over a small number of parameters (such as a priori knowledge and learning rate) to construct a pool of student instances [6].

The population model needs to be extended and deepened in many ways. One direction of improvement would be to vary not only learning-specific parameters but also to include a more holistic model of the learner that considers not only learning, but the context in which learning occurs. This goal, however, requires further discussion of components within implementations of simulated students.

Figure 2 illustrates the conceptual components within a typical simulated student. Here we generalize from our current simulated-student representation to a more abstract one intended to include more thoroughly developed implementations of simulated students, such as SimStudent [2, 5, 8]. The training environment delivers training content to the student. For the purposes of this analysis, we assume the content could range from basic lesson material (e.g., as one might encounter in a traditional computer-based trainer or CBT) to interactive practice, such as one would encounter in a simulation-based training environment. The content can include assessment as well, such as direct questions the systems asks the learner about learning content. Interpretation of the content and (in some cases) reasoning about it leads to an action or reaction (moving to the next frame; taking an action in a simulation; answering a question). An important category of content is feedback, ideally timely and targeted [21], as the learner engages with the content and takes action. In this way, interaction with the learning environment proceeds.

The learning process is modeled within the simulated student. In some cases, there may be an explicit perceptual component that maps from the content representation (e.g., words of text) to a more meaningful internal representation (semantically-meaningful concepts). This internal representation is interpreted by the simulated student, while also drawing on knowledge and skills relevant to the content. The simulated student eventually takes an action, which may require some decoding. For example, a decision to turn a simulated vehicle might get decomposed into a series of individual steering wheel and deceleration/acceleration adjustments.

Learning can take place across any/all of the encoding, interpretation, and action stages, resulting in various kinds of learning. For example, the progressive composition of the "turning action" from its various individual adjustments would be a kind of cognitive skill learning [22–24]. The specific types of learning embedded in a simulated student are likely to reflect the goals of the users. For example, if skill compilation is not a particular emphasis of the learning environment, then a pre-programmed macro for "turning" might be functionally sufficient for the simulated student. Importantly, some simulated students also distinguish between what is learned, how it was learned and where (under what context) some information was learned [8]. Such distinctions are important for generalization and transfer, as knowledge gained in one context may not be readily accessible in another without additional learning.

In addition to what components are present within a simulated student, it is also important to note what is omitted. Simulated students are usually "idealized" students. They attend to and process the content as it is presented; motivation and attention are not typically modeled. Similarly, they remain on task, and will repeatedly attempt to respond to content (solve a problem; answer a question) without tiring or growing frustrated.

Simulated students also typically lack "how to learn" knowledge that many human students bring to bear on problems. Learners use various learning techniques to build up knowledge of a domain or to break down problems. A simple and pervasive example is the use of rubrics as a shortcut for a more complex concept (HOMES to recall the Great Lakes; "every good boy..." to recall the notes on the treble clef; etc.) A population of simulated students that was used to model the time course of learning of conceptual knowledge would likely be inaccurate without considering such learning strategies: the speed at which these concepts are reliably recalled oftentimes depend on such rubrics.

These shortcomings in simulated students represent opportunities for research and improvement. Cognitive architectures, such as ACT-R [25] and Soar [26], could provide a substrate for the development of more comprehensive simulated students, embedding richer models of human reasoning and psychology that could improve the realism and completeness of simulated students. However, given vast breadth in instructional choices and native capabilities that impact learning [27, 28], what kinds of capabilities might go into such a model and with what priority? To explore these questions, we turn to the study of self-regulated learning, in order to attempt to identify which general learning strategies might have large, determinate effects in future implementations of simulated students.

4 Self-regulation Strategies for Learning

Self-regulation in learning refers to the ability of learners to control and to modulate attentional, affective, cognitive and behavioral processing during a learning experience [15]. Self-regulation generally implies a goal-oriented mindset that reflects a desire on the part of learners to reach some level of performance (e.g., a grade). We hypothesize that creating models of self-regulated learning within simulated students will lead to more useful and more realistic populations of simulated students that, in turn, will better support training-systems verification and design. Rationales include:

1. The goal orientation of theories and mechanisms of self-regulation maps directly onto the goal-focused theories of mind of cognitive architectures. This shared orientation both provides a shared common conceptual framework for mapping self-regulation to an architecture, it also points to specific, existing mechanisms for implementation within an architectural framework.
2. Theories of self-regulation have strongly shaped the learning sciences and its understanding of how learners interact with and adapt within a learning environment. Self-regulation is already a "lens" through which learning sciences is viewed. Thus, simulated students may benefit from also adopting this commonplace point of view.
3. It is increasing common for adult learning to take place outside of highly structured learning environments. Two or three decades ago, an employee who wanted to learn a new, job-relevant skill might seek to attend company-sponsored training. Today,

employees are much more likely (and, at least implicitly, be expected) to find and to complete that training themselves (e.g., via YouTube or Khan Academy). Self-regulation is much more important in these less structured learning contexts [14].

The meta-analysis of Sitzmann and Ely [15] identifies and unifies various constructs that comprise self-regulated learning. Table 1 enumerates and describes the three major categories of constructs identified in their analysis. In all, they identify sixteen distinct constructs that appear in the literature as examples of self-regulation. Thus, while self-regulation could be useful for simulated students, a full-scale integration of all 16 constructs within a simulated-students framework would require significant time and resources.

Table 1. Categories of constructs in Sitzmann and Ely [15].

Construct category		Description
1	Regulatory agent	The regulatory agent initiates self-regulated learning. Goals are the primary regulatory agent identified. The establishment of goals (such as successful performance) provides criteria for other regulatory mechanisms and appraisals
2	Regulatory mechanism	These mechanisms directly influence outcomes because they are under the control of the learner and help determine if and what progress is made toward the goal. Examples of specific constructs include: metacognition, attention, learning strategy, help seeking, motivation, and (expenditure of) effort
7	Regulatory appraisal	Appraisals evaluate progress toward learning goals and condition continuing attempts at progress. Self-efficacy is the most widely understood construct: a learner's belief that they will succeed results in better goals, better choice of learning strategies, and increased effort and persistence

Their analysis does attempt to assess the relative effect of the various constructs on learning. Table 2 summarizes some of the findings reported in the meta-analysis. The table lists various constructs, the reported correlation of the construct with learning outcomes, and then some comments. The constructs are listed from strongest to weakest effect; only a subset of the constructs is reported here.

Of the 16 constructs identified, only goal-level and self-efficacy have strong effects on learning according to the meta-analysis. Interestingly, although a number of constructs for self-regulation of learning have been explored in simulated students (Table 2 lists two of these), neither of these constructs with stronger effects are typically included. There is an implicit goal in all simulated students that they should perform as well as possible; this "perform to perfection" goal is another example of how idealized simulated students often are in their realization. Similarly, self-efficacy (nor self-doubt) are explicitly represented in a simulated student. In subsequent analysis within Sitzmann and Ely, the baseline cognitive ability, prior knowledge and the four constructs with the largest effect size (1–4 in Table 2) account for greater than 50% of the variance in learning outcomes [15].

The constructs that are more typically encountered in simulated students, such as support for various learning strategies and help-seeking behavior, actually have much weaker impacts on learning. Help-seeking is of note especially because of the affordance of help-seeking in computer-based training environments is different than help-

Table 2. Various regulatory constructs and their impact on learning (adapted from [15]).

Construct		Correlation with learning outcomes	Observations
1	Goal level	$\rho = .44$	The goal level has a moderate-to-strong effect on observed learning. Goal level requires a learner to establish a performance standard, whether applied to the next item or to the overall learning goal. A performance standard is not typically explicit in simulated students
2	Self-efficacy	$\rho = .35$	Self-efficacy has a moderate-to-strong effect on observed learning. Self-efficacy measures learner confidence and expectation in being able to succeed. Like goal level, this feature is not typically included in simulated students
3	Effort	$\rho = .28$	Learner effort had a moderate-to-weak impact on observed learning. Effort is a measure of time on task. Time on task is typically unconstrained for simulated students (unless time on task is fixed in advance)
4	Persistence	$\rho = .27$	Persistence had a moderate-to-weak impact on observed learning. Persistence measures a willingness to continue to work on training material, regardless of boredom or frustration. Simulated students are typically implemented with unlimited persistence
5	Attention	$\rho = .24$	Attentional regulation had a moderate-to-weak impact on observed learning. Attention relates to effortful concentration and mental focus. Simulated students typically manifest unrealistically full attention (no distraction)
11	Learning strategy	$\rho = .12$	Learning strategy had a weak impact on observed learning. Learning strategy measures the learner's knowledge and use of techniques to elaborate learning content and to integrate materials. Simulated students typically encode, at best, a few narrow learning strategies
15	Help seeking	$\rho = .06$	Help-seeking behavior had a weak impact on observed learning. Help-seeking measures the learners' willingness to seek outside assistance in learning new material. Simulated students do often encode help-seeking behaviors in ITS environment (e.g., ask for a hint)

seeking in traditional courses (the source of most of the data used in the meta-analysis). It is an open question whether help-seeking has a greater effect in computer-mediated environments where the cost of requesting a hint or scaffolding is negligible. However, it has been observed that human students readily abuse hint functions in Intelligent Tutoring Systems when they are made available [29] suggesting differences in help seeking behavior.

5   Improving Simulated Students via Models of Self-regulation

Do these observations about the effects of specific constructs of self-regulation inform how to improve simulated students? We recommend the following courses of action for future research and development:

1. **Integrate goal level in future simulated students.** This construct has the largest effect on learning outcomes and is absent in simulated students. It also should be relatively simple to incorporate in architecture-based models, as goals are explicitly represented in these systems. Explicit performance goals could also mitigate the unrealistic persistence and effort evinced by simulated students, as outlined previously. For example, a simulated student that sets lower performance goals may reduce its effort once those lower goals are reached. Open questions include understanding and encoding various levels of goals in the population model, as well as estimating the distribution of these various levels of goals. However, implementation and exploration could begin more or less immediately.

2. **Research mechanisms to represent and to process self-efficacy within cognitive architectures.** Because self-efficacy impacts other constructs (such as goal setting, choice of learning strategy, persistence and effort), a good model of self-efficacy could lead to simulated students that more easily demonstrate the effects of these other constructs. Unlike goal level, basic research is needed to explore how to represent the effects of self-efficacy for learning within a cognitive architecture. Integrated computational models of other kinds of self-efficacy [30] and appraisal [31] provide a starting point for incorporating this construct in simulated students.

3. **Shift focus from learning strategies to learning context.** There has been relatively more work in representing various learning strategies and help-seeking behavior in simulated students than toward the context-setting constructs outlined above. Based on this analysis, a more productive course for future effort would be to deemphasize the process-focused constructs. These appear to have a relatively modest impact on learning, especially in comparison to the contextual ones.

A final recommendation is to conduct additional learning sciences research to determine if the effects of self-regulation in learning differ greatly in computer-based learning environments. The recommendations above assume that the effects of self-regulation in learning in computer-based environments will roughly follow those seen in traditional instructional environments (the primary focus of the studies reviewed in the Sitzmann and Ely meta-analysis). Understanding if and when the effects of various self-regulatory mechanisms differ in role and impact in a computer-based medium will aid in the development of improved simulated students. More importantly, it will also serve the ultimate goal of delivering learning experiences more attuned to the abilities and needs of individual learners.

References

1. Van Lehn, K.: Two pseudo-students: applications of machine learning to formative evaluation. In: Lewis, R., Otsuki, S. (eds.) Advanced Research on Computers in Education. Elsevier, Amsterdam (1991)
2. Matsuda, N., et al.: Cognitive anatomy of tutor learning: lessons learned with SimStudent. J. Educ. Psychol. **105**, 1152–1163 (2013)
3. MacLellan, C.J., Koedinger, K.R., Matsuda, N.: Authoring tutors with SimStudent: an evaluation of efficiency and model quality. In: Trausan-Matu, S., Boyer, K.E., Crosby, M., Panourgia, K. (eds.) ITS 2014. LNCS, vol. 8474, pp. 551–560. Springer, Cham (2014). https://doi.org/10.1007/978-3-319-07221-0_70
4. Kieras, D.E.: The role of cognitive simulation models in the development of advanced training and testing systems. In: Diagnostic Monitoring of Skill and Knowledge Acquisition, pp. 51–73. Lawrence Erlbaum Associates Inc, Hillsdale (1990)
5. Matsuda, N., Cohen, W.W., Sewall, J., Lacerda, G., Koedinger, K.R.: Predicting students' performance with SimStudent that learns cognitive skills from observation. In: Luckin, R., Koedinger, K.R., Greer, J. (eds.) Proceedings of the International Conference on Artificial Intelligence in Education, pp. 467–476. IOS Press, Amsterdam (2007)
6. Wray, R.E., Stowers, K.: Interactions between learner assessment and content requirement: a verification approach. In: Andre, T. (ed.) AHFE 2017. AISC, vol. 596, pp. 36–45. Springer, Cham (2018). https://doi.org/10.1007/978-3-319-60018-5_4
7. Wray, R.E., Bachelor, B., Jones, R.M., Newton, C.: Bracketing human performance to support automation for workload reduction: a case study. In: Schmorrow, D.D., Fidopiastis, C.M. (eds.) AC 2015. LNCS (LNAI), vol. 9183, pp. 153–163. Springer, Cham (2015). https://doi.org/10.1007/978-3-319-20816-9_16
8. MacLellan, C.J., Harpstead, E., Patel, R., Koedinger, K.R.: The apprentice learner architecture: closing the loop between learning theory and educational data. In: Proceedings of the 9th International Conference on Educational Data Mining-EDM 2016 (2016)
9. Harpstead, E., MacLellan, C.J., Aleven, V., Myers, B.A.: Replay analysis in open-ended educational games. In: Loh, C.S., Sheng, Y., Ifenthaler, D. (eds.) Serious Games Analytics. AGL, pp. 381–399. Springer, Cham (2015). https://doi.org/10.1007/978-3-319-05834-4_17
10. Elliot, A.J., McGregor, H.A., Gable, S.: Achievement goals, study strategies, and exam performance: a mediational analysis. J. Educ. Psychol. **91**, 549–563 (1999)
11. Chi, M.T.H., Wylie, R.: The ICAP framework: linking cognitive engagement to active learning outcomes. Educ. Psychol. **49**, 219–243 (2014)
12. Bandura, A.: Self-Efficacy: The Exercise of Control. W H Freeman/Times Books/Henry Holt & Co, New York (1997)
13. Noe, R.A., Schmitt, N.: The influence of trainee attitudes on training effectiveness: test of a model. Pers. Psychol. **39**, 497–523 (1986)
14. Brown, K.G., Sitzmann, T.: Training and employee development for improved performance. In: APA Handbook of Industrial and Organizational Psychology, vol 2: Selecting and Developing Members for the Organization, pp. 469–503. American Psychological Association, Washington, DC (2011)

15. Sitzmann, T., Ely, K.: A meta-analysis of self-regulated learning in work-related training and educational attainment: what we know and where we need to go. Psychol. Bull. **137**, 421–442 (2011)
16. Tolk, A. (ed.): Engineering Principles of Combat Modeling and Distributed Simulation. Wiley, Hoboken (2012)
17. Fletcher, J.D.: Education and training technology in the military. Science **323**, 72–75 (2009)
18. Fletcher, J.D., Chatham, R.E.: Measuring return on investment in military training and human performance. In: O'Connor, P.E., Cohn, J.E. (eds.) Human Performance Enhancements In High-Risk Environments, pp. 106–128. Praeger/ABC-CLIO, Santa Barbara (2010)
19. Wray, R.E., Woods, A., Haley, J., Folsom-Kovarik, J.T.: Evaluating instructor configurability for adaptive training. In: Schatz, S., Hoffman, M. (eds.) Advances in Cross-Cultural Decision Making, pp. 195–206. Springer, Cham (2016). https://doi.org/10.1007/978-3-319-41636-6_16
20. Leibowitz, N., Baum, B., Enden, G., Karniel, A.: The exponential learning equation as a function of successful trials results in sigmoid performance. J. Math. Psychol. **54**, 338–340 (2010)
21. Shute, V.J.: Focus on formative feedback. Rev. Educ. Res. **78**, 153–189 (2008)
22. Fitts, P.M., Posner, M.I.: Learning and Skilled Performance in Human Performance. Brock-Cole, Belmont (1967)
23. Taatgen, N.A., Lee, F.J.: Production compilation: a simple mechanism to model complex skill acquisition. Hum. Factors: J. Hum. Factors Ergon. Soc. **45**, 61–76 (2003)
24. Newell, A., Rosenblum, P.S.: Mechanisms of skill acquisition and the law of practice. In: Anderson, J.R. (ed.) Cognitive Skills and their Acquistion. Erlbaum, Hillsdale (1980)
25. Anderson, J.R., Bothell, D., Byrne, M.D., Douglass, S., Lebiere, C., Qin, Y.: An integrated theory of the mind. Psychol. Rev. **111**, 1036 (2004)
26. Laird, J.E.: The Soar Cognitive Architecture. MIT Press, Cambridge (2012)
27. Koedinger, K.R., Corbett, A.T., Perfetti, C.: The knowledge-learning-instruction framework: bridging the science-practice chasm to enhance robust student learning. Cogn. Sci. **36**, 757–798 (2012)
28. Koedinger, K.R., Booth, J.L., Klahr, D.: Instructional complexity and the science to constrain it. Science **342**, 935–937 (2013)
29. Aleven, V., Koedinger, K.R.: Limitations of student control: do students know when they need help? In: Gauthier, G., Frasson, C., VanLehn, K. (eds.) ITS 2000. LNCS, vol. 1839, pp. 292–303. Springer, Heidelberg (2000). https://doi.org/10.1007/3-540-45108-0_33
30. Pirolli, P.: A computational cognitive model of self-efficacy and daily adherence in mHealth. Transl. Behav. Med. **6**, 496–508 (2016)
31. Marinier, R., Laird, J.E.: A cognitive architecture theory of comprehension and appraisal. In: Agent Construction and Emotions Conference (2006)

Author Index